Computer-Aided Molecular Design: Theory and Applications

Computer-Aided Molecular Design: Theory and Applications

Jean–Pierre Doucet
Institut de Topologie et de Dynamique de Systèms
Université de Paris 7–Denis Diderot
1 rue Guy de la Brosse
75005 Paris
France

Jacques Weber
Départemeut de Chimie Physique
Université de Genève
30 quai Ernest-Ansermet
1211 Genève 4
Suisse

Academic Press
Harcourt Brace & Company, Publishers
London   San Diego   New York   Boston   Sydney   Tokyo   Toronto

ACADEMIC PRESS LIMITED
24–28 Oval Road
LONDON NW1 7DX

U.S. Edition Published by
ACADEMIC PRESS INC.
San Diego, CA 92101

This book is printed on acid free paper

A catalogue record for this book is available from the British Library

ISBN 0-12-221285-1

Typeset by Phoenix Photosetting, Chatham, Kent, UK
Printed and bound in Great Britain by Hartnolls Limited, Bodmin, Cornwall

Contents

Preface

The originality of this book is to train undergraduates, who have grown up during the computer revolution, in the core parts of those computer tools intended to help chemists, with graphics assistance, in handling molecular representations. The goal is complex, as there is an important synergy in the creative design of molecules or properties between the fundamental theories of chemistry and their computational extensions.

This duality, felt constantly through the last thirty years, has led to immense progress thanks to major developments in both chemical informatics and computational chemistry.

The authors, well known actors in this transformation of chemistry, have identified in a single book the foundations of the modelling and design successes in chemistry. Their aim is to teach some crucial components both in the computer field and in theoretical chemistry, often closely linked in CAMD. The two authors, who have participated in the deployment of CAMD, have identified and selected, with talent and efficiency, its fundamental elements. Their teaching is clear with alternate explanations and applications. The task was not simple, as these elements are shadowed by the progress of the information revolution and the rapid advances of molecular modelling. The synergy of these actions often obscures the identification of those basic components essential for an up to date course.

Thus, **Computer Aided Molecular Design** (CAMD) is identified as a mature discipline. The theoretical and practical aspects of its potential and of its basic tenets are described. Applications are chosen to underline the power of various CAMD strategies.

There are many reasons for such a course at the undergraduate level. Students are usually familiar with the information revolution, and they know that the rapid evolution of science and technology makes it imperative to acquire generalized training enabling them to deal with many changes in their active life. They are sensitive to the radical changes around them in communication fields, and they are aware that equivalent transformations and mutations are taking place in science and technology. These affect our vision of science but also modify our working methods both in our chosen field of activity and in the methods we use. A brief sample enumeration of such trends helps to justify special training such as CAMD. Let us cite, for instance: the dramatic improvement expected in data and information sharing; CAMD collaboration (online or offline) between researchers and engineers separated by physical distance; the increase use of visual tools and methods for viewing, animating, interacting with CAMD mechanisms; the development of heuristic design through intelligent agents endowed with browsing capabilities.

For chemists to take advantage of all this, it is essential to point out that chemistry is based on numerous concepts conferring more or less precise 2D or 3D shapes on 'invisible' molecules.

The imaginary world of chemistry is thus geometrical, shape oriented and thereby open to graphics. Molecular species lead to either property maps or volumes incorporating all information derived from the molecular paradigm. Graphics presentations are used to handle the conceptual nature of bonds, the conformational flexibility of entities and chemical transformations.

This knowledge will be necessary for the future chemist if he is to make free use of modern tools.

J. P. Doucet and J. Weber have woven together the powerful tools of classification and correlation that help conduct similarity searches. Here the basic blocks are those derived from fragmentation or topological procedures. They are used to build adequate working 'spaces of states' suited to correlation searches.

Molecular similarity is presented in a drug design application to illustrate the complexity involved in the search of a 'lead' drug and the difficulties encountered in estimating the embedding of a drug on a biological receptor. The book maintains its homogeneity on structural design, but its authors successfully tackle CAMD tools and applications in bioinformatics.

In short, this book presents an excellent package of complementary ideas and tools and should greatly help students to master their use of molecular software. Instead of blindly applying programs, their knowledge of CAMD will give them the freedom required in creative molecular design research. Finally, this work should be kept by them as a long term classical reference volume.

Jacques–Emile Dubois
Professor, ITODYS, University Paris 7.
(President of CODATA)

Acknowledgements

The following have kindly granted permission to reprint the illustrations cited.

Scheme on pp. 20–21   From W.M. Newman and R.F. Sproull *Principles of Interactive Computer Graphics.* 2e, 1979. McGraw-Hill Inc. Ed, Copyright 1979 McGraw-Hill, reproduced with permission of McGraw-Hill Inc.

Figure 2.2   Reproduced from K. Kronlof and M. Tamminen *The Visual Computer.* 1:1985;24–36. Copyright Springer Verlag 1985. By permission of Springer Verlag Gmbh & Co. *A Viewing Pipeline for Discrete Solid Modeling*, p. 25.

Figure 2.5   Adapted from W.M. Newman and R.F. Sproull *Principles of Interactive Computer Graphics.* 2e, 1979. McGraw-Hill Inc. Ed, Copyright 1979 McGraw-Hill, reproduced with permission of McGraw-Hill Inc.

Figure 2.6   Adapted from W.M. Newman and R.F. Sproull *Principles of Interactive Computer Graphics.* 2e, 1979. McGraw-Hill Inc. Ed, Copyright 1979 McGraw-Hill, reproduced with permission of McGraw-Hill Inc.

Figure 2.7   Adapted from W.M. Newman and R.F. Sproull *Principles of Interactive Computer Graphics.* 2e, 1979. McGraw-Hill Inc. Ed, Copyright 1979 McGraw-Hill, reproduced with permission of McGraw-Hill Inc.

Figure 2.8   Adapted from W.M. Newman and R.F. Sproull *Principles of Interactive Computer Graphics.* 2e, 1979. McGraw-Hill Inc. Ed, Copyright 1979 McGraw-Hill, reproduced with permission of McGraw-Hill Inc.

Figure 2.11   Reproduced from G. Wyvill and T.L. Kunii *The Visual Computer.* 1:1985;3–14. Copyright Springer Verlag 1985. By permission of Springer Verlag Gmbh & Co. *A Functional Model for Constructive Solid Geometry*, p. 3.

Figure 2.22   Reproduced from D.F. Rogers *Mathematical Elements for Computer Graphics.* French Translation by J.J. Lecoeur, 1988. McGraw-Hill Inc. Ed, Copyright 1938 McGraw-Hill Inc. reproduced with permission of McGraw-Hill Inc.

Figure 2.26   Adapted from W.M. Newman and R.F. Sproull *Principles of Interactive Computer Graphics.* 2e, 1979. McGraw-Hill Inc. Ed, Copyright 1979 McGraw-Hill, reproduced with permission of McGraw-Hill Inc.

Figure 2.31   Reproduced from D.F. Rogers *Mathematical Elements for Computer Graphics.* French Translation by J.J. Lecoeur, 1988. McGraw-Hill Inc. Ed, Copyright 1988 McGraw-Hill Inc. reproduced with permission of McGraw-Hill Inc.

Figure 2.34   Adapted from W.M. Newman and R.F. Sproull *Principles of Interactive Computer Graphics.* 2e, 1979. McGraw-Hill Inc. Ed, Copyright 1979 McGraw-Hill, reproduced with permission of McGraw-Hill Inc.

Figure 2.35   Adapted from W.M. Newman and R.F. Sproull *Principles of Interactive Computer Graphics.* 2e, 1979. McGraw-Hill Inc. Ed, Copyright 1979 McGraw-Hill, reproduced with permission of McGraw-Hill Inc.

Figure 2.36   Adapted from W.M. Newman and R.F. Sproull *Principles of Interactive Computer Graphics.* 2e, 1979. McGraw-Hill Inc. Ed, Copyright 1979 McGraw-Hill, reproduced with permission of McGraw-Hill Inc.

Figure 3.2   Reprinted with permission from G.M. Smith and P. Gund *J. Chem. Inf. Comput. Sci.* 18:1978;207–210. Copyright 1978 American Chemical Society, and *Computer-Generated Space-Filling Molecular Models*, p. 208.

Figure 3.5   From L.H. Pearl *J. Molecular Graphics.* 6:1988;109–111, and *Calculating CPK images on a UNIX workstation*, p. 110.

Figure 3.7   From D.S. Goodsell, I. Saira Mian and A.J. Olson *J. Molecular Graphics.* 7:1989;41–44, and *Rendering Volumetric data in Molecular Systems*, p. 43.

Figure 3.9   From M. Gwilliam and N. Max *J. Molecular Graphics.* 7:1989;54–59, and *Atoms with Shadows- an Area-based Algorithm for Cast Shadows on Space-filling Molecular Models*, p. 56.

Figure 3.10   Reprinted with permission from K. Anjyo, T. Ochi, Y. Usami and K. Kawashima *The Visual Computer*. **3**:1987;4–12. Copyright Springer Verlag 1987, by permission of Springer Verlag Gmbh & Co. *A Practical Method for Constructing Surfaces in Three-Dimensional Digitized Space*, p. 5.

Figure 3.16   From A. Koide, A. Doi and K. Kajioka *J. Molecular Graphics*. **4**:1986;149–155, and *Polyhedral Approximation Approach to Molecular Orbital Graphics*, p. 150.

Figure 3.17   From G.D. Purvis and C. Culberson *J. Molecular Graphics*. **4**:1986;88–92, and *On the Graphical Display of Molecular Electrostatic Force-Fields and Gradients of the Electron Density*, p. 98.

Figure 3.19   Reproduced with permission from W. Heiden, M. Schlenkrich and J. Brickmann *J. Comput.-Aided Mol. Design*. **4**:1990;255–269. Copyright 1990 ESCOM Science publishers B.V.

Figure 3.20   From R. Lavery and B. Pullman *International J. of Quantum Chemistry*. 1981;259–272. Copyright 1981 John Wiley. Reprinted by permission of John Wiley and Sons Inc. *Molecular Electrostatic Potential on the Surface Envelopes of Macromolecules*: B-DNA.

Figure 4.2   Reproduced from P.J. Wheatley *The determination of Molecular Structure*, Dover Publications Inc, New York, 1968, p. 18, by permission of Oxford University Press.

Figure 4.3   Reproduced from P.J. Wheatley *The determination of Molecular Structure*, Dover Publications Inc, New York, 1968, p. 101, by permission of Oxford University Press.

Figure 4.8   Reproduced from *Introduction to NMR Spectroscopy* by J. Abraham, J. Fisher and P. Loftus, John Wiley (ed), 1988. Copyright 1988 John Wiley & Sons. Reprinted by permission of John Wiley and Sons Ltd.

Figures 4.9 and 4.10   Reprinted with permission from S. Cheatham *J. Chem. Educ.* **66**:1989;116–117, and *Nuclear Magnetic Resonance Spectroscopy in Biochemistry*, pp. 114–115.

Figure 4.11   Reprinted with permission from S.W. Fesik *J. Med. Chem.* **34**:1991;2937–2945. Copyright 1991 American Chemical Society. *NMR Studies of Molecular Complexes as a Tool in Drug Design*, p. 2938.

Figure 4.12   Adapted from G.M. Clore and A.M. Gronenborn *Progress in NMR Spectroscopy*. **23**:1991;43–92. J.W. Emsley, J. Feeney and L.H. Sutcliffe (eds). Copyright 1991 with kind permission from Elsevier Science Ltd, the Boulevard, Langford Lane, Kidlington OX5 1GB UK. *Application of Three- and Four-Dimensional Heteronuclear NMR Spectroscopy to Protein Structure Determination*.

Figure 4.14   Reproduced by courtesy of F.H. Allen.

Figure 4.15   Reproduced from F.H. Allen, O. Kennard, D.G. Watson, L. Brammer, A.G. Orpen and R. Taylor *J. Chem. Soc. Perkin Trans II* 1987, S1–S19. Reproduced with permission of the Royal Society of Chemistry. *Tables of Bond-Lengths Determined by X-Ray and Neutron Diffraction. Part I. Bond Lengths in Organic Compounds*, p. S2.

Figure 4.18   Reprinted with permission from A. Cosse-Barbi and J.E. Dubois *J. Am. Chem. Soc.* **109**:1987;1503–1511. Copyright 1987 American Chemical Society. *Anomeric Orbital and Steric Control in Static Conformations and System Dynamic Rotations of Methoxy Groups in 2,2 Dimethoxy Propane and Similar Crystallographic COCOC Fragments*, pp. 1509 and 1510.

Figure 4.19   Reproduced from A. Cosse-Barbi and J.E. Dubois *Tetrahedron Let.* **27**:1986; 3501–3504. Copyright Elsevier, 1986; with kind permission from Elsevier Science Ltd the Boulevard, Langford Lane, Kidlington OX5 1GB UK. *Conformational dynamics: Association of Correlated and Non-correlated Rotations of Methoxy Groups in Anomeric Structures. Convergences of a Theoretical Study of 2,2-dimethoxypropane and of Some Crystallographic Data of Acyclic Analogues of Pyranoses.*

Figure 4.20   Reproduced from A. Cosse-Barbi and J.E. Dubois *Tetrahedron Let.* **27**:1986; 3501–3504. Copyright Elsevier, 1986; with kind permission from Elsevier Science Ltd the Boulevard, Langford Lane, Kidlington OX5 1GB UK. *Conformational dynamics: Association of Correlated and Non-correlated Rotations of Methoxy Groups in Anomeric Structures. Convergences of a Theoretical Study of 2,2-dimethoxypropane and of Some Crystallographic Data of Acyclic Analogues of Pyranoses.*

Figure 4.22   Reprinted with permission from C. Gilli, V. Bertolasi, F. Bellucci and V. Ferretti *J. Am. Chem. Soc.* **108**:1986;2420–2424. Copyright 1986 American Chemical Society. *Stereochemistry of the R1(X)=C(sp²)-N(sp³) R2R3 Fragment. Mapping of the Cis-Trans*

Isomerization Path by Rotation Around the C-N Bond from Crystallographic Structural Data, p. 2424.

Figure 4.24   Reprinted with permission from E. Bye, W.B. Schweizer and J.D. Dunitz *J. Am. Chem. Soc.* **104**:1982;5893–5898. Copyright 1982 American Chemical Society. *Chemical Reaction Paths. 8. Stereoisomerization Path for Triphenylphosphine Oxide and Related Molecules: Indirect Observation of the Structure of the Transition State*, p. 5898.

Figure 4.26   Reproduced from G. Häfelinger, C.U. Regelmann, T.M. Krygowski and K. Wozniak *J. Comput. Chem.* **10**:1989;329–343. Copyright 1989 John Wiley. Reprinted by permission of John Wiley and Sons Inc. *Basis Set Dependence, Precision and Accuracy of Full Ab Initio Gradient Optimizations of Molecular Structures of Nonstrained Hydrocarbons. I: CC Bond Lengths.*

Figure 5.10   From J.L.M. Dillen *J. Comput. Chem.* **11**:1990;1125–1138. Copyright 1990 John Wiley, by permission of John Wiley and Sons Inc. *An Improved Empirical Force Field for Saturated Hydrocarbons.*

Figures 5.14 and 5.15   Adapted from J.E. Dubois, J.A. McPhee and A. Panaye *Tetrahedron.* **36**:1980;919–928. Copyright 1980, with kind permission from Elsevier Science Ltd, the Boulevard, Langford Lane, Kidlington OX5 1GB UK. *Steric Effects III. Composition of the Es parameter. Variation of Alkyl Steric Effects with Substitution. Role of Conformation in Determining Sterically Active and Inactive Sites.*

Figure 5.18   By courtesy of A. Cossé-Barbi.

Figure 5.19   Adapted from E. Osawa and H. Musso *Angew. Chem. Int. Ed. Engl.* **22**:1983;1–12. By permission from VCH. *Molecular Mechanics Calculations in Organic Chemistry. Examples of the Usefulness of This Simple Non-Quantum Mechanical Model*, p. 6.

Figure 5.20   From P.A. Kollman, G. Wippff and U.C. Singh *J. Am. Chem. Soc.* **107**:1985;2212–2219. Copyright 1985 American Chemical Society. Reprinted with permission. *Molecular Mechanical Studies of Inclusion of Alkali Cations into Anisole Spherands*, p. 2213.

Figure 5.21   From J.R. Damewood Jr, W.P. Anderson and J.J. Urban *J. Comput. Chem.* **9**:1988;111–124. Copyright 1989 John Wiley. Reprinted by permission of John Wiley and Sons Inc. *A Molecular Mechanics Study of Neutral Molecule Complexation with Crown Ethers.*

Figure 5.22   Part a and b from G. Wipff, P. Weiner and P. Kollman *J. Am. Chem. Soc.* **104**:1982;3249–3258. Copyright American Chemical Society. Reprinted with Permission *A Molecular Mechanics Study of 18-Crown-6 and its Alkali Complexes: Ar. Analysis of Structural Flexibility, Ligand Specificity and the Macrocyclic Effect*, p. 3253.

Figure 5.22   Part c from J.R. Damewood Jr, W.P. Anderson and J.J. Urban *J. Comput Chem.* **9**:1988;111–124. Copyright 1989 John Wiley. Reprinted by permission of John Wiley and Sons Inc. *A Molecular Mechanics Study of Neutral Molecule Complexation with Crown Ethers.*

Figure 5.24   From A. Bouraoui, M. Fathallah, B. Blaive and R. Gallo *J. Chem. Soc. Perkin Trans. II.* 1990;1211–1214. With permission of the Royal Society of Chemistry. *Design and Molecular Mechanics Calculations of New Iron Chelates*, pp. 1212–1213.

Figure 6.1   Reprinted with permission from W.H. Press, B.P. Flannery, S.A. Teukolsky and W.T. Vetterling *Numerical Recipes, the Art of Scientific Computing*, Cambridge University Press, Cambridge, 1986. Copyright 1986 Cambridge University Press.

Figure 6.2   Reprinted with permission from J.L. Yarnell, M.J. Katz, R.G. Wenzel and S.H. Koenig *Phys. Rev. A*, **7**:1973;2130. Copyright 1973 American Physical Society.

Figure 6.3   Reprinted with permission from J.A. Barker and R.O. Watts *Chem. Phys. Lett.* **3**:1969;144. Copyright 1969 Elsevier Science.

Figure 6.4   Reprinted with permission from G.C. Lie, E. Clementi and M. Yoshimine *J. Chem. Phys.*, **64**:1976;2314. Copyright 1976 American Institute of Physics.

Figure 6.5   Reprinted with permission from S. Swaminathan, S.W. Harrison and D.L. Beveridge *J. Am. Chem. Soc.* **100**:1978;5705. Copyright 1978 American Chemical Society.

Figure 6.7   Reprinted with permission from G. Alagona, C. Ghio and P.A. Kollman *J. Am. Chem. Soc.* **107**:1985;2229. Copyright 1985 American Chemical Society.

Figure 6.9   Reprinted with permission from J. Chandrasekhar, S.F. Smith and W.L. Jorgensen *J. Am. Chem. Soc.* **107**:1985;154. Copyright 1985 American Chemical Society.

Figure 6.11   Reprinted with permission from M. Wojcik and E. Clementi *J. Chem. Phys.* **84**:1986;5970. Copyright 1986 American Institute of Physics.

Figure 6.12 Reprinted with permission from A.K. Nowak, A.K. Cheetham, S.D. Pickett and S. Ramdas *Mol. Simulation*. **1**:1987;67. Copyright 1987 Gordan and Breach.

Figure 6.13 Reprinted with permission from C.R.A. Catlow, C.M. Freeman, B. Vessal, S.M. Tomlinson and M. Leslie *J. Chem. Soc. Faraday Trans*. **87**:1991;1947. Copyright 1991 The Royal Society of Chemistry.

Figure 7.1 From A.J. Holder and D.L. Wertz *J. Comput. Chem*. **9**:1988;684–688. Copyright 1988 John Wiley; by permission of John Wiley and Sons Inc. *Conformational Energetics of 1,3 Dichloropropane as Predicted by Several Calculation Methods*.

Figure 7.4 From G.M. Crippen *J. Comput. Chem*. **10**:1989;896–902. Copyright 1989 John Wiley; by permission of John Wiley and Sons Inc. *Linearized Embedding. A New Metric Matrix Algorithm for Calculating Molecular Conformations Subject to Geometric Constraints*, p. 898.

Figure 7.7 From A.P. Tonge, P. Murray Rust, W.A. Gibbons and L.K. McLachlan *J. Comput. Chem*. **9**:1988;522–538. Copyright 1988 John Wiley; by permission of John Wiley and Sons Inc. *Determination of the Major Solution Conformation of Tyrocidine A Using Molecular Mechanics Energy Minimization and NMR-Derived Distance and Torsion Angle Constraints*.

Figures 7.8 and 7.9 From D.M. Ferguson, W.A. Glauser and D.J. Raber *J. Comput. Chem*. **10**:1989;903–910. Copyright 1989 John Wiley, by permission of John Wiley and Sons Inc. *Molecular Mechanics Conformational Analysis of Cyclononane Using the RIPS Method and Comparison with Quantum-Mechanical Calculations*.

Figure 7.12 From M. Randic, B. Jerman-Blazic and J. Trinajstic *Computers Chem*. **14**:1990;237–246, with permission. *Development of 3-Dimensional Molecular Descriptors*, p. 341.

Figures 7.14 and 7.15 Reprinted with permission from D.P. Dolata and R.E. Carter *J. Chem. Inf. Comput. Sci*. **27**:1987;36–46. Copyright 1987 American Chemical Society. *WIZARD. Applications of Expert System Techniques to Conformational Analysis. 1. The Basic Algorithms Exemplified on Simple Hydrocarbons*, pp. 43–44.

Figure 7.16 From W.T. Wipke and M.A. Hahn, *Tetrahedron Comput. Methodology* **1**:1988;141–167. Copyright 1988. With kind permission from Elsevier Science Ltd, the Boulevard, Langford Lane, Kidlington OX5 1GB UK. *AIMB: Analogy and Intelligence in Model Building. System Description and Performance Characteristics*.

Figure 8.2 From B. Lee and F.M. Richards *J. Mol. Biol*. **55**:1971;379–400, with permission. *The Interpretation of Protein Structures: Estimation of Static Accessibility*.

Figure 8.4 From M.L. Connolly *J. Appl. Crys*. **16**:1983;548–558. By permission of the International Union of Crystallography. *Analytical Molecular Surface Calculation*, pp. 550–551.

Figure 8.5 From M.L. Connolly *J. Am. Chem. Soc*. **107**:1985;1118–1124. Copyright 1985 American Chemical Society, with permission. *Computation of Molecular Volume*, p. 1120.

Figure 8.6 From M.L. Connolly *J. Appl. Crys*. **16**:1983;548–558. By permission of the International Union of Crystallography. *Analytical Molecular Surface Calculation*, pp. 550–551.

Figure 8.7 From J.L. Pascual-Ahuir and E. Silla *J. Comput. Chem*. **11**:1990;1047–1060. Copyright 1991 John Wiley, by permission of John Wiley and Sons Inc. *An Improved Description of Molecular Surfaces. I: Building the Spherical Surface Set*.

Figure 8.12 From T.R. Stouch and P.C. Jurs *J. Chem. Inf. Comput. Sci*. **26**:1986;4–12. Copyright 1986 American Chemical Society, with permission. *A Simple Method for the Representation, Quantification and Comparison of the Volumes and Shapes of Chemical Compounds*, p. 9.

Figure 8.14 From M.Y. Pavlov and B.A. Fedorov *Biopolymers*. **22**:1983;1507–1522. Copyright 1991 John Wiley; by permission of John Wiley and Sons Inc. *Improved Technique for Calculating X-Ray Scattering Intensity of Biopolymers in Solution. Evaluation of Form, Volume and Surface of a Particle*, p. 1509.

Figures 8.23, 8.24 and 8.25 From M. Lewis and D.C. Rees *Science*. **230**:1985;1163–1165. Copyright 1985 by the AAAS. *Fractal surfaces of Proteins*, p. 1164.

Figure 10.2 Reprinted with permission from N. Thalmann and J. Weber *Chimia*. **31**:1977;361. Copyright 1977 Schweizer Chemiker-Verband (A, B,); W.L. Jorgensen and L. Salem *The Organic Chemist's Book of Orbitals*, Academic Press, New York (1973). Copyright 1973 Academic Press Inc.

Figure 10.7 Reprinted with permission from K. Angermund, K.H. Claus, R. Goddard and C. Kruger *Angew. Chem. Int. Ed. Engl*., **24**:1985;237. Copyright 1985 VCH Verlagsgesellschaft.

Figure 10.8 Reprinted with permission from M. Roch, J. Weber and A.F. Williams *Inorg. Chem.* **23**:1984;4571. Copyright 1984 American Chemical Society.

Figure 10.10 Reprinted with permission from R. Bonnaccorsi, E. Scrocco J. Tomasi and A. Pullman *Theor. Chim. Acta* **36**:1975;339. Copyright 1975 Springer-Verlag.

Figure 10.11 Reprinted with permission from J.P. Doucet, S.Y. Yue, J.E. Dubois, M. Roch and J. Weber *J. Chim. Phys.* **84**:1987;647. Copyright 1987 Elsevier.

Figure 10.12 Reprinted with permission from S.L. Price and N.G.L. Richards *J. Comp. Aided Mol. Des.* **5**:1991;41. Copyright 1991 ESCOM Science Publishers B.V.

Figure 11.2 From P. Gund, J.D. Andose, J.B. Rhodes, G.M. Smith *Science.* **208**:1980;1425–1431. Copyright 1980 by the AAAS, by permission of *Science. Three-Dimensional Molecular Modeling and Drug-Design*, p. 1429.

Figure 11.15 From P.M. Dean *Molecular Foundation of Drug-Receptor Interaction*, Cambridge University Press, 1987, p. 122. Copyright 1987, Cambridge University Press, with permission.

Figure 12.3 From L.F. Kuyper, B. Roth, D.P. Baccanari, R. Ferone, C.R. Beddell, J.N. Champness, D.K. Stammers, J.G. Dann, F.E.A. Norrington, D.J. Baker and P.J. Goodford *J. Med. Chem.*, **25**:1982;1120–1122. Copyright 1982, American Chemical Society, with permission. *Receptor-Based Design of Dihydrofolate Reductase Inhibitors: Comparison of Crystallographically Determined Enzyme Binding with Enzyme Affinity in a Series of Carboxy-Substituted Trimethoprim Analogues*, p. 1121.

Figures 12.4 and 12.5 From I.D. Kuntz, J.M. Blaney, S.J. Oatley, R. Langridge and T.E. Ferrin *J. Mol. Biol.*, **161**:1982;269–288, by permission. *A Geometric Approach to Macromolecule-Ligand Interactions*, pp. 272–273.

Figure 12.6 From K.D. Stewart, J.A. Bentley, M. Cory *Tetrahedron Computer Methodology.* **3**:1990;713–722. Copyright 1991, with kind permission from Elsevier Science Ltd, the Boulevard, Langford Lane, Kidlington OX5 1GB UK. *Docking Ligands into Receptors: the Test Case of a Chymotrypsin.*

Figure 12.7 From R.L. DesJarlais, R.P. Sheridan, J.S. Dixon, I.D. Kuntz and R. Venkataraghavan *J. Med. Chem.* **29**:1986;2149–2153. Copyright 1986 American Chemical Society, with permission. *Docking Flexible Ligands to Macromolecular Receptors by Molecular Shape*, p. 2151.

Figures 12.8 and 12.9 From R.L. DesJarlais, R.P. Sheridan, G.L. Seibel, J.C. Dixon, I.D. Kuntz and R. Venkataraghavan *J. Med. Chem.* **31**:1988;722–729. Copyright 1988 American Chemical Society, with permission. *Using Shape Complementarity as an Initial Screen in Designing Ligands for a Receptor Binding Site of Known Three-Dimensional Structure*, pp. 724 and 726.

Figure 12.10 From R.P. Sheridan, A. Rusinko III, R. Nilakantan and R. Venkataraghavan *Proc. Ntl. Acad. Sci. USA.* **86**:1989;8165–8169. *Searching for Pharmacophores in Large Coordinate DataBases and its Use in Drug Design*, pp. 8166 and 8168.

Figure 12.11 From M.G. Bures, C.W. Hutchins, M. Maus, W. Kohlbrenner, S. Kadam and J.W. Erikson *Tetrahedron Computer Methodology.* **3**:1990;681–696. Copyright 1990, with kind permission from Elsevier Science Ltd, the Boulevard, Langford Lane, Kidlington OX5 1GB UK. *Using 3D Substructure Searching to Identify Novel, Non-Peptidic Inhibitor of HIV-1 protease.*

Figure 12.12 From A.E. Weber, T.A. Halgren, J.J. Doyle, R.J. Lynch, P.K.S. Siegl, W.H. Parsons, W.J. Greenlee and A.A. Patchett *J. Med. Chem.* **34**:1991;2692–2701. Copyright 1991 American Chemical Society, with permission. *Design and Synthesis of P2-P1 Linked Macrocyclic Human Renin Inhibitors*, p. 2693.

Figure 12.13 Adapted with permission from D.J. Danziger and P.M. Dean *Proc. R. Soc. London.* **B236**:1989;101–113. Copyright 1989 The Royal Society, with permission. *Automated Site-Directed Drug Design: a General Algorithm for Knowledge Acquisition About Hydrogen-Bonding regions at Proteins Surfaces.*

Figure 12.14 From R.A. Lewis and P.M. Dean *Proc. R. Soc. London.* **B236**:1989;141–162. Copyright 1989, The Royal Society, with permission. *Automated Site-directed Drug Design: the Formation of Molecular Templates in Primary Structure Generation.*

Figure 12.15 From D.J. Danziger and P.M. Dean *Proc. R. Soc. London.* **B236**:1989;101–113. Copyright 1989, The Royal Society, with permission. *Automated Site-Directed Drug Design: a General Algorithm for Knowledge Acquisition About Hydrogen-Bonding regions at Proteins Surfaces.*

Figure 12.16 From R.A. Lewis and P.M. Dean *Proc. R. Soc. London.* **B236**:1989;125–140. Copyright 1989, The Royal Society, with permission. *Automated Site-directed Drug Design: The Concept of Spacer Skeletons for a Primary Structure Generation.*

Figure 12.17 From R.A. Lewis and P.M. Dean *Proc. R. Soc. London.* **B236**:1989;141–162. Copyright 1989, The Royal Society, with permission. *Automated Site-directed Drug Design: the Formation of Molecular Templates in Primary Structure Generation.*

Figures 12.18 and 12.19 From R.A. Lewis *Proceedings of the Sixth European Seminar and Exhibition Computer-Aided Molecular Design.* London, 1989, with permission.

Figure 12.20 From V.J. Gillet, W. Newell. P. Mata, G. Myatt, S. Sike, Z. Zsoldos, and A.P. Johnson *J. Chem. Inf. Comput. Sci.* **34**:1994;207–217. Copyright 1991, American Chemical Society, with permission. *SPROUT: Recent Developments in the de Novo Design of Molecules.*

Figure 12.21 From V.J. Gillet, A.P. Johnson, P. Mata and S. Sike *Tetrahedron Computer Methodology*, **3**:1990;681–696. Copyright 1991 with kind permission from Elsevier Science Ltd, the Boulevard, Langford Lane, Kidlington OX5 1GB UK. *Automated Structure Design in 3D.*

Figure 12.22 From H.J. Böhm *J. Comput.-Aided Mol. Design.* **6**:1992;61–78. Copyright 1999 ESCOM Science Publishers B.V., with permission. *The Computer Program LUDI: A New Method for the de Novo Design of Enzyme Inhibitors.*

Figure 12.23 From R.A. Lewis *J. Comput.-Aided Mol. Design.* **3**:1989;133–147. Copyright 1989 ESCOM Science publisher B.V., with permission. *Determination of Clefts in Receptor Structures.*

Figures 12.24, 12.25 and 12.26 From N.C. Cohen *J. Med. Chem.* **26**:1983;259–264. Copyright 1983 American Chemical Society, with permission. *β Lactam Antibiotics: Geometrical Requirements for Antibacterial Activities*, pp. 260 and 261.

Figure 12.27 From J.M. Schulman, M.L. Sabio and R.L. Disch *J. Med. Chem.* **26**:1983;817–823. Copyright 1983 American Chemical Society, with permission. *Recognition of Cholinergic Agonists by the Muscarinic Receptor. 1. Acethylcholine and Other Agonists with the NCCOCC Backbone.*

Figures 12.28, 12.29, 12.30 and 12.31 From R.P. Sheridan, R. Nilakantan, J.S. Dixon and R. Venkataraghavan *J. Med. Chem.* **29**:1986;899–906. Copyright 1986 American Chemical Society, with permission. *The Ensemble Approach to Distance Geometry: Application to the Nicotinic Pharmacophore*, p. 904.

Figures 12.33 and 12.34 From G.M. Crippen *J. Med. Chem.* **8**:1979;988–997. Copyright 1979 American Chemical Society, with permission. *Distance Geometry Approach to Rationalizing Binding Data*, pp. 990 and 992.

Figure 13.1 From L. Stryer, *Biochemistry* (2e). Copyright 1981 by Lubert Stryer. Used with permission of W.H. Freeman and Company.

Figure 13.3 From L. Stryer, *Biochemistry* (2e). Copyright 1981 by Lubert Stryer. Used with permission of W.H. Freeman and Company.

Figure 13.4 From L. Stryer, *Biochemistry* (2e). Copyright 1981 by Lubert Stryer. Used with permission of W.H. Freeman and Company.

Figure 13.5 From L. Stryer, *Biochemistry* (2e). Copyright 1981 by Lubert Stryer. Used with permission of W.H. Freeman and Company.

Figure 13.7 From L. Stryer, *Biochemistry* (2e). Copyright 1981 by Lubert Stryer. Used with permission of W.H. Freeman and Company.

Part (a) after A.E. Edmundson, *Nature.* **205**:1965;883 and H.C. Watson, *Prog. Stereochem.* **4**:1969;299–333.

Part (b) after R.E. Dickerson, *The Proteins*, H. Neurath (ed), (2e) vol. 2, Academic Press, 1964, p. 634.

Part (c) modified from J.M. Widom and S.J. Edelstein *Chemistry: an Introduction to General, Organic and Biological Chemistry.* Copyright © 1981 by W.H. Freeman and Company, used with permission.

Figure 13.8 From J.M. Widom and S.J. Edelstein *Chemistry: an Introduction to General, Organic and Biological Chemistry.* Copyright © 1981 by W.H. Freeman and Company, used with permission.

Figure 13.9 From M. Carson, *J. Mol. Graphics.* **5**:1987;103–106. *Ribbon Models of Macromolecules*, p. 104.

Figure 13.10   From D.J. Barlow and J.M. Thornton *J. Mol. Graphics.* **4**:1986;97–100. *Interactive Map Projection Algorithm for Illustrating Protein Surface*, p. 98.

Figure 13.11   By courtesy of J. Gharbi.

Figure 13.12   From *Introduction to NMR Spectroscopy* by J. Abraham, J. Fisher and P. Loftus, John Wiley (ed), 1988, copyright 1988 John Wiley & Sons. By permission of John Wiley & Sons Ltd.

Figure 13.14   From G.N. Ramachandran and V. Sasisekharan *Adv. Protein Chem.* **23**:1968;283–438. From C.R. Cantor and P.R. Schimmel *Biophysical Chemistry Part 2*, p. 259. Copyright © 1980 by W.H. Freeman and Company, used with permission. After J.P. Flory *Statistical Mechanics of Chain Molecules*, New York Interscience 1969 and G.N. Ramachandran et al. *J. Mol. Biol.* **7**:1963;95.

Figure 13.16   With permission from W.G.J. Hol *Angew. Chem. Int. Ed. Engl.* **25**:1986;767–778. *Protein Crystallography and Computer Graphics. Toward Rational Drug Design*, p. 771.

Figure 13.17   With permission from W.R. Taylor and J.M. Thornton *Nature.* **301**:1983;540–542. Copyright 1983, MacMillan Magazines Limited. *Prediction in the Super-Secondary Structure in Proteins*, p. 541.

Figure 13.20   From D.A. Clark, G.J. Barton, and C.D. Rawlings, *J. Mol. Graphics.* **8**:1990;94–107. *A Knowledge-Based Architecture for Protein Sequence Analysis and Structure Prediction*, p. 96.

Figure 13.21   From J. Greer *J. Mol. Biol.* **153**:1981;1027–1042, with permission. *Comparative Model Building of the Mammalian Serine Proteases.*

Figure 13.22   From T.L. Blundell, B.L. Sibanda, M.J.E. Sternberg and J.M. Thornton, *Nature* **326**:1987;347–352, with permission. *Knowledge-Based Prediction of Protein Structures and the Design of Novel Molecules*, p. 349.

Figure 13.23   From J. Novotny, R. Bruccoleri and M. Karplus *J. Mol. Biol.* 1984;787–818, with permission. *An Analysis of Incorrectly Folded Protein Models. Implications for Structure Predictions.*

Figure 13.24   From G.G. Ferenczy and G.M. Morris *J. Mol. Graphics.* **7**:1089;206–211. *The Active Site of Cytochrome P-450 Nifedipine Oxidase: a Model Building Study*, p. 209.

Figure 13.25   From C.R. Cantor and P.R. Schimmel *Biophysical Chemistry Part 1.* Copyright 1980 by W.H. Freeman and Company, used with permission. After M.O. Dayhoff (ed) *Atlas of Protein Sequence and Structure.* Vol. 5, Silver Spring Md: National Biomedical Research Foundation, 1972.

Figure 13.26   From G. Vriend and C. Sander *Proteins: Structure, Function and Genetics.* **11**:1991;52–58. Copyright 1991 John Wiley, with permission of John Wiley and Sons Inc. *Detection of Common Three-Dimensional Substructures in Proteins.*

Figure 13.27   With permission from M.G. Rossmann and P.A. Argos *J. Mol. Biol.* **105**:1976;75–95. *Exploring Structural Homology of Proteins*, p. 77.

Figure 13.28   From T.L. Blundell, B.L. Sibanda, M.J.E. Sternberg and J.M. Thornton *Nature, London.* **326**:1987;347–352. Copyright 1987, MacMillan Magazines Limited, with permission from Nature. *Knowledge-Based Prediction of Protein Structures and the Design of Novel Molecules*, p. 349.

Figure 13.29   From J.S. Richardson *Nature, London.* **268**:1977;495–500. Copyright 1977 MacMillan Magazines Limited, with permission from Nature. *B-Sheet Topology and the Relatedness of Proteins*, p. 496.

Figures 13.30 and 13.31   From P.J. Artimyuk, D.W. Rice, E.M. Mitchell and P. Willett *Proteins Engineering.* **4**:1990;39–43, by permission of Oxford University Press. *Structural Resemblance Between the Families of Bacterial Signal-Transduction Proteins and of G-proteins Revealed by Graph Theoretical Techniques.*

Figure 13.32   From K. Ooi and K. Nishikawa *Conformation of Biological Molecules and Polymers.* Bergmann and B. Pullman (ed) *Jerusalem*, 1973, Israel Academy of Sciences and Humanities.

Figure 13.33   From G.D. Rose, *Computational Molecular Biology*, A.M. Lesk (ed), copyright 1988, by permission of Oxford University Press.

Introduction

The computer-aided design of novel molecular systems has undoubtedly reached the stage of a mature discipline offering a broad range of tools available to virtually any chemist. Indeed, as various programs, known as model builders, enable chemists to readily calculate molecular geometries, it is now possible to perform structure-based design applications that give a practically unlimited range of novel systems exhibiting specific properties. To this end, recent progress in computational chemistry play an important role by leading to a realistic description of molecular shape and motion, solvent effects, binding free energies, etc. However, it is fair to say that computer-aided techniques have in most cases to be used in combination with X-ray, NMR and other experimental tools to get a complete design of novel systems.

The basic concepts of computer-aided molecular design (CAMD) are generally taught in different courses such as computer science (numerical methods and graphics), physical chemistry (quantum mechanics, statistical mechanics and simulation techniques), organic chemistry (molecular mechanics and conformational analysis) and biochemistry (protein structure and molecular recognition), rather than in a specific course devoted to CAMD. A similar statement is probably also applicable to books, as the corresponding material is often spread out in various monographs presenting more general subjects. The authors found it useful to collect, in a single book, a description of the theoretical foundations of CAMD together with applications illustrating their practical implementation on computers, with a special emphasis on drug design and protein modelling. Both authors have a long experience in teaching various topics of CAMD, and have found it useful and timely to put together, harmonize and update various teaching material on CAMD. The main purpose of this book is to cover all the techniques in a single volume using a language that can be understood by second or third year students or by chemists with a limited knowledge of theoretical chemistry.

This book is intended for undergraduate students with some basic knowledge of general chemistry, or for course instructors of chemistry who can use certain chapters to illustrate the role of computing in conformational analysis or molecular recognition, for example. As most of the subjects are presented at a rather elementary level, chemists or, more generally, molecular scientists who want to refresh their knowledge in CAMD will also find this book useful. Finally, is could also be used as a source of reference by professionals in their quest for information on the actual possibilities of CAMD.

Since the generation of synthetic images, using a computer, plays an outstanding role in molecular modelling and, ultimately, in CAMD, the first chapter presents an introduction to computer graphics by reviewing the main aspects of the discipline, ranging from hardware characteristics to software features such as graphics primitives, standards, libraries and geometrical transformations. The second chapter focuses on the algorithms used to generate realistic images of three-dimensional (3D) objects such as surface modelling, constructive geometry, viewing operations, hidden lines, and surfaces removal and rendering. After the first two technical chapters, devoted to graphics in general, Chapter 3 looks at molecular graphics and reviews the possible modes of representation of chemical objects. It describes the procedures and algorithms leading to the display of molecular shapes generated either from structures or properties. As molecular geometry is an indispensable piece of information for numerous modelling applications, Chapter 4 presents the most common experimental approaches leading to it such as X-ray diffraction, neutron scattering and NMR. This chapter gives general examples, with special emphasis on the possibilities and limitations of these techniques on polypeptides, proteins and DNA. Also it looks at structural databases, such as the Cambridge and Brookhaven, with emphasis on their organization, possible strategies for searches, and applications to nucleosides and structure correlations.

As modelling is the other alternative to obtain structural information, Chapter 5 is devoted to empirical force fields methods and molecular mechanics. The main underlying principles of these methods are reviewed together with minimization techniques. Also several applications are presented, dealing with anti-Bredt olefins, solvolysis rate constants, and host-guest systems. A natural extension of these techniques to Monte Carlo and molecular dynamics is described in Chapter 6. After making the necessary distinction between modelling and simulation, this chapter reviews the basic features of simulation techniques by resorting to numerous examples which illustrate their importance in CAMD. Relevant topics such as umbrella sampling of the configuration space, simulated annealing, free-energy pertubation calculations and coupled NMR – molecular dynamics structural refinements are presented. Keeping with structural properties, Chapter 7 focuses on techniques that explore the conformational space, namely distance geometry and model builders. These methods are very important to describe molecular flexibility by using a sampling of the conformational space. The various algorithms on which distance geometry is based are presented, together with applications devoted mainly to peptides. Also several well-known model builder packages that allow flexibility to be taken into account are introduced such as WIZARD, COBRA, CONCORD, CORINA, etc. in this chapter. Chapter 8 is devoted to the calculation of molecular surfaces and volumes, which represent important properties in CAMD, as they are strongly involved in topics such as intermolecular interactions, drug design and protein folding. The various definitions of molecular surfaces are examined, with special emphasis on numerical methods that allow their evaluation, and the concept of voxel is introduced as a useful tool to characterise and compare molecular surfaces and volumes.

Quantum chemistry methods are another indispensable component of CAMD and Chapter 9 present their key features. Starting from the time-independent Schrödinger equation, the usual approximations leading to the

Hartree-Fock-Roothaan equations are introduced by keeping the mathematical developments at a minimum level. The main features of *ab initio* SCF and post-SCF are also described, with the concern that CAMD practitioners probably do not need to be acquainted with all the subtleties of these techniques. As they can be found in practically all the program packages used in CAMD, both semi-empirical and density functional theory methods are also included with several examples illustrating their capabilities. Chapter 10 is devoted to the derivation and visualisation of molecular properties calculated from quantum chemical methods. This entails molecular orbitals, electron densities, molecular electrostatic potentials, electric fields and reactivity indices. The basic procedures used to derive and visualize these properties on computer displays are reviewed, together with numerous applications stressing on their important role in CAMD.

The comparison of structural characteristics within a set of molecules that display common features is a frequent problem in CAMD, relying on the general assumption that similarity in behaviour implies similarity in structure. It is therefore important to quantify molecular similarity and this is the subject of Chapter 11. The most common methods leading to geometrical comparisons are presented, followed by a description of the various tools used to perform common substructure searches. Also, the process of identifying a pharmacophore within large files of structures is tackled and similarity indices used are also reviewed. Proceeding along these lines, Chapter 12 focuses on drug-receptor interactions, i.e. receptor mapping and the pharmacophore approach. The different techniques allowing to search for 'lead' compounds are presented, ranging from the pharmacophore hypothesis to receptor-based design and the automated detection of receptor-binding regions. A broad range of examples are included, i.e. recent applications of drug design. Chapter 13 is devoted to the modelling of proteins and outlines the various techniques known today that perform the following applications on these systems: structural analysis, 2D, 3D and 4D NMR, model building from homology and similarity evaluation. Various examples are presented from recent important investigations.

CAMD is an exciting and burgeoning field, both from a methodological and an application point of view. The authors sincerely hope that the readers will share their enthusiasm about the discipline when using this book.

Acknowledgements

Chapter 5, 7, 11 and 12 have been prepared in collaboration with Professor A. Panaye, whereas Dr. V. Fabart contributed to Chapters 3 and 8.

The authors are very grateful to all their colleagues and students, too numerous to be quoted here, for their support and for their own excitement about CAMD, that helped in their task. They are also grateful to the editors and authors who have agreed to reproduce several figures in this book.

1 Computer graphics: an introduction

Generating synthetic images from a computer plays an outstanding role in *molecular modelling*. So it seemed to us useful to introduce some basic but key features of these techniques. Obviously, our aim is by no means to present a comprehensive study or to examine the algorithms in depth. We wish only to give a short introduction to the main aspects of the discipline, to present some of the problems commonly encountered, and to sketch some general solutions, bearing in mind their applications in molecular modelling. Special emphasis will be put on *interactive graphics*, i.e. techniques which allow the user to generate on-line synthetic images and modify them in real-time. Indeed, the chemist often has to display representations generated by heavy simulation programs for hours of Central Processing Unit (CPU) time (as *ab initio* MO calculations) but he also frequently needs to know rapidly how a chemical system accommodates small structural perturbations. Such questions occur, for instance, when examining molecular flexibility (conformational changes, sterically induced geometrical distortions from standard values, etc.) or dynamic processes, etc.

For more complete detailed presentations of computer graphics, it is suggested that the reader consults specialized textbooks in this field [1–5].

1.1 DISPLAY AND INPUT DEVICES

To carry on a dialogue with images, one needs a material support which is both flexible and very interactive. Any review of chemical graphics is doomed to be quickly outdated because of the rapid evolution in graphics support (the annual growth of graphics stations was predicted to be about 30% by 1992). This is, of course, a matter of hardware development, but it also implies a symbiosis between the computer hardware functions and largely integrated application programs to fully exploit the machine's capabilities. These mutual influences and interactions between software packages and hardware result in a rapidly changing field, made still more complex by the lack of defined standards.

Basically, three main types of graphics hardware systems can be distinguished:

1. *Workstations*: self-governing devices with important storage memory are often provided with a mathematical coprocessor and dedicated processors to perform graphics operations. Stations now offer both high computational power (120 MIPS, 20 MFLOPS are now currently available, and some stations even reach 300 MFLOPS) and efficient graphics treatments, with fair screen resolution (typically 1280×1024 pixels). These devices generally have a user-friendly approach, making it easier to fully exploit the qualities of the machine.
2. *Microcomputers* ("personal computers") are now available with faster clocks, 32 bit processors, mathematical coprocessors and cards to give enlarged graphic resolution on appropriate display monitors. The future of this type of device, however, is not clear, since the additional elements needed to enhance performance significantly increase the cost, giving capabilities still lower than those offered by true workstations.
3. *Graphics terminals* have to be connected to a host computer. However, they now offer opportunities for local treatments. This may be an attractive feature for molecular modelling: complex simulations can be carried out on the host computer, while results are displayed locally (and therefore rapidly) on the terminal.

This evolution is quite rapid. Various (limited) molecular modelling packages are now available at the PC level. However, despite their extended capabilities, PCs still have some difficulties carrying out heavy MO methods for medium-size molecules. With their RISC (reduced instruction set computing) architecture and more efficient buses, workstations offer ever increasing power. Graphics terminals integrate more sophisticated output primitives, and offer better assistance. They still maintain their general purpose bias thanks to the computational resources of the host computer. In any case, the changes tend to offer to non-specialists a more efficient and largely integrated graphics support, so as to free the end-user from the technical problems of representation.

Relevant to the same concern for increasing efficiency, let us note the rapid growing up of information and communication networks. At internal level (within a laboratory) this allows for an optimized use of the resources (CPU power, graphics capabilities, etc.) from the various machines available. At national or international level, particularly thanks to the Internet, it makes possible easy and rapid information exchange (data-file transfer, communication of softwares, access to databases, etc.).

1.1.1 The cathode-ray tube (CRT)

Most display devices now use the standard *cathode-ray tube* (CRT) (Figure 1.1). However, an increasing number of portable terminals are appearing which are provided with a *plasma panel display, light-emitting diodes* (LEDs) or *liquid crystal display* (LCD). In a CRT, a beam of electrons (cathode rays) emitted by an electron gun is directed (after focusing and deflection) towards a phosphor-coated screen. Points reached by this electron bombardment emit a brief spot of light with an intensity depending upon the kinetic energy of the incoming electrons. Deflection plates, set to appropriate voltage levels, allow one to direct the beam towards specified points of the screen. The light emitted by the phosphors is very brief. To maintain an image on the screen, it is necessary to periodically excite the phosphors: this refresh cycle reconstructs the image several times a second (about 30 times/s) so the viewer's eye only perceives a continuous sensation.

According to the way in which the electron beam deflection is monitored, two techniques can be distinguished: *random scan*, used in vector (or calligraphic) systems, and *raster scan*.

1.1.2 Random scan monitors

The screen is considered as a 2D space provided with an orthogonal coordinate system (x,y). To draw a line segment between two points $P_1(X_1, Y_1)$ and $P_2(X_2, Y_2)$,

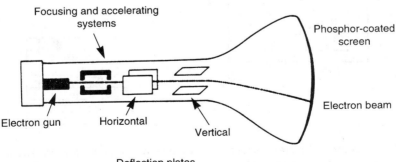

Figure 1.1 Schematic representation of a CRT (adapted from Hearn and Baker [1]).

the electron beam is gradually moved from position P_1 to position P_2, leading to a continuous line on the screen. Graphics patterns are reproduced as line drawings, i.e. as if they are composed of line segments, arcs of curves, etc.

An essential part of all graphics systems is the display processor. Basically, it exploits the digital information from the central processing unit (CPU), converting it into analogue signals to generate pictures or characters on the display screen, and ensuring screen refreshment through a refresh storage area, where the picture definition is kept. Graphics commands are gathered in a display file, where the line style (dashed, dotted, solid, etc.) and the end point coordinates for each line to be represented are defined. These are drawn one line at a time during each refresh cycle (Figure 1.2).

Random scan systems imply only a limited storage area. However, memory requirements directly depend upon the picture's complexity. For complex drawings with many elementary vectors, the refresh cycle duration is too short to refresh the whole picture: this causes "flickering". On the other hand, owing to the very compact mode of storage for picture definition, high interactivity is achieved: drawings can be modified between two refresh cycles. Hardware implementation of varied transformations (rotation, translation, zooming) speeds up image modification, and is quite useful for true real-time animation sequences. Another advantage is the high quality of the drawing produced. Obviously, though, vector systems are unable to represent 3D shapes as solid areas ("solid images").

1.1.3 Raster scan systems

Images here are made up by a matrix of discrete cells, each of which can be made bright or not. The refresh storage area now contains attributes (intensity, colour) for each screen position rather than a list of graphic commands. It is

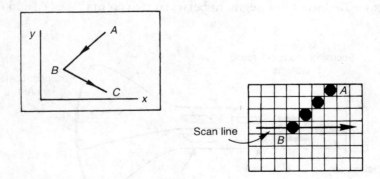

Figure 1.2 Vector and raster displays. Vector system: arrows represent the motion of the electron beam on the screen to draw the AB and BC line segments. Raster display: the status of each pixel along the scan line is systematically explored, one line after the other. Segment AB is represented as a series of enlighted cells on the screen.

usually called a *frame buffer* or *bit map*. As in TV technology, pictures are painted on the screen through a systematic traversal of the frame buffer, along horizontal lines (scan line), one line after the other (possibly with interlacing). Within each scan line the intensity varies according to the content of the corresponding element of the frame buffer: "picture element" or *pixel*. In other words, a pixel may be defined as the smallest element of a display that can be independently assigned a colour or an intensity. By contrast with calligraphic devices, the frame buffer is always examined in its entirety, so raster images do not suffer from any flickering, provided the refresh time is not too long (about 1/60 s). Another interesting point is the ability to represent surfaces filled with colours and shading patterns, well suited to the display of solid objects.

The major drawbacks are:

- its slowness to modify images, since the frame buffer has to be updated (some tricks do exist to speed up this step);
- a somewhat minor precision in line drawing, owing to the definition of a line as a succession of pixels. Aliasing (stairstep effects) may occur if some special treatment is not carried out, or with lower resolution systems (Figure 1.3). Raster systems are thus more naturally devoted to the representation of static images and 3D solid shapes, where rendering effects (hidden part removal, shading, etc.) contribute towards giving more realistic displays.

1.1.4 Resolution

The quality of the image obviously depends upon the number of points which can be displayed on the CRT screen. High precision CRTs provide about 4000 × 4000 points. However, it also depends upon the capability of the upstream

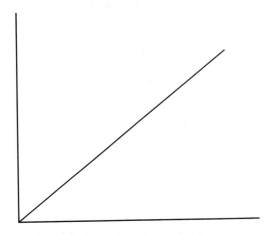

Figure 1.3 Aliasing causes the stepping appearance of oblique lines on a low resolution raster screen.

computer (resolution of the display processor or frame buffer). For raster systems, resolution – initially low at the beginning of the technique – gradually becomes more respectable; for PCs, typical values range from 300 × 200 to 1024 × 1024. For more sophisticated devices, values of 1280 × 1024 are commonly proposed, and can reach up to 2048 × 1568. Note that a resolution of 1024 × 1024 implies handling more than a million differing pixels.

1.1.5 Colour screens and colour models

Most video monitors generate colours by combining various amounts of three primary components. Green, red, blue have been chosen as "primary colours" in the so-called *RGB additive system*. Three separate electron beams are used, each exciting corresponding sets of phosphors. Three primary colours, on a binary status only (on/off), allow for displaying eight colours, as schematized in the RGB cube model (Figure 1.4), where each pure colour is defined by its components along the three axes R, G or B. Combining different intensity levels allows us to generate a wider range of colours, as in *half toning* procedures: expanding each position in the scene to a 2 × 2 pixel grid (giving a five-level setting per primary colour) leads to 125 possible colours.

More flexibility is attained with a *look-up table*: the frame buffer is considered to be formed by a packet of several successive bit maps (Figure 1.5). Each pixel is described as a series of bits. This series defines a binary number used as an address in the look-up table. At this address, the display processor will find the percentages of the elementary colours (Red, Green, Blue). This technique provides a large number of colours to be displayed, but it is a function of the number of bits allowed for the screen description: 4 bitplanes define 16 entries in the look-up table. With 9 bits per entry (three for each of the three components R, G, B), the 16 colours may be selected from among a pallette of 512 possible. These numbers depend upon the display device, but these capabilities are more or less efficiently used by the application programs.

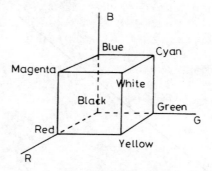

Figure 1.4   The RGB additive colour model. Each colour point within the unit cube is represented by a triple (*r,g,b*) where intensities of the primary colours *r*, *g* and *b* are defined in the range 0, 1. Black is at origin; shades of grey correspond to the main diagonal (black, white).

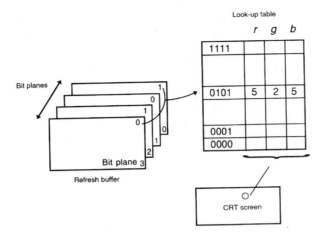

Figure 1.5 Look-up table. Each pixel is decribed as a series of bits characterizing its status in each bit plane (represented here by 0 or 1 in the upper right corner of the bit plane). The binary number so formed defines an address in the table. At this address the intensities of the r, g, b primary colours for that pixel are indicated.

What characteristics [6] can be used to define more precisely what a colour we perceive may be?

Colour (or *hue*) directly relies on the wavelength of the corresponding electromagnetic radiation (Blue: about 400 nm; Green: 530 nm; Red: 700 nm). Let us note also that when a white light is reflected by a coloured object, some frequencies are absorbed, and therefore are missing in the reflected beam.

Other important properties are *brightness*, related to the intensity of the source, and *purity* (or *saturation*). Pastels or pale colours look more "washed" than pure colours Usually, the light coming to the observer encompasses a large frequency range with a dominant frequency over a continuous background (white light). The hue is fixed by the dominant frequency, and purity describes how much the dominant frequency exceeds the background.

Derived from the RGB model, the HSV colour model not only takes into account the **hue** but also its **saturation** (amont of white) and its **value** related to its black component (1 for pure colour, 0 for black). The hexagon (at the basis of the hexcone) corresponds to the RGB cube viewed along its diagonal (representing white light). Saturation is measured on the horizontal radius and value on the hexcone axis (Figure 1.6).

For printers or plotters giving hard copies where we perceive colours by reflection, the Cyan, Magenta, Yellow (CMY) system operates by the subtraction of components within these three primary colours (whereas the RGB system acts by addition)[1].

According to Hearn and Baker [1], the human eye may distinguish about 128 hues and 130 different saturation levels (tints). Discernible shade levels depend upon the colour (about 16 in the blue area, 23 in the yellow), amounting to about 380 000 different "colours". Such a diversity is not easy to handle: treating 128

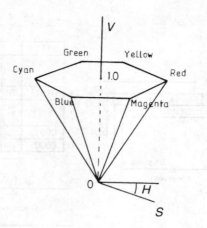

Figure 1.6   The HSV hexcone. Values $V = 0$ and $V = 1$ on the vertical axis (value) correspond respectively to Black and White. Saturation (S) is measured by the distance to the vertical axis (for pure colours, $V = 1$ and $S = 1$). Hue (H) is defined by the angle about the vertical axis (0 for Red to 360°).

hues, 8 saturation levels and 15 value settings amounts to nearly 16 K levels and would require 14 bits per pixel. However, a look-up table can reduce storage requirements. As for black and white pictures, 16 grey intensity levels at least are necessary. 256 levels (on a logarithmic scale) give results that are nearly as good as a photograph [7].

1.1.6 Printer and plotter

Hard copy pictures can be obtained by directing graphics output to a *plotter* or *printer*. Dot matrix printers, commonly available on PCs, for instance, allow for typing pre-defined alphanumeric characters by pressing an inked ribbon onto a sheet of paper. They can also accommodate any dot pattern to be reproduced thanks to appropriate interfaces. Ink jet, laser and electrostatic methods constitute non-impact techniques, where electric field or electrostatic effects are used to transfer a toner or direct a stream of ink towards a sheet of paper. Plotters reproduce line drawings, mainly using ink pens, although other techniques (ink jet, laser) are available for both line drawings or filled area representations.

1.1.7 Interactive input devices

The most familiar device to input data and communicate with the computer is the alphanumeric *keyboard*. Additionally, for graphics applications, cursor control keys can be used to specify the position of a cursor on the screen.

However, to handle the graphics information, some other devices are often more convenient, among which are:

- dials and potentiometers directly monitoring pre-defined graphics functions,
- joystick or trackball,
- lightpen or touch panel for LED units,
- graphics tablet (probably the most accurate method for selecting coordinate positions), and
- mouse, widely used on PCs or workstations for packages provided with pop-up menus.

1.2 ELEMENTARY GRAPHICS PRIMITIVES

Synthetic images are generated thanks to *output primitives*: routines which allow the display processor to generate those elements constituting the picture on the screen. The screen is considered as an orthonormal coordinate system in which the motions of the spot and the end points of the line segments or curves to be drawn are defined. Among the very basic elementary instructions (existing in all systems, with perhaps small syntax differences) is the generation of some simple geometric features: point, line, circle, etc.

Instructions more or less similar to MOVE (x,y) and DRAW (x,y) allow us to locate the spot at the specified position (x,y), or to draw a line segment from the current spot position to point (x,y) (Figure 1.7).

POLY (n,x,y) draws a polygon with the n points (x,y) as successive vertices.

ARC (r,a_1,a_2) draws an arc of a circle centred on the cursor position, with r as the radius. The arc is drawn between angular positions a_1,a_2.

More sophisticated routines include filling areas delimited by polygons with colours or encompass sets of elementary instructions to carry out, as pre-

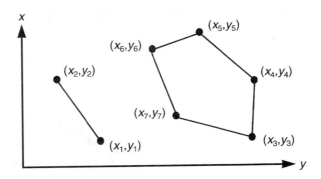

Figure 1.7   Instructions MOVE (x_1,y_1) and DRAW (x_2,y_2) bring the cursor to point (x_1,y_1) and draw a line segment from (x_1,y_1) to (x_2,y_2). The polygon is drawn by instruction POLY (5,array), where array contains the coordinates of the vertices of the polygon.

defined functions, more complex operations: hidden part removal, lighting, shading, display of strings of characters, etc.

According to the capability of the display device and its degree of integration, some of these graphics primitives, as well as subroutines for image manipulation (geometrical or viewing transformations), can be implemented in hardware for faster operation and greater efficiency.

1.2.1 Line drawing

A point plotting routine takes coordinate information, as input, and selects the phosphors to be turned bright as output. In a CRT random-scan device, this is accomplished by setting appropriate deflection voltages on the electron beam. In raster systems, the corresponding screen pixels are enlightened (i.e. the phosphor is submitted to the electron beam) when scanning the frame buffer one line after the other by the display controller. A line is then drawn as a succession of pixels between its two end points (Figure 1.8).

For a line segment between two points (X_1, Y_1) and (X_2, Y_2), the equation is:

$$Y = aX + b$$

where:

$$a = (Y_2 - Y_1)/(X_2 - X_1) \quad b = Y_1 - aX_1$$

Rather than the brute line equation, differential expressions are generally preferred in order to use incremental methods, which are more rapid. In the digital differential analyser (DDA), the pixel position is calculated with:

$$\Delta Y = a\Delta X$$

The current-pixel abscissa X_i is changed by unit steps, starting from the first point ($i=1$) of the segment. The corresponding Y_i value is calculated, and then rounded off:

$$Y_{(i + 1)} = Y_i + a$$

(if the slope is greater than 1, the role of x and y are reversed. If $X_2 < X_1$, ΔX is taken as -1).

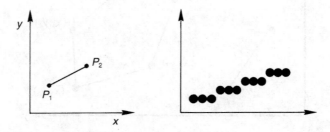

Figure 1.8 In a vector system (left), a line is drawn by continuously moving the electron beam from P_1 to P_2. In a raster system (right), a line is formed by a succession of enlightened pixels.

The *Bresenham's line algorithm* [8] also uses a differential expression: at each iteration step, where one coordinate changes by ± 1, the problem now is to determine if the other coordinate has to be modified or not. Given the current point, the choice between its possible neighbours is set by examining their distance to the real line (distance measured perpendicular to the axis of greatest movement) (Figure 1.9).

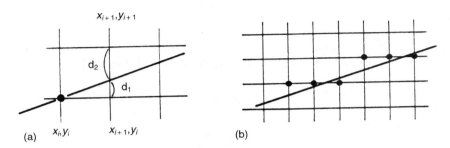

(a) x_i, y_i (b)

Figure 1.9 Bresenham's line algorithm. (a) The pixel centred at x_i, y_i being plotted, the choice between pixels (x_{i+1}, y_i) or (x_{i+1}, y_{i+1}) depends upon the distances d_1 and d_2. Here pixel (x_{i+1}, y_i) is preferred. Only the centre of each pixel is represented. (b) Centres of the enlightened pixels (dots) and the true line to be drawn (solid line).

1.2.2 Antialiasing

Since they are made up of discrete pixels, lines may suffer from a stepping appearance. Antialiasing routines improve the look of the image [9, 10]. In the *sampling approach*, one takes advantage of the fact that the pixels and the line drawn have finite dimensions: each pixel traversed by the line (in fact a rectangular area of about one pixel in width) is given an intensity proportional to the relative surface overlapped by the "line" (Figure 1.10).

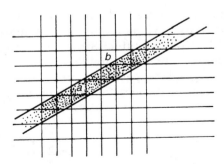

Figure 1.10 Antialiasing. The intensities of pixels a and b are set to about 75% and 10% of the maximum (adapted from Hearn and Baker [1]).

1.2.3 Circle generation

Circles or derived shapes (ellipses, etc.) are largely used in computerized images, and are therefore often proposed as output primitives in graphics systems.

Although Cartesian or polar parametric equations are available:

$$(X-X_c)^2 + (Y-Y_c)^2 - r^2 = 0$$

and:

$$X = X_c + r \cos \theta$$
$$Y = Y_c + r \sin \theta \quad \text{(centre } C(X_c, Y_c) \text{, radius } r \text{) (Figure 1.11)}$$

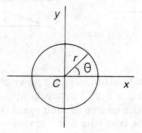

Figure 1.11   Polar coordinates to represent a circle.

The Bresenham's circle algorithm [11] largely reduces the computational task. As for straight line drawing, given one pixel already plotted, the next one is selected from among the neighbours, choosing the closest to the circle.

Similar methods can be used for several other common functions (including polynomial expressions, splines etc.). Otherwise, the individual data points are connected by line segments to give a continuous curve.

1.2.4 Characters

In raster systems characters are easily defined thanks to a dot matrix (usually of 5×7 to 9×14 positions). When required, this matrix is copied at the selected position of the frame buffer. Strings of characters can thus be incorporated in pictures (Figure 1.12). Standard fonts are generally stored in the memory, but some graphics packages provide the capability for user-defined symbols. In vector devices, characters may be drawn as line segments.

Figure 1.12   Generation of a character from vectors or pixels of a dot matrix.

1.2.5 Attributes

All these graphics primitives may be given attributes to vary their appearance on the display: colour, style (dashed, dotted, etc.), width for lines; size, orientation for characters, etc.

Attributes can be defined to be used with a given graphics device, so other devices may be unable to reproduce them. Building a bundled table allows for specifying which attributes are generated by different output devices, and can so activate diverse workstations, thus increasing portability.

1.2.6 Area filling

Area filling intervenes when displaying representations of solid objects limited by coloured faces. For this we need to know what pixels on each scan line are internal to the polygonal contour limiting the face.

Interior stretches can be defined by determining the intersections of the current scan line with the polygon edges (Figure 1.13).

Figure 1.13 For each scan line, the pixels interior to the polygon have to be enlightened.

Note that if a convex polygon is described counterclockwise, an observer moving along the edges always has his left-hand inside the polygon (Figure 1.14).

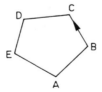

Figure 1.14 The polygon (A,B,...,E) is described counterclockwise.

A convex polygon is described counterclockwise if, for three successive vertices A,B,C, the cross product $AB \times BC$ is positive. Similarly, to determine if a point P is at the left of the vector P_1-P_2, it suffices to examine the sign of $P_1P_2 \times P_1P$:

$$C = (X_2-X_1)(Y-Y_1) - (Y_2-Y_1)(X-X_1)$$

If $C > 0$, P is on the left side of the vector P_1P_2 (Figure 1.15).

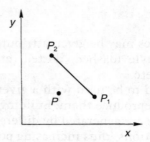

Figure 1.15   Determining the position of point P.

Some caution is necessary when the scan line encounters a vertex: examination of the Y variations along the edges removes the ambiguity (Figure 1.16). Taking into account coherence in the object (a scan line is generally not very different from the preceding one) may largely reduce the computation time. For a detailed presentation of the algorithm and techniques of antialiasing, see Hearn and Baker [1].

1.2.7 Graphics standards and graphics libraries

Portability is an essential condition for designing simulation packages that can be widespread and easily transferred from one installation (workstation, graphics terminal) to another. It is also important to aid the understanding of graphics methods by application programmers, and to make the use of graphics

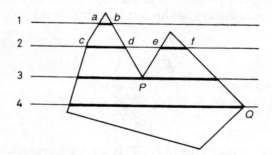

Figure 1.16   Determination of interior stretches. Given a polygon described in a continuous order:

    scan lines 1 and 2: pairing successive intersection points, *a-b* or *c-d*; *e-f* defines the internal stretches;

    scan line 3: at vertex P one edge has decreasing Y, the other increasing Y when moving sequentially on the polygon contour. P is considered as two (or zero) intersection points limiting two internal segments;

    scan line 4: at vertex Q the two edges correspond to variations of Y in the same direction. Only one intersection point is set in Q.

in applications easier. Here are the goals of many efforts for the standardization of graphics software, a task of prime interest but perhaps not yet completely fulfilled. Once defined and accepted as standard, a set of basic graphics functions can be collected in a *library* to be used when developing applications. This avoids writing repeatedly the same instructions for carrying out basic and frequently called graphics operations.

A good graphics library must offer:

- restricted number of basic primitives, easy to use,
- existence of high level primitives,
- non device-dependence, and
- clear definitions, non language-dependent.

As for the last criterion, it can be met by defining a set of functions performing graphics operations independently of any programming language. For their implementation, *language bindings* (language dependent layers) were defined, which allow us to access the various graphics functions within specific languages such as Fortran, Pascal, C, etc.

The possibility to write applications indendent of the physical device can be achieved thanks to the concept of "workstation", in fact an abstract graphics device (not an actual physical device) with typical characteristics (simple logical interfaces controlling the physical device). Of great help in the field are also *inquiry functions*, which allow the user to retrieve information about the capability and state of the graphics system.

The more basic elements which can be easily implemented on any material include 2D (and 3D) tracing. More refined functions correspond to:

- changing the reference system,
- parameterizing the display,
- graphics input,
- error handling,
- utilitarians,
- segments or structures (for local storage of subobjects, formed from several geometrical primitives, and better interaction),
- metafiles, allowing us to store and recall an audit of the calls to graphics functions in a picture generation session. A metafile can be interpreted so as to reproduce the picture created by the original application, or used for long-term graphics data storage.

The Graphical Kernel System (GKS) [12], developed since 1976 and resulting from wide international cooperation, became an ISO (International Standard Organization) standard in 1985. It may be considered as the first international standard in the field, giving a methodological framework, and a source of common understanding and terminology. Originally devoted to 2D graphics, GKS was subsequently extended to 3D.

As for output primitives, although only six types of object are supported by GKS, they allow for building any type of drawing (Figure 1.17). They are

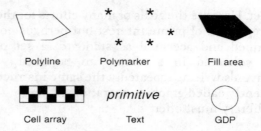

Polyline   Polymarker   Fill area

primitive

Cell array   Text   GDP

Figure 1.17 Examples of GKS primitives [13].

polyline (polygonal line drawing), *marker* (to put small symbols on the vertices, without drawing edges), bit *map* (cell array), *polygon* (filling areas), *text* (for strings of characters) and call to *generalized drawing primitives* (GDP) (workstation dependent). These objects may be given attributes. The library is organized in hierarchical levels of increasing complexity or layers. The lowest level contains only a minimal set of output functions; the highest includes the more complex functions (input, workstation, attributes, segments, metafiles, etc.).

Some related areas, however, are not considered by GKS, and separate standards have been developed. Standardization for graphics interface to output devices is proposed in the computer graphics interface (CGI) system. Standards for archiving and transporting pictures are gathered in the computer graphics metafile (CGM). Methods for real-time graphics are considered in the Programmer's Hierarchical Interactive Graphics Standard (PHIGS), another ISO standard, of largely and rapidly increasing broadcasting [14].

Among its functions, PHIGS allows the user to dynamically edit graphics data, manipulate geometrically-related objects, and modify the relationships between graphics data. It mainly operates by:

- creating structures, a collection of elements (basic building blocks), which specify the information about the graphics objects being created, and allows one to draw the output primitives;
- organizing structures in a structure network: hierarchical tree with an inheritance capability; and
- posting them to a workstation.

Complex output primitives include, for example, the creation of multifaceted surfaces (quadrilateral mesh, triangle strip) and advanced rendering (shading, lighting, invisibility control).

Besides official international standards, some graphics libraries involved in largely widespread commercial software acquired an outstanding importance, and became kinds of non-official standards gathering the efforts of manufacturers. As an example (particularly in the field of molecular modelling) OPEN GL from Silicon Graphics, now licensed to many hardware vendors, is supported by the majority of graphics workstations and PCs.

1.3 GEOMETRICAL TRANSFORMATIONS

The creation and manipulation of computerized images frequently involves a set of geometrical operations to modify the appearance of the display. For example, it may be useful to experiment with different viewing positions, so that certain parts of the scene are more visible. Sometimes, the size of the pictures has to be reduced or magnified according to the regions of interest. In animated sequences, the relative location of the objects constituting the scene reproduced have to be changed continuously. Other applications directly rely on visualization techniques: deriving a 2D image on a monitor screen from a 3D scene usually implies several changes in the coordinate systems defining the objects or their images.

Elementary transformations encompass translation, scaling, rotation, etc. Since such operations frequently occur in computer-image generation, efficient handling methods are needed. Easier handling is attained if these transformations can be formulated to satisfy the following conditions:

- denotation as a single mathematical entity, and
- concatenation: the successive application of single transformations can be expressed as a unique transformation from the starting point A to the final transform D:

$$A \xrightarrow{T_1} B \xrightarrow{T_2} C \xrightarrow{T_3} D$$

$$A \xrightarrow{T} D$$

1.3.1 Matrix representation and homogeneous coordinates

For the sake of clarity, we first consider only a 2D space. It is clear that elementary transformations such as translation by a vector (T_x, T_y), scaling by a vector (S_x, S_y), and rotation of θ about the origin, transform the current point (X, Y) into a new one (X', Y'), according to the following expressions:

Initial point

$$P \ \begin{vmatrix} X \\ Y \end{vmatrix}$$

Transform

$$P' \ \begin{vmatrix} X' \\ Y' \end{vmatrix}$$

Translation

$$T_x$$
$$T_y$$

$$X' = X + T_x$$
$$Y' = Y + T_y$$

Scaling

$$S_x$$
$$S_y$$

$$X' = S_x X$$
$$Y' = S_y Y$$

Rotation

$$\theta$$
measured clockwise

$$X' = X \cos\theta + Y \sin\theta$$
$$Y' = -X \sin\theta + Y \cos\theta$$

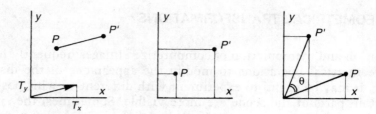

All these operations (including translations) are readily expressed in a more convenient matrix form, thanks to the introduction of *homogeneous coordinates*, representing point (X,Y) in the 2D space by the triple $(X,Y,1)$[1]:

$$P \longrightarrow P' \quad (X',Y',1) = (X,Y,1)\,\mathbf{T}$$

where $\mathbf{T}$ is the transformation matrix expressed as:

Translation	Scaling	Rotation

$$
\begin{vmatrix} 1 & 0 & 0 \\ 0 & 1 & 0 \\ T_x & T_y & 1 \end{vmatrix}
\qquad
\begin{vmatrix} S_x & 0 & 0 \\ 0 & S_y & 0 \\ 0 & 0 & 1 \end{vmatrix}
\qquad
\begin{vmatrix} \cos\theta & -\sin\theta & 0 \\ \sin\theta & \cos\theta & 0 \\ 0 & 0 & 1 \end{vmatrix}
$$

Using this matrix formalism (and here is one of its main advantages) sequential transformations are easily concatenated. The matrix representing the overall transformation is simply the product of the matrices of the individual transformations.

Note that the order of the transformations must be strictly preserved when multiplying matrices. For example:

$$
p_2 = p\mathbf{T} \qquad \mathbf{T}
\begin{cases}
\begin{array}{l}
P \\
\downarrow\ \mathbf{T}_1 \\
P_1 \\
\downarrow\ \mathbf{T}_2 \\
P_2
\end{array}
\end{cases}
\qquad
\begin{array}{l}
p_1 = p\mathbf{T}_1 \\[2mm]
p_2 = p_1\mathbf{T}_2 = (p\mathbf{T}_1)\mathbf{T}_2 \\
\quad = p(\mathbf{T}_1\mathbf{T}_2)
\end{array}
$$

So: $\mathbf{T} = \mathbf{T}_1\,\mathbf{T}_2$

where p_i represents the row vector $(X_i,Y_i,1)$ associated with point P_i.

Various other elementary transformations – reflection (symmetry) in one of the reference axes or in the origin, X or Y shear – can similarly be expressed simply by matrices.

Note that adding a third component to the duple (X,Y) does not weigh down the calculations: the third column of the matrices is always $(0,0,1)$ so for any

---

[1] In homogeneous coordinates, any point $P(X,Y)$ of the 2D space is represented by the triple (fX,fY,f) where $f\ (\neq 0)$ can be considered as a scale factor. It comes to map the space of ordinary coordinates (dimension n) to a space of dimension $n+1$: the space of homogeneous coordinates. Conversely, the n space may be considered as a projection of the $n+1$ space. These homogeneous coordinates are very useful for manipulating graphics in operations such as projections, size changes and geometrical transformations.

2D transformation only the usual (2×2) matrix needs to be stored, the full matrix being easily restored simply by attaching the third column. Similarly, rather than using the general algorithm for multiplication, one can speed up the process, taking into account the nature of the third column, and avoiding the evaluation of non-intervening elements.

With raster systems, some tricks, such as copying blocks (bit block transfer) for some simple transformations as translation or 90° rotation, rather than carrying out the usual matrix operations, give extended capabilities in, for example, animated sequences.

1.3.2 Inverse transformations

One sometimes needs to know what initial point a given (transformed) point comes from:

$$P'\ (X',Y') \quad \xleftarrow{\ \ \mathbf{T}\ \ } \quad P\ (X,Y)$$
$$\xrightarrow{\ \ \mathbf{T}^{-1}\ \ }$$

This inverse transformation just corresponds to the inverse matrix $\mathbf{T}^{-1}$, as can easily be seen:

from $p' = p\mathbf{T}$ it comes
$p'\mathbf{T}^{-1} = (p\mathbf{T})\mathbf{T}^{-1} = p(\mathbf{T}\mathbf{T}^{-1})$
i.e. $p'\mathbf{T}^{-1} = p$

Such inverse transformations intervene, for instance, to return to an original coordinate system, temporarily changed to more easily perform some process. Similarly, moving an object in a direction can be treated as moving the reference axes in the opposite direction.

As another example, suppose a polygon is created with a pattern filling it. Transforming the polygon also requires transforming the pattern inside. An easy way to determine the attributes of a point P' in the new display is then to determine to what initial point P, P' corresponds, and to examine the status of point P in the pattern table (intensity, grey level, colour attributes). This is easy using the concept of inverse transformation.

1.3.3 3D transformations

The matrix formalism, presented above in 2D, is easily extended to 3D space. The possibility of concatenating sequences of transformations so as to represent the result as a product of matrices is, of course, maintained.

A point (X,Y,Z) is now represented in homogeneous coordinates by $(X,Y,Z,1)$. The matrices associated with the more frequent transformations are gathered in Table 1.1. Let us note that rotations are now more complex, since the rotation axis has to be specified.

Table 1.1 Common transformation matrics.

| Translation | $\begin{vmatrix} 1 & 0 & 0 & 0 \\ 0 & 1 & 0 & 0 \\ 0 & 0 & 1 & 0 \\ T_x & T_y & T_z & 1 \end{vmatrix}$ | |

Scaling
(centre in the origin)

$$\begin{vmatrix} S_x & 0 & 0 & 0 \\ 0 & S_y & 0 & 0 \\ 0 & 0 & S_z & 0 \\ 0 & 0 & 0 & 1 \end{vmatrix}$$

Rotation: (angles measured clockwise when looking along the rotation axis from its positive part towards the origin

$$\begin{vmatrix} \cos\theta & -\sin\theta & 0 & 0 \\ \sin\theta & \cos\theta & 0 & 0 \\ 0 & 0 & 1 & 0 \\ 0 & 0 & 0 & 1 \end{vmatrix}$$

$$\begin{vmatrix} \cos\theta & 0 & \sin\theta & 0 \\ 0 & 1 & 0 & 0 \\ -\sin\theta & 0 & \cos\theta & 0 \\ 0 & 0 & 0 & 1 \end{vmatrix}$$

$$\begin{vmatrix} 1 & 0 & 0 & 0 \\ 0 & \cos\theta & -\sin\theta & 0 \\ 0 & \sin\theta & \cos\theta & 0 \\ 0 & 0 & 0 & 1 \end{vmatrix}$$

1.3.4 Rotation about an arbitrary axis

Rotation around an axis not aligned with the coordinate axes can be performed by a sequence of primitive transformations.

Given an axis passing through point (X_o, Y_o, Z_o) and with (a,b,c) as direction cosines, the following sequence is carried out:

a   Translation defining a new reference system with origin in (X_o, Y_o, Z_o): matrix **T**.
b   Rotation around the new x and y axes to bring the unit vector (a,b,c) on to the z axis: matrices $\mathbf{R}_x$ and $\mathbf{R}_y$.

c   Rotation θ around the z axis of the new coordinate system: matrix $\mathbf{R}_\theta$.

d   inverse of step b } to return to the original axis system.

e   inverse of step a

So the overall transformation may be written as:

$$\mathbf{T}\,\mathbf{R}_x\,\mathbf{R}_y\,\mathbf{R}_\theta\,\mathbf{R}_y^{-1}\,\mathbf{R}_x^{-1}\,\mathbf{T}^{-1}$$

$$\mathbf{T} = \begin{vmatrix} 1 & 0 & 0 & 0 \\ 0 & 1 & 0 & 0 \\ 0 & 0 & 1 & 0 \\ -X_o & -Y_o & -Z_o & 1 \end{vmatrix} \qquad \mathbf{R}_x = \begin{vmatrix} 1 & 0 & 0 & 0 \\ 0 & c/v & b/v & 0 \\ 0 & -b/v & c/v & 0 \\ 0 & 0 & 0 & 1 \end{vmatrix}$$

$$\mathbf{R}_y = \begin{vmatrix} v & 0 & a & 0 \\ 0 & 1 & 0 & 0 \\ -a & 0 & v & 0 \\ 0 & 0 & 0 & 1 \end{vmatrix} \qquad \mathbf{R}_\theta = \begin{vmatrix} \cos\theta & -\sin\theta & 0 & 0 \\ \sin\theta & \cos\theta & 0 & 0 \\ 0 & 0 & 1 & 0 \\ 0 & 0 & 0 & 1 \end{vmatrix}$$

where:
$$v = (b^2 + c^2)^{\frac{1}{2}}$$

rotation about the x axis until the axis of rotation is in the xz plane

$$\cos\alpha = c/v$$
$$\sin\alpha = -b/v$$

situation after the R_x rotation and rotation until the axis of rotation corresponds to the z axis

$$\cos\beta = v$$
$$\sin\beta = a$$

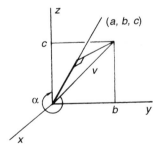

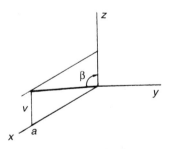

(From Newman and Sproull [4] with permission).

REFERENCES

1. D. Hearn and M.P. Baker *Computer Graphics*. Prentice Hall, London, 1986.
2. D.F. Rogers *Procedural Elements for Computer Graphics*. McGraw Hill, New York, 1985.
3. S. Harrington *Computer Graphics, A Programming Approach*. McGraw Hill, New York, 1987.

4. W.M. Newman and R.F. Sproull *Principles of Interactive Computer Graphics*. McGraw Hill, New York, 1979.
5. J. Foley and A. Van Dam *Fundamentals of Interactive Computer Graphics*. Addison Wesley, Reading, MA, 1981.
6. G.W. Meyer *The Visual Computer*. **2:** 1986; 278–290.
7. P. Schweizer *Infographie*. II. Presses Polytechniques Romandes, Lausanne. Switzerland, 1987, p. 824.
8. J.E. Bresenham *IBM Syst. J.*, **4:** 1965; 25–30.
9. F.C. Crow *IEEE Computer Graphics and Applications*, **1:** 1981; 40–47.
10. A. Fujimoto and K. Iwada *IEEE Computer Graphics and Applications*, **4:** 1984; 11–23.
11. J.E. Bresenham *Comm. ACM*, **20:** 1977; 100–106.
12. G. Enderle, K. Kansy and G. Pfaff *Computer Graphics Programming. GKS The Graphics Standard*, Springer-Verlag, Berlin, 1984.
13. *DEC GKS User Manual*, Digital Equipment Corporation, April 1989.
14. *DEC PHIGS*, Digital Equipment Corporation, August 1990.

2 Computer graphics: towards realistic images

Computer generated representation of 3D objects (here molecular features) basically involves a series of operations not so different to those needed when taking a photograph of a scene in the real world. Indeed, in the generation of synthetic images, the computer has often been compared to a "synthetic camera".

First the photographer has to select his location with respect to the scene and direction of vision. This determines what part of the scene will be represented (that appearing in the viewfinder). An image is formed on the planar surface of the film. Then the development generates the print. Similarly, to create computerized images, we first have to select a *viewpoint*, a *viewing direction* and fix a *window* demarcating what will be reproduced. Finally, the drawing will be displayed on a given area of the screen: the *viewport*. This will be carried out using graphics functions (output primitives) to draw line segments, curves or paint filled areas.

Before examining these viewing operations and the basic graphics functions, we first introduce the most common representations used in computer graphics to reproduce objects of the 3D space. Owing to some specific features of the chemical "objects", either defined as real bodies (the usual wood or plastic molecular models) or as more conceptual, abstract entities (for displaying electronic properties), we will deal in more detail with the representation of molecular shapes in a later chapter.

2.1 REPRESENTATION OF 3D OBJECTS

Solid objects are often represented by shaping their outer surface like a skin around them: this is *surface modelling*, very popular for applications involving free form curves. In raster systems, surfaces are commonly described as polyhedra, limited by a network of polygon-shaped faces [1]. These polyhedra either exactly represent the surface or are selected to give a good approximation of complex objects, provided the number of faces is sufficient. In calligraphic displays, complex objects may be defined from a mesh of polygonal surface patches: succesive sections by sets of parallel planes along orthogonal directions give packets of contours (cross section outlines) leading to "chicken wire" models [2].

Such serial sectioning methods can also be useful in raster systems to derive a network of facets limiting a solid, thanks to triangulation [3]. Some procedures have been proposed for triangulation between successive contours, or for analytically defined molecular surfaces [4–6] (Figure 2.1).

Other techniques involve *solid geometry* [7–9], which combines simple shapes through logical operations (union, intersection). Discrete modelling assembles cells [10] to represent the body of the object. Rather than using a set of similar elementary cells, octree encoding often gives a very convenient data structure to make easier this edification [11,12] (Figure 2.2).

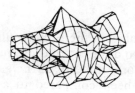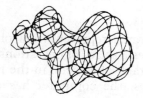

Figure 2.1   Representation of a solid by a planar faced polyhedron or by a mesh of contour outlines.

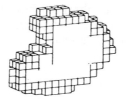

Figure 2.2 Discrete modelling: representation of an object by a set of elementary cells (volume elements or voxels) (from Kronlof and Tamminen with permission [10]).

2.1.1 Polyhedra

Polygons defining the faces are identified by the coordinates of the vertices and the sequential order to join them. For easier display, it may be efficient to organize the data in tables containing the geometrical characteristics and the rendering attributes (colour, shading, etc.). For instance, one table may contain the coordinates of the vertices, another the list of the edges with their end points, and a third the polygons with their constituting edges (Figure 2.3).

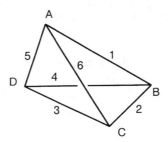

Figure 2.3 Identification of a polyhedron.

Vertices			
B	X_B	Y_B	Z_B
C	X_C	Y_C	Z_C

Edges		
3	C	D
4	B	D
5	A	D

Faces		
c	A	B	D
d	A	B	C

2.1.2 Plane equation

The plane equation for a face can be written:

$$aX + bY + cZ + d = 0 \qquad (1)$$

Coefficients a,b,c,d are readily obtained from three non-colinear vertices by writing that their coordinates (X_i,Y_i,Z_i) satisfy equation (1) and solving the system obtained. It is easily checked that these coefficients can be derived from the determinant Δ:

$$\begin{vmatrix} X_i & Y_i & Z_i \\ X_j & Y_j & Z_j \\ X_k & Y_k & Z_k \end{vmatrix} = \Delta$$

So, $d = -\Delta$ and a,b,c are equal to the determinants formed by replacing in Δ, respectively, the first, second or third column elements by 1. Note also that a,b,c are the components of a vector normal to the face. Plane equation and normal coefficients a,b,c will be used later for hidden part removal or shading problems.

2.1.3 Curved surfaces

Curved surfaces may be approximated by a polyhedron, with a sufficient number of planar faces. Otherwise, one can use a set of curved paths to define the surface. A parametric representation is often the most convenient when an analytical form is available.

So a spherical surface can be described with the equations:

$$X = r \sin u \cos v$$
$$Y = r \sin u \sin v$$
$$Z = r \cos u$$

where r = radius and u and v specify the azimuth and longitude of the current point (Figure 2.4).

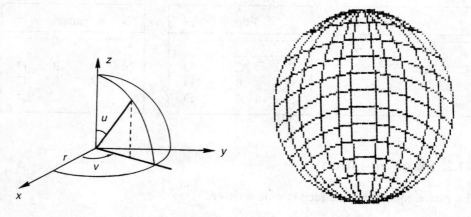

Figure 2.4   Sphere described as a set of curved paths. u = constant, v = constant.

Complex situations may require the joining of curve segments. Smooth transitions are obtained ensuring continuity conditions of increasing order: the curves meet at the junction point (zero order continuity), with the same tangent direction (first order) and same curvature (second order) (Figure 2.5).

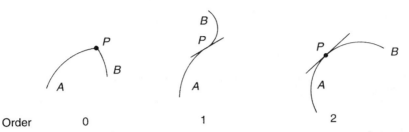

Order   0   1   2

Figure 2.5   Curve junction at various degrees of continuity (from Newman and Sproull with permission [16]).

2.1.4 Splines and Bezier curves and surfaces

We will not develop these points, since until now such methods have not been widely used in molecular graphics, whereas they are largely widespread in computer assisted design to interactively adjust the shape of a curve according to the user's wishes.

The principle is to define a curve thanks to a sequence of polynomial functions, formed from the coordinates of user-defined *control points*. These control points monitor the general shape of the curve, and allow for modifying it interactively [13–15]:

(a) Bezier curve

$$\text{Current point } \boldsymbol{P(u)} \quad \begin{vmatrix} X(u) & \text{is defined thanks} \\ Y(u) & \text{to } (n+1) \text{ control} \\ Z(u) & \text{points } \boldsymbol{p_i} \end{vmatrix}$$

$$\boldsymbol{P(u)} = \sum_{i=0}^{n} \boldsymbol{p_i} B_{i,n}(u) \quad 0 \le u \le 1$$
$$B_{i,n}(u) = C(n,i) u^i (1-u)^{n-i}$$
$$C(n,i) = (n!) / i!(n-i)!$$

A Bezier curve always passes through the end control points and lies within the convex hull of the control points (Figure 2.6).

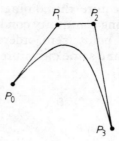

Figure 2.6 Bezier curve defined by four control points (order $n = 3$) (from Newman and Sproull with permission [16]).

(b) B-splines

B-spines (order $k-1$): current point $P(u)$ is defined from the $n + 1$ control point p_i, (with i varying from 0 to n) (Figure 2.7):

$$P(u) = \sum_{i=0}^{n} p_i N_{i,k}(u) \quad \text{with } 0 < u < n - k + 2$$

$$N_{i,1}(u) = \begin{cases} 1 & \text{if } u_i \le u < u_{i+1} \\ 0 & \text{otherwise} \end{cases}$$

$$N_{i,k}(u) = \frac{(u - u_i)N_{i,k-1}(u)}{u_{i+k-1} - u_i} + \frac{(u_{i+k} - u)N_{i+1,k-1}(u)}{u_{i+k} - u_{i+1}}$$

$$\left(\text{convention } \frac{0}{0} = 0 \right)$$

breakpoints u_j define $n + k$ subintervals for u

$$\begin{aligned} u_j &= 0 & \text{if } j < k \\ \text{breakpoints } u_j \qquad\quad j - k + 1 & & \text{if } k \le j \le n \\ 0 \le j \le n + k \qquad\quad n - k + 2 & & \text{if } j > n \end{aligned}$$

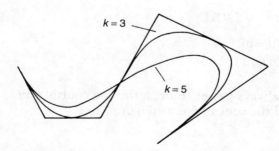

Figure 2.7 B-splines of order 2 and 4, from 6 control points (from Newman and Sproull with permission [16]).

As seen with other curves (for instance, circles to represent a sphere), sets of such splines or Bezier functions (in a biparametric form) can be used to generate curved surfaces defined through control points (Figure 2.8).

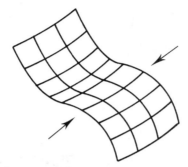

Figure 2.8 Bezier surface created by joining two Bezier patches at a boundary (indicated by arrows). First order continuity is ensured by the choice of the control points (from Newman and Sproull with permission [16]).

2.1.5 Octree structure

Solid objects can be represented as being formed by the summation of identical small elementary volumes (typically small cubes) called "voxels" (volume elements) contained in the object. This is the principle of discrete solid modelling, largely used to represent free-form objects and also in medical applications ("computer assisted tomography") [10].

Rather than using any arbitrary, but fixed, resolution, the octree method prefers a hierarchical-tree organization through recursive subdivision. The data structure adopted takes advantage of the space-coherence of the object. It reduces memory space requirements and allows for easy downstream treatments (hidden part removal, logical operations, etc.).

An octree is created by recursively dividing space into eight octants. Each node (which corresponds to a region of space) is assigned eight data elements for storing the characteristics of the eight octants generated. If an octant is homogeneous (same colour, location either totally internal or external to the object) its attributes are encoded into the corresponding data element. Otherwise, the subdivision is continued until each region of space becomes homogeneous. A system of pointers ensures the correspondence between the divided octant and the resulting next nodes in the octree structure. Algorithms have been proposed to built an octree structure for objects defined as a mesh of polygons or resulting from solid constructive geometry (this can be performed by testing, octant by octant, a *3D box* containing the object) (Figure 2.9).

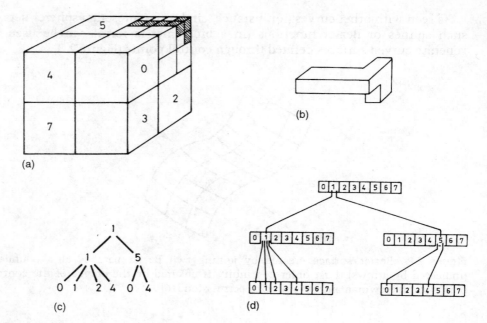

Figure 2.9 Octree encoding of an object. (a) Subdivision into octants; (b) object; (c) octree; (d) data elements for the octree nodes.

For the representation of an object on a video screen, a projection is made onto a plane, leading to a similar *quadtree structure*: each node now corresponds to four data elements. The quadtree structure is then mapped on to the frame buffer (Figure 2.10).

2.1.6 Solid constructive geometry

Objects are generated by combining simpler shapes, considered as primitives, through Boolean operations of union, intersection and difference. For easy welding and cutting, a convenient data structure has been proposed by Wyvill and Kunii [7]. It is provided by a directed acyclic graph, where each vertex represents a primitive or a combination of them. A directed acyclic graph is used instead of a tree structure, since the same subobject can be used several times in the description. Data processing is then performed thanks to a modified octree (Figure 2.11).

Such an approach may be interesting in chemistry, since the representation of usual space filling CPK (Carey, Pauling, Koltan) or ball and stick models only needs a few simple primitives: spheres (for atoms), cylinders (for bonds), etc. Note also that, for building such elementary shapes, one can take advantage of the existing symmetry, and generate the whole shape by rotational or translational sweeping of a simpler figure (great circle or generatrix, for instance) through the appropriate region of space.

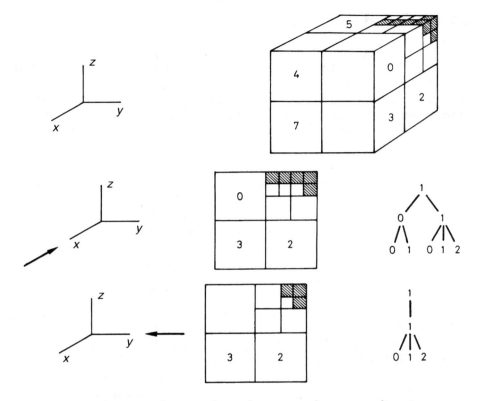

Figure 2.10 Resulting quadtrees for viewing from x or y directions.

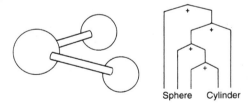

Figure 2.11 Solid constructive geometry: building a molecular model from simple primitives (sphere and cylinder) (from Wyvill and Kunii with permission [7]).

2.2 Viewing, windowing and clipping

Mapping a three dimensional scene on a planar viewing surface involves several operations: projection, change of the coordinate system, clipping, etc., usually gathered within the generic term of *viewing operations*. We briefly examine them in this section.

2.2.1 Projection

To get a planar representation of a scene on a screen, the main operation required is obviously projection. We only consider here *parallel projection*. Points of the object are projected along parallel lines, so that the relative dimensions are maintained (Figure 2.12). The projection is said to be orthographic when the projection lines are perpendicular to the projection plane (otherwise it is an oblique projection).

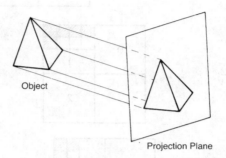

Figure 2.12 Oblique parallel projection.

The transformation matrix for an oblique projection along a direction defined by its director cosines (a,b,c) is (see also Figure 2.13):

$$\begin{vmatrix} 1 & 0 & 0 & 0 \\ 0 & 1 & 0 & 0 \\ -a/c & -b/c & 0 & 0 \\ 0 & 0 & 0 & 1 \end{vmatrix}$$

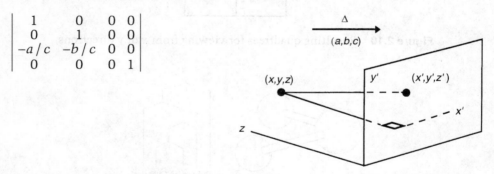

Figure 2.13 Transformation matrix for an oblique projection (adapted from Hearn and Baker [1]*) *reference from Chapter 1.

Perspective projection, where projection lines converge towards a common centre, leads to more realistic pictures, since distant objects appear smaller, but as a counterpart, the relative dimensions of objects are not maintained.

2.2.2 Viewing transformation

To locate on the display screen the projections of the objects constituting the scene, we first have to convert their position from the *world coordinate-*

system in which they are defined to a *viewing coordinate system* (eye coordinate system): its origin is the *viewpoint* and its z axis corresponds to the direction of view (Figure 2.14). This is performed using transformation matrices (see p 17). Let us note that the eye coordinate-system is generally chosen as *left handed* (the z axis pointing forward from the viewpoint), whereas the world coordinate system is right-handed. So, at some step z must be changed to $-z^1$.

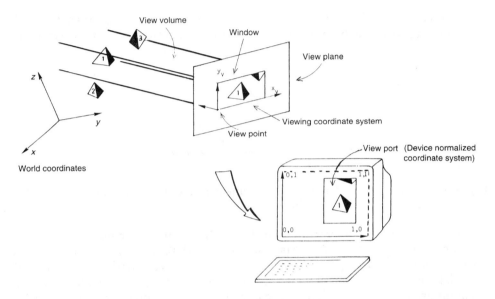

Figure 2.14 Various coordinate systems. Only objects within the view volume are displayed.

2.2.3 Clipping

On a photograph, only that part of the scene observed in the viewfinder will appear on the print. Similarly, that part displayed on the screen corresponds to a *window* in the view plane. This window, with its projection lines, defines a *viewing volume* which limits the portion of the world the viewer can see. Futhermore, this volume can be reduced by near and far planes: this allows for eliminating objects far away from the viewer or near objects that can mask those located behind (Figure 2.15). In molecular modelling, for example, we can thus eliminate the nearest parts of a molecular surface to preserve some view of the molecular framework.

---

[1] A trick (among others) to remember the orientation of axes in a left-handed system is: for the left hand, if the thumb and the first finger align with the x and y directions, the second finger points towards the z direction.

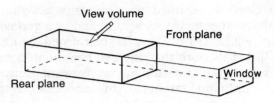

Figure 2.15   The view volume can be limited by front and rear planes.

We have now to discard all line segments (or parts of them) located outside the viewing volume before performing the projection on the view plane (Figure 2.16).

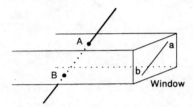

Figure 2.16   3D clipping. Only the line segment inside the view volume (AB) will be visible (and is projected as ab on the window).

This 3D clipping operation can be seen as an extension of the two dimensional clipping operations (see section 2.2.4). Basically, one examines the position of each line segment of the object successively with respect to each boundary plane limiting the viewing volume: if the two end points of the segment are on the inside face of the current boundary plane, the full segment is maintained. If the two points are outside it is discarded. If the segment intersects the boundary plane, one calculates the coordinate of the intersection point to determine that part of the segment that will be saved (Figure 2.17).

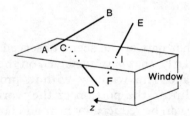

Figure 2.17   Clipping by the upper plane of the view volume: CD and FI saved, AB and IE discarded.

After the clipping operation, the window is mapped on a projection viewport (rectangular) part of the screen, where the window's content will be represented. As stressed by Newman and Sproull [16]: "the window defines what we want to display, the viewport specifies where on the screen to put it". This corresponds to the "windowing transformation."

To allow for using varied display devices within a same system, it is more convenient to describe the viewport with *normalized device coordinates* (varying from 0 to 1 for the full screen area). Finally, the normalized coordinates of the displayed features are converted to *device coordinates* for specific display on the monitor used to get the picture on screen. For the sake of efficiency, actual implementation of this clipping treatment prefers a somewhat differing scheme: the viewing volume (in the eye coordinate system) is transformed in a viewbox (in a normalized screen coordinate system). Clipping is more easily performed at this step. Then data can be transformed into device (screen) coordinates for display. They also have the appropriate form to be passed to hidden-part removal routines for more realistic pictures.

With VLSI (Very Large Scale Integration) technology, viewing operations can be hardware implemented using a pipeline of chips performing trans-formations, projections, clipping and conversion to device coordinates.

Let us also note that the geometrical transformations the user wishes to apply can be performed either before or after the viewing operations: when intervening before, on some of the segments defining the scene, they allow one to modify the relative location of some objects. If applied after, they only change the X, Y coordinates in the screen system: the objects' location on the screen is modified, but the angle along which they are seen is not changed.

2.2.4 2D clipping

To discuss the clipping operations in more depth, we prefer, for the sake of clarity, to present the approaches proposed in 2D graphics, since they give a good introduction to the processes used in 3D.

As previously said, clipping eliminates parts of the scene (in 2D the "scene" is a picture defined in some world coordinate system) which will not be represented on the display screen, i.e. points of the picture outside the window (one can alternatively eliminate screen points outside the viewport). An essential part of clipping operations corresponds to *line clipping*, since line segments are used to define polygons limiting complex objects either in raster or calligraphic graphics systems. Line clipping is performed by examining if the end points are within or external to the window (respectively the viewport), as indicated for 3D clipping. An essential element has been proposed by Cohen and Sutherland [17]. They suggest defining the location of each line end point with respect to the window using a four digit code: so the first (left) bit is set to 1 if the point is above the window (otherwise zero). The bits (from left to right) shown in Figure 2.18, when set to 1, similarly specify a location below, right or left.

This code allows for a fast comparison of line end-points and window boundaries, through examination of the relevant bits and logical operations. So it appears to be very efficient for determination of the lines to be immediately discarded as totally external to the window, or saved as completely inside. Then, the lines intersecting the window edges are examined to determine

1001	1000	1010
0001	0000	0010
0101	0100	0110

Figure 2.18 Binary codes allow for locating line end-points in the nine regions defined by the window edges. The window corresponds to the area encoded 0000.

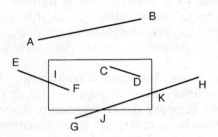

Figure 2.19 Only segments CD, IF, JK have to be saved.

what part will be maintained (Figure 2.19). Other algorithms have also been proposed, using midpoint subdivision of intersecting lines [18] or working on parametric equations such as that of Liang and Barsky [19, 19a].

Polygons can be clipped with these algorithms by successively treating all their edges (Figure 2.20). The outline created is no longer closed, but this causes no problem for line drawings generated from a calligraphic device. On the contrary, when polygons limit filled areas, we need clipping to produce closed outlines (Figure 2.21). Sutherland and Hodgman [20] proposed clipping the polygon against one of the four window edges, then clipping the resulting shape against another edge, and so on. The algorithm operates on the vertices defining the polygon: each vertex is in turn compared to the window boundaries. If the edge between two successive vertices crosses a boundary, the intersection point becomes a new vertex, added to the output polygon, and the (old) external vertex is discarded.

To limit storage requirements, a point (original vertex or intersection) is saved only after comparison with the four window boundaries. At the end, a closing routine is applied for the first and last point clipped against each window edge. Special methods have been proposed for non convex polygons [20].

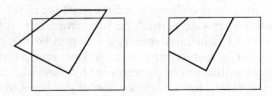

Figure 2.20 Clipping hollow polygons (heavy lines) leads to non-closed outputs.

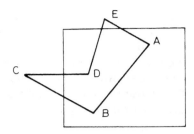 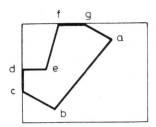

Figure 2.21 Clipping a filled polygon against the window edges. Lower case letters correspond to the vertices or intersections saved (starting from A).

Blanking

Complementary to clipping, which erases objects outside the window, blanking allows for erasing parts of the display which are contained within a selected window area. This may be useful to superimpose pictures, to add labels, etc.

2.3 SEGMENTS

For some applications it may be interesting to *modify only part* of a complex image. Such situations may occur in the display of animated sequences representing dynamic processes, but also when the chemist, in interactive sessions, is looking for a better fit between two interacting molecules, for instance a small drug or ligand binding to a large biomolecule (protein, part of a receptor, etc.). Such selective modifications are made easier if the various objects or groups of objects to be moved are defined as separate modules. This is the role of *segments*.

A segment, considered as a part of an overall display, gathers a set of instructions of the display file representing graphical primitives and which can be manipulated as a single unit. Segments are given attributes monitoring:

- visibility,
- geometric transformations (to change size, position), and
- priority (in their order of display, mainly for raster devices).

The display file must be organized so as to reflect and take advantage of this subpicture structure. Manipulation of segmented display files can be efficiently carried out with a minimal set of functions, not so far in principle from those used in handling sequential disk files: opening, closing, deleting. An important feature is the ability to make segments visible or not. In a calligraphic system, this is obtained by adding the segment to the refresh cycle for its display or removing it for erasing. In raster systems, an internal data structure is organized ("pseudo display file") which allows for updating the frame buffer according to the segment attributes.

The display file may be organized by gathering the properties of segments in a linear array or using a linked list: for a given segment, instructions are not stored in order but a pointer gives access to the successive graphical instructions to execute. Storage through fixed-length blocks is also possible. Each segment is identified by an integer. To handle segments easily, their main attributes are gathered in a segment directory (address, length, visibility), or can be found in headers located at the beginning of each segment block and accessed through pointers.

2.4 HIDDEN LINES AND SURFACES REMOVAL

Removal of hidden parts is essential to produce realistic-looking images. In real life, only front parts of objects are attainable by light and can therefore be visible, the bulk of opaque material hiding on the back faces. On the contrary, in computer generated images, all parts of the objects are displayed with no thought as to how the real objects would appear. Adapted algorithms are therefore necessary to maintain only those lines or volumes which are visible for a given observer position.

The same problem occurs in wire frame drawings: faces of objects are replaced by their bounding contours, but it is difficult from these wire outlines to judge which parts or lines lie in front and which belong to the back. Suppressing non-visible lines removes the ambiguity, as can be seen in the scheme shown in Figure 2.22.

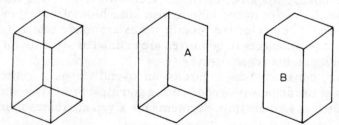

Figure 2.22 Without hidden line removal, the same wire frame drawing may be interpreted as a view from down left (A) or top right (B) (from Rogers with permission [2]*).
*Reference from Chapter 1.

As an essential part in producing realistic images, hidden part removal problems have received continual interest and prompted numerous algorithms. However, no single answer is now proposed and has even to be expected: the efficiency of the methods proposed depend upon the complexity of the scene being represented or the image being displayed, and also on the goals in mind. Handling highly interactive images needs different requirements than producing still-life looking shaded surfaces.

Despite a great diversity, some general features may, however, be stressed:

- Hidden part removal relies on a *geometrical sorting* of what is near the observer, and what is further away (the order in which depth and lateral sorting are carried out may vary, but is not of prime importance).
- *Coherence*: the fact that a scene generally exhibits locally some regularity is capitalized at various degrees to limit the calculations and speed up the process. For reviews, see, for instance, Sutherland *et al.* [21] and Clark [22].

Algorithms work either on the *object space* (on the real location of the objects in the scene), or in the *image space* (looking only at what must appear on the screen). These later applications are more clearly relevant to raster devices (examining each pixel of the image) and are largely dedicated to hidden surface removal, while the former applications can also treat hidden lines for calligraphic pictures. When working in the object space, the computer cost increases with the number of objects, whereas for image space methods it depends upon the complexity of the image. Note also that image space algorithms are implemented in the screen coordinate and carried out with the (limited) precision of the screen. Large zooming on such images may result in some shortcomings in the visual result. On the contrary, object space algorithms (working in the space where objects are defined) can be carried out with a high precision so that images can be enlarged without any problems.

Although algorithms working on curved surfaces have been proposed (see below), we mainly consider here objects approximated by polyhedra, i.e. volumes limited by planar faces (polygons). Treatment is easier for convex polygons: if it is not the case it is still possible to arrive at convex shapes by subdivision of the original faces.

2.4.1 Preliminary treatment: back face elimination

Hidden part removal is an onerous task, and any method allowing for limiting the calculations or speeding up the process is of great interest. Back faces of the object (those situated at the rear with respect to the observer) are obviously non-visible, and can be safely eliminated to limit the number of faces submitted to further treatment. The process gives only partial solutions: *faces potentially visible*. In fact, front faces may be visible but can also be obscured by other parts of the object (for non-convex shapes), or by other objects of the scene. However, such pre-processing can save about 50% of the computer time required, and is therefore widely used.

The determination of front and back faces is easily performed through the polyhedron approximation of the object. Planar equations of the limiting faces can be directly used, otherwise the normals to the faces are considered (normals are also used in shading operations).

Let us first consider the planar polygonal faces limiting a polyhedral object. We can distinguish for each plane a side facing the interior part of the object, with the other looking outside. We shall call these faces *internal* or *external*. For a face of equation

$$aX + bY + cZ + d = 0$$

the expression

$$S(P') = aX' + bY' + cZ' + d$$

is zero for any point P' (X', Y', Z') located in the plane: for other points (not in the plane), its value is non-zero and changes sign according to the side of the plane on which the point is lying.

Remark that for plane:

$$aX + bY + cZ + d = 0$$

coefficients a,b,c correspond to the direction of a vector (generally a non-unit vector) perpendicular to the plane ("normal" to the plane). We adopt the convention that the plane is specified by three vertices described according to a counterclockwise motion when viewing the outer side of the face, in a right-handed system (or CW in a left-handed one). Recall that in a plane a polygon is described counterclockwise if for three successive edges Pi,Pj,Pk the third point Pk is always on the left side of the oriented direction Pi,Pj.

Given such three points:

$$P_1 (X_1,Y_1,Z_1), \quad P_2 (X_2,Y_2,Z_2), \quad P_3 (X_3,Y_3,Z_3)$$

$$a = \begin{vmatrix} 1 & Y_1 & Z_1 \\ 1 & Y_2 & Z_2 \\ 1 & Y_3 & Z_3 \end{vmatrix} \quad b = \begin{vmatrix} X_1 & 1 & Z_1 \\ X_2 & 1 & Z_2 \\ X_3 & 1 & Z_3 \end{vmatrix} \quad c = \begin{vmatrix} X_1 & Y_1 & 1 \\ X_2 & Y_2 & 1 \\ X_3 & Y_3 & 1 \end{vmatrix} \quad d = - \begin{vmatrix} X_1 & Y_1 & Z_1 \\ X_2 & Y_2 & Z_2 \\ X_3 & Y_3 & Z_3 \end{vmatrix}$$

vector (a,b,c) is normal to the plane and points from inside to outside.

In a right-handed system (viewing direction along negative z axis), normal has component c along the z axis. If c is negative, it means that the normal points away from the viewing position: the plane is a back face. Similarly, in a left-handed system (viewing along positive z axis), with points P_i described clockwise, back faces correspond to $c>0$ (Figure 2.24).

As to the position of point P' by respect to plane $aX + bY + cZ + d = 0$, with the convention adopted for deriving the equation of the plane, it comes that expression $S(P') = aX' + bY' + cZ' + d$ is positive for a point situated on the external side of the face.

Such a simple test in fact implies that the coefficients a,b... are given a correct and fixed sign (as just indicated above). It is easy to check the correctness of these signs by testing the value of S for a point with a well known position with respect to the object faces. Otherwise coefficients a,b... have to be multiplied by -1.

In a right-handed system, the viewing direction is along the negative z axis. Let us consider the sign of S for point $(0,0,-\infty)$, at infinity along the viewing direction. If S is positive (that is in fact if c is negative), it means that the point at infinity along the negative z axis is on the external side of the face. Therefore, for the observer, the plane is a back face, hidden by the bulk of the object. Similarly, when working in a left-handed system (viewing direction along positive z axis) a back face corresponds to a positive c coefficient.

Let us note that giving a sign to the coefficients $a,b,\ldots$ means that it is not equivalent to write for instance a plane equation:

$$z + 1 = 0 \quad \text{or} \quad -z - 1 = 0$$

In a right-handed system, the first case ($c > 0$) would correspond to a front face (with origin on the external side), the second ($c > 0$) indicates a back face (with the origin on the internal side). See Figure 2.23.

An equivalent formulation considers two adjacent (non-colinear) edges which meet at a convex vertex and are described counterclockwise (in a right-handed system). Their vector cross product yields a vector directed towards the external side, provided the two edges make a convex angle. This face is visible

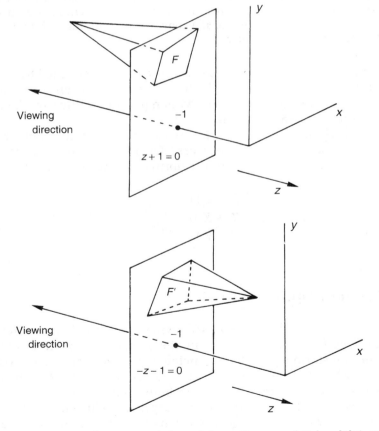

Figure 2.23 Back face elimination (adapted from Hearn and Baker [1]*). *Reference from Chapter 1.

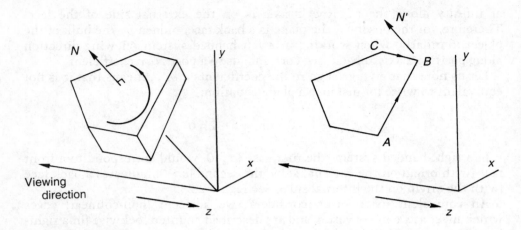

Figure 2.24 Normal to a face. The face is described as counterclockwise when viewing the outer side of the face. Normal points towards the external side (the axis system is right-handed). Alternatively, the normal direction can be determined from the cross product of vectors AB and BC meeting at a convex vertex (B) (adapted from Hearn and Baker [1]*). *Reference from Chapter 1.

if the normal points towards the viewer (this can be determined by evaluating the dot product of the normal with the direction of vision, or more easily, looking at its z component for an observation point at infinity on the negative z axis, as in the preceding method).

To avoid searching for adjacent edges with a convex angle, another method considers only sums over all vertices. Quantities a', b', c', calculated as indicated below, are proportional to the normal coefficients (a, b, c):

$$\left. \begin{array}{l} a' = \Sigma \ (Y_j - Y_i)(Z_j + Z_i) \\ b' = \Sigma \ (Z_j - Z_i)(X_j + X_i) \\ c' = \Sigma \ (X_j - X_i)(Y_j + Y_i) \end{array} \right\} \quad \begin{array}{ll} \text{with } j = 1 & \text{for } i = n \\ \text{otherwise} & j = i + 1 \end{array}$$

2.4.2 z-buffer (depth buffer)

The z-buffer method [23] is widely used owing to its simplicity to implement and its good performance. It works in the screen coordinate system (after clipping and transformations). The principle is to give each pixel of the frame buffer the colour of the surface element closest to the viewer thanks to a geometrical sorting along the z axis. The z-buffer is made up of two arrays with an entry for each pixel (X, Y): intensity and depth (Z). The buffer is first cleared with the background colour. Then polygons are entered one by one by the display file interpreter (scan conversion algorithm) (Figure 2.25). The depth of the current point treated is compared to the depth of the point already stored

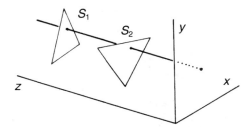

Figure 2.25   Pixel (X, Y) is given the colour of the nearest surface (S_2).

in the buffer. If the current point is in front, the buffer is updated with this new Z value; otherwise (point behind the point stored) it is ignored and the older Z value is maintained. Calculations may be accelerated using interpolation or incremental routines through the scan line.

The main drawbacks of the z-buffer method are the execution time and the cost in memory: a second buffer is required in addition to the refresh buffer, and each pixel has to be examined: this amounts to over one million values for a 1024×1024 resolution. In fact, the execution time depends upon the resolution of the screen and is proportional to the number of polygons (whereas other methods depend upon the square of this number): so, although relatively slow for treating a few polygons, it is more competitive for complex scenes where many faces are to be considered. As to the need for a large memory, this can be avoided by processing only one section of the scene at a time, or by using the derived *scan line* method.

2.4.3 Scan line method

In this approach, the scan line z-buffer [24], the z-buffer is limited to only one line (the current scan line) (Figure 2.26). This comes to process all polygons for one scan line rather than processing all the scan lines for one polygon (as in the usual z-buffer). The method is an extension of the scan line algorithm for filling polygon interiors, but now working to multiple surfaces. The method is greatly accelerated by treating not individual pixels but spans, i.e. sequences of pixels on a scan line lying within the same polygon: only end points or intersection points (when polygons interpenetrate) need to be considered [25–27].

This scan line approach is well suited to hardware realization. It is also adapted to downstream rendering treatments such as giving the picture grey shade depending upon z to get highly realistic displays.

2.4.4 Priority algorithms: painter's algorithm

This class of method [28] relies first on priority relationships on the depth between polygons in the object space, the XY calculations and scan

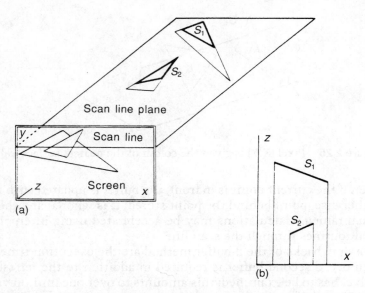

Figure 2.26 (a) Scan-line z-buffer; (b) projection of spans within polygons S_1 and S_2 in the scan line plane (from Newman and Sproull with permission [16]).

conversion being performed afterwards in the image space (starting with the surface of greatest depth). The term *painter's algorithm* suggests some analogy with the way in which a painting is created: first the background is painted, then the more distant objects are added, and finally, the nearer objects are represented. New paint covers the background ones, and only the newest layers are visible.

Polygons are arranged in a priority order based on their depth (those nearer the viewer having a higher priority). Then they are "painted" on to the frame buffer (through scan conversion) starting with the furthest polygons. Rather than a pixel-by-pixel comparison, as in the z-buffer, visibility is calculated using geometrical criteria, taking advantage of the coherence of polygons in depth.

The method looks very simple, but only at first glance, since each polygon cannot be entered independently, and for each one, intersections with all other polygons are to be investigated. This is necessary to determine which are situated behind the others, and to fix the order of their display.

Simple tests can greatly help in examining the key relation: "Is polygon P_i obscuring polygon P_j?" This is the *MINIMAX* (or boxing) test: clearly, two polygons cannot intersect if boxes just containing them do not overlap (Figure 2.27).

This is immediately determined by examination of the extreme values of X and Y. If the test is inconclusive, it does not mean that the polygons interpenetrate. A more refined test is necessary: for instance, comparison

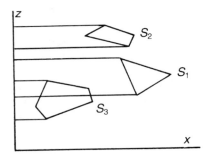

Figure 2.27 Depth test indicates that S_1 and S_2 cannot overlap. The test is inconclusive for S_1 and S_3, but a test on X rules out any overlap between them (adapted from Hearn and Baker [1]*). *Reference from Chapter 1.

between an edge of one polygon to all the other edges of the other one. Sequences of increasing complexity have been proposed:

- minimax test in Z,
- minimax test in XY,
- comparison of all vertices of P_i with respect to the plane of P_j,
- same test (vertices of P_j, plane of P_i),
- full overlap test in XY.

Other priority algorithms have been proposed, such as the Encarnacao method [29]: all faces are first pre-processed and decomposed to triangles. This minimizes storage requirements and execution time, and also avoids trouble occurring with non-convex polygons. Concave structures can thus be treated and cyclic overlap problems solved (Figure 2.28).

2.4.5 Space partition

Ordering polygons in priority algorithms (such as the painter's) may take advantage of space partition techniques. Planes are selected so as to separate clusters of polygons, but not to intersect any face within a cluster. Once

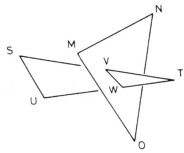

Figure 2.28 Considering separately polygons USVW and WVT allows for treating alternate visibility.

these groups are formed, sorting is faster (ordering between clusters is straightforward and within a cluster priority is more readily determined because a few faces are considered). Interestingly, this first classification into clusters is maintained when the position of the observer changes. So, if wanted, a modification of the viewpoint can be allowed. This would appear attractive in some animation applications [30].

Another approach [31] to order polygons first compares each vertex of one polygon against the plane of other polygons: if the planes intersect, the polygon is split in two. Two groups are constituted: one for the faces in front, the other with faces behind the current polygon. For each subgroup the separation is repeated, choosing one polygon to fix the new comparison plane, and so on until all polygons have been sorted. Results appear as a binary tree where nodes represent polygons: one branch gathers the front faces, the other the faces behind.

2.4.6 Warnock's algorithm

Basically, the main idea of the *area subdivision methods* is that frequently parts of a scene or landscape are globally immediately analysed, whereas some elements need a more refined description. Let us consider, for instance, the case of a tree in a large meadow. The eye will get a quick perception of the stretch of the meadow, whereas it will focus more attention on the tree to get more details about its branches, and then the leaves on them. In terms of computer graphics, as the scope of interest becomes narrower, more precision must be sought. But some advantages (in priority algorithms) can be gained from area coherence within polygons. It is therefore easier to deal with areas representing parts of a single surface.

In the Warnock's method [32, 32a], working in the image space, the screen is recursively divided within smaller windows. A window bearing one or no polygons is easy to solve: if there is no face within the window, this is given the colour of the background. If a face covers the window entirely, it is filled with the corresponding colour. Otherwise, this window is subdivided into four smaller parts until a window that is easy to solve is found, or the pixel size is reached (Figure 2.29).

For this method we need a test identifying clearly whether the window corresponds to a single surface or is still too complex. A classification is made between polygons not overlapping the window ("disjoint polygons"), polygons completely surrounding the window ("surrounder") or falling wholly ("interior") or partly ("intersector polygons") within the window. The main step is the treatment of surrounders. Polygons behind a surrounder have to be eliminated. This is carried out by first comparing the Z values (Z minimum of the polygon, Z maximum of the surrounder).

The recursive character of the subdivision allows for a faster traversal of the model. So, a polygon disjoint to a window is disjoint to all the subwindows created thereafter, and does not have to be considered. In fact, this method seems to be the first presentation of a data structure not so far from the

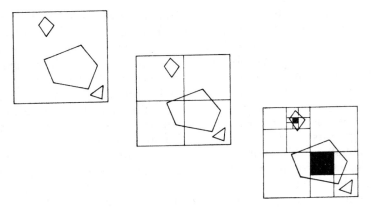

Figure 2.29 Windows are subdivided until they are easy to solve (the shadowed windows need no further subdivision). Not all the subdivisions necessary to treat the whole picture are indicated.

quadtree technique. The computer time varies roughly with the complexity of the display, and not with the complexity of the scene.

Rather that subdividing the window into four parts, as an alternative solution one can use boundaries of polygons to partition the screen [33].

2.4.7 Octree method

Hidden surface removal is fast and simple when an octree structure is used for encoding objects: octree nodes are projected onto the viewing surface in a front to back order, creating a quadtree of visible areas (Figure 2.30).

2.4.8 Hidden line algorithms

Wire frame objects are defined only by the edges of their polygonal faces, as if they were totally transparent, but for the sake of clarity, the hidden portions of lines have not to be drawn. Hidden line removal algorithms can sometimes be derived from hidden surface algorithms, or one can alternatively directly treat

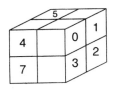

Figure 2.30 Node representing octants 0,1,2,3 are traversed before nodes 4,5,6,7.

the object edges. z-buffer and painter's (or derived) methods, which use the frame buffer or similar devices, are inadequate for calligraphic drawings.

The Roberts' algorithm [34, 34a] was the first solution proposed (Figure 2.31). Initially, within each object, edges of polygons are examined to see if they are hidden by the object to which they belong ("self-hidden" parts). Then each surviving line of the object tested is compared to other objects (test objects) to determine the segments of edges eclipsed by them. In the third step, penetration of edges of the tested object into the test volumes is investigated. Junction segments are drawn between all pairs of penetration points and their visibility checked. All visible segments remaining are then displayed. Of course, adequate depth or minimax tests and priority lists can limit the volume of calculations.

Rather than comparing all lines of the tested object to all other volumes, we can compare contour edges of all objects to each line. This is the method chosen in the *quantitative invisibility model* [35]. One follows a line and examines its intersections with edges to determine its invisibility when being hidden by an object lying in front of or penetrating it.

2.4.9 Treatment of curved surfaces

A common way to treat objects represented by curved surfaces is to approximate their shape by a polyhedron with a sufficient number of planar facets. The octree method, which can represent any type of object, is also usable without modifications. A surface subdivision algorithm which directly considers curved surfaces has been proposed [36, 36a, 36b]. Other methods

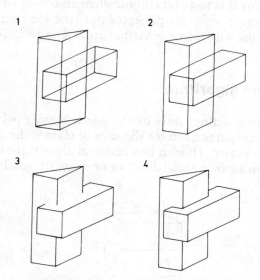

Figure 2.31 Successive steps in the Roberts' hidden line removal algorithm (adapted from Rogers with permission [2]*). *Reference from Chapter 1.

derive from the scan line algorithms [37–40]. The intersections with the curved surface (which are no longer segments, as for planar faces) are evaluated from the curves defining the surface, mainly from numerical techniques.

The originality of Encarnacao's scan grid method [29] relies on the treatment of any arbitrary curved surface defined by a grid of lines limiting curved patches. A scan grid is superimposed on the projection of the surface on the image plane. A mimimax test is used to determine which surface patches may potentially overlap each scan grid area. After this presorting, one has to determine visibility only for those patches lying within the same scan grid area. For surface patches, the edges are approximated by straight lines and visibility is tested by selecting equally spaced points along them and looking to see if they are obscured by other patches (within the same scan grid area). At this step, curved patches are replaced by planar facets obtained by inserting its two diagonals into the patch. As for Warnock's algorithm, recursive subdivision of the grid scan can be carried out for regions of increased complexity.

2.4.10 Ray tracing

In contrast with preceding methods that take advantage of coherence, using ray tracing to determine the visible part of a scene is a "brute force" technique which will be discussed with shading models (see Section 2.5) [41–43]. Some scan line adaptation significantly speeds up the method [44].

2.4.11 Efficiency

The performance of the hidden part removal modules depends upon the efficiency of the sorting steps (nature of the sorting algorithm), and also on the number of items to be treated. For instance, a scene with horizontally well separated surfaces favours a scan line method, whereas a distribution in depth is better approached through a z-buffer. A detailed discussion of the execution time, specifying the relative weight of the different steps, can be found elsewhere [45] (see also [16 p. 387, 21, 22]). The priority algorithm of Newell *et al.* seems very attractive when few polygons are considered, but it slows down for a large number of faces, whereas the z-buffer has uniform performance but suffers from high memory requirements. Octree and area subdivision look very attractive if many surfaces are considered.

2.5 RENDERING

Whereas calligraphic devices are very attractive for interactive image modification or animation effects, one of the main advantages of raster devices is their ability to represent objects as space-filling solids.

The *shading model* (determining the light intensity and colour on the points of the surface displayed) constitutes the ultimate and more refined operation in the search for realistic images [46].

2.5.1 Shading

This problem involves characteristics of the surface of the object and properties of the light falling on it. When a ray of incident light falls on the surface of an object, one (small) part is absorbed, the other is reflected or some light may be (partly) transmitted through the object (transparency) (Figure 2.32).

The scene may receive light from some *point sources* (such as a lamp, a candle) giving highlights on the surface, either directly or by reflection on neighbouring objects. There is also a *diffuse illumination*, constant for all directions. It arises from multiple reflections of light from nearby objects, walls of the room, sky, etc., constituting the ambient or background light. *Shadows* complete the display.

A model giving visually satisfactory results was given by Bui-Tuong Phong [47] (see also [48]). It stresses that the colour or illumination attributes of each point of the object (that is, the total energy of light coming from each point (p) of the object) can be calculated by summing up the contribution of diffuse illumination (E_{pd}), point light sources (E_{ps}) and taking into account transparency effects (E_{pt}):

$$E_p = E_{pd} + \sum_s E_{ps} + E_{pt}$$

For a *diffuse illumination* of intensity I_d falling on the object, the shading on point P is:

$$E_{pd} = R_p I_d$$

The *reflective factor* R_p $(0 \leq R_p \leq 1)$ relates the reflected energy leaving P to that arriving on it. For coloured surfaces, R_p depends upon the light

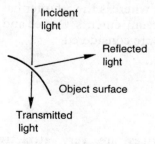

Figure 2.32 Light falling on an object's surface may be absorbed, reflected or transmitted through the object.

wavelength, and three values (for green, red, blue colours, for instance) are usually incorporated in the treatment to reproduce the effects.

Since changing the orientation of a face does not modify its shading, diffuse illumination does not generally lead to very realistic pictures. In fact, such a situation rarely occurs in the real world, where point sources nearly always intervene.

When a *point source* is present, the shade varies with the orientation. According to Lambert's law, the energy of reflected light from a surface varies as the cosine of the incidence angle.

In *diffuse reflection*, arising from the surface roughness, the light is uniformly scattered in all directions:

$$E_{ps} = (R_p \cos i)I_{ps}$$

(I_{ps} is the intensity of the incoming radiation from the point source.)

In fact, the intensity falling on point P depends upon the distance (d) to the light source S. An expression like

$$E_{ps} = R_p \frac{I_{ps}}{d + d_o} \cos i$$

where d_o may be adjusted, gives satisfactory shading.

Note that $E_{ps} \equiv 0$ if $i > 90°$, i.e. if the surface point is hidden from the light sources.

In addition to diffuse reflection, bright spots (highlights) are created by *specular reflections* more intense on shiny than on dull surfaces. In this process, the reflected light has the same colour as the incident light (a white light illuminating a red object causes a white spot on the surface).

Specular reflection depends upon the position of the observer with respect to the source and the surface. According to elementary optics, for a perfect reflector, the principal reflected ray corresponds to an reflection angle (r) equal to the incidence one (i) (Figure 2.33).

For other directions the intensity of the reflected ray depends upon the angle of observation (θ), and upon a specular reflection coefficient $W(i)$, a function of the incidence angle (i) varying with the nature of the surface (Figure 2.34).

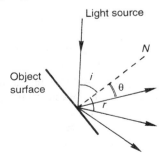

Figure 2.33 Shading. i = incidence angle, N = normal to the surface.

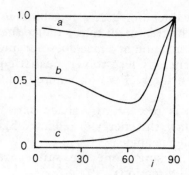

Figure 2.34 Specular reflection coefficient $W(i)$ as a function of the incidence angle i. a = silver, b = gold, c = glass (from Newman and Sproull with permission [16]).

According to the Phong model, the specular reflection is expressed by:

$$E_{ps} = W(i)\cos^n (\theta{-}r) I_{ps}$$

Generally speaking, $W(i)$ increases towards a value near to the theoretical maximum of 1 at grazing angle $(i = 90°)$. For shiny metallic surfaces such as silver, $W(i)$ is high for all i values whereas for glass, for instance, $W(i)$ is low, except for values near 90° where it takes nearly the maximum value of 1. Similarly, for metallic surfaces, exponent n is large $(n \sim 100)$: the reflection range is narrow about the direction $r = i$. On the contrary, a dull surface (such as paper) is given a low n value (down to 1) corresponding to a larger extent of reflected light.

So, for a point source illumination:

$$E_{ps} = \frac{I_{ps}}{d + d_o}\left[R_p \cos i + W(i)\cos^n(\theta - r)\right]$$

The last term of the Phong model (E_{pt}) corresponds to *transparency*. Coefficient T_p determines what part of the energy arriving at P from behind (E_{pb}) is transmitted:

$$E_{pt} = T_p E_{pb}$$

Background objects seen through a transparent face may be treated by adapting some hidden surface method with depth-sorting.

Refraction effects can also be included in the shading model. Although both diffuse and specular refraction can take place, shading models usually consider only specular refraction to limit calculation time.

The preceding formulae give a straightforward evaluation of shading on each point, but for a reasonable resolution of 1024 × 1024 pixels this would lead to a large number (1 million) of calculations. This can be alleviated by taking advantage of *shading coherence*: the intensity of adjacent pixels generally varies only a little. So, for an object represented by planar polygons, the intensity on a face is constant and can be calculated from the normal vector to the face (provided the light source and the viewer are far enough apart, angular terms do not vary). For a rapid calculation of the angles involved see elsewhere [16, p. 393].

2.5.2 Gouraud and Phong smooth shading of polyhedra

For a polyhedron formed of planar facets, shading, as explained above, leads to an image somewhat rough looking with straight line segments limiting areas of constant shading. These shortcomings are largely lessened but not totally erased in Gouraud smooth shading [49], which restores a smoother appearance to objects. Shading is varied across polygon surfaces so that the shade at the edges matches that of the neighbours (Figure 2.35).

Normals are calculated at the vertices defining the facets, either, directly from the surface model (if known), or by interpolating the normals at the facets surrounding the vertex. By interpolation from the shades calculated at the vertices, shade values along edges of the facet are then obtained. Finally, for points internal to the facet, interpolation between the edge shades at the end points of the scan line is carried out.

While this is satisfactory to get smooth pictures when the normals on two neighbouring facets have nearly the same direction, discontinous shading results from abrupt changes in orientation for adjacent facets. This relies on the Mach band effects, related to the inability of the human eye to accommodate discontinuous changes of illumination. The eye then perceives light or dark bands at the discontinuity (dark areas appear darker and light are lighter at the boundary). Adding more polygons decreases the discontinuities and lessens this effect. On the other hand, if averaged normals have nearly the same direction (as for sheets gently folded), a misleading constant shading is obtained. Here also this can be removed by adding more polygons near to the boundaries, so that the average normal orientation varies (Figure 2.36).

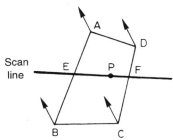

Figure 2.35 Gouraud's interpolation. Intensities at E and F are interpolated from intensity values at A and B (for E), and C and D (for F). Along the scan line intensity at P is interpolated between E and F values. Normals at vertices are represented by arrows. In Phong's model, normals rather than shades are interpolated (from Newman and Sproull with permission [16]).

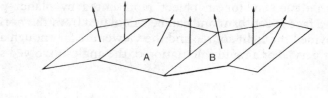

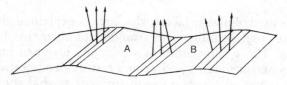

Figure 2.36 If the average normals point in the same direction, a constant shade is calculated for A and B, rubbing out relief perception (above). Adding supplementary polygons removes this effect (below) (from Newman and Sproull with permission [16]).

The Phong's algorithm interpolates normal vectors rather than shades, and shading is applied to each pixel displayed. This technique remedies the main problems encountered with Gouraud's algorithm, and gives very realistic pictures, but requires more calculations. Sampling requirements to obtain highly realistic displays and problems encountered on the edges are detailed elsewhere [16, p. 402].

2.5.3 Transparency

Adding the transparency contribution in equations implies knowledge of the light intensity coming at a surface point from behind it. This can be easily implemented in frame buffer algorithms. When successively adding polygons of a higher priority, the old intensity represents the light arriving from behind. In scan line algorithms, depth sorting also gives a way in which to cope with transparency effects.

2.5.4 Ray tracing

Ray tracing is a very powerful technique that is easy to implement and able to give highly realistic displays, since in the same step it also treats the problems encountered in hidden surface removal, shadows, transparency, etc. [50–53]. The main drawback is, however, its slowness, since, as can be seen from the following description, many intersections of light rays with surfaces have to be calculated.

The principle of ray tracing is to follow the pathway of a light ray from the light sources to each pixel of the display. In fact, the reverse way is easier: going back from the viewpoint through each displayed pixel into the object space [54] (Figure 2.37).

Intersections with all the surfaces encountered are examined. If more than one surface is on the ray's path, only that closest to the observer is retained (unless transparent), performing hidden part removal at the same time [41]. The shadow problem is similarly solved, examining if the ray from the surface to the light source is interrupted by another surface, thus obscuring the light source [42, 43]. Various solutions have recently been proposed to increase the speed of ray tracing algorithms [55–58].

The shading (intensity value) for each visible point of the surface can then be evaluated considering the diffuse illumination and the light received from other objects through specular reflection. For this, the surface point is now considered as the viewpoint. One recursively examines the light coming from a given direction, taking into account, if necessary, intersections with other objects and related effects (reflection, transparency, etc.). For reflection, for instance, one can search for a reflected ray giving light along the ray followed, etc.

2.5.5 Shadows and special effects

More realism can be gained by displaying shadows. Their treatment implies knowing what faces (or parts of faces) are attained by light, i.e. are visible from the light sources. The problem is therefore closely related to hidden part elimination. One way is to repeat hidden surface calculation using the light source as the viewpoint. Surfaces which are visible but which are hidden from the light source are displayed with only diffuse illumination. The others (visible from both the actual viewpoint and the light source) in addition receive direct light source illumination.

A possible solution is to define *shadow polygons* [59] constructed from the outline of the object when viewed from the light source. These shadow polygons are added to the polygons representing the faces of the objects, and all

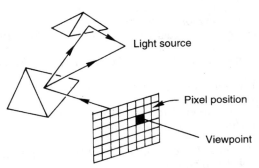

Figure 2.37 Following the ray trace backwards from the viewpoint to the light source (adapted from Hearn and Baker [1]*). *Reference from Chapter 1.

these polygons are submitted to hidden part removal algorithms. The operation is easily incorporated into z-buffer techniques: a first z-buffer determines the foremost visible surfaces, and a second one (a shadow z-buffer) establishes if these points are in the shadowed volume, i.e. between the back and front faces of the shadow volume (Figure 2.38).

Texture and surface patterns allow for more accurate representations of real world objects. We shall not develop these points, since such techniques are not used largely in molecular modelling: it is sufficient to say here that texture can be simply introduced by giving some appropriate modulation to the reflection coefficients in the shading model. This changes the colour without modifying the flat appearance of the surface.

Changing reflection coefficients determined by sampling was also used to simulate highlights arising from an image mapped on to the surface of the object displayed (see elsewhere [16, p. 408] for more details).

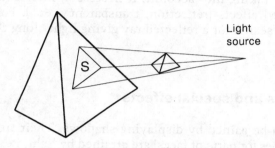

Figure 2.38 Shadow polygon (S) does not receive direct illumination from the light source (adapted from Hearn and Baker [1]*). *Reference from Chapter 1.

REFERENCES

1. J.E. Scott *Introduction to Interactive Computer Graphics*. Wiley, New York, 1982.
2. J.E. Dubois, S.Y. Yue and J.P. Doucet *The Visual Computer*. **2:** 1986; 367–378.
3. M. Roch, PhD Thesis. Geneva, 1986.
4. M.L. Connolly *J. Appl. Cryst.*, **18:** 1985; 499–505.
5. E. Keppel *IBM J. Res. Develop.*, **19:** 1975; 2–11.
6. H.N. Christiansen and T.W. Sederberg *Computer Graphics*, **12:** 1978; 187–192.
7. G. Wyvill and T.L. Kunii *The Visual Computer*, **1:** 1985; 3–14.
8. A.A.G. Requicha and H.B. Voelcker *IEEE Computer Graphics and Applications*, **2:** 1982; 9–24.
9. C.L. Jackins and S.L. Tanimoto *Computer Graphics and Image Processing*, **14:** 1980; 249–270.
10. K. Kronlof and M. Tamminen *The Visual Computer*, **1:** 1985; 24–36.
11. T. Pavlidis *Algorithms for Graphics and Image Processing*, Computer Science Press, Rockville, MD, 1982.
12. K. Yamaguchi, T.L. Kunii, K. Fujimura and H. Toriya *IEEE Computer Graphics and Applications*, **4:** 1984; 53–59.
13. P. Bezier *Numerical Control, Mathematics and Applications*, (A.R. Forrest and A.F. Pankhurst, trans) J. Wiley, London, 1972.

14. P. Bezier *Emploi des Machines à Commande Numérique*, Masson, Paris, 1970.
15. W.J. Gordon and R.F. Riesenfeld in *Computer Aided Geometric Design*, R.E. Barnhill and R.F. Riesenfeld (eds.), Academic Press, New York, 1974.
16. W.M. Newman and R.F. Sproull *Principles of Interactive Computer Graphics*, Mac Graw Hill, New York, 1979.
17. D. Cohen and I.E. Sutherland, see for instance: Principle of interactive computer graphics, W.M. Newman and R.F. Sproull, McGraw-Hill, New York 1979, p. 65.
18. R.F. Sproull and I.E. Sutherland *AFIPS Fall Joint Computer Conference*, **33:** 1968; 765–776.
19. Y.D. Liang and B.A. Barsky *Comm. ACM*, **26:** 1983; 868–877.
19a. Y.D. Liang and B.A. Barsky *ACM Trans*, **3:** 1984; 1–22.
20. I.E. Sutherland and G.W. Hodgman *Comm. ACM*, **17:** 1974; 32–42.
21. I.E. Sutherland, R.F. Sproull and R.A. Schumacker *ACM Computing Surveys.*, **6:** 1974; 1–55.
22. J.H. Clark *Comm. ACM*, **19:** 1976; 547–554.
23. E. Catmull *Proc. IEEE Conf. Computer Graphics, Pattern Recognition and Data Struct.*, 1975; 11–17.
24. L. Carpenter *Proceedings of NW76 ACM*, Seattle, WA, 1976.
25. G.S. Watkins University of Utah Computer Science Department Technical Report UTEC CSC 70-101, NTIS AD 762 004, June 1970.
26. A.J. Myers *Report to the NSF*, Ohio State University, Computer Graphics Research Group, July 1975.
27. W.J. Bouknight *Comm. ACM*, **13:** 1970; 527–536.
28. M.E. Newell, R.G. Newell and T.L. Sancha *Proc. ACM Annual Conference*, 1972; 443–450.
29. J. Encarnacao and W.K. Giloi *Proc. AFIPS FJCC*, **33:** 1972; 985–998.
30. R.A. Schumacker, B. Brand, M. Gilliland and W. Sharp U.S. Air Force Human Resource Lab. Tech. Rep. AFHRL-TR-69-14, NTIS AD 700 375, September 1969.
31. H. Fuchs, Z.M. Kadem and B.F. Naylor *Computer Graphics*, **14:** 1980; 124–133.
32. J.E. Warnock University of Utah, Computer Science Dept. Tech. Rep., TR 4-15, NTIS AD 761 995; May 1968.
32a. J.E. Warnock University of Utah, Computer Science Dept. Tech. Rep., TR 4-15, NTIS AD 753 671; June 1969.
33. K. Weiler and P.R. Atherton *Computer Graphics*, **11:** 1977; 214–222.
34. L.G. Roberts in *Optical and Electro-Optical Information Processing*. J.T. Tippet (ed.) MIT Press, Cambridge, MA 1964, pp. 159–197.
34a. L.G. Roberts, MIT Lincoln Lab. Rep. TR 315, May 1963.
35. A. Appel *Proceedings of the ACM National Conference*, 1967; 387–393.
36. E. Catmull PhD University of Utah, 1974.
36a. E. Catmull, UTEC CSc, 74–133.
36b. E. Catmull, NTIS A004 968.
37. J.F. Blinn *Computer Graphics*, **12:** 1978; 286–292.
38. J.M. Lane, L. Carpenter, T. Whitted and J.F. Blinn *Comm. ACM*, **23:** 1980; 23–34.
39. J.M.Lane and L. Carpenter *Computer Graphics and Image Processing*, **11:** 1979; 290–297.
40. J.H. Clark *Computer Graphics*, **13:** 1979; (supplement to *Proc. SIGGRAPH 79*), 174.
41. R.A. Goldstein and R. Nagel, *Simulation*, **16:** 1971; 25–31.
42. S.D. Kay and D.P. Greenberg *Computer Graphics*, **13:** 1979; 158–164.
43. T. Whitted *Comm. ACM*, **23:** 1980; 343–349.
44. P.R. Atherton *Computer Graphics*, **17:** 1983; 73–82.
45. W.K. Giloi *Interactive Computer Graphics – Data Structures, Algorithms Languages*, Prentice-Hall, Englewood Cliffs, NJ, 1978.
46. R.A. Hall *The Visual Computer*, **2:** 1986; 268–277.
47. Bui-Tuong Phong *Comm. ACM*, **18:** 1975; 311–317.

48. J.F. Blinn *Computer Graphics*, **11:** 1977; 192–198.
49. H. Gouraud *IEEE Trans. Comput.*, 1971; C-20, 623–629.
50. S.D. Roth *Computer Graphics and Image Processing*, **18:** 1982; 109–144.
51. R.A. Hall and D.P. Greenberg *IEEE Computer Graphics and Applications*, **3:** 1983; 10–20.
52. T. Nishita and E. Nakamae *Computer Graphics*, **19:** 1985; 23–30.
53. R.L. Cook, T. Porter and L. Carpenter *Computer Graphics*, **18:** 1984; 137–144.
54. A. Appel *AFIPS Spring Joint Comput. Conf.*, 1968; 37–45.
55. J. Amanatides *Computer Graphics*, **18:** 1984; 129–135.
56. P.S. Heckbert and P. Hanrahan *Computer Graphics*, **18:** 1984, 119–127.
57. D.J. Plunkett and M.J. Bailey *IEEE Computer Graphics and Applications*, **5:** 1985; 52–60.
58. H. Weghorst, G. Hooper and D.P. Greenberg *ACM Trans. Graphics*, **3:** 1984; 52–59.
59. F.C. Crow *Computer Graphics*, **11:** 1977; 242–248.

3 *Displaying molecular shapes*

Creating molecular images may be considered as a part of *generative graphics*: the creation of synthetic representations of real or conceptual objects. The general method is to first define an abstract description of the scene or objects that is understandable by the computer. This model is then transformed into a picture on a display device. However, owing to the diverse nature of chemical representations, various approaches have been considered, according to whether interest is focused on either geometrical and structural descriptions or electronic features.

For the representation of objects in the real world on a computer screen, it is important in a first step to search for a data structure or for simpler constituent parts, *graphics primitives*, to make picture generation easier. This structuralization can be achieved either on the object itself ("sketchpad") or on the space to be embodied (voxel subdivision) (Figure 3.1).

Such a situation is sometimes also found in chemistry. When creating molecular models reproducing the usual wire frame or plastic CPK ("space filling") representations on the display screen, the structure of the chemical object already exists. The basic laws of valence theory, for example, express the organization of the atomic framework, and provide the generation algorithms required to build the images. Atomic spheres (for atoms) and cylinders (for bonds) constitute a set of primitives for the molecular body (possibly supplemented with a few other simple shapes for macromolecules such as a ribbon, helix, etc.), and the problem mainly relies on geometric or analytical handling.

On the other hand, the problem is more complex when representing property shapes related to electron distribution. For these more conceptual objects, the overall shape is neither known *a priori*, nor easily defined through an analytical expression. The first task is therefore to generate the information to be represented in the 3D space and give it the appropriate structuralization.

From a quite different point of view, the type of display device favours certain representations. Of course, calligraphic (vector) systems only allow line drawings, whereas raster devices make it possible to display "solid" images. So, vector systems are particularly interesting for wire frame models stressing molecular topology and atomic positions. Their advantages lie in the high resolution and image quality attainable, and the possibility of (quasi-) real-time manipulation.

Rapidly increasing capabilities of microcomputers and raster display devices prompt interest in shaded images, more static but superior for the presentation or publication of 3D models. Many recent efforts focus on improving rendering and creating, within this context, more aesthetic images, able to approach photorealism. In fact, realism provides good clues for a more rapid and complete understanding of 3D features such as depth, size and shape. However, some authors put a caveat on the excessive use of transparencies, reflections, etc. able to produce images that are visually confusing and difficult to interpret.

Another distinction also appears between 2D and true 3D images. In the first case, one only builds a projection of a 3D object on the screen, and the calculations have to be repeated when the object or the point of view is moved. With true 3D treatments, motions are allowed without the need for a complete new calculation; only the projection of the scene has to be updated.

Among general trends (according to the widespread use of raster devices), one can note the frequent use of *templates*, calculated only once and "stamped" at the appropriate position, an idea already used by Basch [2] as early as 1983. The *z-buffer* technique (possibly with some adjustments) is also largely generalized for hidden-part removal.

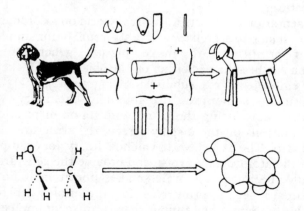

Figure 3.1 Representation of real beings or objects first implies the identification of primitives and any knowledge of their combination laws. In some chemical applications, shape primitives and structuralization rules are already known (from Dubois *et al.* [1]).

We shall first detail some representations of structural shapes before coming
to the more complex problem of property shapes.

3.1 REPRESENTATION OF STRUCTURAL SHAPES

The simplest images, such as "stick" (or "wire frame") models, present a "see-
through" representation focusing on molecular topology (relative positions of
atoms and existing bonds between them) [3], which can be made interactive up
to a few thousand atoms. The simplest (and perhaps the oldest) computerized
images in chemistry only use a single line segment for each bond, and can be
modified in real time [4]. In more recent displays, for a better visual perception,
bonds are colour encoded along half their length, according to the atom type
from which they are issued.
   A better insight into the molecular body is provided by space filling models
built by the union of the van der Waals spheres of the constituting atoms.
These can be treated either as line drawing representations (the only one
attainable on vector devices) or as solid images (raster systems) (Plate I).
   Dot representations also deserve special emphasis in view of their
widespread use. The Connolly's Molecular Surface algorithm [5, 5a] spreads
points with a constant density onto the molecular envelope, built from atomic
spheres, possibly completed with reentrant parts (see Chapter 8). These dots
give a very pleasant representation of the molecular body, since they allow for
some perception of the atom framework inside. They also provide a data
structure either for triangulation processes, leading to the display of the
molecular surfaces as solid images, or for deriving structured surfaces (4D
images), where dots are colour encoded with the value of a property: MEP,
hydrophobicity, etc. [6, 7]. Owing to their large involvement in molecular
surface and volume evaluations, these dot representations will be developed in
a separate section (see Chapter 8 and Plate II).
   Space structuralization by a lattice of nodes (used either directly or as a basis
for further refinement) is also a possible avenue to surface representation, as
detailed in Chapter 8. Such applications are still in close relation to surface and
volume calculation owing to their ease of implementation and their capability
for easy logical (Boolean) operations. We will only present space
stucturalization for the display of electronic properties here, since it is the
most general approach of these features.

3.1.1 Calligraphic representations

Starting from the pioneering work of Warme [8], in 1978 Smith and Gund [9]
proposed a line drawing algorithm for the display of CPK representations on
vector devices. Atom spheres project onto the screen plane as one of their great

circles. Intersection between spheres is a circle (in a plane perpendicular to the line of their centres) that projects as an elliptical arc. Thanks to appropriate scaling, rotation and translation, all circles to be drawn are derived from a standard circle approximated by a 24-edge polygon. Hidden-part elimination is carried out by comparing each polygon edge with each relevant sphere. For these two operations (hidden-line removal and determination of intersections), analytical solutions are carried out. Time is saved, for each current atom, by sorting among the other atomic spheres those that are intersecting or eclipsing it (Figure 3.2) Adding some parallels or meridians to the spheres, or drawing hatchings to parts facing away from a light source, as in the earlier options of the PLUTO package[1], gives images greater realism. Largely widespread in crystallographic papers, the ORTEP algorithm draws ball and stick models, where the atomic spheres are replaced by thermal ellipsoids [10] (Figure 3.3).

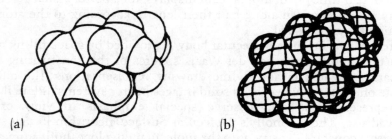

(a) (b)

Figure 3.2 Calligraphic display of space-filling models (5Me-5Et-5,6 dihydro-2(1H)-pyridone) from the SPACEFIL algorithm. The outline drawing (a) is completed in (b) with cross-hatching from meridians and parallels (with permission from Smith and Gund [9]). (See also Plate I a.)

3.1.2 Molecular representations as solid objects

In calligraphic representations, such as those obtained from vector devices, the drawing is built from line segments characterized by their start- and end-points. For solid object images (attainable with raster systems), the attributes of each pixel constituting the image on the screen have to be determined and stored in the frame buffer.

Before presenting more refined programs to generate CPK images on raster devices, let us first recall that the *painter's algorithm* suggests a very simple way in which to feature a molecule. It paints on the frame buffer a filled disc for each of the atoms sorted from the rear to the front. No sphere intersection is represented, but the overall shape of the molecule is fairly well suggested. Note that this very simple and cheap algorithm is used on some low level modelling programs running on personal computers (Figure 3.4).

Not so far from the painter's algorithm is the treatment proposed by

[1] PLUTO, a program for plotting molecular and crystal structures; University Chemical Laboratory, Lensfield Road, Cambridge CB2 1EW, UK (S Motherwell).

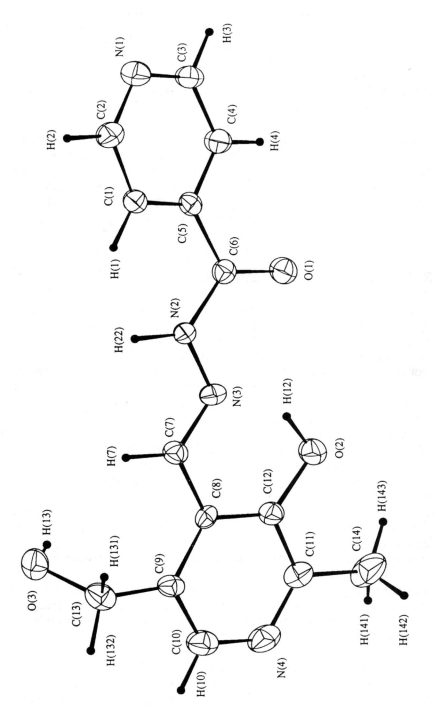

Figure 3.3 ORTEP representation of a crystal structure, including thermal ellipsoids.

Figure 3.4   The painter's algorithm allows for an immediate rough perception of the molecular shape, without treating intersections between atom spheres.

Thomson *et al.* [11] from the well-known Connolly's dot representation (the Connolly's Molecular Surface program generates dots on the molecular envelope and specifies the direction of its normals). To get a nearly complete solid surface, dots are replaced by discs (in fact octagons) tangential to the surface. Sorting by depth (with respect to the screen plane), as in the painter's algorithm, virtually ensures hidden-part removal.

3.1.3 Some aspects of z-buffer techniques

Owing to their simplicity of implementation, the z-buffer techniques have been widely used as standard tools [12, 12a]. For each pixel of the screen plane (defined by its x,y coordinates), it suffices to determine which atom is the closest to the observer. In principle, spheres need not to be sorted, but are rendered one by one. For each pixel, the z value of the current atomic sphere is compared to the value stored in the buffer. If the new value is lower than the previous one, the point is visible and therefore the depth and colour are updated.

However, such a comparison, involving all the atoms of a molecule, may become cumbersome when carried out on each pixel of the image (a medium resolution image of 512×512 pixels would require 256 K comparisons ...) [13]. Sorting the pixels to test dramatically improves the situation. On the array of pixels forming the image, the vertical and horizontal extents of each atom are determined. For each (let us say horizontal) pixel line, one selects only those atoms whose vertical extent spans this line. A similar sorting can be made on the horizontal extents so as to substantially limit the depth comparisons needed on each pixel (Figure 3.5).

One advantage of the z-buffer, when compared to other hidden-part removal methods using depth sorting on the scene to be drawn, is the ability to add new components at any stage of completion. We have only to update the depth and colour of the relevant pixels. In the study of molecular interactions, for example, a small ligand can be added to a large biomolecule without modifying the whole image.

Variants requiring less memory have been proposed, such as the line buffer ("scan line of the z-buffer") proposed by Porter [14]. Only the scan line is examined each time, and its contents are sent on the frame buffer (incremental methods rather than time consuming calculations involving square-root evaluations speed up the process [12, 12a].

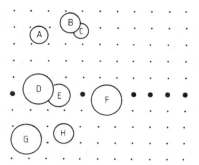

Figure 3.5   Circles, projections of the atom spheres on to the array of screen pixels, can be sorted according to their vertical extension. For the current scan line investigated (bold dots), only spheres D, E and F have to be considered (from Pearl [13]).

In the scan line plane, intersections of atomic spheres are circles, and substantial time is saved by taking advantage of some scan line coherence. Intersections of spheres determine on the scan line stretches along the x axis where only one sphere is visible. For each of these intervals, visibility is calculated on an arbitrary position of each arc. Assigning priorities to circles within the scanning plane and making use of some correlation between the current scan line and the preceding one increase speed still more [19].

Another interesting adaptation is the "3D depth buffer" [15]. Usually, a 2D depth buffer stores only one pixel for each (x,y) location, that nearest to the observer. On the contrary, in the 3D depth buffer, several pixels (in fact voxels) are stored, each at a different depth. As stated by Connolly [15], "hidden surfaces are not eliminated but rather stored in a linked list sorted by z coordinate, so they can be revealed by clipping and translucency." This approach constitutes an attempt to transfer to raster systems some of the capabilities of vector graphics: the ability to clip the image so that the interior is visible, to display a transparent molecular surface, etc.

3.1.4. Templates

Among the current trends for both constructing and rendering operations is the increasing use of templates calculated once and stamped at the appropriate location in the image. The representation of molecular bodies easily falls in with that type of solution. Assuming orthographic parallel projection and a light source at infinity, the size and shading of a sphere become independent of its position. Thanks to this [3], the expensive shading calculations are made only once, which allows for providing high quality images for systems up to 1000 atoms within a reasonable time. Although interesting in terms of speed, templates are memory expensive, and image manipulation requires recalculation of the pixels' attributes. An elegant way in which to ensure geometrical transformation is to come back temporarily to a wire frame representation, allowing for more rapidly defining the new atomic positions. Templates are then stamped at the updated locations (Figure 3.6).

Interestingly, there is an obvious analogy between a processor array architecture and the pixel array of a raster screen. This connection has been taken advantage of to generate molecular images on a highly parallel computer. A 512×512 pixel image is organized by blocks of 64×64 pixels and treated using an array of 64×64 single bit processors (4096 in all) [16]. No doubt promising solutions are to be expected in this field opening the way towards truly interactive raster graphics. Parallel processing facilities have also been presented [17], where four different memory buffers are simultaneously used to update the pixels' position and visibility.

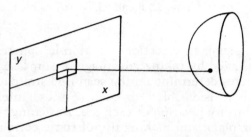

Figure 3.6 Atom templates are made up by a front-facing hemisphere of van de Waals spheres. Each pixel (x, y) of the frame buffer is encoded with the corresponding z value and shade value (given the light source direction) (Palmer and Haussher [3]).

Let us now look in more detail at image generation by templates. Palmer and Hausheer's "context free" spheres [3] designed for improved rendering, or Johnson's device-independent model [18] are relevant such approaches. The template is stored in a matrix of cells containing both shading and depth information, organized for instance as a $4 \times n$ array (n depending upon the size of the atom) containing the (relative) x,y location of the pixel, its surface z value and shading value (for a preset direction of illumination).

Given the atomic positions, this information will be transferred to the image data structure: a matrix of cells equal in size to the output image generated. Stamping templates at the appropriate position results in a viewer-visible shell hanging between the background and the image plane. Image space comparison of the depth values then determines visibility with a z-buffer technique.

Typical values for template definition (indicated by Basch [2]) correspond to about 20 pixels per Å (i.e. 35 pixels per carbon atom), which allows for an image about 45×45 Å on a screen displaying 900×900 pixels. Such a use of templates would imply that the atomic centres always coincide with centres of pixels, which does not always occur. However, the resulting distortions of molecular geometry remain small on static imges, although they may create unpleasant "jump" effects in animated sequences.

3.1.5. Rendering: shadowed images and ray tracing

Adding shadows provides supplementary visual clues for understanding the 3D structure. One possibility is to repeat the visibility calculation taking the light source as the point of view, according to an idea proposed by Williams [20]. "Visible" regions are the illuminated ones, whereas if an atom is not seen by the light source, then it is in shadow and may be shaded as such [21–23] (Figure 3.7).

To generate this second image with the light source as point of view, Goodsell *et al.* [22] introduce a second template: the shadow template. It contains the z value of all points of the sphere visible from the light source and projected onto a plane coplanar with the image plane. These data are stored in a second z-buffer (shadowing z-buffer) with modified atom positions. To determine points in shadow, the two buffers are compared pixel by pixel if the depth of the surface pixel is greater than the z value of the shadow map, the point is shadowed.

A priori, ray tracing seems very attractive owing to its powerful possibilities for rendering treatments, the simplicity of calculations for sphere/ray intersections and its easy implementation. However, for large molecules it would rapidly become inefficient in view of the large number of spheres to test.

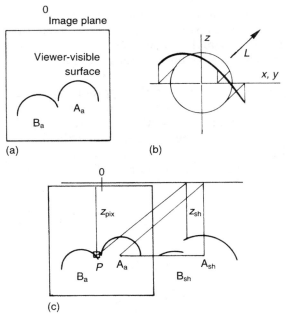

Figure 3.7 In the scan plane, the visible atomic surface is delimited by a circle. Templates are stamped at positions Aa and Ba (a). For a set direction of illumination, a shadow template can be drawn (b), L being the light vector. To determine the shadowed points, shadow templates are mapped into a second z-buffer. Comparison of z values in the z and shadow buffers determines points in shadow; here, for instance, point P is shadowed. The viewer is at the top of the figure with z axis oriented towards him (with permission from Goodsell *et al.* [22]).

Methods have been proposed to reduce this number, allowing for rendering of complex molecular systems [24–26].

Derived from ray tracing and z-buffer techniques, special rendering effects were proposed by Goodsell *et al.* [22, 23] for molecules embedded in space-filling coloured clouds or transparent surfaces, representing electrostatic potential, electron density, etc. and sampled on a grid surrounding the molecule. A ray is cast from each pixel away from the viewer through the data grid. Its shade is incremented as it passes through more or less opaque areas of the grid until it strikes the molecular surface or the background.

3.1.6 Spherical lune templates

Most of the preceding methods only generate 2D images taking advantage of z-buffer capabilities. As a part of our concern for a flexible and transferable molecular modelling system, we have developed a true 3D representation of molecular bodies using only spherical lunes as templates [27]. Rather than storing a full sphere, the template is constituted by one spherical lune, approximated by a set of quadrangular planar facets and two triangles at the ends.

Although dedicated to a raster device, the method is suited for both calligraphic or solid displays, since the atom spheres are first defined by a set of planar polygonal facets which can be treated by looking only at their contour lines (chicken-wire drawings) or considered as colour filled surfaces (Figure 3.8 and Plate III).

The atomic spheres are drawn at the appropriate location by successive rotations of this graphics primitive; an operation easily and rapidly performed using hardware rotation and translation capabilities now offered by most workstations. Shading needs to be recalculated for the newly generated part of the image, but this is not a problematic task (with hardware functions), since normals are determined by the rotation process. The method is very cheap in

Figure 3.8   Spherical lune template.

terms of memory requirements and rendering time. For a good quality image, constituted from about 300 facets, a template of only 12 elementary facets needs to be stored in memory.

The user can *interactively modify* the number of facets, choosing either more facets for realistic displays, or only few for a rough representation, which may be animated in nearly real-time. This capability looks quite interesting in docking or shape adjustment studies, since the user is able to maintain some perception of the bulk of the molecular body during the session (such "steric" features are, of course, completely lost with the common use of wire frame models at this step).

Portability

In the past, device dependence seriously hampered a wide diffusion of modelling packages. Portability requires that specific hardware functions are avoided, so that the speed of the algorithms becomes essential. In this context, as stressed by Johnson [18], although a direct comparison is difficult from the published data, in view of the diversity of the devices and representations chosen, the simple use of templates and of the z-buffer algorithm offers quite interesting capabilities. For better portability, the image is generated in the memory of the host computer as a colour matrix (each element corresponding to a pixel on the display screen). Once this is achieved, the image is transferred on to the memory of the raster display, limiting the number of code lines to be re-written when changing the display device or graphics library.

Stick-type models

For more complex molecules, where it would be too long to display all the atom spheres, a stick model, where bonds are represented by shaded cylinders, may more easily give some (approximate) insight into the molecular body and its steric requirements [18]. However, cylinder templates are not as easy to treat as spheres, since their shape and shading depend upon their location. As a compromise to rapidly generate such shapes, a circle is first drawn and rotated so as to be normal to the cylinder axis direction. Shading is calculated for each point of the circle. The corresponding generatrix is drawn with this constant shading and mapped onto the buffers.

3.1.7 Space subdivision

Other hidden-surface algorithms rely on space subdivision [28]. Each sphere image is divided into a list of regions bounded by vertical line segments or arcs of circles. For the sake of simplicity, intersections (circles projected as ellipses) are treated on the screen as circles passing through three points of the ellipse, and are used to define new points for subsequent subdivision. Non-visible parts are removed from the list. Once all intersections have been examined, atom by atom, the remaining parts in the atom list are displayed as trapezoids and rendered one-by-one along vertical scan lines (Figure 3.9).

Shading and highlights have been added by Max [21, 29, 29a]. *Ray tracing* and *z-buffers* require time proportional to the number of pixels. With an *area based algorithm*, the visible portions of each object are calculated to an arbitrary precision (usually that of the computer) and rendered at the desired resolution. Shading cost is resolution-dependent, but not the visibility calculation, which depends upon the number of atoms and the complexity of the image [21]. To add shadows, the subdivision method can be applied twice, to produce first one view from the observer and then one from the light source [21].

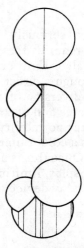

Figure 3.9 Successive steps of the space subdivision method. Each sphere is subdivided into two trapezoids. Intersecting and overlapping spheres create additional trapezoids (from Knowlton and Cherry [28]).

3.2 REPRESENTATION OF PROPERTY SHAPES

The representation of shapes related to electronic properties poses a double problem. First, for these properties, neither an easy analytical solution nor predefined primitives are known, so we have to generate data and define for them a convenient data structure. Second, given the property to display, we have to select the type of representation. Indeed, in our 3D space, representation of functions of one or two variables, $y = f(x)$ or $z = F(x, y)$ does not cause any problem. In the later case, for instance, one possibility is to draw, in a 2D map, contour lines associated with a given value: $F(x,y) =$ constant. Another is to represent a perspective drawing of the surface $z = F(x,y)$, the z coordinate giving the value of the function for each couple of x,y coordinates [30]. Unfortunately, the situation is more complex for the commonly invoked electron properties (electrostatic or hydrophobic potential, electron density).

They are scalar functions of 3D coordinates, $F(x,y,z)$, taking a value for each point of the space. Some methods have to be found for an easy visual perception of such 4D entities. Among them, using colour proves to be very useful for representing this supplementary dimension.

The problem is still a little more complex for the display of a property such as the electrostatic field, which is defined as a vector with a modulus and a direction: small arrows (with appropriate orientation and length) are generally used to schematize the field values on selected points in the space. Although not commonly used, until now, the electrostatic field has been proposed as another "electric image of a molecule", giving more importance to charges near the probe point than the electrostatic potential does, owing to a variation in r^{-2}, in place of r^{-1}. We will not discuss this point any more.

For the representation of scalar properties, $P(x,y,z)$, a first solution consists in displaying an *isovalued* envelope gathering all points where the property takes the same preset value (constant valued surface: $P(x,y,z) = Cst$. This relies on surface modelling where objects are described by shaping their envelope like a skin around them. Images are then presented as a mesh of curved surface patches or a network of polygon shaped facets (Figure 3.10) (Plates V, VI).

However, for those properties we are concerned with, the envelopes corresponding to diverse values recover each other, as the skins of an onion, and so partially mask each other. Even with line drawings (chicken-wire models) or using transparencies, only a few surfaces can be drawn without resulting in a very intricate and confusing image.

Another avenue is to restrict the representation space: either in a given *plane*, leading to maps of *isovalued contour lines* (as common geographical maps), or along a *molecular surface* or layers derived from it. One can even display only a projection of this encoded surface on the screen plane [32] (Plate VI).

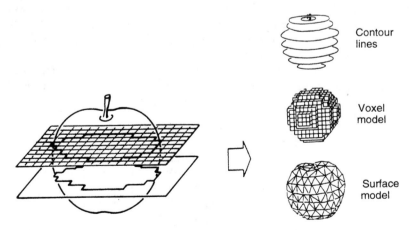

Contour lines

Voxel model

Surface model

Figure 3.10   Serial sectioning by parallel planes and creation of 3D images: packets of cross-section surface outlines or isovalued contour lines, $P(x,y,z) = Cst$, can be used to directly schematize the shape or make up the basis of a model built by triangulation for volume reconstruction. Alternatively, a voxel model can be derived (with permission from Aniyo *et al.* [31]).

Colour encoding molecular surfaces with a property value (giving a visual indication of this property value on points of the 3D space) corresponds to *structured surfaces* [6], or *4D images*. Note that, as stressed by Heiden *et al.* [33], such 4D images constitute a very powerful representation. Displaying the molecular surface and colour encoding it (according to, say, molecular electrostatic potential – MEP – values) focus attention on both topological and electrostatic complexity, allowing for easier identification of recognition sites. This global perception also appears very useful for stressing differences regarding the activity of structurally similar molecules. (See also Plates II b and XII.)

As previously pointed out, for those properties which can be easily attained only through numerical evaluation, one must first generate a structured array of data and specify its organization (particularly the relation between the localization of the observation points and the values of the property sought).

A common approach is to use a structured space maintaining the image. For instance, the system under investigation is immersed in a lattice ("3D grid"). Its nodes or elementary cuboids ("voxels" or volume elements) will constitute the basic elements from which information will be sought. Another possible data structuralization method comes from the very popular Connolly algorithm, scattering dots on to the molecular surface.

3.2.1 Lattice embedding and determining isovalued points

Let us now look in more detail at the derivation of such isovalued envelopes. Suppose the problem is to represent the contour corresponding to a preset value of the MEP or any other one electron property. First, the molecule is immersed in a 3D lattice constituting the calculation and representation space, and we evaluate the MEP on each node of the lattice. Then, along each edge, points corresponding to the set value ("isovalued points") are determined by interpolation between adjacent nodes (Figure 3.11).

Once the isovalued points are determined, joining them forms polygonal contour-lines, and representation relies on *serial sectioning*. From cross-section

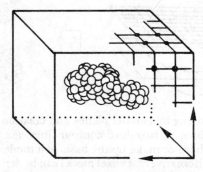

Figure 3.11 Lattice embedding.

outlines drawn for packets of parallel planes, various strategies are possible, according to the type of display selected (see preceding section).

Chicken wire drawings

A combination of two (or better, three) packets of such contour-lines in orthogonal planes provides a mesh of surface patches representing the 3D surface in chicken-wire models [34–37] (Figure 3.12).

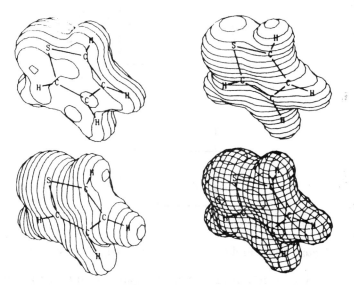

Figure 3.12 Chicken wire representation by combination of cross-sections: 50 kcal/mol repulsive MEP (featuring the molecular shape) for thiopene. Three packets are drawn (athough two would in principle be sufficient for a 3D perception) to give more aesthetic displays when rotating the image (Yue [36]).

As to graphics aspects, we stress two features:

1. Given the points with the appropriate property value, how to sequentially join them along a polygonal contour? The procedure relies on the Freeman code. From the current point, one explores the adjacent edges successively to form a linked list of isovalued points (Figure 3.13).

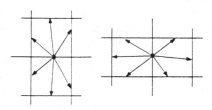

Figure 3.13 Searching for the next isovalued neighbour points located on either a vertical or horizontal edge of the node lattice: 2D schematization.

2. Some ambiguity can appear when the same voxel is traversed by two contour lines (four intersection points on the edges). For simple shapes (assuming monotonic variations of the property), examination of the property gradient allows for specifying how points must be joined [36,37].

Note also that some uncertainties arise in situations where two intersections would exist on a single edge (Figure 3.14): part (a) on the scheme will not be represented. Fortunately, this would only occur in regions where

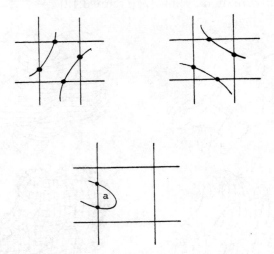

Figure 3.14   Problems in connecting edge intersections.

the object to be displayed has an extent comparable in size to that of a voxel, so that the resulting error on the overall shape remains small.

Finally, to avoid confusing images that are difficult to interpret, a hidden-line removal process has to be carried out. Possible processes will be indicated in Chapter 8. One can, for instance, examine successively the visibility of each polygon edge owing to its location with respect to the planes of the other polygons.

Triangulation and solid images

With raster systems (chemical) objects can be presented as polyhedral solids limited by coloured facets. Various approaches have been suggested to define the solid surface thanks to a mosaic of triangles connecting suitable data points on adjacent contours in a slice (Figure 3.15).

Triangle (the simpler planar polygon) interpolation is highly attractive, since it makes the calculation and use of rendering refinements easier to derive more aesthetic images with smooth shaded and shadowed surfaces (see Chapter 8). It can also be extremely efficient in terms of speed, provided some care is devoted to optimize the generation process [33]. To build such a model it is necessary to get topological information about the relation between two

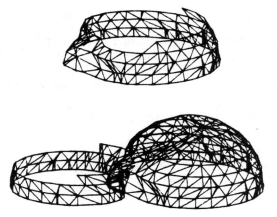

Figure 3.15 Generating triangular facets from cross-section outlines (intermediate steps in the representation of the MEP of a water molecule) (from Roch [38]).

adjacent cross-sections, and several solutions have been proposed [31, 38–42]. In the approach of Koide *et al.* [43], data are organized through linked lists between hierarchical levels characterizing the surface to be represented, the constituting elementary triangles, their vertices and normals (Figure 3.16).

Although more naturally devoted to raster displays, owing to their capability of providing solid images, triangulation algorithms were also recently used in vector devices, and look well-suited to plotters because of the simplicity of the drawing [43]. Serial sectioning can be also be used in discrete solid modelling. The volume included between successive contour lines is divided into small cuboids considered as basic cells of the solid body.

Polyhedron-ID	Surface constant C	Molecular orbital ID	Molecule ID
*	*	*	*
*	*	*	*
*	*	*	*

(a)

Polyhedron-ID   Triangle list

3 vertices (pointers to vertices list)		
J	K	L
*	*	*
*	*	*

Vertex list

Coordinates			Normal vectors		
x	y	z	u	v	w
*	*	*	*	*	*
*	*	*	*	*	*

(b)

Figure 3.16 Successive levels of data organization in the tessellation approach of Koide *et al.* [43].

Subvoxel treatment

In the same way, elementary cells defined by the embedding lattice can be spanned into smaller units that are easier to treat: cutting off corners of a voxel (plus the central part) creates five tetrahedrons. As in the previous methods, intersections of the contours sought for with the edges of these tetrahedrons are determined by interpolation from node values. They constitute the vertices of triangular or tetragonal facets limiting the shape [43, 44] (Figure 3.17).

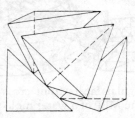

Figure 3.17   Subvoxel treatment of a node lattice. The elementary cell is expanded into five smaller tetrahedrons (the upper right-hand one is not shown). Isovalued points on edges and diagonals of the cell are interpolated between node values (from Purvis and Culberson [44]).

3.2.2 Reducing the representation space

Introducing geometrical constraints leads to *subimages*: simplified views, but allowing for simultaneous display of a larger number of property values. Several options have been proposed.

Map of isometric lines

The same generation strategy is applied: evaluation of the property on a pre-defined grid in a given plane, interpolation to find the isovalued points then drawing the contour lines. Note that, with a raster display, one can lay colour on areas between isocontours corresponding to preset values [39, see also 45, 46] (Figure 3.18; see Plate VII).

Representation of a property on van der Waals type surfaces

Thanks, for instance, to the Connolly's algorithm, dots are scattered with a regular density on to the molecular surface. The property is evaluated on these points[2], and colour encoded. To increase speed, an octree technique has been introduced [47]. The molecular surface is produced by recursive spatial

[2] Evaluation of one or one-electron properties on the van der Waals surface and derived layers, or on a preset grid, is currently provided by the Gaussian *ab initio* or other semi-empirical M_o packages.

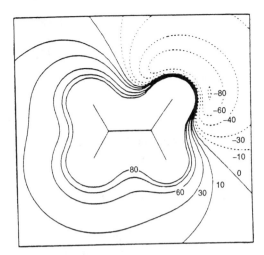

Figure 3.18 Planar map of isovalued contours: MEP STO 4-31 G of formamide in the molecular plane. Contours are given in kcal/mol (from Roch [38]).

subdivision. The cubic space area investigated is divided into eight smaller cubes. Those completely empty or situated inside the molecular body are discarded. The remaining ones (containing some part of the surface) are further subdivided into eight, down to pixel resolution.

Although pleasant, since they allow for perceiving the molecular skeleton, dotted surfaces in some cases may give complex images on screen, camouflaging the relevant information by the interference of background and foreground dots [33]. For simpler displays, triangulation can be used to generate solid representations from these dotted surfaces (a surface defined by point coordinates). This also appears useful for quantitative evaluation of molecular areas and volume. Connolly [48, 49] proposed a triangulation algorithm working on analytically defined solvent-accessible surfaces: recursive subdivision of edges limiting the curved faces (parts of spheres or tori) determines the elementary triangles. Sequences for building such images are illustrated in Plate II.

A special emphasis on speed for interactive manipulation and applicability to large biomolecules leads Brickmann [33] to propose a hierarchy of triangulation strategies, applying different conditions to different surface regions. In the *tmesh strategy*, the propagation phase generates each elementary triangle from an already existing edge, so that each new piece is defined by only one new point (Figure 3.19). Although proposed in the context of a Connolly dot-representation of the molecular surface, the triangulation algorithm of Brickmann is also applicable to isovalued points derived from interpolation on a node lattice. (See also Koide *et al.* [43] for a propagation algorithm to generate triangles from a set of isovalued points.)

According to the authors, the capabilities of actual super workstations, able to generate roughly 10^5 Gouraud shaded triangles per second, allows for the interactive manipulation of large biomolecules (about 30 000 dots for trypsin).

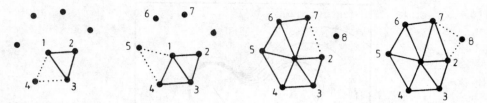

Figure 3.19 Growing a triangle tmesh. Starting from point 1, the first triangle is built with the two nearest neighbours (points 2 and 3). The last edge drawn (1–3) is the basis of the next triangle (with permission from Heiden *et al.* [33]).

Generation of a 2D view by a planar projection outside the molecular body

This method is somewhat similar to that of Greer and Bush to determine the molecular surface (see Chapter 8): from each node of a (x, y) grid, a line is cast along the z direction until it reaches the surface. The property sought (say, the MEP) is calculated at this point, and the returned value is used to colour encode the corresponding part of the screen [32] (Figure 3.20).

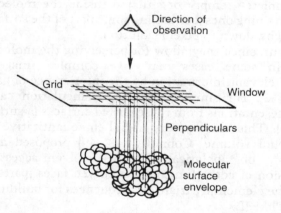

Figure 3.20 Generation of a 2D view by a planar projection outside the molecular body (from Lavery *et al.* with permission [32]).

3.3 CONCLUDING REMARKS: SYMBOLIC PICTORIAL PRIMITIVES

Large biomolecules are far too complex for the representation of all atoms in all cases. The structural formulae of polypeptides are, for example, mainly described using one (or three) letter symbols identifying the constituting amino acids. Similarly, for display of their 3D organization, a hierarchy of levels with increasing precision has been managed.

In a first step, to feature only the overall shape, sets of larger graphics primitives (prefabricated standard fragments) can suggest the essential morphological characters. On-line modelling of proteins often uses such schematic representations to define the secondary structure: cylinders or regular helices depict α helices, broad arrows or ribbons represent β strands, etc., randomly coiled chains being only indicated by their C_α–C_α bonds. More details are sometimes necessary for limited parts of the macromolecule, for instance when looking at interactions between a ligand and an active site. In such cases, one can turn to the usual representations at atomic scale, and display, for instance, part of the molecular surface as previously described. (See also Chapter 13.)

REFERENCES

1. J.E. Dubois, J.P. Doucet and S.Y. Yue In *Molecules in Physics, Chemistry and Biology*, J. Maruani (ed.), Kluwer Academic Publishers, 1988, vol 1, 173–204.
2. P.A. Basch, N. Pattabiraman, C. Huang, T.E. Ferrin and R. Langridge *Science*, **222:** 1983; 1325–1327.
3. T.C. Palmer and F.H. Hausheer *J. Mol. Graph.*, **6:** 1988; 149–154.
4. C. Levinthal *Sci. Am.*, **21:** 1966; 42–52.
5. M.L. Connolly *Science*, **221:** 1983; 709–713.
5a. M.L. Connolly *Science*, **211:** 1981; 661–666.
6. H.R. Karfunkel and V. Eyraud *J. Comput. Chem.*, **10:** 1989; 628–634.
7. H.R. Karfunkel *Match*, **19:** 1986; 67.
8. P.K. Warme *Comput. Biomed. Res.*, **10:** 1977; 75.
9. G.M. Smith and P. Gund *J. Chem. Inf. Comput. Sci.*, **18:** 1978; 207–210.
10. C.K. Johnson *ORTEP II program*. Oak Ridge National Laboratory, Tennessee, 1976.
11. C. Thomson, D. Higgins and C. Edge *J. Mol. Graph.*, **6:** 1988; 171–177.
12. N.L. Max *J. Med. Sys.*, **6:** 1982; 485–499.
12a. N.L. Max *J. Mol. Graph.*, **2:** 1984; 8–13.
13. L.H. Pearl *J. Mol. Graph.*, **6:** 1988; 109–111.
14. T. Porter *Computer Graphics*, **12:** 1978; 282–285.
15. M.L. Connolly, *J. Mol. Graph.*, **3:** 1985; 19–24.
16. R.E. Hubbard, D. Fincham *J. Mol. Graph.*, **3:** 1985; 12–14.
17. P. Schultze and K. Wüthrich *J. Mol. Graph.*, **4:** 1986; 108–111.
18. B.A. Johnson *J. Mol. Graph.*, **5:** 1987; 167–169.
19. J. Wu, Y. Guan and Q. Zheng *J. Mol. Graph.*, **5:** 1987; 190–192.
20. L. Williams *Computer Graphics*, **12:** 1978; 270–274.
21. M. Gwilliam and N.L. Max *J. Mol. Graph.*, **7:** 1989; 54–59.
22. D.S. Goodsell, I.S. Mian and A.J. Olson *J. Mol. Graph.*, **7:** 1989; 41–47.
23. D.S. Goodsell *J. Mol. Graph.*, **6:** 1988; 41–44.
24. A. Fujimoto and T. Tanaka *IEEE Computer Graphics and Applications*, **6:** 1986; 16–26.
25. A.S. Glassner *IEEE Computer Graphics and Applications*, **4:** 1984; 15–22.
26. J. Arvo and D. Kirk *Computer Graphics*, **21:** 1987; 55–61.
27. J.P. Doucet and V. Fabart (to be published).
28. K. Knowlton and L. Cherry *Comput. Chem.*, **1:** 1977; 161–166.
29. N.L. Max *Computer Graphics*, **12:** 1978; 348.
29a. N.L. Max *NCGA 81* Conference Proceedings. National Computer Graphics Association, Washington, DC, 1981; 495.
30. M.M. Gilbert, J.J. Donn, M. Peirce, K.R. Sundberg and K. Ruedenberg *J. Comput. Chem.* **6:** 1985; 209–215.

31. K. Aniyo, T. Ochi, Y. Usami and Y. Kawashima *The Visual Computer*, **3**: 1987;
    4–12.
32. R. Lavery and B. Pullman *Int. J. Quantum Chem.*, **20**: 1981; 259–272.
33. W. Heiden, M. Schlenkrich and J. Brickmann *J. Computer-Aided Molecular Design*,
    4: 1990; 255–269.
34. J.E. Scott *Introduction to Interactive Computer Graphics*, Wiley, New York, 1982.
35. W.L. Jorgensen *Q.C.P.E. Program 340*, Indiana University.
36. S.H. Yue, Thesis, University of Paris, 1987.
37. J.E. Dubois, S.Y. Yue and J.P. Doucet *The Visual Computer*, **2**: 1986; 367–378.
38. M. Roch, Thesis, University of Geneva, 1986.
39. H.N. Christiansen and T.W. Sederberg *Computer Graphics*, **12**: 1978; 187–192.
40. S. Ganapathy and T.G. Dennely *Computer Graphics*, **16**: 1982; 69–75.
41. H. Boissonat *ACM Trans. Graph.*, **3**: 1984; 266.
42. R.J. Zauhar and R.S. Morgan *J. Comput. Chem.*, **9**: 1988; 171–187.
43. A. Koide, A. Doi and K. Kajioka *J. Mol. Graph.*, **4**: 1986; 149–155.
44. G.D. Purvis and C. Culberson *J. Mol. Graph.*, **4**: 1986; 88–92.
45. J. Weber and M. Roch *J. Mol. Graph.*, **4**: 1986; 145–148.
46. J. Weber, P.Y. Morgantini, J.P. Doucet and J.E. Dubois, New trends in computer
    graphics, *Proceedings of CG international*, N. Magnenat–Thalmann and D.
    Thalmann (Eds.), Springer–Verlag, Berlin, 1988; 499–508.
47. P. Quarendon, C.B. Naylor and W.G. Richards *J. Mol. Graph.*, **2**: 1984; 4–7.
48. M.L. Connolly *J. Appl. Crystallogr.*, **18**: 1985; 499–505.
49. S.D. Kahn, Iris Universe Spring, 1989; 24–30.

4 *Access to experimental geometrical parameters*

Molecular geometry is of course the starting and virtually indispensable information for numerous modelling problems: display of the molecular architecture, determination of the electronic structure, evaluation of the interaction mechanisms, etc. Detailed structural information provided by atom coordinates has an enormous impact in chemistry and biology, and appears essential for the understanding of many biological or chemical processes at the molecular level [1].

Of course, a "theoretical" estimation of molecular geometry is attainable through molecular or quantum mechanics programs, but access to experimental determinations for the molecule under investigation, or (no less interesting) for a set of closely related structures, appears to be valuable. Various spectroscopic techniques can give some information (more or less comprehensive) about interatomic distances, depending upon the size of the system under investigation (small molecules or large biomolecules) and the physical state of the sample (crystal, liquid or gas). Among the various possible methods, two emerge in the field with which we are concerned regarding mainly medium-size (drugs) or large (biomolecule) organic systems, and deserve a brief presentation here: X-ray diffraction and NMR. Other approaches (neutron or electron diffraction, microwave or far infrared spectroscopies) are of very limited use in the field.

Although in fact restricted to crystalline samples, structure analysis using X-ray diffraction has become an essential tool in the investigation of molecular geometry, probably owing to its wide applicability to small or large molecules (proteins). Improvement of experimental devices, and increased computer capabilities (largely reducing the time necessary to "solve" a structure), rapidly yield a huge and exponentially growing amount of data. Faced with the large number of atom coordinates now available, computer-based files appear to be the only practical (and virtually indispensable) tool to store and broadcast the structural information. To manage efficient access to these large files, and to make them available to the scientific community, large efforts have been devoted to building extensive and reliable databases which are now standard sources for molecular geometry information.

We will first briefly present experimental approaches (mainly X-ray diffraction and nuclear overhauser enhancement – NOE – measurements in NMR) before some introductory comments on the crystallographic Cambridge and Brookhaven databases.

4.1 CRYSTALS AND X-RAY DIFFRACTION

4.1.1. Crystal lattice

Crystal state is characterized by the fact that the crystal components (atoms, molecules or ions) tend to pack together as closely as possible, and consequently arrange themselves periodically in a three-dimensional ordered array where the environment of each unit is identical. This arrangement, the *crystal lattice*, can be described as the packing of identical elementary units, each containing the same constituents (one or more atom, ion or molecule) in an identical position. In the repetition of such a unit cell by translations along three directions of the space to generate the whole crystal, the translation vectors, their modules a, b, c and angles between them (α, β, γ) define the crystallographic axes and the size of the elementary cell. For easier representation, we consider the constitutuent elements as points, and use a lattice of points as a picture of the actual crystal (Figure 4.1).

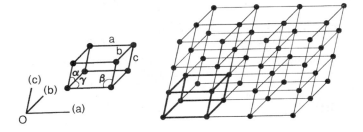

Figure 4.1 The crystal lattice can be considered as being generated by repetition of a unit cell.

To fill the whole space without leaving gaps by packing identical blocks, only certain shapes are allowed (let us recall, for instance, that in 2D a regular pavement cannot be obtained with pentagonal blocks). Symmetry considerations show that only seven different shapes of unit cell can match. They are known as the *crystal systems* (Figure 4.2).

For any crystal lattice it is always possible to define a primitive triclinic cell, containing only one motif (in fact, one motif per corner shared between eight adjacent cells). However, to better display the symmetry of the edifice, possibly hidden when considering primitive cells, it is often more convenient to choose non-primitive cells, i.e. cells of greater dimensions and containing more than one motif: *body centred* with one extra lattice point at the centre of the cell, *side centred* with two extra points on one pair of opposite faces, *face centred* with extra points on all faces. These various possibilities define 14 *Bravais lattices* within the seven crystal systems (Figure 4.3). It must be noted that the choice of a Bravais lattice is not always unique for a given crystal (although some conventions are accepted), and that this choice does not modify the interpretation of X-ray scattering experiments. In addition, one can also distinguish 32 crystal classes related to symmetry within the unit cell: for the nomenclature of the 32 crystallographic point groups, see, for instance Wheatley [2] and Dean [3].

Special interest is devoted to planes containing a high density of lattice points, because these points (in fact atoms or molecules in the actual crystal) are the scattering centres in X-ray diffraction studies.

Given the geometrical parameters of the unit cell, one can gather sets of parallel planes into families. Each family is defined by the intersections of one plane of the family with the three crystallographic axes. They correspond to three multiples of the cell parameters a, b, c. To describe the orientation of the family of planes, we can retain integers proportional to the reciprocal of these multiples: so are defined the *Miller indices*. Miller indices characterize the orientation of a family of planes, and from them the equidistance between successive planes of the family can be easily derived (Figure 4.4).

System	Unit cell	
Triclinic	$\alpha \neq \beta \neq \gamma \neq 90°$	$a \neq b \neq c$
Monoclinic	$\alpha = \gamma = 90°$ $\beta \neq 90°$	$a \neq b \neq c$
Orthorhombic	$\alpha = \beta = \gamma = 90°$	$a \neq b \neq c$
Trigonal		
Rhombohedral axes:	$\alpha = \beta = \gamma \neq 90°$	$a \neq b \neq c$
Hexagonal axes:	$\alpha = \beta = 90°$ $\gamma = 120°$	$a = b \neq c$
Tetragonal	$\alpha = \beta = \gamma = 90°$	$a = b \neq c$
Cubic	$\alpha = \beta = \gamma = 90°$	$a = b = c$

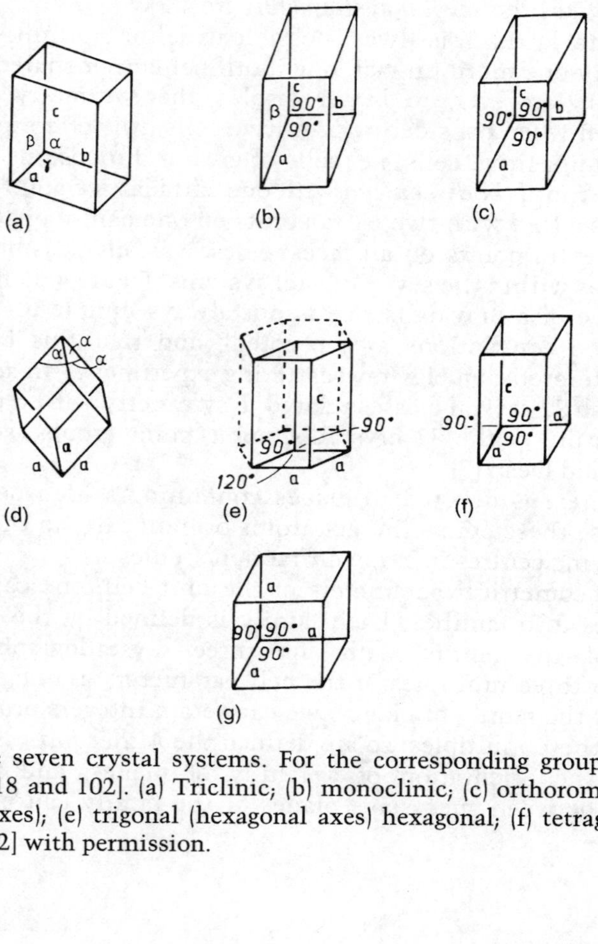

Figure 4.2 The seven crystal systems. For the corresponding group symmetry see Wheatley [2, p. 18 and 102]. (a) Triclinic; (b) monoclinic; (c) orthorombic; (d) trigonal (rhombohedral axes); (e) trigonal (hexagonal axes) hexagonal; (f) tetragonal; (g) cubic. From Wheatley [2] with permission.

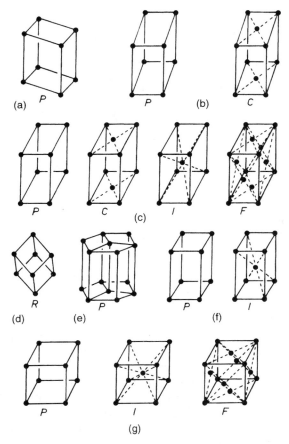

Figure 4.3 The 14 Bravais lattices. (a) Triclinic; (b) monoclinic; (c) orthorhombic; (d) trigonal (rhombohedral axes); (e) trigonal (hexagonal axes) hexagonal; (f) tetragonal; (g) cubic (from Wheatley with permission [2] p. 106).

4.1.2 Bragg equation

We now consider the interaction between a crystal lattice and an X-ray beam. The energy absorbed by the atoms is reemitted in all directions. Owing to the comparable magnitude for the wavelength of the incoming radiation and the periodicity of the lattice, the waves diffracted by the various atoms may build an interference pattern. In the Bragg approach, beams *diffracted* from lattice nodes can be considered as beams *reflected* by a family of lattice planes. Only for certain incidence angles can reflected beams reinforce each other, giving rise to sizable intensity, according to the Bragg relationship:

$$n \lambda = 2 \, d \sin \theta$$

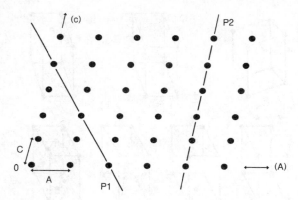

Figure 4.4 Identifying lattice planes by Miller indices: 2D schematization (B axis is supposed perpendicular to the figure). For plane P1 the intercepts with the crystal axes are 2(A), ∞ (B) and 4(C). Taking the reciprocals and multiplying them by the lowest factor to get integers (4), we obtain for the Miller indices (hkl) 201. Similarly, for P2 the Miller indices are 100.

λ wavelength of the radiation. The integer n corresponds to the order of diffraction, (90–θ) is the angle of incidence of the X-rays and d is the spacing between adjacent lattice planes in the family (Figure 4.5).

This fundamental relationship shows that X-ray reflection is selective, and occurs only for some discrete incidence angles. In a diffraction experiment, where X-rays of a given wavelength interact with a crystal lattice, each family of planes gives rise to a reflected beam in a direction corresponding to the Bragg angle, and depending upon the interplanar spacing. As there are many families of lattice planes, the diffraction pattern (as observed on a photographic plate) is composed of a large number of discrete spots (corresponding to the directions where the Bragg condition is met for one family of planes). Conversely (given the X-ray wavelength), measuring the Bragg angles from the location of the spots allows for determining the

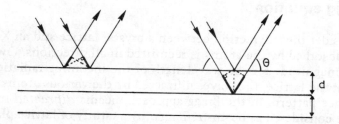

Figure 4.5 Reflection from a family of equispaced lattice planes. Parallel rays reflected from the same plane are in phase; parallel rays reflected from two partially-reflecting planes are in phase only when the path difference is a multiple of the wavelength (Bragg relation). The components of the path difference are shown by bold lines.

spacing of lattice planes, and gives access to the size of the repeat unit (the unit cell) in the crystal.

4.1.3 Structure factor and electron densities

Intensities of the spots are another interesting factor, since they allow for determining atom positions within the unit cell (expressed as a fraction of the repeat unit).

X-ray scattering is almost entirely due to external electrons, and the intensity of the scattered radiation depends upon the electron distribution within the atoms. The amplitude of the wave scattered by an atom is called *the atomic scattering factor f*. At small angles of diffraction ($\theta \# 0$), *f* is equal to the atomic number (possibly minus the charge for ions), but for larger angles *f* is reduced because of interferences between the different scattered rays.

Going now to a molecule, to evaluate the amplitude of the wave diffracted by atoms of planes *hkl* (or according to Bragg presentation, reflected by planes *hkl*), we have to combine the waves diffracted by the individual atoms. The resulting amplitude is given by the structure factor F [4][1]. For a plane *hkl* passing through atom *j* (of fractional coordinate x_i, y_i, z_i.) and a parallel plane passing through the origin:

$$F_{hkl} = \sum_{j}^{N} f_j \exp\left(2\pi i (hx_j + ky_j + lz_j)\right)$$

f_j represents the scattering factor (at the value of ($\sin \theta / \lambda$) for atom *j* whose fractional coordinates are x_j, y_j, z_j in the unit cell:

$$x_j = \frac{x}{a}; \ y_j = \frac{y}{b} \cdots$$

Except for some particular cases (cubic crystal, for instance), F is a complex number characterized by an amplitude and a phase.

In fact, electrons are not strictly localized, and we have to consider explicitly their distribution in space, characterized by an electron density function $\mu(x, y, z)$: number of electrons in the element of volume dv at point (x, y, z). The structure factor (amplitude of the wave diffracted by dv is therefore:

$$F_{hkl} = \sum_{j=1}^{N} \iiint \mu f_j \exp\left(2\pi i (hx/a + ky/b + lz/c)\right) dv$$

---

[1] Taking into acount some correction factor related to the geometry of the measurement device, absorption of the X-ray beam by the crystal, etc.

So, the structure factor F and the density of scattering elements (in fact, the electron density) are linked via a Fourier transformation, and we can write:

$$\mu(x,y,z) = (1/V)\iiint F_{hkl} \exp -(2\pi i(hx/a + ky/b + lz/c))dv$$

or using only a discrete summation:

$$\mu(x,y,z) = (1/NV)\sum\sum\sum F_{hkl} \exp -(2\pi i(hx/a + ky/b + lz/c))$$

where NV is the volume of the entire array considered NV being the volume of the unit cell [5, p. 719].

Electron density could therefore be calculated if the cell parameters and the structure factor for a sufficient number of h,k,l values (i.e. beams diffracted by different families of planes) are known. From electron densities $\mu(x,y,z)$ calculated for various x,y,z points of the cell (generally on the nodes of a grid), a density map can be drawn, where peaks indicate the location of the atoms.

Solving a structure by X-ray diffraction can therefore be formulated as *Given the intensities of the diffracted beam, determine the positions of the atoms.* The trouble is that the only attainable quantity is the intensity of the diffracted beam, proportional to the square of the amplitude [4]. This fundamental difficulty is known as the *phase problem.* In other words, the amplitudes of the diffracted beams are known but not their relative phases, and so we cannot combine them to get the electron density map. Among the main methods used to help in solving the phase problem, we note [4]:

Use of a good trial structure obtained via a "Patterson synthesis". It uses a similar Fourier series but with intensities rather than amplitudes. The result is a map giving not the actual atomic positions, but the vector distances of individual atoms (all translated to a common origin). This method is very convenient when the structure contains heavy atoms ($Z > 25$) for which the scattered intensity is important.

When a pair of isomorphous crystals (differing by one atom) can be studied, differences in structure factors selectively reflect the contribution of the differing atoms.

In the direct method, no structural information has to be input. The criterion is that electron density cannot be negative. This electron density is obtained from summation of waves of frequency (h,k,l) and amplitudes varying from $-F$ to $+F$. Their phase angle has to be adjusted so that the best map is obtained.

Once a chemically plausible trial molecular structure is obtained, a refinement step is performed to flatten the differences between experimental data and calculated structure factors. This can be done using least squares procedures or difference synthesis (working on the difference between observed and calculated F), which allows for adjustments of the position of heavy atoms and location of hydrogens.

The quality of the proposed structure is generally expressed through a reliability index, the residual factor R:

$$R = \Sigma \left| \left| F_o \right| - \left| F_c \right| \right| / \Sigma \left| F_o \right|$$

where indices o and c refer to observed (respectively calculated) values, and the summation is over all h,k,l reflections.

For a random distribution R would be 0.586 for a non-centrosymmetric structure (and 0.828 if a centre of symmetry exists). R of about 0.4 is correct for a trial solution. Values of about 0.05 are now considered as good for a refined structure, whereas higher values correspond to partially incorrect structures or to less precise determinations. $R = 0.25$ suggests that atoms are correctly located within about 0.1 Å [5, p. 763].

What precision may be attainable to fix the atom location in a crystal?

- First, as μ is calculated from a finite Fourier summation, it is clear that the more h,k,l reflections are considered, the more details in the crystal organization would be derived (this is similar to truncation effects in Fourier IR or NMR spectroscopies). Resolution is therefore increased if high scattering angles are considered, but the number of data to process increases very rapidly: Cantor and Schimmel [5] quote the example of a macromolecule with a cell of about 150 000 Å³ (linear dimensions about 50 Å): going from a 4 to 1 Å resolution implies treating from 1200 to 75 800 diffraction spots [5, p. 752].
- Another important feature is that hydrogen atoms are difficult to locate (accuracy is often limited to 0.1 Å). This is a result of the fact that X-ray scattering is almost entirely due to electrons (the scattering factor varies roughly as the atomic number Z), and so is largely greater for heavy atoms than for hydrogen. Computer programs exist for automatically adding hydrogen atoms to structures when crystallographic data do not specify their location (by means of a table of standardized bond lengths) [3].
- Other limitations come from the wavelength of the X-ray radiation, the quality of the crystal (disorder) and mainly thermal motion of atoms: atomic displacements of ca. 0.05 Å are not unusual at room temperature for organic molecules. This can lead to an apparent shortening of bond lengths up to several hundredths of an Å unit.

Small molecules are generally analysed with a resolution of about 1 Å (where each atom is clearly distinct). For macromolecules, the resolution is seldom better than 1.5 Å: however, a resolution of 2.5 Å still allows for seeing the protein side chains or salient groups (such as carbonyls) of peptide moieties.

4.2 NEUTRON SCATTERING AND MISCELLANEOUS TECHNIQUES

4.2.1 Neutron scattering

Although not so common as X-ray experiments, neutron diffraction studies can give comparable (and often complementary) information [2,6]. Small elementary particles (such as neutrons or electrons) may exhibit behaviour comparable to transverse electromagnetic waves. The associated wavelength depends upon their energy according to the de Broglie relation:

$$\lambda = h/mv$$

where m is the mass of the particle and v its velocity. If the wavelength is of a magnitude comparable to interatomic distances, diffraction effects from the atom arrangements in crystals or molecules are to be expected, similar to those observed in X-ray diffraction studies.

Nuclear reactors are the common neutron source: fast neutrons generated during fission can be slowed down by collision with atoms of a moderator in a pile. So, one can get a beam of thermal neutrons corresponding to associated wavelengths of about 1 Å; that is a convenient value for molecular diffraction studies.

The root mean square velocity of the neutrons produced in a pile of temperature T is given by:

$$v = (3kt/M)^{0.5}$$

(M = mass of the neutron), or according to the de Broglie relation:

$$\lambda = h/(3MkT)^{0.5}$$

This corresponds to $\lambda = 1.33$ Å at 100°C. A monochromatic beam (necessary for diffraction experiments) can be generated by filtering using a Bragg reflection over a large single crystal: this allows for selecting a rather small band of wavelengths from the incident beam.

Unlike X-rays, which are mainly scattered by electrons, neutrons are diffracted by nuclei. Neutron scattering factors are roughly independent of both the nature of the atom (within a factor of 2 or 3) and the scattering angle. As a consequence, hydrogen atoms, which are difficult to cope with in X-ray experiments, are now easily located with an accuracy comparable to that of other elements (ca. 0.001 Å). Absorption of neutrons is generally quite weak (because of their electrical neutrality) and, unless an exception, does not require careful corrections.

However, some disadvantages as compared to X-rays do exist: neutrons are available from only few centres, in contrast to the more ancillary and widespread X-ray instruments. Larger crystals are required (ca. 4 mm³ or 5 mg in place of 0.01 mm³ or 0.01 mg for X-rays), and data collection times (typically a few weeks) are much longer, since neutron beams are weak as compared to X-ray beams from usual devices. Detectors (only counters and not photographic films) are low in resolution (owing to the size of the beam) and need to be efficiently shielded against parasite radiations.

Location of hydrogens is, of course, quite interesting, particularly regarding the paramount importance of hydrogen bonds in the conformational preferences of biomolecules and in the stabilization of their complexes with drugs or ligands [7, 7a, 7b]. Nevertheless, as a counterpart, the number of unknown parameters increases, and there may even be a need for deuteration in more complex situations.

Some other methods (particularly electron diffraction and microwave or far infrared spectroscopies) are able to give access to geometrical parameters. Although they can attain very high accuracy, their use is restricted to small molecules, and so is of limited interest for direct application in the scope of this book. However, it must be recalled that precise geometrical information is of prime importance in various fields of molecular modelling: defining structural primitives for model builders, identifying geometrical distortions characteristic of specific structural arrangements, setting parameters in empirical force field methods for fragments or bonds not yet considered, and so on.

4.2.2 Electron diffraction

As for neutrons, diffraction experiments can be carried out with electron beams. According to the de Broglie relation, for an acceleration voltage of 40 kV, the associated wavelength is about 0.06 Å, small enough with respect to interatomic distances to generate diffraction patterns. Although some results have been reported on solid samples (mainly for surface studies), typical experiments are performed on gas phase. Diffraction therefore occurs from randomly oriented molecules. A further difficulty results from the fact that the diffraction pattern is the summation of coherent molecular scattering (the only one of interest for the derivation of molecular geometry) and atomic (both coherent and incoherent) scattering, which complicates the analysis [2,8].

Until now, diffraction experiments which can attain very high accuracy (few thousandths of an Å) have been restricted to small molecules (unless symmetry considerations reduce the number of parameters).

4.2.3 Microwave spectroscopy

The same is true for microwave spectroscopy, which seems up to now devoted to gas phase molecules less than 10 non-hydrogen atoms. The principle can be approached by the simpler case of a diatomic molecule, where transitions between successive rotational levels $(J{\rightarrow}J+1)$ are given by:

$$h\nu = \Delta\,E_{\text{rot}} = 2hB(J+1)$$

where J is the rotational quantum number. Rotational constant B is related to the inertial moment I calculated by reference to the centre of mass, where x_i is the distance of atom i (mass m_i) from the centre of mass:

$$B = h/8\,\pi^2\,I \quad \text{and} \quad I = m_1\,x_1^2 + m_2\,x_2^2$$

So, the internuclear distance can be calculated from the rotational frequencies observed.

For a polyatomic molecule, the rotational spectrum (now located in the microwave region and no longer in the far infrared, as for lighter diatomic molecules) allows determination of the three principal moments of inertia of the molecule. This is generally not sufficient to derive all the interatomic distances involved, but this difficulty can be overcome thanks to isotopic substitutions (changing the mass of the nuclei does not modify the internuclear distances, but changes the inertial moments and the spacing of rotational lines). Even for smaller molecules (the only ones that can be studied), the synthesis and computational efforts are important, but high accuracy can be obtained.

4.2.4 Comment on distances from various techniques

Although molecular graphics applications generally do not require the precision attainable by some of the above techniques, it may be useful to specify that, what is called "interatomic distance" does not actually always represent the same thing, and that minute differences appear from one approach to another [9]. Although very small, such variations may be confusing when discussing structural distortions from data of various origins. Such minute analyses are, however, obviously restricted to the smaller molecular systems, where various approaches could be carried out to determine the molecular geometry. In particular, it must be noticed that X-ray diffraction experiments "yield values which refer to distances between centres of electron densities, which need not correspond exactly to the positions of the nuclei, and yield only rather imprecise hydrogen positions whereas neutron diffraction values refer to distances between positions of the nuclei, which are accurate even for hydrogen" [10].

4.3 NMR: A SOURCE OF GEOMETRICAL DATA IN SOLUTION

X-ray crystallography has long remained the privileged (and often the only) way in which to get molecular geometries from solid samples, hence comes the success and widespread use of crystallographic databases such as the Cambridge Database or the Brookhaven Protein Data Bank, which still constitute the main source of information for geometrical data about organic substrates and proteins. However, since 1978 a new approach was being developed using NMR data, and taking advantage of the spectacular developments of this technique. Improvements of NMR equipment now deliver, on commercial spectrometers (using high magnetic fields), working frequencies of 500 and 600 MHz (for ^{1}H), in place of the 60 MHz during the 1960s, dramatically increasing sensitivity. The Fourier transform, and new pulse sequences particularly regarding 2D (and more recently 3D and 4D) NMR offer extended capabilities [11]. The role of NMR as a source of geometrical data has grown rapidly and this approach now provides new avenues for determining stereochemistry and interatomic distances. The NMR approach has become an indispensable tool in the determination of the 3D structures of proteins in solution (this important field of application will be detailed in Chapter 13), that of conformational forms of DNA etc.

Although not comprehensive, NMR-derived information is, nevertheless, of prime interest, since it directly concerns molecules in solution, that is in conformations which may be assumed to be not so far from that adopted in the living media. Indeed, if there are numerous cases where the crystal structure of proteins well fits measurements in solution (indicating that side chains in the interior of the molecule were locked in a rigid position), other examples showed large differences between crystal and solution [12–14]. At last, whereas crystal structure provides static representations, NMR gives access to some dynamic information. It is able to reflect both low frequency exchange processes or at the molecular scale much faster motions (about to the picosecond) in the solution (see Chapter 13). Some reviews on NMR structural studies and the dynamics of solid-state bio-macromolecules have also recently appeared [15, 16].

In the tremendous advances attained during the last few years, the Nuclear Overhauser Effect (NOE) deserves special interest in view of its outstanding role [13,17] in 3D structure determination, and the great efficiency offered for its measurement by 2D NMR. A single NOESY (Nuclear Overhauser Enhancement SpectroscopY) experiment indicates spatial proximities between individually assigned H atoms, and provides a network of H–H distances which spans the entire biomolecule under investigation.

Although NOE is more widely used, other NMR information may also give supplementary insight about the spatial location of atoms, particularly in the protein field.

Identifying *internal H-bonds* in polypeptides may be derived from amide-proton exchange rates: the exchange is slow for protons involved in H-bonds or buried in the internal part of the protein. *Torsion angles* may be estimated from J coupling constants in Karplus–Pople type laws (Figure 4.6). For instance, in proteins, NH–C$_\alpha$H couplings give dihedral angle restraints from $^3J_{NH\alpha}$. This, plus energy considerations, constrains the backbone torsion angle Φ to values between -80 and $-160°$ for residues where $^3J > 8$ Hz and -60 to $+40°$ if $^3J < 5.5$ Hz [18]. (For other examples, see also Moore *et al.* [19].)

4.3.1 Nuclear Overhauser Enhancement and 2D NMR

NOE corresponds to a selective fractional enhancement of a given NMR line by the irradiation of another resonance in dipolar coupled spin systems [20].

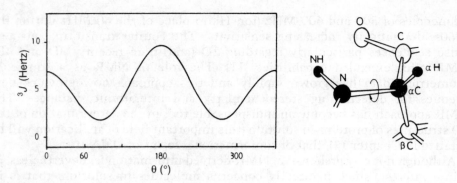

Figure 4.6 Karplus curve for $^3J_{HN\alpha}$ coupling vs. the torsion angle H–N–C_α–H (θ) for an amino acid residue (see Chapter 13). For L amino acids, $\theta = |\Phi - 60|$, Φ being the torsion angle around the C_α–N bond. $\Phi = 180°$ for an s-trans arrangement of the main chain.

For sensitivity reasons, here we will only consider interactions between protons. However, 3D and 4D heteronuclear experiments were recently proposed using ^{13}C and ^{15}N labelled compounds to extend the capabilities. NOE is a consequence of the effects of dipole–dipole relaxation on the spin population of the individual nuclei, and it suffers the strong distance-dependence of the cross relaxation rates among proton pairs. Indeed, the intensity of the NOE enhancement depends upon the distance between interacting nuclei (r) and an effective correlation time (τ) for rotational motion of the vector between the nuclei [20]:

$$I \alpha r^{-6} f(\tau)$$

In practice, NOE enhancements will therefore be observed only for protons in close spatial proximity (< 5 Å), and the preceding relationship offers a way in which to estimate interproton distances [21, 22].

NOE and distance determination

For small, roughly spherical proteins (MW about 6000), only tumbling motions are important. NOE between protons can be observed for sites distant by less than 4–5 Å (otherwise the effect is below the common detection level). Besides instrumental limitations (due to the poor sensitivity of NOESY experiments), dynamic factors introduce intrinsic limits for interpreting NOEs.

Motions on the picosecond timescale introduce averaging effects that decrease the cross relaxation rate, and consequently the NOE, by a scale factor relative to rigid models [23]. However, because of the r^{-6} term, motional errors of a factor of 2 lead to only a 12% uncertainty in the distances [23]. Internal calibration (by reference to pairs of protons whose distances are known) may overcome this difficulty. However, some caution must be exercised with specific NOEs altered by internal motions: local mobility influences the correlation time, which may not be considered as uniform over all the molecule. For example, in the external regions of proteins, peripheral segments of side chains may undergo rapid local motions which quench the NOE. For this reason, only identified NOEs are considered (absence is not viewed as characteristic of long distances, since it may be due to local mobility). In any event, NOE experiments are better considered as giving a network of distance extremes rather than exact values.

In a "uniform averaging model", NOEs are classified as *strong* (associated with interproton distances less than 2.5 Å), *medium* (between 2.5 and 3.0 Å) and *weak* (for distances less than 5 Å) – recall that a minimum distance for two hydrogen atoms is given by the sum of their van der Waals radii, about 2.0 A.

NOE measurements and 2D NMR spectroscopy

Although most of the data needed have first been investigated through usual (one-dimensional) NMR, 2D sequences correspond to definite improvement, and prompted the development of practical distance determination via NMR, leading to efficient protein structure assignments in about 1977 (20 years after the first report of a protein NMR spectrum). One advantage results from *improved resolution* spreading the resonances over a plane rather than along an axis. Indeed, for complex molecules, resonance peaks may overlap and some broadening may occur due to molecular motions slower than in small molecules. Another main advantage of 2D spectra is *efficiency*, since a single experiment gives all spin-spin interactions in the entire molecule.

In a conventional (1D) NMR, the spin system is first submitted to a RF pulse. Immediately, or after a given delay, the free induction decay is recorded with time $M(t_2)$, and then Fourier transformed to yield a signal in the frequency domain, giving the usual spectrum display $S(f_2)$.

In a 2D experiment, the signal is not recorded immediately after the preparation time (during which the spin system is submitted to a sequence of RF pulses), but only after a delay called the "evolution time (t_1)" and a "mixing period" according to the scheme:

preparation, evolution (t_1), mixing, detection (t_2)

A set of measurements is carried out where this evolution time is sequentially incremented: so two time domains $(t_1$ and $t_2)$, are acquired in a single experiment. The data matrix obtained in such multipulse experiments, $M(t_1,t_2)$, is Fourier transformed in one dimension, transposed and transformed with respect to the second time variable (the second dimension) to yield a grid of resonance intensities vs. two frequencies, f_1 and f_2.

Typically, in a 2D spectrum, diagonal peaks correspond mainly to the conventional (one dimension) spectrum, and display the positions of the resonance lines. Cross peaks establish correlation between diagonal peaks [24]: these cross peaks determine nuclei interconnected by some interaction. For

NOESY, such interactions involve a mutual *dipolar relaxation* pathway, a through-space mechanism, our main concern here. Other pulse sequences concern *spin coupling*. They correspond to through-bond mechanisms. In COSY (COrrelation SpectroscopY), cross peaks reflect scalar spin-spin couplings between protons separated generally by no more than three bonds (to get appreciable coupling). HETCOR (HETeronuclear chemical shift CORrelation) specifies direct ^{1}H–^{13}C connectivity. A second type of 2D experiment separates chemical shifts and couplings along the two axes [13,24,25].

Typically, a NOESY spectrum contains a large number (about 100) of cross peaks which express Nuclear Overhauser Enhancement, and indicate a close proximity ($d < 5$ Å) in the 3D space. Such stringent constraints on the conformation can be used in modelling programs to derive the 3D structure. (For a simple example, see Chapter 13, p. 417.)

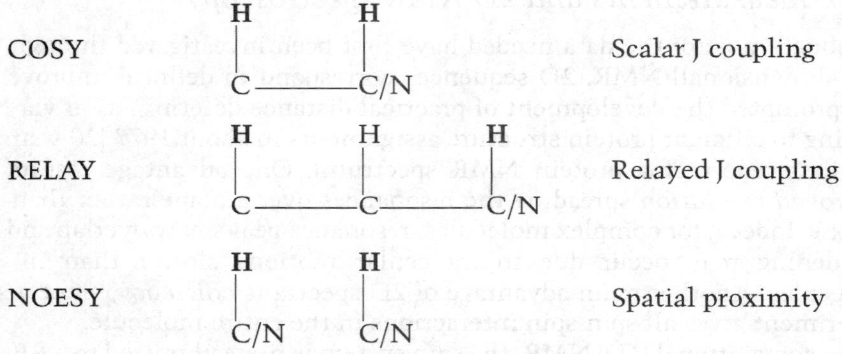

COSY

$$\begin{array}{cc} \text{H} & \text{H} \\ | & | \\ \text{C}\!-\!\!\!-\!\!\!-\!\!\!-\!\!\!-\!\!\!-\!\!\!-\!\!\text{C/N} \end{array}$$ Scalar J coupling

RELAY

$$\begin{array}{ccc} \text{H} & \text{H} & \text{H} \\ | & | & | \\ \text{C}\!-\!\!\!-\!\!\!-\!\!\!-\!\text{C}\!-\!\!\!-\!\!\!-\!\!\!-\!\text{C/N} \end{array}$$ Relayed J coupling

NOESY

$$\begin{array}{cc} \text{H} & \text{H} \\ | & | \\ \text{C/N} & \text{C/N} \end{array}$$ Spatial proximity

(Correlated protons are indicated by bold letters.)

For example, in solving protein structures, valuable information regarding the sequence of amino acids in the chain is sought by examining NOESY through-space proximity between αCH$_i$ and NH$_{(i+1)}$: $d_{\alpha N}$ distance, and/or between NH$_i$ and NH$_{(i+1)}$: d_{NN} distance (Figure 4.7).

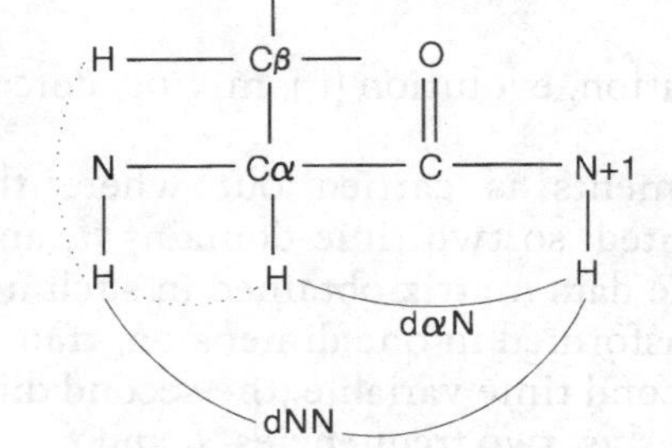

Figure 4.7 Determining the amino acids in a polypeptide. The spin set for each amino acid is established through C$_\alpha$H, NH coupling (COSY) and possibly by C$_\beta$H,NH correlation (RELAY) shown by dotted lines. The sequence is attainable with NOESY experiments (distances $d_{\alpha N}$ and d_{NN}, shown by full lines).

In a first step, the constituent amino acids present in a polypeptide chain can be identified by COSY spectra, since for the 20 naturally occurring amino acids, COSY connectivity patterns (protons within three bonds of each other) are quite characteristic. Also of great help in this step is RELAYED COSY (relayed coherence transfer). The J connectivity possessed by two spins A and M is passed to X (through AM and MX couplings) even though A and X are not scalar coupled. For example, in threonine, COSY peaks indicate ^3J normal connectivity between αH and the βH, and between βH and γCH$_3$. In RELAYED COSY a cross peak exists between this γCH$_3$ and αH (through their common coupling to βH). The intensity of the RELAY cross peaks depends upon the J coupling creating the original coherence, and also on the coupling through which it is transferred. It is therefore possible to optimize the experiment for better identifying selected amino acids (Figure 4.8). The more recent TOCSY sequence (transfer of magnetization through ^3J coupling between NH, C$_\alpha$H$_j$, C$_\alpha$H, C$_\beta$H and C$_\beta$H, C$_\alpha$H directly gives the entire spin system for each residue).

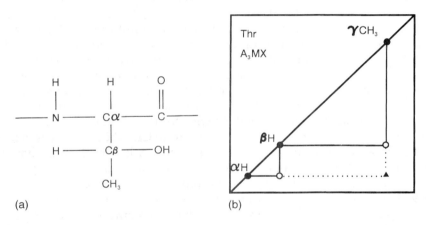

(a)   (b)

Figure 4.8 J-connectivity pattern from 2D NMR spectroscopy. COSY (o) and RELAYED COSY (▲) crosspeaks for the non-exchangeable protons of the threonine residue. Spectra run in D$_2$O (from Abraham *et al.* with permission [25]).

4.3.2. NOE and model building

On account of its strong distance dependence, NOE tells us nothing about long-range information, and a direct biomolecule-structure determination is not possible without the help of some other knowledge (for example, in the protein field, indication of the motifs constituting the chain). One can also introduce additional information provided by empirical energy functions [23, 26].

Generally, NMR-derived constraints are incorporated into the usual methods, where they appear to be of definite help, as shown with distance geometry approaches [24] or molecular dynamics simulations [27] – these methods have been discussed elsewhere, and only few examples will be given. NMR information was also introduced for exploring the conformational space of cyclic molecules [28] or proposed as a pseudo-energy term in molecular mechanics [29]. The principle is to introduce in the calculation some supplementary penalty terms: they tend to destabilize any geometry where the calculated distance between two protons is different from the value assumed on the basis of NOE effects. For instance, in the model of Tonge *et al.* [34], a conformation where distances deviate by more than 1% of a given NOE constraint is penalized by about 2 kcal/mol[2].

Polypeptides and proteins

As an example of the utility of NMR-derived interproton distances, one can note the test carried out by Havel and Wüthrich [30] on a small protein (58 amino acid residues) a basic pancreatic trypsin inhibitor (BPTI) of known crystal structure [31]. Experimental data correspond to 73 strong, 135 medium and 299 weak NOEs. These distance constraints have been introduced in a distance geometry approach, where the complete embedding process is decomposed into two successive, more tractable, calculations on "sub-structures". The authors were able to reproduce the crystal structure within 1.3 Å root mean square coordinate difference for the backbone atoms.

Similarly, the structure of the small protein Crambin was searched for, introducing 240 interproton distances less than 4 Å in a molecular dynamics simulation. Calculations from different starting structures converge to the known crystal structure, with rms deviations of 1.3 Å for the backbone atoms and 1.9 Å for the side-chain atoms [23, 32]. For other examples, see for instance, Lautz *et al.* [33] where molecular dynamics simulations were used to study the immunosuppressive drug cyclosporin A, either as an isolated molecule in solution, or in a set of four crystal cells embedding 16 cyclosporin A and 22 water molecules (introducing in the calculation crystalline periodic-boundary conditions). Interestingly, this study points out significant conformational differences between crystal and apolar solution.

NMR constraints were also used for traversing the conformational space of cyclic molecules [28]. This has been illustrated recently using the example of a cyclic decapeptide tyrocidine A (presumed as existing in solution as a single conformation) [34]. The strategy is detailed in Chapter 7. It first involves splitting the molecule into two open halves (terminated by a pair of common overlapping terminal groups). Then, examination of the conformational space available for each individual part (taking into account MM and NOE requirements) is carried out using a penalty function on NOE constraints.

---

[2] The penalty function is: $E(NOE)_{kcal/mol} = K \times (R-Ro)^2/Ro^3$ with $K = 800/R$Å.

Finally, a few representative conformers of each of the halves are assembled by superimposition of common terminal atoms and classified. Adding side chains and minimizing energy give the proposed final structure.

Conformational analysis of DNA

Nucleic acids are key molecules in the storage and transmission of genetic information carried by chromosomes, and play a preeminent role in the cellular function of living organisms. Deoxyribonucleic acid (DNA) in solution is a flexible macromolecule composed of a series of nucleotide units: each one is formed by a phosphoryl group ensuring the junction with other units, a furanose ring (sugar) and one of the four bases: adenine, thymine, cytosine and guanine. Two polynucleotide chains, associated via hydrogen bonding between adequate bases, form the well-known double helix.

The DNA strand may take several conformations, called A,B or Z. The A and B forms correspond to right-handed helices. A-form favours C3' endo puckering, whereas C2' endo puckering is associated with B-form. The other form, Z DNA (left handed helix), corresponds to sequences with alternating purines and pyrimidines. Two conformational parameters are of interest in the description of such units: the sugar pucker, corresponding to envelope or twist forms, and the glycosidic bond angle, leading to syn or anti conformations [35] (Figure 4.9).

NOE can be used to identify and sequence DNA oligomers [35]. The analysis starts from the 5' nucleotide, which is the only one to bear only one NOE peak. The basic remark is that the H_1 sugar proton is approximately equidistant (4Å) from H_8 or H_6 of its own base (respectively, depending on whether the base is a purine or a pyrimidine), and the H_8 or H_6 of the 5' but not the 3' nucleotide. In right-handed DNA helices, the sugar H_1,H_2/$H_{2'}$ give NOEs to H_8 or H_6 of the 3' neighbouring base, but not to the 5' one. In contrast, NOE occurs between thymidine methyl protons and H_6/H_8 of the 5' residue, but not the 3' one [35]. Differentiation between A and B forms is readily achieved by comparing internucleotide and intranucleotide NOE between $H_{2'}$ and H_8 or H_6 [35].

In Z DNA (left-handed helix of alternating purines and pyrimidines), pyrimidines have an anti-configuration and $C_{2'}$ endo sugar pucker, purines a syn glycosidic bond and $C_{3'}$ endo sugar pucker. The NOE pattern is therefore very different, and is easily detected. Slow exchange between B and Z forms has also been characterized [35] (Figure 4.10).

4.3.3 New trends and recent developments

The determination of the actual 3D structure of a drug in its active conformation (that bound to its receptor) is of course a fascinating challenge for the design of novel active analogues. Recent improvements of NMR experiments now give promising avenues for the direct study of drug-receptor

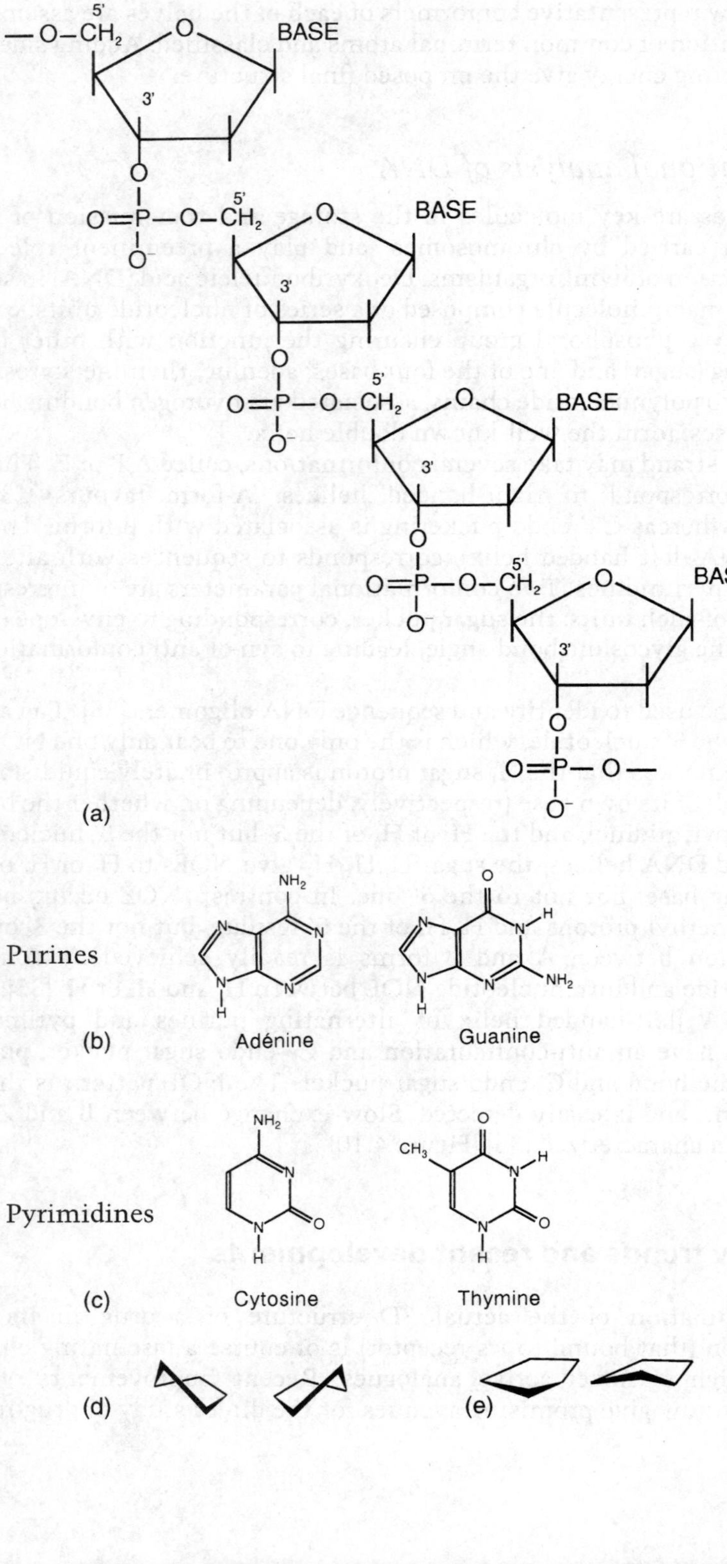

(a)

Purines

(b) Adénine   Guanine

Pyrimidines

(c) Cytosine   Thymine

(d)   (e)

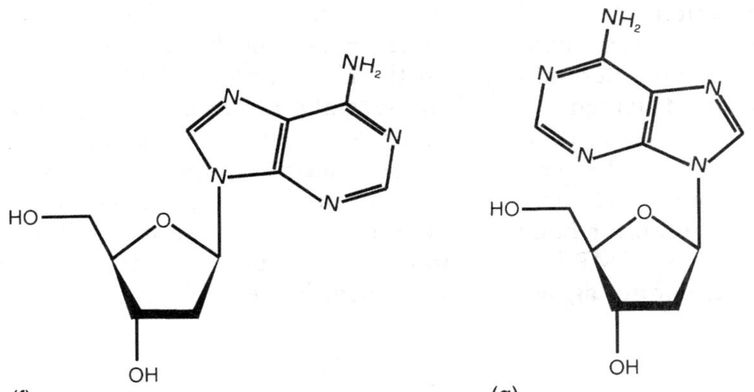

Figure 4.9 DNA structure analysis. (a) Polynucleotide chain; (b) purines; (c) pyrimidines; (d), (e) twist and envelope conformations of a cyclopentane ring and anti (f) syn (g) conformers around the glycosidic bond here for desoxyadenosine (from Cheatham [35] with permission).

complexes: various methods have been proposed to simplify the complicated proton NMR spectra of large receptor-ligand complexes [36]. Identifying those portions of the ligand that interact with the target site, and determining their 3D structure, are the main objectives of such studies.

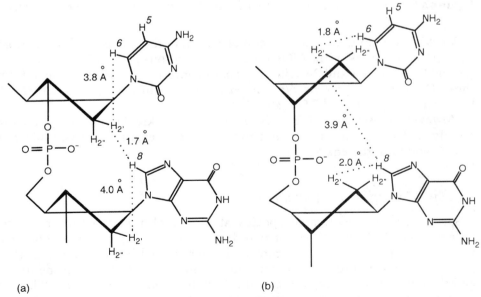

(a)   (b)

Figure 4.10 Differing proton-proton distances in (a) A form DNA (C3′ endo sugar pucker) and (b) B form DNA (C2′ endo sugar pucker) (from Cheatham with permission [35]).

The conformation of ligands weakly bound to a macromolecule and exchanging rapidly from bound to free state can be attained through *transferred NOE*. The method relies on the observation of NOEs between a target spin (on the exchanging ligand) and several protons on a rigid part of the system. These values lead to a probability density to find the target spin in given regions of space that can be used as a starting point in a subsequent model building procedure. An application example of transferred NOE was presented on an inhibitor bound to cytidine-5'-monophosphate-3-desoxy-D-*manno*-octulosonate, (CMP-KDO) synthetase [36]. Such inhibitors may be of clinical use as antibiotics against Gram-negative bacteria (Figure 4.11).

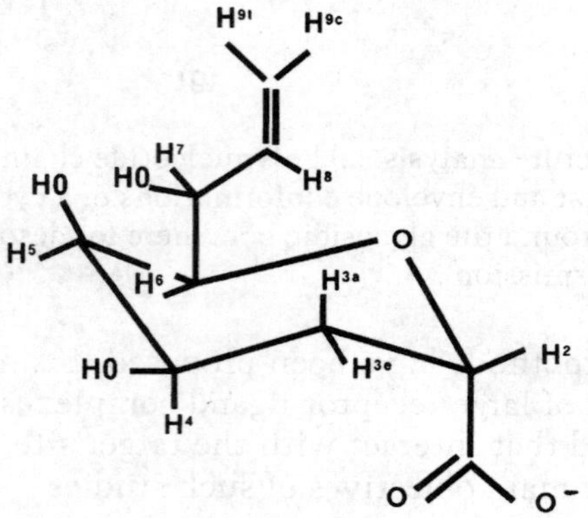

Figure 4.11   Application example of transferred NOE (from Fesik with permission [36]).

For the bound inhibitor, the $^5C_2$ conformation of the ring is established from the equal intensities of the NOE observed for protons pairs 2/3a and 2/3e, and the NOE between 4 and 3e. The side chain conformation is established from the NOE between protons 6/8 and 9t/7 and the low (or inexistent) NOE between 7/6, 7/5, 7/8 and 7/9c, [36].

However transferred NOE only applies to weakly bound and therefore rapidly exchanging ligands and not to the (more interesting) active inhibitors tightly bound to the enzyme [36].

Isotope-edited proton spectra

Spectra run on deuterated species allow for eliminating proton peaks. *Subtraction 2D NOE spectra* of two drug-receptor complexes (one obtained with a protonated ligand, the other with a deuterated ligand) affords an easy way for differentiating NOEs within the ligand and NOEs between the ligand and the enzyme, providing structural information on the active site. The use of *perdeuterated receptors* has been also proposed for a rapid investigation of the bound conformations of the ligands. A selective detection of protons attached to isotopically labelled nuclei (^{13}C or ^{15}N) is also possible [36].

Towards higher dimensions: 3D and 4D NMR

For larger systems, chemical shift degeneracy, peak overlap and increased linewidths limit the interest of usual 2D sequences relying on small H–H couplings. One attractive solution to overcome these problems is provided by 3D and 4D heteronuclear experiments [37]. Such developments seem very promising for extending the range of applicability of the NMR method to proteins in the 15–30 kDa range. Experiments are carried out on uniformly labelled (> 95%) ^{15}N and/or ^{13}C substrates.

Resolution is improved by increasing the dimensionality of the spectrum, since the proton resonances are now sorted along a third dimension: the cross peaks are spread into different planes according to the shifts of the heteroatoms (^{13}C or ^{15}N). Efficient magnetization transfer, ensured through large one-bond heteroatom-proton coupling, is less sensitive to linewidth influence. Furthermore, additional information can be gained through the shifts of these heteroatoms ^{15}N and ^{13}C. This leads to a massive simplification in analysing the spectra.

A classical 2D experiment processes a double Fourier transform on a data matrix where the data collected are a function of both a detection period (as in the 1D spectrum) and an (independently incremented) evolution period. Higher dimension (3D or 4D) NMR spectra are obtained by combining two or three 2D sequences. For example, in a 3D experiment, two evolution periods are incremented independently giving a data matrix $s(t_1, t_2, t_3)$.

Conventional 2D spectra try to identify NOE connectivities between adjacent residues involving NH, $C_\alpha H$ and $C_\beta H$ protons. Similar data are obtained in a 3D experiment, except that the spectrum is spread out in a series of slices according to the ^{15}N or the ^{13}C shifts [37] (Figure 4.12).

4.4 THE CAMBRIDGE STRUCTURAL DATABASE

The continuous interest in structural information, coupled with a dramatic expansion of crystallographic data, prompts an increased interest in crystallographic databases (mainly the Cambridge and Brookhaven databases, which have become standards for organic materials and proteins). The reliability and integrality of the data stored in these archives make such databases indispensable starting points, or essential components, of many molecular modelling applications.

At this step, it seems important also to point out that, besides the resolution of individual cases (i.e. providing a suitable geometry for a given target), storing geometrical information about a large set of structures provides new tools for structural investigation. Screening the database and carrying out statistical studies on selected candidates allows for detecting common trends, or conversely, attributing unusual behaviours to specific substructures or atom

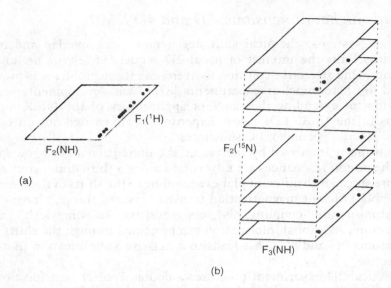

Figure 4.12   From (a) 2D to (b) ^{15}N edited 3D spectra, for NH/HC pairs. NOE cross peaks (dots) not differentiated along F_2 (NH) in a 2D NOESY experiment are separated and sorted according to ^{15}N shifts in a 3D experiment. Sorting can be pursued (in 4D) along ^{13}C shifts of the attached carbon (adapted from Clore and Gronenborn with permission [37]).

arrangements. Extracting dynamic information about interconversion or reaction pathways is also a spectacular application of such database screenings. As pointed out by Kennard [38], "The whole is greater that the sum of parts".

In this text, being largely devoted to modelling medium-size molecules (rather than large biomolecules), interest will be focused on the Cambridge Structural Database (CSD). Established as early as 1965, the CSD stores information derived from X-ray (mainly) and neutron diffraction studies on organic, organometallic compounds and metal complexes. Some purely inorganic structures (carbides, carbonates, cyanates, etc. are not considered; nor are high molecular-weight polymers. Originally, proteins were also outside the field of the Cambridge database. Resulting from a recent agreement with the producers of the Brookhaven Data Bank for proteins (PDB, see below), the Cambridge Structural Database now deliver bibliographics and sequence PDB entries. On the Cambridge October 1995 release, 3528 entries are available.

The Cambridge Crystallographic Data Centre (CCDC) not only produces and maintains the CSD database, but also ensures its distribution to academic centres and industrial companies, develops software and performs its own research projects. Distribution to the academic community is performed through "national affiliated centres", which contribute to the maintenance of the database by paying an annual national subvention.

In October 1995, the CSD files contained about 146 000 structures, with an update rate of about 20 000 new entries in the last year. More than half of the entries have been published since 1980, and correspond to an R factor (mean) of about 6%, characterizing high quality determinations. The consistency of the data is carefully checked, from the chemical name to the consistency of atomic positions, so errors are reduced down to 1.5%, whereas it was indicated that nearly 16% of the original data submitted contained at least one error. The database gathers primary literature data, collected from over 500 journals and their supplementary materials, or originating directly from authors or from a depository [39–42].

4.4.1 Organization of the database

The system is organized around a key module, the ASER database, and is provided with a set of modules for treatment, statistical analysis or information display.

In the ASER database, each entry is identified by a REFCODE, an eight-letter reference code, originally an acronym of the chemical name. NBS identification numbers (whenever cell parameters are given), or CAS registry numbers (unfortunately only for about 10% of all entries) are sometimes indicated.

Three levels of information are managed:

1. "One-dimensional" information gathers bibliographic data or textual comments.
2. "Two dimensional" information describes the chemical structure (atom type and connectivity) and bond characteristics (nature, cyclicity).
3. "Three-dimensional" information specifies the atom coordinates, the unit cell. These data are those used for molecular graphics representations or evaluation of molecular geometries.

For almost 99% of entries, cell parameters and information about the connectivity of the chemical moieties within the crystal are given ("chemical connectivity"). Atomic coordinates are indicated for 85% of all entries. To make information more readily usable by chemists, these coordinates refer to the crystal chemical unit (a covalently bonded structure), and not to the crystallographic asymmetric unit (which is a structural motif) (Figure 4.13).

To manage an efficient exploitation, the ASER database stores three records for each entry:

1. mandatory integers and bit screens.
2. 2D chemical connectivity and text.
3. 3D atomic coordinates and connections.

Various strategies are available for 1D or 2D information searches. Queries may, of course, concern refcodes, author(s) or compound names, but one can also ask for chemical formulae or chemical classes (among 86 organic classes). Besides the identification of particular molecules, a highly interesting feature

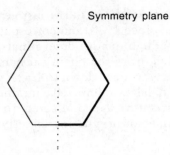

Symmetry plane

Figure 4.13 Crystal chemical unit and crystallographic unit: benzene molecule (only carbon atoms are shown). The complete molecule is generated from the asymmetric unit (bold lines) by the addition of symmetry-related atoms (from Kennard *et al.* [39] with permission).

for structural studies is the possibility of performing searches for chemical fragments (substructures), according to the nature of atoms, the existence and nature of bonds etc. Atoms can be fully defined or within a given class (any metal, alkali metal, and so on).

The search query is defined by tests, screens or a combination of the two. A large variety of tests, defined by key words, can be selected, including 19 text and 38 integer-field types. Tests may be combined with logical operators (and/or/not). The introduction of 682 bit screens appears very useful, allowing for saving a large percentage of the search time. Searches are more efficiently performed, comparing first screens generated from the query to the bit screens of the ASER base (created on the building of the base). Candidates issued from this first selection are subsequently submitted to the more time-consuming step of in-depth searches (via atom by atom, bond by bond fragment matching). The 682 bit screens proposed, generated as search keys, may ask for an organic material, an *R* factor smaller than a threshold value, an absolute configuration determination, etc. They can also select element groups, entry information, text screens or connectivity screens (the presence or not of 434 pre-defined subfragments). Tests based on elements, formulae, connectivity of special fragments or cell parameters are also provided.

Large facilities are offered for the input and output thanks to the QUEST module, allowing for alphanumeric or graphic man/machine communication. Menu-driven graphics interfaces provide an easy, user-friendly access. Note that the 2D edition causes some specific problems, and surprisingly appears more complex than 3D wire frame representations of the molecular framework. For 3D representations, the atom coordinates give a well-defined representation basis, and only limited choices have to be made about angles of vision, overall size, etc. By contrast, usual (2D) drawings of structural formulae implicitly obey some well-accepted conventions: a naphthalene molecule is always represented with the two rings side by side, and not one above the other. This means that no algorithm procedures can generate satisfactory graphics directly from connection tables for a wide range of compounds. To get suitable graphics (implying some standardization in the presentation of chemical classes), coordinates relating to a backlog of 30 000 connection tables are inserted into the CSD files.

Basically, the QUEST module carries out the following functions:

- construct search queries,
- display search results as text, summary tables, simple histograms or scattergrams,
- permit manipulation of geometrical parameters, and
- prepare subfiles for link with other modules (VISTA, PLUTO) or external software packages.

Other modules (VISTA and PLUTO) process and represent the 3D information. VISTA ensures a wide variety of numerical, statistical and graphical analyses on the retrieved geometrical data. It calculates intra- and inter-molecular geometrical parameters and performs diverse processes based on these parameters (selection of fragments by geometrical criteria, tabulation of their geometry, output of coordinates, etc.). It also allows for 3D queries such as those involved in pharmacophoric-pattern search. In the treatment of these numerical multi-parameter features, statistical methods play an important role. Modules propose varied approaches of data analysis, including analysis of variance, principal component analysis and cluster analysis [43, 44].

The plotting package PLUTO is used to prepare, from the numerical files in CSD, a variety of 3D displays and illustrations of crystal and molecular structures. A single molecule or an assembly of molecules (packing diagram) can be represented under various styles of illustration: stick or ball-and-stick diagrams, space filling models, with some control capabilities about size, colour, rendering, and so on (Figure 4.14).

Among the current developments are the integration and extension of the 3D search capabilities, and improvement of the connectivity description. Indeed, the "crystallographic connectivity" representation lacks bond-type information and details on atomic charges explicited in the "chemical connectivity". The absence of some hydrogens in the reported crystal structure, the presence of more than one bonded residue in the crystal cell, etc., may constitute other sources of difficulty in the matching of these two representations [41].

4.4.2 Applications

The CSD system, gathering extensive data files and user-friendly software, has become an important tool in many fields of chemistry. 3D information is, of course, an important starting point for the modelling of chemical systems or similarity searches between a set of related molecules. Nevertheless, an interesting feature is the ability to carry out systematic analysis via statistical treatments. These allow for "waking up the dormant information" that

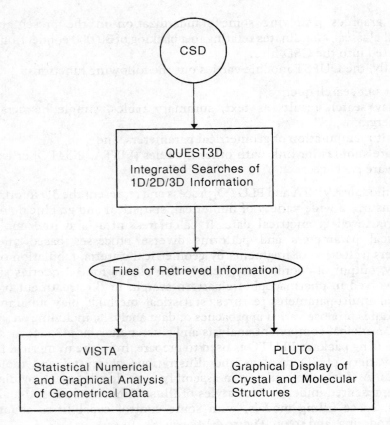

Figure 4.14 General organization scheme of the Cambridge Structural Database (reproduced with permission from F.H. Allen).

reveals, from the published data, some hindered information, underlying but not yet perceived.

Systematic analysis of molecular geometry

Mean geometries and model builder primitives

X-ray structures of course represent solid-state conformations which are not necessarily identical either with those in solution, nor with the reactive species in a living organism (because of crystal packing, solvent effects or possible conformational changes during adaptation processes) [45]. However, knowing the crystal structure is a unique source for the easy construction of molecular models. Of course, it directly gives the answer for rigid substrates, but it is also of great help for flexible molecules where crystal structures provide a "reasonable" starting geometry which can be further refined. In model building calculations, standard bond lengths or angles can constitute a first set of input data, but more realistic proposals can be attained from an average of the geometries observed experimentally. This approach thus provides structural fragments closer to reality, since more subtle environment effects, associated with specific structural frameworks, are taken into account. Assembling such predefined fragments speeds up the elaboration of complex systems.

A first step consists of the determination of mean interatomic distances, depending upon the molecular environment in which the bond is embedded (Figure 4.15). Tables are now available gathering the average bond lengths involving the most common elements, corresponding to nearly 700 bond types [46].

For the derivation of fragment libraries, particularly for applications in molecular graphics, estimating average molecular dimensions, the Cambridge database can constitute a major source of data. However, estimated standard deviations on the atom coordinates are not quoted, and the scattering of the results encompasses both experimental errors and environment effects. Practical guidelines to derive the best average distances, depending upon the relative importance of these effects, have been proposed [47].

Nucleoside and nucleotide studies

Besides a better estimate of bond lengths, depending upon their environment in the molecule under investigation, standardization of larger molecular

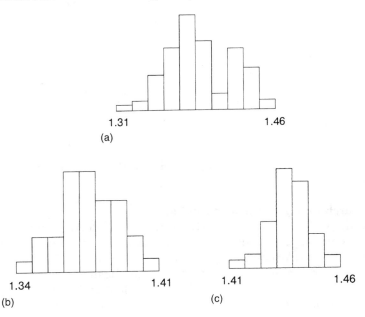

Figure 4.15 Distribution histogram of the $C_{ar}-N$ bond lengths in a $C_{ar}-N-(Csp^3)_2$ fragment. (a) Complete distribution; (b) (c) resolution into two subsets, (b) for planar nitrogen (mean valence angle at $N > 117.5°$) and (c) for pyramidal N (mean valence angle at N in the range 108–114°) (from Allen *et al.* with permission [46]).

fragments, considered as molecular building bricks, have been outlined. Using a least square minimization procedure on nearly 90 crystal structures, Taylor and Kennard [48] proposed a set of orthogonal coordinates for nucleic acid residues as close as possible to the average dimensions observed in crystals subject to the necessity for ring closure. For these primitives, the residues are planar and the maximum discrepancies are no more than 0.001 Å for bond lengths and 0.1° for angles. An example is given in Figure 4.16 where the geometry is specified in internal coordinates (more easily perceived for a chemist than Cartesian values). Apart from model building, other applications may be the refinement of oligonucleotide structures or the parameterization of empirical force fields.

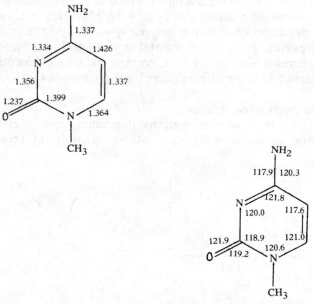

Figure 4.16 Average dimensions of the cytosine motif (Å and degrees) (Taylor and Kennard [48]).

Systematic search for a given pharmacophore
Screening of the CSD files has recently been used to derive putative ligands complementary in shape to a receptor site of a known structure, or to propose new ligands if a pharmacophore geometry is already established and the geometry of receptor-bound conformations of some ligands are known [49–51]. This point will be discussed in more detail in Chapter 12, dedicated to the pharmacophore approach.

Systematic analysis of intramolecular interactions

Apart from deriving more realistic geometries (either to be used directly or submitted to energy minimization), such systematic studies are essential for fundamental approaches of structure and bonding. Often, effects are small and close to uncertainty limits of individual structure determinations. Statistical analysis on a large set of samples is therefore necessary to safely detect trends [42].

Large variations due to hybridization changes are well known:

C——C——C——   C===C——C——   C≡≡≡C——C——

1.538 Å   1.507 Å   1.464 Å

but more subtle effects also modify bond lengths to a lesser extent.

As a well-known example, one can note substituent-induced deformations in benzene rings, which have been widely documented. An electron-withdrawing substituent increases the *ipso* α angle and slightly decreases β by about one half of Δα. Adjacent α bonds are shortened (opposite effects being observed with electron-donating groups). Statistical treatment through factor analysis rationalized the observations (more than 100 derivatives investigated) on the basis of two independent mechanisms involving σ and π interactions [42, 52, 53] (Figure 4.17).

Figure 4.17 Substituent-induced deformations in benzene rings.

Hydrogen-bonding

Hydrogen-bonding is involved in many chemical and biochemical processes (stabilization of drug-receptor complexes, anaesthesic properties, etc.) Analysis of the crystallographic environment of about 700 hydrogen atoms show that short ($d < 2.4$ Å) CH $\cdots$ O contacts (significantly lower than the sum of the van der Waals radii of H and O: 2.7 Å) are not unusual in amino acid crystal structures. In contrast, short CH $\cdots$ C or CH $\cdots$ H contacts are extremely rare. So, it was established that CH $\cdots$ O hydrogen-bonding may be a significant factor in determining the conformations and the crystal packing arrangement of amino acids, and also in stabilizing certain DNA-drug complexes [42].

Distribution of O $\cdots$ H distances and O–H $\cdots$ O angles have also been investigated in O–H $\cdots$ O arrangements (from neutron diffraction data). The mean values are respectively, 1.818 Å and 167.1°, in good agreement with MO predictions. Shorter hydrogen bonds tend to be more linear than longer ones. In hydrogen-bonding to etheroxides, it appears that the proton is in the plane of the oxygen lone pairs, but with no privileged direction. In contrast, H-bonding to a sp² oxygen (in a carbonyl group, for instance) shows a definite directional influence; the incoming hydrogen statistically prefers the region of the "rabbit ears" of the oxygen atom:

$$O\text{———}H\text{--------}O\begin{array}{c}\diagup R\\[6pt]\diagdown R\end{array}$$

Chemical dynamics from static crystal structures: structure correlation

Extracting dynamic information from crystallographic data has been proposed as the "structure correlation method" by Burgi and Dunitz [54]. When a molecule is embedded in a crystal, it suffers forces from the crystal environment, which can more or less distort its geometry, so that its structure is not necessarily identical with the equilibrium structure attained for the isolated species, for instance in the gas phase. Similarly, the structure for a given molecular fragment depends to some extent upon the molecular environment in which the fragment is embedded, as well as on crystal lattice influences.

Examining a large set of compounds containing a given fragment provides a series of snapshots of this fragment, each in a particular molecular environment. These data can be ordered so as to depict a gradual deformation of the fragment. In a "many-dimensional" conformational space (one dimension for each geometrical variable defining the substructure), each of these fragment geometries may be associated with a sample point. The basic hypothesis is that these interactions with the molecular or crystal environments may be considered as small perturbations to the total molecular potential energy. One may therefore reasonably expect that the sample points (corresponding to the structures observed) will tend to concentrate in the low-lying energy regions of the potential energy hypersurface. The distribution of sample points in the conformational space will thus delineate the shape of the energy valleys. It gives information about the preferred regions on this surface, and on the probable interconversion pathways between stable conformers (or even the energy profile of related chemical reactions).

So, structure correlation method allows for mapping low-energy paths for stereoisomerization processes, and proposing an approximate structure for the transition state involved.

One may question, however, what the scattergrams of experimental structural data actually represent? The fragments investigated are also submitted to environment effects from the molecule in which they are embedded. Their distribution therefore involves the structural constraints imposed (for instance, ring closure conditions may fix more or less rigidly some angle values). If such interactions are large, the scatter can be viewed as a response path and not strictly as a minimum-energy path. However, if the potential surface possesses narrow, steep-sided energy valleys, whatever the constraints may be, the sample points are more likely to indicate the minimum-energy pathway.

Conformational interconversions of anomeric systems

The anomeric effect (observation that in sugars the percentage of axial conformers is greater than in the corresponding hydrocarbons) is related to the

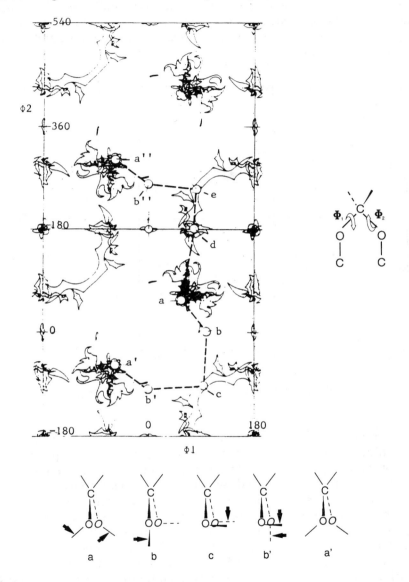

Figure 4.18 Interconversion pathways from crystallographic data of acyclic and cyclic fragments C–O–C–O–C. The lines outline clusters of sample points with their density. The "peripheral pathway" a,b,c,b',a' involves several steps: correlated disrotations (a-b and b'-a'), non-correlated ones (b-c and c-b'). The "lateral pathway" a,d,e,b"a" at first maintains the constancy of one angle Φ (a,d,e) and in a second step follows the above peripheral pathway (e,b",a") (from Cossé-Barbi and Dubois with permission [55]).

existence of two electronegative atoms in gem position on a single carbon. Analysis of 85 structures which have a common C–O–C–O–C fragment was carried out in a conformational space built on two torsion angles Φ_1 ($C_1O_2C_3O_4$) and Φ_2 ($O_2C_3O_4C_5$). Screening experimental data indicates that numerous acyclic fragments are gathered in areas close to 60° for both Φ_1 and Φ_2 corresponding to a C_{2v} symmetry and associated with a double anomeric effect [55] (Figures 4.18, 4.19).

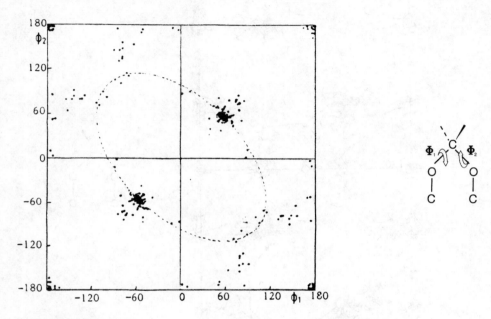

Figure 4.19 Crystallographic data for acyclic C–O–C–O–C fragments (from Cossé-Barbi and Dubois with permission [55]).

Extension of this study to cyclic and acyclic fragments gives some insight about conformational energy minima, and suggests two possible inter-conversion pathways, both avoiding the constrained structure ($\Phi_1 = \Phi_2 = 0$). These conclusions are supported by a conformational study of dimethoxy-propane, considered as a model compound, and carried out by quantum chemistry methods (at the semi-empirical INDO level), which reveals the same probable interconversion pathways [55] (Figure 4.20).

For other applications of the structure correlation method, see also Nachbar *et al.* and Cossé-Barbi *et al.* [56] on structure- and internal dynamics-implications of the *gem-6 effect* (a widely documented mechanism in systems containing two tert-butyl groups in close proximity).

Cis-trans isomerization around a Csp²–Nsp³ bond
Another example, providing a good indication of the relevance of the structure correlation method to foresee interconversion pathways and transition state

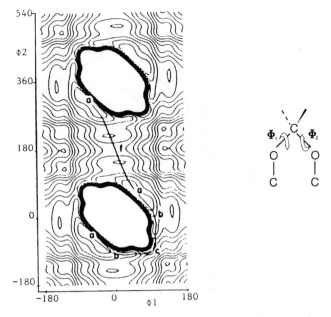

Figure 4.20 Theoretical (INDO) conformation map for 2,2 dimethoxypropane. Separation of lines 0.627 kcal/mol (from Cossé-Barbi and Dubois with permission [55]).

geometry, is afforded by the recent study of the cis-trans isomerization process by rotation around a Csp^2-Nsp^3 bond, as occurring in amides, enamines, etc. [57] (Figure 4.21). (For other examples see Wilson and Huffman [45].)

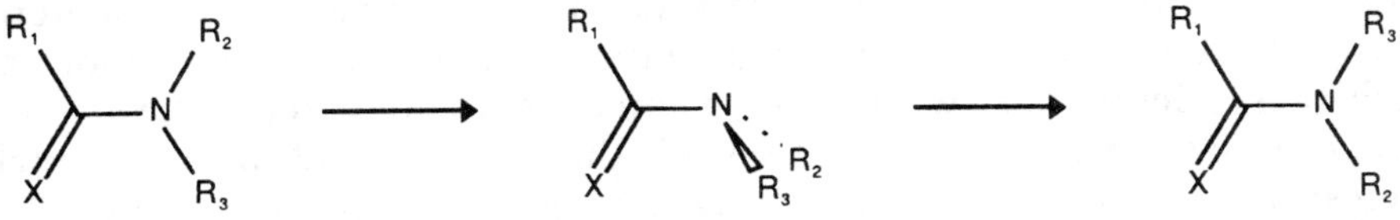

Figure 4.21   Cis-trans isomerization process example.

Such systems tend to be planar owing to the partial double bond character of the C–N bond. However, in crystal environments, distorted geometries are encountered. The deformations have been described by the torsion angle around the C–N bond and out-of-plane bending of the amino nitrogen, from planar sp^2 to tetrahedral sp^3 (the Csp^2 atom resisting much more strongly to deformation and remaining nearly undistorted). Out-of-plane deformation origins from two different mechanisms: a "butterfly" bending of sp^2 nitrogen, and a complex motion combining both the twisting around the C–N bond and the nitrogen bending.

From crystallographic data, the structure correlation method suggests that this complex mechanism satisfactorily maps the cis-trans isomerization process. Distorted O–C–N fragments reflect the geometrical changes occurring in the group along the reaction pathway (Figure 4.22).

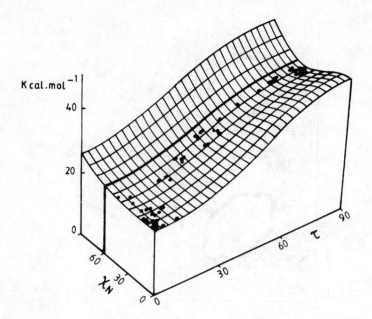

Figure 4.22 Sample points mapped on to the calculated energy surface (function of τ = twisting angle around the C–N bond, and $\varkappa_N$ = out-of-plane bending angle of N) (from Gilli *et al.* with permission [57]).

Mapping from crystallographic data has been compared to a theoretical study of the reaction profile on the potential energy surface. From molecular mechanics calculations (with adapted force field parameters), the potential energy is evaluated as a function of the torsion and bending angles. Interestingly, it appears that the experimental geometries retrieved for distorted fragments are located along the valley leading from the planar initial conformation to a rotated and pyramidalized transition state. Note, however, that several MO calculations propose another pathway beginning with N pyramidalization, and only after rotation around the C–N bond [57].

Reaction pathways from structure correlation
As to the use of crystallographic information for the inference of a chemical reaction, one can note the earlier work of Burgi *et al.* [58] about the addition of a nitrogen nucleophile to a carbonyl group:

N $\cdots$ C $=$ O contact distances were examined in a set of crystals corresponding to C $\cdots$ N distances ranging from 2.91 Å (non-bonded contact) to 1.49 Å (covalent N–C–O bond). Analysis of the geometry of numerous fragments shows that the N $\cdots$ C distance (d_1) is inversely correlated to the C $=$ O distance (d_2) and the deviation from coplanarity (Δ) (Figure 4.23). Or, in terms of reaction pathways; as the nitrogen atom approaches the carbon, the CO distance increases and the carbon progressively pyramidalizes. The N–C–O angle suggests an attack from a constant direction at ca. 107° of the CO bond [54, 58, 59].

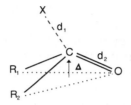

Figure 4.23   Geometry of N $\cdots$ C $=$ O and O $\cdots$ C $=$ O contacts (X=N or O).

Transition state inference
The method can be extended one stage further. It is clear that small displacements along a valley cost less in energy than motions perpendicular to it. The paths along the valley would therefore correspond to soft vibrational modes of the molecule. So, large amplitude internal motions of a molecule in a crystal environment can be inferred from the equilibrium crystal structures of a set of related molecules [60,61]. In favourable cases where extensive crystallographic observations are available, putative structures of the transition states can be proposed from experimental data (an uncommon situation, since transition states are more often only suggested from calculations) [60].

For triphenylphosphine oxide, the equilibrium structure of an isolated Phe$_3$P=O fragment is close to a symmetric propeller shape with the three phenyl rings rotated in the same sense by about 40° (Figure 4.24). The stereo-isomerization path proceeds through a transition state with one phenyl ring

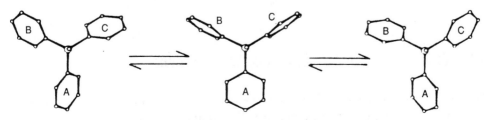

Figure 4.24   Transition state inference example (from Bye *et al.* with permission [60]).

nearly perpendicular to its C–P=O plane and the other two nearly coplanar (within 10°) with their own C–P=O planes, and leads to the enantiomer species (a mechanism related to the "two-ring flip mechanism" proposed by Mislow *et al.* from dynamic NMR studies of triaryl-boranes, -methanes, etc. [62]. For triphenylphosphine oxide, appreciable internal rotations of the phenyl groups about their respective P–C bonds are observed. The phenyl ring nearly perpendicular to the P=O bond (ring A) shows a much larger libration amplitude than the other two [61].

Multidimensional problems and statistical treatments
In more evolved problems, where the systems present numerous degrees of freedom, preventing interpretation from visual inspection, it is often useful to decrease the complexity of the problem.

One way in which to reduce dimensionality uses techniques such as *principal component analysis*. It substitutes the correlated (initial) variables with a few synthetic variables (referred to as *principal components*) more easily representing the variability of the system. The new variables are a linear combination of the original set, and are derived in decreasing order of importance. Selecting only the first ones allows for accounting for as much as possible of the variance of the original set, reducing the dimensionality of the problem without losing too much information. Another advantage is that this treatment avoids any *a priori* hypothesis on the significant parameters to examine, which could bias the analysis. On the other hand, some effort is necessary to connect the factors obtained (synthetic variables) to geometrical characteristics directly understandable by the chemist. In terms of molecular geometry, one can hope that a certain linear combinations of parameters can be interpreted in chemical terms: for example, a linear combination of ring torsion angles was identified as representing a puckering parameter in a pseudo-rotation itinerary [63,64].

Another possibility is to reduce the number of data points. In *clustering techniques* neighbouring points are grouped together and the relationships between clusters are analysed. In the field of conformational analysis of molecular fragments, statistical clustering methods have been adapted to cope with highly symmetrical and periodic distributions of sample points in the conformational space. The method was illustrated in a study of the molecular fragment $M(P-Phe_3)_2$ with eight torsional degrees of freedom. The conformational interconversion model proposed implies gearing motions of the two $P-C_3$ fragments alternating with stepwise inversions of the helicities of the $P-Phe_3$ propellers [65].

4.5 THE BROOKHAVEN PROTEIN DATA BANK

This project was started in 1971, with the constitution of standard format files providing the structural information for macrobiomolecules as the aim. Originally devoted to proteins, the database also distributes data on DNA, tRNA and polysaccharides. Moreover, besides experimental determinations it includes some structures generated by computer simulation or derived by analogy with chemically homologous molecules. NMR data have been also incorporated [1,3]. Data are mainly gathered deposits (since, owing to the huge volume needed, atom coordinates are generally not published for large macromolecules). About 3500 atomic coordinate entries (in October 1995) are held in the database, with an increase of about 600 structures during the last year.

4.5.1 Organization of the database

For each structure, entry identifiers include the name and the source of the protein, bibliographical references and some comments. Then the sequence of residues is given, with information about the secondary structure: helices, sheets or turns. They are followed by geometrical data: a first line summarizes crystal cell parameters with the position of the origin and the scale factor (coordinates are stored in an orthonormal axial system in Å units, whereas crystallographers commonly use axes based on the unit-cell vectors). Then each atom corresponds to one line of information gathering the atom name, its type, the residue name, chain identifier, residue sequence number, the 3D atomic coordinates (x,y,z), occupancy and temperature factor. Hydrogen bonds or disulfide linkages can also be identified thanks to connectivity records (Figure 4.25).

$$N(1) \quad C_\alpha(2) \quad C(3) \quad O(4) \quad C_\beta(5)$$

Figure 4.25 Ordering of atoms within amino acid residues.

For some cases, structure factor data are also stored. Although these primary experimental data imply massive memory requirements, they possibly could be used in future to reexamine and refine the structures.

As with the Cambridge Structural Database, various molecular modelling packages offer capabilities for easy interfacing to the Brookhaven Data Bank. They allow for varied displays of the molecule (or parts of it) and derivation of geometrical characteristics (torsion angles, interatomic distances, etc.). Frequently, for new structures entering the database, only the C_α coordinates are given (before the full structure is solved), and various programs attempt to locate the side chains in energetically favoured conformations [66–68].

4.5.2 Main applications

Until now, the main interest of the protein database has been to provide chemists with atomic coordinates of biological macromolecules. Each year, a large number of papers cite the Brookhaven database as a source of information, particularly regarding active sites of selected enzymes. Docking varied inhibitors to an enzyme of a known structure and looking at protein-ligand interactions through energy calculations and computer graphics gives some clues to a rational approach of drug design.

Another fascinating field of applications (but perhaps a long range objective) would be forecasting the 3D organization of a protein (tertiary structure) from the knowledge of its amino acid sequence. As a consequence of the rapid advance of sequencing techniques, the chemical sequence is known for many more proteins than their 3D structure, and the gap is growing, sequencing being much faster than structure solving. Known protein structures may constitute the root for predictive rules thanks to the homology concept or structural correlations. These points will be developed further in Chapter 13.

Let us remark that the PDB is access free on the Internet.

4.6 DATABASES OF CALCULATED STRUCTURES

Such databases of calculated structures are generated thanks to various "model builders", that is, basically, programs able to propose 3D structures from connection tables indicating only the molecular topology. Model builders are generally based on a set of rules and some rough energy minimizers.

This point is developed in Chapter 7.

APPENDIX: INTERRELATION OF SOME EXPERIMENTAL DISTANCE PARAMETERS

For more details see elsewhere [9, 10].

- Equilibrium nuclear position (r_e), attainable by quantum mechanics optimization, corresponds to the nuclear separation at the potential minimum (hypothetical vibrationless state).
- Average distance r_g (averaged over thermal vibrations for different thermally occupied states) is related to distances reported in gas phase electron measurements. From this value, harmonic correction and temperature extrapolation yield r_α^0, which refers to the vibrational ground state.
- From microwave spectra, average position r_z is averaged over zero point vibration (average nuclear position at the ground vibrational state), and can be determined from rotational constants (after appropriate corrections).

r_z and r_α^0 are close to r_{av} and r_e, which respectively correspond to experimental or theoretical distances at minima of potential energy or energy hypersurfaces. Other values r_m, r_o, r_s derived from moments of inertia refer to specific states.

(r_o is based on a simple analysis disregarding the effects of zero-point vibration; r_s results from a combination of isotopic rotational constants in microwave spectroscopy, from which r_m (modified mass dependence) can be derived.

A bond length is better represented by r_g, which is a real vibrational average (with small isotopic dependence), whereas non-bonded distances are better described by r_z or r_α, which refer to average nuclear positions.

Some of these interrelations are summarized in Figure 4.26.

MICROWAVE   ELECTRON DIFFRACTION

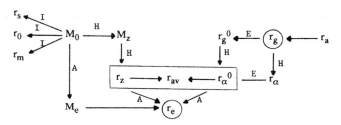

Figure 4.26 Interrelation of experimental distance parameters. H = harmonic corrections; A = anharmonic corrections; I = isotopic substitution; E = temperature-dependent extrapolation (from Haëfelinger *et al.* with permission [10]).

REFERENCES

1. E.E. Abola, F.C. Bernstein and T.F. Koetzle in *The Role of Data in Scientific Progress*, P. Glaeser (ed.), Elsevier Science Publishers, 1985; 139–143.
2. P.J. Wheatley *The Determination of Molecular Structure*, Dover Publications, New York, 1968.
3. P.M. Dean *Molecular Foundations of Drug-Receptor Interaction*, Cambridge University Press, Cambridge, UK, 1987; 34–69.
4. J.P. Glusker X ray crystallography, an introduction, *International School of Crystallography, 11th course: Static and Dynamic Implications of Precise Structural Information*, Erice, Italy, 1985; 23–60.
5. C.R. Cantor and P.R. Schimmel *Biophysical Chemistry: Part II–Techniques for the Study of Biological Structures and Functions*, W.H. Freeman, New York, 1980; p. 719.
6. G.A. Jeffrey Precise structure analysis by neutron diffraction, *International School of Crystallography, 11th course: Static and Dynamic Implications of Precise Structural Information*, Erice, Italy, 1985; 133–148.
7. R. Taylor, O. Kennard and W. Versichel *J. Am. Chem. Soc.*, **106**: 1984; 244–248.
7a. R. Taylor, O. Kennard and W. Versichel *J. Am. Chem. Soc.*, **105**: 1983; 5761–5766.
7b. R. Taylor, O. Kennard and W. Versichel *Acta Crys.*, **B40**: 1984; 280–288.
8. I. Hargittai Gas-phase electron diffraction, *International School of Crystallography, Static and Dynamic Implications of Precise Structural Information*, Erice, Italy, 1985; 107–132.
9. K. Kuchitsu The potential energy surface and meaning of internuclear distances, *International School of Crystallography, 11th course: Static and Dynamic Implications of Precise Structural Information*. Erice, Italy, 1985; 351–371.

10. G. Häfelinger, C. U. Regelmann, T.M. Krygowski and K. Wozniak *J. Comput. Chem.*, **10:** 1989; 329–343.
11. R. Benn and H. Gunther *Angew. Chem. Internat. Ed.*, **22:** 1983; 350–380.
12. J.S. Cohen, L.J. Hughes and J.B. Wooten in *Magnetic Resonance in Biology*, vol 2, J.S. Cohen Ed., J. Wiley, 1983; 130–247.
13. K. Wüthrich *Acc. Chem. Res.*, **22:** 1989; 36–44.
14. A.D. Kline, W. Braun and K. Wüthrich *J. Mol. Biol.*, **189:** 1986; 377–382.
15. A.E. Derome and S. Bowden *Chem. Rev.*, **91:** 1991; 1307–1320.
16. T.M. Alam and G.P. Drobny *Chem. Rev.*, **91:** 1991; 1545–1590.
17. K. Wüthrich *NMR of Proteins and Nucleic Acids*, J. Wiley, New York, 1986.
18. D. Marion and K. Wüthrich *Biochem. Biophys. Res. Commun.*, **113:** 1983; 967–974.
19. J.M. Moore, D.A. Case, W.J. Chazin, G.P. Gippert, T.F. Havel, R. Powls and P.E. Wright *Science*, **240:** 1988; 314–317.
20. J.H. Noggle and R.E. Shirmer *The Nuclear Overhauser Effect*, Academic Press, New York, 1971.
21. G. Wagner and K. Wüthrich *J. Mol. Biol.*, **160:** 1982; 343–361.
22. E.T. Olejniczak, F.M. Poulsen and C.M. Dobson *J. Am. Chem. Soc.*, **103:** 1981; 6574–6580.
23. M. Karplus in *Computer Simulation of Chemical and Biomolecular Systems*, D.L. Beveridge and W.L. Jorgensen Ed. New York Academy of Sciences, New York, 1986; 255–266.
24. T.F. Havel and K. Wüthrich *J. Mol. Biol.*, **182:** 1985; 281–294.
25. R.J. Abraham, J. Fisher and P. Loftus *Introduction to NMR Spectroscopy*, J. Wiley, 1988; 217–232.
26. B.R. Brooks, R.E. Bruccoleri, B.D. Olafson, D.J. States, S. Swaminathan and M. Karplus *J. Comput. Chem.*, **4:** 1983; 187.
27. G.M. Clore, A.M. Gronenborn, A.T. Brünger and M. Karplus *J. Mol. Biol.*, **186:** 1985; 435–455.
28. G.M. Smith and D.F. Weber *Biochem. Biophys. Res. Comm.*, **134:** 1986; 907–914.
29. J.N. Scarsdale, P. Ram, J.H. Prestegard and R.K. Yu *J. Comput. Chem.*, **9:** 1988; 133–147.
30. T.F. Havel and K. Wüthrich *Bull. Mathematical Biology*, **46:** 1984; 673–698.
31. J. Deisenhofer and W. Steigemann *Acta Cryst.*, **B31:** 1975; 238–250.
32. A.T. Brünger, G.M. Clore, A.M. Gronenborn and M. Karplus *Proc. Nat. Acad. Sci. USA*, **83:** 1986; 3801–3805.
33. J. Lautz, H. Kessler, R. Kaptain and W.F. van Gunsteren *J. Computer-Aided Molecular Design*, **1:** 1987; 219–241.
34. A.P. Tonge, P. Murray-Rust, W.A. Gibbons and L.K. McLachlan *J. Comput. Chem.*, **9:** 1988; 522–538.
35. S.D. Cheatham *J. Chem. Educ.*, **66:** 1989; 111–117.
36. S.W. Fesik *J. Med. Chem.*, **34:** 1991; 2937–2945.
37. G.M. Clore and A.M. Gronenborn *Progress in NMR Spectroscopy*, J. Emsley, J. Feeney and L.H. Sutcliffe (Eds.) Pergamon Press **23:** 1991; 43–92.
38. O. Kennard *6th European Seminar on Computer-Aided Molecular Design*, October 1989; London, UK, 1–17.
39. O. Kennard, D.G. Watson, F.H. Allen, W.D.S. Motherwell, W. Town and J.R. Rodgers *Chem. Britain*, **11:** 1975; 213–216.
40. F.H. Allen, S. Bellard, M.D. Brice *Acta Cryst.*, **B35:** 1979; 2331–2339.
41. F.H. Allen, J.E. Davies, J.J. Galloy *et al. J. Chem. Inf. Comput. Sci.* **31:** 1991; 187–204.
42. F.H. Allen, O. Kennard and R. Taylor *Acc. Chem. Res.*, **16:** 1983; 146–153.
43. G.W. Snedecor and W.G. Cochran *Statistical Methods*, Iowa State University Press, 1980; 73.
44. C. Chatfield and J.A. Collins *Multivariate Analysis*, Chapman & Hall, London, 1980; 57.
45. S.R. Wilson and J.C. Huffman *J. Org. Chem.*, **45:** 1980; 560–566.

46. F.H. Allen, O. Kennard, D.G. Watson, S. Bellard, M.D. Brice *J. Chem. Soc. Perkin Trans.* II, 1987; S1–S19.

47. R. Taylor and O. Kennard *J. Chem. Inf. Comput. Sci.*, **26:** 1986; 28–32.

48. R. Taylor and O. Kennard *J. Am. Chem. Soc.*, **104:** 1982; 3209–3212.

49. R.P. Sheridan and R. Venkataraghavan *J. Computer-Aided Molecular Design*, **1:** 1987; 243–256.

50. R.P. Sheridan, A. Rusinko III, R. Nilakantan and R. Venkataraghavan *Proc. Nat. Acad. Sci. USA*, **86:** 1989; 8165–8169.

51. R.L. DesJarlais, R.P. Sheridan, G.L. Seibel *et al. J. Med. Chem.*, **31:** 1988; 722–729.

52. A. Domenicano, A. Vaciago and C.A. Coulson *Acta Cryst.*, **B31:** 1975; 221–234 and 1630–1641.

53. A. Domenicano and P. Murray–Rust *Tetrahedron Lett.*, 1979; 2283–2286.

54. H.B. Burgi and J.D. Dunitz *Acc. Chem. Res.*, **16:** 1983; 153–161.

55. A. Cossé-Barbi and J.E. Dubois *J. Am. Chem. Soc.*, **109:** 1987; 1503–1511 (and *Tetrahedron Lett.*, **27:** 1986; 3501–3504).

56. R.B. Nachbar Jr, C.A. Johnson and K. Mislow *J. Org. Chem.*, **47:** 1982; 4829–4833 (and J.E. Dubois and A. Cossé-Barbi *J. Am. Chem. Soc.*, **110:** 1988, 1220–1228).

57. C. Gilli, V. Bertolasi, F. Bellucci and V. Ferretti *J. Am. Chem. Soc.*, **108:** 1986; 2420–2424.

58. H.B. Burgi, J.D. Dunitz and E.J. Shefter *J. Am. Chem. Soc.*, **95:** 1973; 5065–5067.

59. J.D. Dunitz, *X-ray Analysis and the Structure of Organic Molecules*, Cornell University Press, London, 1979; 366–379.

60. E. Bye, W.B. Schweizer and J.D. Dunitz *J. Am. Chem. Soc.*, **104:** 1982; 5893–5898.

61. C.P. Brock, W.B. Schweizer and J.D. Dunitz *J. Am. Chem. Soc.*, **107:** 1985; 6964–6970.

62. K. Mislow *Acc. Chem. Res.*, **9:** 1976; 26–33.

63. P. Murray–Rust and W.D.S. Motherwell *Acta Cryst.*, **B34:** 1978; 2534–2546, 2518–2526.

64. R. Taylor, *J. Mol. Graphics*, **4:** 1986; 123–131.

65. L. Norskov–Lauritsen and H.B. Burgi *J. Comput. Chem.*, **6:** 1985; 216–228.

66. P.A. Corea, *Proteins, Structure, Function and Genetics*, **7:** 1990; 366–377.

67. L.S. Reid and J.M. Thornton *Proteins, Structure, Folding and Design*, **2:** 1987; 93–102.

67a. L.S. Reid and J.M. Thornton *Proteins, Structure, Function and Genetics*, **5:** 1989; 170–182.

68. L. Holm and C. Sander *J. Mol. Biol.*, **218:** 1991; 183–194.

5 *Empirical force field methods and molecular mechanics*

Obviously, molecular geometry is the necessary starting point for most modelling treatments. Although modern NMR techniques can give some indications about interatomic distances for liquid samples, geometrical information, for large or medium-size molecules, is up to now mainly derived from crystallographic data (either from direct measurements or through retrieval from databases such as the well known Cambridge Database).

However, such information only concerns crystals, where the geometry may be somewhat affected by packing effects and is not automatically the same as for the isolated molecule in the gas phase, or for the reacting species in biological media. Furthermore, such data are not attainable for samples that are difficult to obtain as good crystals, for hypothetical structures (conformers differing from the most stable one) or for molecules not already synthesized. Finally, if X-ray data give the geometry of the more stable form, they say nothing about energetics and the possible existence of other low-energy structures. Suitable computational approaches are obviously needed for such applications.

The methods of quantum chemistry are of course quite suited to predicting the geometric, electronic and energetic features of known or unknown molecules. Nevertheless, they remain, until now and probably in the foreseeable future, too expensive in terms of computer time and nearly intractable, even at the simplest, semi-empirical level, for many organic molecules or biological macromolecular structures. In this context, increased interest has focused on models able to quickly give energy favoured conformations for large systems. An important role is played in the field by techniques known as *molecular mechanics* (MM) or *empirical force field* (EFF) methods. On another hand, "molecular dynamics", or at a more pragmatic level, "model builders" (introduced in the following chapters) propose alternate avenues towards a non-experimental access to molecular geometry.

Basically, molecular mechanics treats molecules as being composed of masses and springs (according to an idea traced back to Andrews as early as 1930 [1]). It uses the laws of classical mechanics to treat the diverse interactions occurring in the real molecule, according to a model that is empirically parameterized. Such a presentation in terms of atoms and bonds is not so far from the usual way of thinking of chemists with their ball-and-stick molecular models. However, it offers extended capabilities as to the treatment of non-standard situations that cannot be represented easily with such naive models (bond lengths or angles differing from usual values, small geometrical distortions due to steric crowding, structures departing from the common valence rules such as transition states, intermolecular associations, etc.). Molecular mechanics also gives important information about energetics: more stable or low-energy conformers, interconversion pathways, etc. Although it tells us nothing about electronic characteristics, its computational speed and its ability to deal with large systems make it very attractive whenever electronic properties are not needed or in other cases to get quickly optimized geometries to subsequently submit to the heavier quantum chemistry methods.

These methods are often considered as the other "facet" of the Born Oppenheimer separation of nuclear and electron motions. In quantum mechanics calculations, one starts with a given position for the nuclei and searches for the best repartition of electrons in the potential generated by the nuclei (the calculation can be continued with modified positions of the atoms to get the conformer of best stability in an energy minimization process). In molecular mechanics, on the contrary, one studies the position of the nuclei in the field generated by the electrons. Electrons are not explicitly considered (with some exceptions, however), and the field they generate is not actually calculated, but rather represented by an "effective" potential treated according to classical mechanics.

Several well-documented reviews on empirical force field methods and molecular mechanics have been published recently [2–10] so we only present here a brief survey of the main underlying principles, and few indications about some common types of applications. The very popular molecular mechanics approach from Allinger *et al.* [2, 10–16] will be largely used as a basis for this presentation.

Looking at the internal energy of a molecular system, we can represent it, about the point of minimum energy, by a Taylor series expansion involving atomic coordinates. This is more easily carried out by use of internal coordinates (more familiar to the chemist): bond lengths, bond and torsion angles, together with interaction terms between non-bonded atoms, expressing van der Waals repulsions.

This can be illustrated on the simple example of a diatomic molecule treated by the "classical mechanics" model of two balls joined by a spring (Figure 5.1): for any small variation of the internuclear distance (r) around the equilibrium position (r_o) the potential energy is expressed by:

$$V = V_o + \frac{\mathrm{d}V}{\mathrm{d}r}(r - r_o) + \frac{1}{2}\frac{\mathrm{d}^2V}{\mathrm{d}r^2}(r - r_o)^2 + \dots \tag{5.1}$$

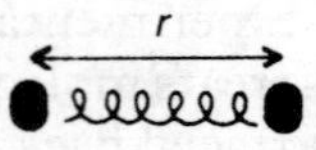

Figure 5.1   "Classical mechanics" model.

If we assign to the energy a null value at the minimum (V_o) and since at the equilibrium position the first derivative is zero:

$$V = \frac{1}{2}\frac{\mathrm{d}^2V}{\mathrm{d}r^2}(r - r_o)^2 + \dots = \frac{1}{2}k(r - r_o)^2 + \dots \tag{5.2}$$

This corresponds to the well-known Hooke's law, familiar to spectroscopists, k being the stretching force constant.

A similar expression can be written for more complex systems:

$$V = \frac{1}{2}\Sigma k_{ij}\Delta r_i.\Delta r_j + \dots \tag{5.3}$$

where the k_{ij} are the quadratic force constants and Δr_i the *displacement coordinates* of nucleus i (changes of the coordinates defining the location of the nuclei from their reference values).

The problem is now to define the k_{ij} coefficients that are the force constants. Their set constitutes the *force field*, which allows us to define the changes in energy depending upon the molecular geometry.

The fundamental assumptions of molecular mechanics, largely justified by the countless successes of the method, are:

- The potential energy of a molecule can be represented as a sum of terms associated respectively with the various types of molecular deformations (changes in bond lengths, valence or torsion angles) or atom-atom interactions. The *steric energy* calculated from the sum of these terms represents the *additional energy associated with the deviations of the structure with respect to an ideal situation where all geometrical elements would be in a reference state.*
- The parameters needed to calculate this steric energy can be derived from information gained on small molecules (bond lengths, angles), their

transferability to large systems being assumed. In other words, as stated by Allinger: "A large molecule consists of the same features we already know about in small molecules, but combined and strung together in various ways" [2].

5.1 THE FORCE FIELD

The key for successful predictions from molecular mechanics lies in a good representation of these different energetic terms, that is in the choice of the force field.

At this point it must be clearly borne in mind that force field calculations "are performed on a model. This model is assigned properties that reproduce experimental facts, but this does not mean that it is in every respect a faithful reproduction of the molecule under study. It only means that the particular information which has been used to develop the model is reproduced by the model" [2].

Although everyone agrees with the need for a consistent force field, according to Lifson and Warshel [17, 17a], i.e. a force field which should be optimized by fitting all kinds of data and all available data, the practical situation is far more complex, since one can only treat, up to now, a limited amount of data. So various force fields have been proposed, which depend to some extent upon the selected target, that is on the experimental property to be reproduced: geometry, heat of formation, vibrational frequencies, and even on the structural scope where they apply.

However, as force fields become more and more complete, thanks to the inclusion of more exotic compounds (not belonging to their original application area), force fields tend to be more similar as their application areas overlap.

5.1.1 Components of the force field

In molecular mechanics calculations, the force fields generally take the form:

$$V = V(r) + V(\theta) + V(\Phi) + V(\text{nb}) + (\text{specific terms}) \tag{5.4}$$

in which the successive terms are associated with bond stretching, bond angle bending, bond torsion, non bonded interactions plus specific terms allowing for out of plane bending, electrostatic interactions and possible hydrogen-bonding. We shall first mainly focus on the simplest approximation level to emphasize the basic principles of the approach. Refinements of the force field will be detailed later.

Bond stretching

The bond stretching contribution is represented by Hooke's law:

$$V(r) = \frac{1}{2} \Sigma k_r (r - r_o)^2 \tag{5.5}$$

the summation being performed over all bonds.

k_r and r_o are the harmonic force constant and the reference bond length for the corresponding bond. For large deformations or in the study of highly congested molecules, an additional cubic term $1/2 \; k'_r \; (r-r_o)^3$ can be introduced (quartic terms are more rarely considered). However some caveat must be put on the use of such a cubic term: due to the shape of the cubic function, the energy goes to minus infinity as the atoms go further apart. If the starting geometry is too bad, the bond would tend to dissociate.

Alternatively, a **Morse potential** can be used:

$$V(r) = \Sigma D[1 - \exp(-a(r - r_o))]^2 \tag{5.6}$$

D and a being parameters characterizing the bond. The use of such a potential seems useful for elongated hydrogen bonds which otherwise tend to dissociate [18].

Angle bending

A quadratic bending function is usually retained:

$$V(\theta) = \frac{1}{2} \sum_{\substack{bond \\ angles}} k_\theta (\theta - \theta_o)^2 \tag{5.7}$$

(the o underscript still corresponds to the reference value).

Such an expression seems to work well up to deformations of about 10–15°, which covers most of the usual cases. For the large distortions occurring in small cycles, additional terms may be introduced, but one generally prefers to maintain such a quadratic expression with different force constants, specific of the cycles investigated.

Torsion

Bond torsion effects can be well-reproduced with a Fourier series:

$$V(\Phi) = \Sigma \left[\frac{V_1}{2}(1 + \cos \Phi) + \frac{V_2}{2}(1 - \cos 2\Phi) + \frac{V_3}{2}(1 + \cos 3\Phi) + \ldots \right] \tag{5.8}$$

Φ represents the torsional angle, i.e. for a four atom fragment ABCD, the dihedral angle between planes ABC and BCD. In a Newman projection along the central bond BC, Φ also appears as the angle between the projections of the bonds AB and CD. According to the convention proposed by Klyne and Prelog [19], a positive dihedral angle corresponds to a clockwise rotation of the first named bond to bring it on the second one. It takes the same value when one looks along the bond from one side or the other (Figure 5.2).

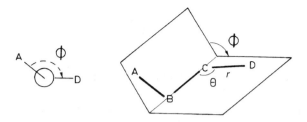

Figure 5.2   Torsion.

In molecular mechanics, thanks to the inclusion of van der Waals repulsions between non-bonded atoms (see below), and owing to the degree of precision required, it appears that three terms are sufficient to cope with all situations (Figure 5.3).

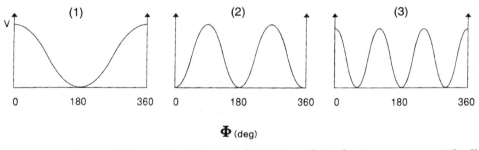

Figure 5.3   Shapes of the torsional potentials V_1, V_2, and V_3: three terms cope with all situations.

Torsion around a highly symmetrical sp³–sp³ bond, as in ethane, is well-reproduced when only the third term corresponding to a three-fold barrier is used. Later it was shown that, in more general cases, inclusion of 1-fold and 2-fold terms (although smaller) improves the results [11]. On the other hand, for a sp²–sp² bond (as in ethylene), the preferred situation is the eclipsed one. V_2 will be then the principal term so as to get a 2-fold barrier (and the other terms neglected). Note that we do not need here explicit mention to π electrons to treat the rigidity of the double bond.

In more complex functionalized molecules, the three terms have to be considered. V_2 accounts for the 2-fold barrier along double bonds, but it also intervenes for sp³–sp³ bonds (in butane). The 1-fold component deals with 1–4 non-bonded atom-atom repulsions (and electrostatic interactions).

Out of plane bending
For systems with sp² carbons, it is necessary to distinguish between in-plane and out-of-plane deformations. In alkenes, for example, out-of-plane bending is described by a quadratic term (Figure 5.4):

$$V = \frac{1}{2}\Sigma k_{ob.} x^2 \qquad\qquad (5.9)$$

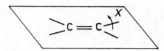

Figure 5.4   Out-of-plane bending.

Such terms are important for small cycles (cyclobutene, cyclobutanone) where valence angles are quite different from the usual 120° value. Without them, the exocyclic atom would tend to move exaggeratedly out of plane to restore a 120° valence angle (Figure 5.5).

Figure 5.5   Without quadratic terms, the exocyclic atom would move exaggeratedly out of plane to restore a 120° valence angle.

Van der Waals interactions

To ensure the transferability of the force field from one molecule to another molecular mechanics takes into account explicitly the interactions between atoms not bonded to each other or to a common atom. These are generally referred as van der Waals interactions. These terms are unfortunately generally omitted in spectroscopic calculations, leading to force constants varying for each individual molecule considered.

Actually, these van der Waals interactions are represented by assuming the additivity of pairwise terms: for atoms i,j:

$$V(nb) = \Sigma \left[A(r_{ij})^{-n} - B(r_{ij})^{-6} \right] \tag{5.10}$$

or, alternatively (6–exp Buckingham potential):

$$V(nb) = \Sigma \left[A \exp(-br_{ij}) - B(r_{ij})^{-6} \right] \tag{5.11}$$

where the summation is taken over all non-bonded pairs of atoms (excluding, of course, atoms bonded to each other or to a common atom, since they are already taken into account in stretching and bending terms). Note that, to correctly reproduce torsion effects, both torsion potential terms and 1–4 non-bonded atom-atom repulsions are necessary.

These expressions, and the other ones used, involve two terms:

1. An attractive part, corresponding to induced dipole-induced dipole interactions, proportional to r^{-6}, where r is the distance between the two atoms.
2. A repulsive part, corresponding to London dispersion terms and rapidly growing at short distances (recall that in a hard sphere model schematizing what happens when two balls are brought close together, the repulsion would become infinite when the distance of the centres becomes smaller

than the sum of the radii). Various expressions have been chosen for this repulsive part, using either an exponential function [20] or an inverse power of r. For the exponent n a value of 12 is correct for expressing interactions in the rare gases or between closed shell molecules considered as a whole (Lennard–Jones 6–12 potential). A softer repulsive term with $n = 9$ is sometimes preferred in molecular mechanics.

The Hill equation:

$$V(nb) = \varepsilon\left[-c_1(r^*/r)^6 + c_2 \exp(-c_3 r/r^*)\right]$$

(5.12)

which works well for rare gases and small molecules, has also frequently been used in larger molecular systems (and is included, for instance, in the MM2 program). In equation (5.12), the c_i coefficients are taken as universal constants. ε is an energy parameter related to the depth of the pair potential well. For a pair of atoms i,j, $\varepsilon = \sqrt{\varepsilon_i * \varepsilon_j}$, where ε_k characterizes the "hardness" of atom k. r^* is defined as the sum of the van der Waals radii of the interacting atoms in the pair [21] (Figure 5.6).

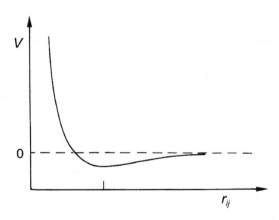

Figure 5.6   Variation of an atom-atom potential vs. internuclear distance.

The c_i coefficients being treated here as universal constants, only two parameters are necessary to adjust van der Waals repulsions for each atom-pair (ε and the sum of the van der Waals radii (r^*)). For numerical values see, for instance, Allinger [2]. Some typical values used in the MM2 parameterization are given in Table 5.1.

Some implicit approximations are to be noticed:

- First, one assumes that terms derived from intermolecular interactions can also represent intramolecular interactions.
- Then, it is considered that these pair interactions are not dependent upon their environment, i.e. of the other atoms present.
- Furthermore, atoms are treated as spheres (with an isotropic electron distribution around them). This is not quite right for atoms involved in a

Table 5.1   Some van der Waals parameters (MM2 1985 force field).

H		ELEMENT i	Lone	
	47	$10^3\ \varepsilon_i$	Pair	16
	1.50	r		1.20
alcohol	36	(kcal/mol, Å)		
	1.20			
amine⌐	34			
imine⌐	1.325			
acid⌐	15			
amide⎪	0.90			
vin alc⌐				

B		C		N	O		F	
trig.	34	sp^3	44	55	alc⌐	50		78
	1.98		1.90	1.82	eth⌐	1.74		1.65
tetr.	34	sp^2	44		$O=$	66		
	1.9		1.94			1.74		
					O_{Fu}	50		
						1.74		

	Si		P		S		Cl	
		140		166		202		240
		2.25		2.18		2.11		2.03

	Ge				Se		Br	
		200				276		320
		2.40				2.25		2.18

	Sn				Te		I	
		270				370		424
		2.55				2.40		2.32

	Pb	
		340
		2.70

vin alc: vinylic alcohols; O_{fu}: sp^2 oxygen in furan

For C $\cdots$ H: $10^3\ \varepsilon = 46$; sum $r^* = 3.34$. Distance reduction = 0.915. Centre of electron density for hydrogen shifted towards carbon and located at 0.915 of the CH distance.

Van der Waals interaction energy between two atoms i, j:
Given

$$P = r^* / r \quad (r^* = r_i + r_j;\ r\ \text{interatomic distance})$$
$$\varepsilon^* = (\varepsilon_i\,\varepsilon_j)^{1/2}$$

if $P \le 3.311$ $E_{ik} = \varepsilon^* (2.9 \times 10^5\ \exp^{(-12.5/P)} - 2.25P^6)$

if $P > 3.311$ $E_{ik} = \varepsilon^* \times 336.176\ P^2$ Kcal/mole molecule, and particularly for hydrogen atoms where the unique electron is engaged in the bond. In that case good results are obtained by shifting the centre of the sphere by about 0.1 Å towards the heavy atom [22]. With atoms bearing lone pairs (such as nitrogen in amines or oxygen in alcohols or ethers, but not carbonyl groups), where non-bonded interactions are expected to be anisotropic, the lone pairs can be explicitly taken into account as dummy (virtual) atoms [23–25] which are given appropriate van der Waals radii.

Note also that for larger biomolecules, some programs provide the possibility of using "united atoms": hydrogen atoms (except those involved in hydrogen-bonding) are omitted. One considers only the heavy atoms with their attached hydrogens as single (spherical) entities, with adapted van der Waals radii.

5.1.2 Parameterization

The selection of appropriate parameters for determining the force field is of paramount importance for the precision to be expected in energy or geometry predictions and the structural scope covered.

Spectroscopic force fields deserve special mention, owing to strong formal similarities between some energetic terms (bond stretching, angle bending) and the expressions used in vibrational spectroscopy (Hooke's law). In vibrational analysis, performed with the *valence force field* approximation, the expression of the potential energy in terms of internal coordinates only involves diagonal terms. This results in force constants *a priori* differing from one compound to another one. However, taking into account some interaction terms (non-diagonal terms) in a *generalized valence force field*, some transferability can be ensured within limited classes of closely related compounds (chemical families as alkanes, alkenes, and so on). A similar result can also be attained by the *Urey Bradley approach*, where 1–3 interactions (interactions between two atoms bonded to a same third one) are introduced [26].

In molecular mechanics, getting transferable parameters is of prime importance, since a single model should be able to apply to a large set of molecules. This leads to different definitions of the force field with, for example, inclusion of the van der Waals terms.

Bond lengths and bond angles are usually available from existing structural information. X-ray crystallography is a privileged source of interatomic distances (supplemented by neutron or electron diffraction, and for smaller molecules by microwave spectroscopy). However, some caution must be exercised. Crystal lattice forces can modify distances derived from X-ray diffraction, when compared to electron diffraction measurements dealing with gas phase isolated molecules. However, these differences are small and can be neglected here. C–H bonds constitute a noticeable exception, since the nuclear position and the centre of electron-density are separated by ca. 0.1 Å (see above).

Bond stretching parameters can be directly derived from vibrational force constants, whereas for angle bending constants, the corresponding values have to be scaled by about 0.8 to 0.5. The coefficients of the torsional barriers can be safely estimated from barrier heights (attainable through microwave spectroscopy or thermodynamic studies, and more recently far infrared and Raman spectroscopies).

More difficult is the evaluation of van der Waals repulsions, a somewhat crucial point since these interactions are of paramount importance in determining the stability of crowded or highly branched molecules. Furthermore, van der Waals parameters are highly correlated with other parameters. "Molecules in which the van der Waals interactions are large and generally complicated, show many degrees of freedom in which the molecule may relax" [2].

A striking example of difficulties encountered in such a parameterization is illustrated by the extensive studies about the destabilization of gauche butane. In an early Allinger force field MM1 [25], this was reproduced with important hydrogen-hydrogen repulsions, but introduction of 1- and 2-fold torsional terms, and comparison with other force fields lead us to prefer softer and smaller hydrogens [11, 26, 27]. Let us note also that, speaking only of van der Waals terms, there is some interdependency between the ε parameter and the radii intervening in the Hill equation, so that different models can work fairly well. Such problems are exemplified in the definition of the AMBER (Assisted Model Building and Energy Refinement) force field [29]. For carbonyl carbons, $r^* = 1.81$ Å and $\varepsilon = 0.184$ kcal/mol are convenient in a 6–9 force field. However, in a 6–12 potential, these values have to be modified to $r^* = 2.175$ Å and $\varepsilon = 0.039$ kcal/mol.

The choice of van der Waals radii is a difficult problem, largely discussed by Allinger [2]. A first source of information comes from the study of contact data on rare gases, or on selected compounds in the crystal phase (for instance, in graphite, the distance between carbon planes results from van der Waals interactions). Indeed, for rare gases, these values actually represent the sum of the van der Waals radii of the atoms. However, for molecules, the influence of other atoms further apart induces some attractive interactions and leads to some interpenetration of the van der Waals radii for the nearest atoms. Such distances of closest approach in crystals are the source of a largely used tabulation [28].

For molecular mechanics, calculations supported by atom spacing in selected alkane crystals allowed for assessing van der Waals radii for carbon and hydrogen. Reasonable estimates of the radii for other most common atoms have been then proposed [2]. It must be noted that these radii (used in Allinger force field for example) are, roughly speaking, 0.3 Å larger than Bondi values.

Optimization of the force field used is difficult work, since it involves many parameters not quite uncorrelated to each other. Some automatic but "blind" procedures have been proposed where the optimization is performed by the computer via multivariate regressions. An interactive approach where the user monitors the successive refinements seems more efficient. Parameterization protocols have been proposed [30,31]. The PEFF program [32, 32a] allows for fitting the force field parameters to a set of experimental data, using a least square method. Conversion of *ab initio* force field to molecular mechanics energy parameters has been also proposed [33, 33a].

Some remarks:

- Owing to the empirical aspect of these methods, it may be that parameters might be lacking for molecules not previously considered in the scope of a particular force field. Parameters taken from another force field cannot be directly introduced, but can only be used as a guide. Any force field is defined as a whole to reproduce experimental facts, but each of its terms has not independently an intrinsic meaning.
- Reference values in equations (5.5)–(5.7) are not the experimentally observed values in simple systems. They are the values which lead to correct predictions within a given force field taking into account the other energy contributions. For example, in the MM2 program, single bond lengths (ca. 1.53, 1.54 Å usually) are well-reproduced provided a reference value $r_o = 1.523$ Å is used (for ethane, experimental value 1.534 Å, calculated 1.532 Å).

5.1.3 Refinements of the force field and cross terms

Beyond the fundamental contributions to the force field (stretching, angle bending, torsion, non-bonded atom-atom repulsions), additional terms may be considered to refine the accuracy of the predictions. For example, the Urey Bradley formulation incorporates 1–3 interactions [26]. Alternatively, cross terms consider simultaneous variations of two parameters, such as changes in bond lengths and variations of the angle they define (this stretch-bend term is important in strained systems where bonds lengthen when constraints compress their angle) (Figure 5.7). Such terms have the general form (for adjacent bonds i and j making angle θ_{ij}):

$$\frac{1}{2}k_{r(i,j,\theta)}[r_i - r_{io} + r_j - r_{jo}][\theta_{ij} - \theta_{ijo}]$$

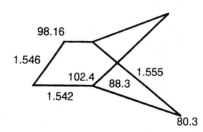

Figure 5.7 Bonds lengthen when the valence angle decreases (for ethane, the calculated C–C distance is 1.532 Å) (Allinger [2]).

Specific terms

Some additional mechanisms must be also taken into account. Bond polarization is weak in alkanes and may be neglected, but it has to be introduced in systems bearing heteroatoms, and has even been considered in alkenes or alkynes. These electrostatic terms are of paramount importance for the evaluation of intermolecular interactions in polar molecules. Such interactions have been first taken into account with bond dipoles. However, for a more accurate representation it seems better to treat electrostatics through a set of partial atomic charges. So, special interest has been devoted to the determination of partial atomic charges or adjustments of bond moments.

To reproduce electrostatic effects, the Coulomb law is used with point charges located at the nuclei:

$$V(e) = \Sigma \frac{q_i q_j}{r_{ij}} \frac{1}{D} \tag{5.13}$$

where q_i and q_j are the atomic charges and r_{ij} the interatomic distance for the pair of atoms i,j. A local dielectric constant D (usually varying from 1 to 4) can be introduced.

Atomic charges may, of course, be derived from quantum chemical calculations. However, such methods (at the *ab initio* and even at the crudest semi-empirical or HMO levels) are not easily available for large molecules such as proteins. Furthermore, the concept of atomic charge is still open to discussion. Mulliken population analysis, often used in MO methods, has been widely questioned and is largely basis-set-dependent. Potential-consistent charges (i.e. point charges centred on atoms and defined as able to reproduce the electrostatic potential on a grid of points surrounding the molecule [34] have also been questioned).

The construction of a practical yet realistic model [30], allowing for the treating of large molecules with reduced computational effort, is therefore a prerequisite for investigating the extended molecular systems of interest for biochemists [31], and particularly for developing reliable molecular mechanics programs.

A possible solution (adopted in AMBER) is to carry out quantum mechanics calculations on fragments and then "patch" them together [29], the transferability of subunits being assumed.

Among other approaches, the concept of *electronegativity equalization* underlies several calculation methods [35–39] and gives promising results: calibration on small test molecules is quite consistent with *ab initio* calculations and reproduces well various physicochemical properties, such as NMR or ESCA shifts, dipole moments, etc. Only one example would be briefly indicated here.

In the model of Abraham and Smith [40], σ effects of polar atoms are separated into one-, two- and three-bond additive contributions. The one-bond effect is proportional to the difference in the electronegativity of the bonded atoms, the other effects being functions of the atomic electronegativity and polarizability.

Charge induced on an atom by the β neighbour is treated as proportional to the polarizability of the former atom and the electronegativity of the latter. The charge induced by the γ atom is directly proportional to the charge induced on the β atom. The actual charge distribution is evaluated by iteration, adjusting polarizability in each step until it reaches zero [41]. The parameterization is calibrated so as to reproduce the experimental molecular dipole moments.

This method, initially proposed for saturated molecules including the most common functionalities, has been extended to conjugated systems in the framework of the *Huckel Molecular Orbital* (HMO) approach with a suitable choice of Coulomb and exchange integrals [42]. A similar approach has been proposed by Gasteiger *et al.* [41] in their *Partial Equalization of Orbital Electronegativity* (PEOE). Here, conjugated systems are included thanks to the concept of π orbital electronegativity.

Another method uses point dipole approximation. Bond dipoles are chosen so as to fit the moments observed in simple molecules. Only permanent dipoles are generally considered, although induced moments have been introduced in the IDME method [43].

Hydrogen-bonds and medium effects

Hydrogen-bonds deserve special interest as a result of their prime importance in the stability of many biomolecules. They can be treated as a part of electrostatic and van der Waals interactions [44]. Otherwise, specific functions have been introduced [45–49]. For a better quantitative agreement, it was proposed to reduce the van der Waals radius of the hydrogens involved in hydrogen-bonding and increase the attraction term toward the electronegative atom by about 1–3 kcal/mol, depending upon the particular atom involved [50]. In AMBER, a different potential (10–12 rather than 6–12) is used for hydrogens involved in hydrogen-bonding [29].

Taking into account the influence of the solvent would also be of prime importance for a safe prediction of conformational energies. The simplest way is to treat these solute-solvent interactions with a continuum model which globally represents medium effects without explicitly considering solvent molecules. For example, dielectric constants of 1–1.5 are generally used for gas phase studies, whereas D = 4, possibly with distance dependence (see below for the AMBER force field) was claimed as more appropriate for crystals or aqueous solutions.

Explicitly adding solvent molecules looks of course a more refined avenue, but at the expense of largely increased computer time, and this solution still does not account for polarizability effects. A common approach, also widely used in Molecular Dynamics simulations (see Chapter 6), is to immerse the molecule under scrutiny in a box containing a substantial amount of water molecules, to explicitly evaluate solute-solvent interactions. For example, in the TIP3P model of Jorgensen [81a], a cubic box (with 18.7 Å sides) contains 216 water molecules.

The box is surrounded in all directions by its 26 images derived from translations along the three axes. With this trick, a constant density of solvent molecules around the studied molecule can be maintained: if during optimization a solvent molecule escapes the box from one side its image enters the box from the opposite site. A convenient cut-off must be chosen to avoid an atom interacting with both another atom and an image of that atom.

A somewhat similar approach was proposed to take into account lattice forces involved in crystal structures (for safer comparisons between molecular mechanics predictions and X-ray data). The crystalline environment around the molecule under study is built up according to the X-ray crystal data, as a block of, say, 5×5×5 unit cells. The energy of the central molecule is first optimized in a restricted environment (thanks to cut-off values). Then the energy of the block of unit cells is optimized using the optimized central molecule and the crystalline environment is adjusted. The process is then iterated until self-consistency is achieved [104].

Diverse other approaches were also proposed. For example, it was also suggested that for a rapid (possibly interactive) evaluation, solvation effects can be considered as roughly proportional to the accessible surface of the constituting atoms of the solute (weighted by a solvation parameter): see Chapter 8.

Finally, a quite different strategy was developed by Gilson and Honig [51], Their model of charge-solvent interactions uses pairwise energy terms which can be easily incorporated into existing force fields. The basic idea is that interaction of a solute atom with the solvent can be described as an interaction of opposite sign with the other atoms of the solute: an atom interacting favourably with the solvent can be viewed as being repelled by the other solute's atoms.

π Systems

Delocalized π systems involve a particular stabilization which had to be treated separately. However, in simple cases one can cope with such situations without taking into account π electrons explicitly. We have already indicated that the rigidity of the double bond can be reproduced by means of an appropriate torsional constant. Similarly, when a conjugated system can be considered as a whole (something like a substituent not directly involved in the property investigated) it may be described with a particular set of parameters. For instance, to maintain the coplanarity of a phenyl ring, one can use large force constants, thus making any out-of-plane deformation costly.

More generally, however, an explicit evaluation of π effects is required and quantum chemistry is invoked. Warshel added to the usual molecular mechanics calculation for the σ system a SCF treatment of π electrons to get the global energy to minimize [52]. Allinger [53] adjusts the bond stretching and torsional constants according to bond orders calculated from an initial geometry in a *VESCF treatment* (Variable Electronegativity Self Consistent Field). The calculation can be iterated to self-consistency if minimization tends to change the geometry of this part of the system.

In a more recent version (MMP2), a SCF π calculation is preferred to the VESCF approach to adjust bond lengths, stretching or torsional constants. The parameterization proposed allows for calculating heats of formation, resonance energies and structures for conjugated hydrocarbons in a way that is quite consistent with the calculations on non-conjugated molecules. The results obtained for a large, diverse group of compounds (aromatic or not, strained or strainless, planar or non-planar) testify to the accuracy of the method [54]. These calculations have been now extended to aromatic heterocycles [55].

5.2 STERIC ENERGY AND DERIVED INFORMATION: HEAT OF FORMATION AND STRAIN ENERGY

The summation of all these energy contributions (see equation (5.4)) defines the *steric energy*. For a molecular system, considered in a given geometry, steric energy represents the difference due to internal deformations and non-bonded atom-atom repulsions with respect to a hypothetical system where all parameters have the reference values and van der Waals interactions are null.

This steric energy is sufficient for investigations on the same molecular system: minimization of steric energy will give the actual geometry ("geometry optimization"); comparison between the steric energy associated with given conformers indicates their relative stability. Variations of steric energy depending upon one or two selected parameters (generally torsional angles) allow for drawing energy profiles or maps from which the inter-conversion processes may be approached. These are the most common uses of molecular mechanics in modelling, and the only ones to be developed here.

However, for a comparison of energetic data with experimental information, as well as in the study of different molecules (i.e. systems with a different number of atoms or even a different topology), other energetic parameters have to be introduced. For example, steric energy is sufficient for comparing cis or trans di substituted alkenes, but not for the corresponding gem-derivative (which has a different topology).

Heats of formation, directly comparable to experimentation, allow for testing the accuracy of the method in terms of energetic predictions, and lead to quantitative indices of thermodynamic stability. *Strain energy* gives a quantitative evaluation of the constraints (steric or geometrical) suffered by the system.

For evaluating heats of formation, steric energy (difference with respect to a reference situation) has to be completed by bond- (and some additional structural-) enthalpy increments and a contribution from partition functions. This latter term, theoretically attainable via statistical mechanics, encompasses the influence of translation, rotation, vibration, internal rotations, etc. However, its evaluation is still controversial. Indeed, the force field parameterization originates from data (geometries) corresponding to a given temperature (which is not 0 K) and so already includes some of these terms.

A more empirical and crude model (such as that used in MM2) seems to work well: the *heat of formation* is defined as the sum of steric energy and bond-enthalpy terms. The latter are evaluated by summation of increments associated with the various types of bonds present in the molecule or, in an equivalent way, to the various types of groups (CH_3, CH_2, $CH...$). Supplementary structural-enthalpy increments take into account the presence of specific structural motifs (branched motifs for instance). The translation/rotation contribution is given a constant value of 2.4 kcal/mole. These increments have been adjusted by comparison with the experimental enthalpies of formation for simple reference systems, and it may be thought that they implicitly contain the other contributions. For some specific situations, additional terms account for the influence of higher energy conformers or low frequency torsional motions (for example 0.36 kcal/mole for a torsional barrier around a bond of less than 7 kcal/mole). The results appear quite satisfactory, since with the early MM2 force field (1977), for example, the heats of formation of alkanes and cycloalkanes are predicted with an accuracy comparable to experimental uncertainty: standard deviation about 0.4 kcal/mol [11]. Heats of formation of C_{60} (buckminsterfullerene, footballene) and C_{70} have also been determined recently [56].

For comparison between non-isomeric systems, strain due to geometrical distortions is better expressed by the *strain energy* (neighbouring terms have been proposed: strain enthalpy, formal steric enthalpy, etc., with nearly equivalent definitions but slightly different numerical values). Strain energy is evaluated as the difference between the heat of formation and a strainless bond energy (which would correspond to a hypothetical isomer, strainless and exclusively considered in its minimum energy conformation). This strainless bond energy (as the bond energy just seen) is calculated by summation of bond increments. These have been chosen so that simple reference compounds have no strain energy at 25°C in the conformational mixture.

Although numerical values may differ among authors as a result of the difficulty of defining what a strainless reference is, strain energy seems a good index of the strain existing in a molecule. However, some extreme situations have been quoted, where according to the force field selected, a comparison of two molecules such as norbornane or 2 tBu-adamantane leads to reverse conclusions as to the relative amount of strain they suffer. For a more detailed discussion, see elsewhere [4, 57].

5.3 SEARCH FOR THE PREFERRED GEOMETRY AND ENERGY MINIMIZATION

For a given geometrical arrangement of the atoms in a molecular system, one can calculate, according to equation (5.4), the steric energy due to distortions of bonds lengths and angles with respect to the reference values and van der Waals interactions (provided the system is not too far from the reference conformation). To determine the actual equilibrium geometry, one then has to minimize this steric energy with respect to all internal degrees of freedom. This locates the minimum of the hypersurface representing the steric energy in the hyperspace defined by the geometrical parameters characterizing the system. The problem may be expressed as searching for the minimum of a function of a vector variable X, the elements of which represent the geometrical parameters of the system (Figure 5.8).

For the minimum, the first derivative (the gradient vector g, with components $\partial V/\partial x_i$) is null and the second derivative (the Hessian matrix G with components $\partial^2 V/\partial x_i \partial x_j$) is positive definite. Minimization methods to be used depend on the information known on the potential energy (hyper) surface: energy values only, first or second derivatives. Methods work either analytically or numerically (finite difference method).

If only energy values are available, there is *a priori* no privileged direction to move on the potential energy surface. A downhill Simplex method, which does not require derivative evaluation, can be used [58], as in the MAXIMIN program [59]. As stated by Press *et al.* [58] *direction-set* methods prefer updating directions of search so as to get "good" directions allowing to go far along narrow valleys, or "non-interfering" directions so that what is gained along one direction is not spoiled by the following one. This is the principle of *conjugate direction* minimization, such as the Powell method (which does not in fact need derivative calculation). Alternatively discrete sampling via the Monte Carlo method is possible (see Chapter 7 devoted to the exploration of the conformational space).

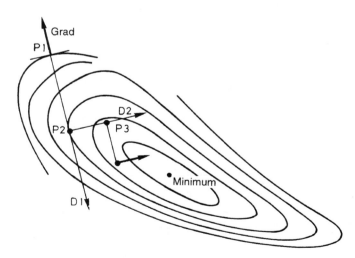

Figure 5.8 Steepest descent method. From the starting point P1, the first direction of displacement D1 is opposite to the gradient at the point. The minimum along D1 is reached at P2. The second displacement is performed along D2 to the minimum P3, and so on.

The gradient, if it is known, indicates the best direction to step the potential energy surface. This is basically the principle of the *steepest descent* approach of Wiberg [60] (Figure 5.8). The iterative process can be presented as follows: starting from a given point of the energy hypersurface, one looks for the direction in which the energy decreases most. It corresponds to the opposite of the gradient vector at this point. This is numerically determined by moving each atom individually along each of the three coordinates and calculating the resulting changes in energy. A new geometry is then proposed by moving simultaneously all atoms over a distance depending on $\partial V/\partial x_i$: the greater the gain in stability, the greater the corresponding atom displacement. The process is repeated until energy decreases no more.

An analytical estimation of the energy derivatives (the gradient coordinates) accelerates the process by a factor of about ten. Such steepest descent methods are very general and seem to never stop hanging on a saddle point, so they are very useful to begin with crude structures far from the minimum, where more refined methods may fail. However, one drawback of the steepest descent method is that, in the current step, the search is made along the gradient direction. Once the minimum along that direction is found the next step is processed along an orthogonal direction, and so on. This may result in a long travel taking many steps to reach the energy valley floor. The method becomes also very slow near the minimum, when the slope is weak. In the "pattern search" of Schleyer *et al.* [61], for each step, the direction of motion from the preceding iteration is added to the current one. This speeds up the minimum search by a factor of two or three.

The *conjugate gradient* minimization (Fletcher–Reeves or Polak–Ribiere

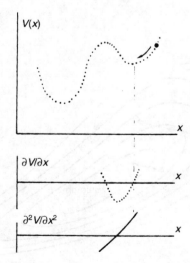

Figure 5.9 The minimization program runs downhill from the starting position (black circle) and stops at the first minimum found. On a minimum, the first derivative is null and the second is positive (schematic representation using a function $V(x)$ of a single scalar variable x).

methods) also gives increased efficiency. For more details about these methods see elsewhere [5, 58].

Apart from these methods working only on first derivatives, the Newton Raphson method [5], using the Hessian matrix (second derivatives) directly determines the minimum for a quadratic function, or with few iterative steps for more complex expressions. Energy variations vs. coordinate changes are better expressed by a quadratic form (therefore involving the Hessian matrix). "The first derivative measures the slope and the second derivative indicates the rate of the change of the slope" [18].

For displacement vector δ around the starting position defined by the vector variable $\mathbf{X}$ (of components x_i ...), a Taylor expansion leads to:

$$g(\mathbf{X} + \delta) = g(\mathbf{X}) + G\delta + \ldots \tag{5.14}$$

where g is the gradient of the energy (components $\partial V/dx_i$) and G the Hessian matrix $(\partial^2 V/\partial x_i \partial x_j)$. At the minimum:

$$g(\mathbf{X} + \delta) = 0 \tag{5.15}$$

leading to the basic expression:

$$G\delta = -g(\mathbf{X}) + \ldots \tag{5.16}$$

The minimum will be found directly if the function is quadratic, otherwise the iteration process is:

$$\delta = -G^{-1}g(\mathbf{X}) \qquad \mathbf{X}_{k+1} = \mathbf{X}_k + \delta \tag{5.17}$$

This Newton Raphson method is very efficient, at least for a guessed conformation neighbour to the solution (about three times faster than methods using only first derivatives), and is now very widely used as the standard energy minimization method. Nevertheless, it may be less efficient (and even not converge) for configurations far from the energy minimum. This may occur at points where the Hessian matrix is not positive definite. In the *restricted step* method, geometrical changes at each iteration are shortened so that energy does decrease. Equation (5.16) is modified as:

$$(G + \lambda I)\delta = -g(\mathbf{X}) + \ldots \tag{5.16'}$$

where I is the unit matrix of dimension $m \times m$ (m = number of parameters to be varied) and γ a scalar chosen so as the new matrix $G + \lambda I$ is positive definite. The method therefore looks like an intermediate between the steepest descent and the Newton Raphson methods. An extension of the Newton Raphson method was proposed that approximately accounts for anharmonicity in bond stretching coordinates, and which provides more rapid convergence [62].

Another handicap is that the demand on computer resources is high since half the Hessian matrix needs to be calculated and stored at each iteration. To avoid such a time-consuming way, in *quasi-Newton* methods, one updates the

Hessian matrix (or the approximate form accepted for it) thanks to information gained during the current iteration [5]. This corresponds to the frequently used Davidon–Fletcher–Powell (DFP) or Broyden–Fletcher–Goldfarb–Shanno (BFGS) methods [58].

One can also use the *bloc diagonal Newton Raphson* method, where only 3×3 submatrices along the diagonal are considered, each related to the coordinates of one atom, i.e. one atom is moved at a time. The geometric improvement is not so fast as with the full matrix method, but the computer time is largely reduced (especially for large molecules).

The ORAL energy minimizer [63] proposes various options such as "floating block" which can either conserve or modify their geometry during minimization, and the possibility to incorporate NOE distance inequality constraints. With this comes added penalty terms which seriously hamper conformations where constraints (interatomic distances in that case) are not satisfied. Minimization under constraints was also tackled by Dillen [32, 32a].

The energy minimization process is generally performed starting from an "initial guess", derived for example from standard geometrical parameters, or realistic data taken from neighbouring systems. However, finding the true energy minimum is not a trivial problem. Usual minimization methods stop at the first minimum encountered near the starting point, even though it may be only a local (secondary) minimum. In other words, they do not go through a potential hill to find the (true) absolute energy minimum. Various solutions to explore the conformational space have been proposed, and will be briefly discussed in Chapter 7, dedicated to the traversal of the conformational space.

Traversing the conformational space is often computer time-consuming in full optimization processes. However for many applications, simplifications can be introduced owing to the relative magnitude of the different force constants intervening. Roughly speaking, it is more costly to relieve strain by stretching a bond rather than opening a valence angle and *a fortiori* modifying a torsional angle. So for large systems it is more convenient to reduce the conformational freedom and adjust only torsions, standard fixed values for bond lengths or angles being assumed. Such a solution has been proposed, for instance, for large polypeptides [46, 64]. It is sometimes better to minimize first on dihedral angles alone, and then from that minimum allow relaxation of all degrees of freedom.

5.4 MOLECULAR MECHANICS: SCOPE, LIMITATIONS AND EVOLUTION

For all problems where the electron distribution is not explicitly needed, (so when only energy or geometry is considered), molecular mechanics is a very attractive approach, working rapidly and efficiently. When compared to quantum chemistry methods, it can be noted that the computer time needed increases as the square of the number of atoms, whereas for quantum methods, time grows as the square or the cube of the number of orbitals (at the semi-empirical level) or as a fourth power for *ab initio* approaches. This gain in computer time (often amounting to some powers of ten) is a notable advantage favouring molecular mechanics as the systems comprise more atoms. So, whereas *ab initio* quantum approaches have been practically limited until now to small or medium size molecules, empirical force field methods can be reasonably applied to large systems. Some versions of molecular mechanics (or empirical force field) programs allow for the treatment of few *hundred* atoms.

In other respects, molecular mechanics involves quite simple concepts, not far from the usual way of thinking of chemists and the elementary reasoning from ball-and-stick models.

However, its empirical character is handicapped by some drawbacks and restrictions. Accurate results are obtained within the scope for which the method has been parameterized. Highly strained molecules or systems with "exotic" geometries (far from the reference systems used to define the force field) cannot be reproduced well with standard force fields, and require the search for new parameters. The same is true for some chemical species or atomic patterns not yet taken into account. This may be performed by fitting available data (experimental or *ab initio*) on relevant small similar systems.

For non-classical situations, specific terms have to be introduced: lone pair, hydrogen bonding and anomeric effects have yet been considered. Others may appear to be necessary for more complex problems. Furthermore, such methods do not explicitly consider π electrons. This deficiency may be biased by a composite approach, adding to the usual MM calculations on the σ framework, a π electron calculation with semi-empirical quantum methods.

Recent refinements of the MM2 force field have been proposed to enlarge the structural scope of the method and give more reliable results. New or modified parameters have been reported, comparisons with *ab initio* results on model compounds being generally used to scale the proposed values and to check the validity of the method in the structural area considered. Besides the numerous chemical families already investigated or revisited (alkanes, alkenes, alcohols, acids, ketones, etc.), imines [65], ketenes [66] peroxides and oxonium ions [67–71], organosilicon compounds [42, 72–74], sulphur heterocycles [55] ethers and anomeric effects [69, 70] have recently been scrutinized. Interestingly, for siloxanes, no torsional potentials are necessary to account for torsional barriers (the effects being reproduced well using only appropriate non-bonding and electrostatic potentials [72–74]. An accurate force field for zeolites was also developed [75]. Slight modifications of the treatment of out of plane deformations in the MMP2 method have also been proposed to improve the underestimated rotational barriers involving ring distortions [76].

While the MM programs of Allinger have nearly become a standard for modelling small or medium size organic molecules, various other programs appeared using somewhat similar expressions of the force field. The MAXIMIN program [59], for example, was presented as performing comparably to Allinger MM1 and proposing special features making it particularly suitable for the treatment of flexible molecules in a pharmacophoric pattern search.

One can note, for example, the "aggregate option" which protects the geometry of molecular fragments (considered as rigid entities) while optimizing their orientation.

In the neighbouring family of the Consistent Force Field programs, Huige *et al.* [77] discussed the determination of the force field parameters for imines and oximes, and provided some insight about the way in which these problems can be approached.

Besides the conventional force fields used in popular molecular mechanics MM2 programs and recent parent approaches such as MM3 or AMBER (see below), a quite different definition of the force field was proposed by Saunders and Jarret [78]: terms involving bond or torsion angles being replaced by two-body central forces between atoms. Promising results have been obtained with increased speed.

In other respects, AMBER [29, 79] or AMBER/OPLS [80, 81], ECEPP [82], CHARM [83, 84] and CFF/VVF [85–87] were mainly devoted to biopolymers, or YETI [88] to small-molecule protein interactions, while other developments aim to incorporate interactions with metal ions [89, 90]. Generally speaking, owing to the relative simplicity of the calculations involved, it is not surprising that most of the modelling packages, even the crudest ones, include some type of energy minimizer with more or less sophisticated force fields.

We will now examine in some detail the recent MM refinements with MM3 and AMBER, as an example of a biomolecule-oriented force field.

5.4.1 MM3 force field

Although, as stated above, empirical force field approaches (and particularly the widespread MM2 method) have provided a lot of successful predictions as to molecular geometries or favoured conformers, typical failures have been identified in some special (and fortunately limited) cases, prompting new attempts to refine the MM2 force field parameterization. These could take advantage of newly available high accuracy experimental data (gas-phase works on relatively simple structures, low temperature crystallographic studies, neutron diffraction information, etc.).

The two models proposed (MM3 by Allinger *et al.* [12–16] and the empirical force field of Dillen [32, 32a] have been adjusted for alkanes considered as the basic element of the molecular architecture (before a further extension to heteroatoms). Although using different solutions to parameterize the force field, the two models largely remove the major flaws of the former MM2 approach: underestimation of high C–C rotational barriers in congested molecules and of bond elongation triggered by eclipsed interactions, overestimation of repulsions in short H · · · H contacts and of sublimation enthalpies of crystalline hydrocarbons.

Rather than modifying only certain values or introducing appropriate correction terms, both research groups prefer to define again a complete and consistent force field reproducing the results of MM2 (in the structural scope where it works well) and correcting its pitfalls for most of the failures identified. Interestingly, as to the empirical character of these molecular mechanics methods, it appears that different sets of parameters and even different analytical expressions for the energy contributions can reproduce former results (in the usual situations) or improve them for specific cases.

As for MM3 parameterization, a quartic term is introduced for bond stretching (to balance cubic terms tending to separate atoms for geometries far from the equilibrium), whereas in EFF a Morse potential is used. Torsion and stretch-bend interactions are treated similarly as in MM2. On the contrary, for angle bending the former quadratic term is completed by higher powers (up to the sixth one), torsion-stretch interactions (in $(r - r_o)(1 + \cos 3\Phi)$) and bend-bend interactions (for angles centred on the same atom) in $(\theta - \theta_o)(\theta' - \theta_o')$ are introduced (the latter for a better description of corresponding vibrational frequencies). However, the previous torsion-bend interaction is now suppressed. Parameterization for the anomeric sequence C–O–C–O–C has been updated. A different solution is adopted in EFF, where anharmonicity in bending or bend-bend interactions are lacking. In MM3, for example, non-bonded atom-atom potentials have been refined from those available in MM2 by examination of crystal parameters and heats of sublimation of some n-alkanes, graphite and benzene. C and H atoms are slightly softer than in MM2 [16]. Different values have been preferred in EFF, corresponding, as for C–C interactions for example, to C atoms larger and harder than in MM3 (Figure 5.10).

An important feature of MM3 is the interest devoted to vibrational frequencies (not considered much in MM2 or EFF). Thanks to a better definition of the force field (with bend-bend and torsion-stretch interactions), vibrational frequencies are evaluated for a selected sample of hydrocarbons with a "chemically acceptable" accuracy of ± 35 cm^{-1}, which allows a possible estimate of entropic terms.

Evaluation of the intensities of the IR bands (from changes of the dipole moments) has been also proposed [91]. Interestingly, it appears that these

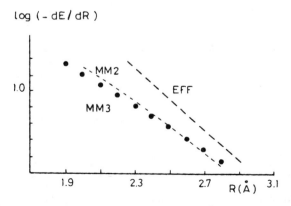

Figure 5.10 Comparison of non-bonded C–C potentials in MM2, MM3 and EFF (from Dillen with permission [32a]).

entropic terms are important in the evaluation of rotational barriers. So, a part of the discrepancies previously observed between experimental NMR data (referring to free energies of activation) and calculated barriers (enthalpic terms) is due to entropic contributions.

For instance, the rotational barrier of 2,2,3,3-tetramethylbutane (exp: 10.2 kcal/mol) is reproduced well by the EFF model (10.01) and MM3 (9.28 kcal/mol, resulting from an enthalpic contribution of 7.77 kcal/mol and an entropic term of −8.57 e.u.), whereas the discrepancy is quite large with MM2 (calculated enthalpy: 5.24).

A detailed discussion of the geometries of representative hydrocarbons is presented elsewhere [14, 32a] compared with either experimental or former MM2 results. Some data for norbornane are gathered in Figure 5.11. Allinger *et al.* [13] discuss particularly the influence of torsion-stretch interactions to reproduce the lengthening of (nearly) eclipsed C–C bonds. From a comparative computational and experimental study of dodecahedrane derivatives, Allinger *et al.* can assess that MM3 calculations are competitive in accuracy with low temperature crystal structures.

In a series of subsequent papers [92–95], the MM3 force field was extended to various classes of compounds: alcohols and ethers, amines, aldehydes and ketones or amides, polypeptides and proteins. It can be noted that, in this new force field, lone pairs which were explicitly included in the preceding MM2 parameterization are no longer utilized. As for alkanes, vibrational spectra are fairly well reproduced (rms about 35–40 cm^{-1}). In the case of amides, particular attention was devoted to electrostatic interactions in an aqueous medium.

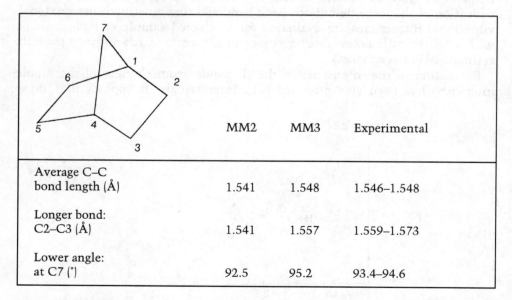

	MM2	MM3	Experimental
Average C–C bond length (Å)	1.541	1.548	1.546–1.548
Longer bond: C2–C3 (Å)	1.541	1.557	1.559–1.573
Lower angle: at C7 (°)	92.5	95.2	93.4–94.6

Figure 5.11 Comparison of some geometrical data for norbornane (from Allinger *et al.* [13]). The agreement is quite good for torsion angles (within 1°). "Experimental" data gather X-ray crystallography, electron diffraction and *ab initio* calculations.

According to the authors, structural results obtained on the small protein crambin "are better than those with MM2 and comparable with the better specialized protein force fields available" [92–95].

5.4.2 AMBER model

The molecular mechanics approaches of Allinger *et al.* [2] and Ermer and Lifson [87] started from the study of saturated hydrocarbons viewed as isolated molecules (in inert solvent or in gas phase). Some authors question the best force field to use for polar or ionic molecules in condensed phases, particularly regarding bioorganic systems such as proteins and peptides [96, 97] on the one hand, and on the other, nucleic acids [98, 99] and proposed various sets of parameters. The AMBER force field, from Weiner, Kollman and co-workers [29, 79], is described as a general force field developed in a consistent way for both proteins and nucleic acids. Basically, although usually using "united atoms" (hydrogens bonded to carbon omitted), it is not very different in its fundamental expression (5.18) from the formula (5.4) previously indicated. Similarly, parameters are first extracted from physical measurements (microwave, neutron diffraction, crystal packing information) and refined through molecular mechanics calculations on model compounds closely related to the structural scope under investigation:

$$E_{\text{total}} = \sum_{\text{bonds}} K_r (r - r_{\text{eq}})^2 + \sum_{\text{angles}} K_\theta (\theta - \theta_{\text{eq}})^2 + \sum_{\text{dihedrals}} \frac{V_n}{2} [1 + \cos(n\phi - \gamma)] +$$

$$\sum_{i<j} \left[\frac{A_{ij}}{R_{ij}^{12}} - \frac{B_{ij}}{R_{ij}^6} + \frac{q_i q_j}{DR_{ij}} \right] + \sum_{\text{H-bonds}} \left[\frac{C_{ij}}{R_{ij}^{12}} - \frac{D_{ij}}{R_{ij}^{10}} \right] \tag{5.18}$$

So, we only comment on some specific features. For the relatively unstrained proteins and nucleic acids, quadratic functions are sufficient for bond stretching or angle bending, joined to a Fourier series for torsional energy. Van der Waals terms are expressed with a 6–12 potential. Electrostatic interactions (treated as a Coulombic expression $(q_i q_j)/DR_{ij}$) deserve more attention, as well as a supplementary 10–12 potential introduced to account for H-bonding effects.

As to the van der Waals term, it is worth noting that, as previously stated, r^* and ε, atomic van der Waals radius and well-depth (intervening here in A_{ij}, B_{ij}...) are strongly interconnected and dependent upon the expression used (6–9 or 6–12 potential). The values retained within this particular method are nearly 0.2 Å larger than the "standard" van der Waals contact radii [28]. For interactions involving hydrogen-bonding hydrogens and heteroatoms, a 10–12 potential is used with a depth of 0.5 kcal/mol to better reproduce H-bond distances (predicted as too short if this term is omitted and too long if one uses the usual 6–12 potential). The evaluation of electrostatic terms involves two interesting features.

First, a distance-dependent dielectric constant $D = R_{ij}$ may be chosen to simulate the fact that in water media, intramolecular electrostatic interactions die off more rapidly with distance than in the gas phase. Then, rather than semi-empirical (CNDO type) Mulliken charges, atomic charges are chosen so as to fit the *ab initio* electrostatic potential surrounding the molecule [34, 100]. These calculations are carried out on subunits and then "patched" together. For example, protein residues were broken into two parts: the bridge containing the α and β carbons and the "chromophore" possessing the remaining side chain atoms. Note also that for hydrogen bonding, sulphur (but not nitrogen or oxygen) requires explicit inclusion of lone pairs.

Another interesting point is the introduction of a cut off distance to avoid calculating interactions between too distant atoms (ca. 9 Å, with a progressive attenuation between 8 and 9 Å to better accommodate the influence of atoms entering (or leaving) the active domain during an energy-minimization process).

Van der Waals hydrogen parameters have recently been revisited [101] to take into account the effect of neighbouring electronegative substituents (electron-withdrawing groups attached to a methyl group pull electron density away from the methyl hydrogens, thus allowing for a closer approach of the surrounding water molecules).

The proposed force field has been satisfactorily tested on diverse examples such as furanose sugar puckering, base stacking, hydrogen-bonding, base paired dinucleoside phosphate refinements, Ramachandran energy contours for dipeptides, and refinement of insulin (a small protein of 500 atoms). These successes confirm that the model is likely to work consistently for both nucleic acids and proteins [29]. It has recently been extended to dehydroaminoacid residues [102].

For the description of proteins in solution or crystalline environments and nucleotide bases, the AMBER force field was modified in OPLS (Optimized Potentials for Liquid Simulations) [80, 81] with the introduction of a set of new potential functions (Coulomb plus Lennard–Jones), bond stretching, angle bending and torsional terms being adopted from AMBER united-atom parameters. This new parameterization was obtained and tested in conjunction with Monte Carlo statistical mechanics simulation on organic liquids or aqueous solutions and optimization of crystal structures for some polypeptides. As for nucleotide bases, *ab initio* calculations for base-water complexes and simulation of the binding energies between pairs of bases were used.

Broadening the structural scope of empirical force field methods is of constant interest in the quest for a universal force field, i.e. a force field which may give good uniform quality predictions over a wide variety of compounds, even if it is not the best one for limited classes of compounds [103]. In this field, one can notice the recent study of the protein crambin by molecular mechanics [92–95, 104]. Such peptides, generally studied with more specific programs (AMBER), constitute an interesting challenge owing to the size of the molecules, the importance of polar and hydrogen-bonding effects, and the influence of an aqueous surrounding medium. The MM2 [104] and more recent

MM3 [92–95] studies of the protein crambin (46 residues, 327 non-hydrogen atoms, 671 atoms) illustrates the capability of the MM methods of treating large systems and to perform energy minimizations (consistent with X-ray data) with a modest cost in computer time (45 minutes on a CYBER 205). Also relevant to the problem of evaluating the quality of the force field is the recent validation of the Tripos force field of the SYBYL package [103] or a comparison of MM2 with various other force fields included in commercial modelling packages (SYBYL, CHEMX) or semi-empirical quantum methods [105]. According to the authors, the conformational energies for hydrocarbons are "reasonably well-reproduced by all the tested methods", the best agreement (even for functionalized compounds) being obtained with MM2 (1985 parameter set).

Recently, Dillen [32, 32a] proposed the PEFF program to assist in the development of empirical force fields. It accepts a wide variety of potential energy functions (with option capabilities) and provides a multidimensional driver to scan energy hypersurface.

5.5 SOME APPLICATIONS

5.5.1 Molecular mechanics and conformational analysis

In molecular modelling, molecular mechanics is now mainly used to predict the favoured geometry, the relative stability of conformers or the ease of their interconversions. More specific applications, such as investigation of steric effects (and related geometrical distortions) on chemical reactivity and determination of the energy profile of a reaction, will not be discussed here.

Numerous studies assess the efficiency and accuracy of empirical force field methods. For hydrocarbons, as quoted by Osawa and Musso [3], the very simple (but well-balanced) Engler force field, with only four terms [61], generally leads to excellent geometry predictions (0.007 Å for lengths, 1–2° for bond angles, 2 kcal/mol for energy). A lot of structures covering a large structural scope have been evaluated by Allinger MM calculations with an accuracy comparable to experimental precision. However, some specific situations are not always reproduced. So, if the stretching of C–C bonds in highly substituted ethane derivatives is predicted (1.68–1.60 Å for the central bond of pentaphenylethane vs. an experimental value of 1.60 Å), the method (with the MM2 force field) fails to reproduce the enlarged bonds induced by through bond coupling of π orbitals (σ bond surrounded by several parallel π orbitals).

Determination of stabilomers

The speed of the calculations makes molecular mechanics very attractive for scrutinizing the more stable isomers in complex polycycles C_nH_m (from C_8 to C_{14}). For each family, the "stabilomers" have been determined through systematic generation of all the possible isomers, elimination of those obviously too severely constrained and MM calculation of the remaining proposals [3]. However, for such complex systems, various local minima of comparable stability may exist on the potential energy hypersurface, and some discrepancies sometimes appear between the rearrangement products obtained and the calculated more stable isomers (Figure 5.12).

Figure 5.12 Among the low energy isomers of the 15 358 possible pentacyclo-undecanes, D_3 tris-homocubane is calculated as the most stable (strain energy = 42 kcal/mol).

Conformational filiations

Numerous structure/activity or structure/property relationships in pharmacology or physical chemistry characterize structures through topological descriptors, i.e. descriptors specifying only the nature of the atoms and bonds linking them, without any information about their actual 3D location (Figure 5.13). The success of such relationships may seem at first glance a little

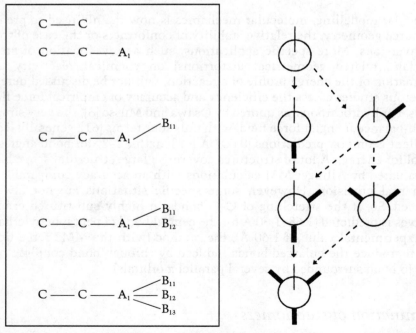

Figure 5.13 One topological adjunction generally corresponds to several possible topographical adjunctions. Projections along the $C-A_1$ bond. B_{1j} positions are located behind the plane of the figure.

amazing, since many properties are known to depend upon conformational or geometrical features. It relies on the fact that, within a limited series of related compounds, topology implicitly reflects the main topographical (3D) aspects: "topology may be described as rubber sheet geometry" [106].

One advantage of topological descriptors is their flexibility for easily deriving relations of formal organization between molecules. For instance, topological models are based on the concept of *filiation* by successive adjunction of sites. Each compound is embedded in a network where it is surrounded by the structures from which it derives or which it generates by adjunction of sites (Figure 5.14).

What does the adjunction of a topological site mean to the actual molecular architecture? In rigid molecules, this corresponds to the occupation of a well-identified spatial location. However for flexible systems, the question is to pass from one given population of conformers to another [107].

To ensure the validity of the models based on 2D descriptors (topological or others) at a predictive or interpretative level, it is therefore of paramount importance to demarcate the areas of structural regularity (where the conformational mixture or the predominant conformer does not change) and their border lines where the models have to be modified.

A striking example is given in the revisitation of Taft steric constants E_s by Dubois and co-workers, from an examination of the accessibility of the carboxyl group in a series of alkylcarboxylic acids [108, 109]. From MM calculations, it was shown that the first terms (up to seven carbon atoms) correspond to eclipsed conformations, whereas more branched structures prefer a bisected conformation. In the first population, successive additions of carbons increase the steric hindrance (as expected), whereas in the second one a levelling of the steric constant is observed. Furthermore, very surprisingly, geometrical distortions in overcrowded skeletons induce decreasing steric effect.

This leads to the concept of "topographically active sites", a site seen by the reaction (or the property) centre (Figure 5.15). This concept has been further extended to cyclic systems, and to other physicochemical properties. In ^{13}C NMR, it allowed rationalization of the surprising variations of α-substituent shifts observed in cyclic derivatives [110].

"Anti-Bredt" olefins

"Anti-Bredt" olefins provide another example where information about the stability of rigid cyclic systems suffering from geometrical constraints is sought for with strain energy calculations.

Amply supported by numerous examples, the Bredt rule can be briefly schematized as: in small bi- or poly-cyclic systems a normal double bond cannot emanate from the bridgehead. Whereas in large bicyclic systems, bridgehead double bonds were known since the late 1940s, the isolation of a first anti-Bredt olefin in 1967 [111] prompted a large amount of work (Figure 5.16) to synthesize similar compounds and to rationalize their stability. The

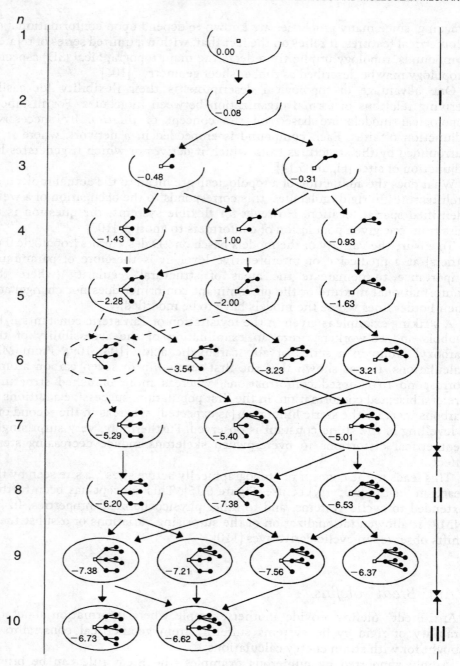

Figure 5.14 Formal generation series of alkylcarboxylic acids. Closed ovals represent bisected conformations, open ovals indicate eclipsed conformations, □ = C–COOH, ● = Me. Numbers indicate the steric parameters E_s' for alkyl groups. The three areas I, II and III correspond to increasing, levelling and decreasing steric effects (from Dubois *et al.* with permission [108]).

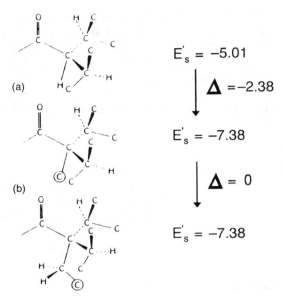

(a) $E'_s = -5.01$

$\Delta = -2.38$

(b) $E'_s = -7.38$

$\Delta = 0$

$E'_s = -7.38$

Figure 5.15   Replacement of H by C on an active site (a) induces a large variation of the steric constant, whereas on an inactive site (b), the variation is negligible (or small). The new carbon introduced is indicated by a circle (from Dubois *et al.* with permission[108]).

key element for interpreting the observations is a clear separation between the strain of the double bond and that of the carbon framework. A definite improvement was the concept of *olefinic strain* (OS) proposed by Schleyer *et al.* [112, 113]. OS represents the difference in strain energy between the bridgehead olefin and the corresponding alkane, the two systems being considered in their most stable conformation. OS values calculated from the MM2 program lead to an empirical loose classification (which up to now has suffered no exception) of bridgehead olefins as [113]:

- *isolatable* (kinetically stable at room temperature at least long enough to permit reactions or spectroscopic observations) when OS < 17 kcal/mol,
- *observable* (detected spectroscopically at low temperature) 17 kcal/mol < OS < 21 kcal/mol,
- *unstable* (generally not observable, except perhaps in matrix isolation and commonly detected by trapping): OS > 21 kcal/mol.

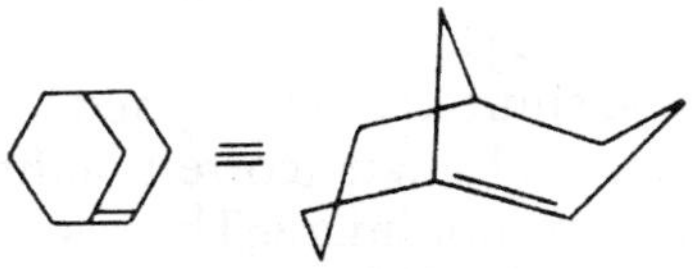

Figure 5.16   Anti-Bredt olefin.

As an example among a lot of systems studied by Schleyer *et al.*, we only mention the two olefins (a) and (b) which have almost identical total steric energy. However, their OSs differ by about 8 kcal/mol in agreement with the fact that (b) but not (a) has been observed. OS values also focus particular interest on *hyperstable* olefins, containing less strain that the parent alkane (their OS values are therefore negative), and which are predicted remarkably unreactive, as for instance (c) which has been already synthesized (OS = −1.5 kcal/mol) (Figure 5.17).

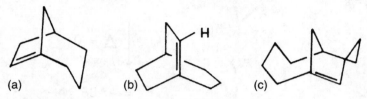

Figure 5.17 Schleyer *et al.* [113] example.

Dynamic conformational analysis

Complementary to the determination of the most stable species, one often needs to know if there are any other low energy conformers, and what the interconversion pathways may be. Indeed, molecular flexibility and shape adaptation processes may be important features in drug-receptor or ligand-protein interactions.

In simpler cases, scanning the conformational space can be reasonably carried out through stepwise variations of only one, or more generally two, geometrical parameters, usually torsional angles. Conformational maps are generally drawn given selected fixed values of two geometrical parameters, on a grid of points. Steric energy is minimized on these points as a function of all other coordinates (some constraint relaxation over the other degrees of freedom thus being allowed for). Such maps are very useful to characterize the possible conformers (those of lower energy) and the probable low energy interconversion pathways (Figure 5.18). Maintaining one or several internal coordinates (geometrical parameters) fixed is easily performed by assessing very high values to the corresponding force constants. In such studies of energy profiles, facilities are given with the "dihedral angle driver" option, which allows one to fix one or two dihedral angles at selected values (a warning about the use of this technique on endocyclic dihedral angles was given by Jaime [114]). Constraints may be also introduced in the Newton Raphson matrix. Some packages allow for maintaining symmetry within groups [115], which can significantly speed up the optimization process or for simultaneously rotating several bonds at a time [116].

Rotational barriers can be similarly determined by minimizing steric energy for selected values of a dihedral angle (corresponding to stepwise variations) acting somewhat as a reaction coordinate. The results are fairly satisfactory as to the trends observed, although MM2 calculated values are lower than those observed. So for the barrier to internal rotation around Csp^3–Csp^3 bonds, in crowded open systems, Tiffon and Lomas [117] found a good linear relationship between the ΔG^+ from NMR and the MM2 results (representing, in fact, enthalpic terms), but with a slope of about 0.64, in agreement with previous results of Osawa and Musso, who quoted that in acyclic compounds, barriers appear as underestimated by a factor of ca. 0.6, the discrepancy increasing with the bulk around the rotating bond [3].

According to more recent work (MM3, EEF) [12–16, 32a], it appears that such discrepancies are largely removed thanks to a refined force field (see above). Note also that a non-negligible part of the deviations may be due to the neglect of entropic terms in previous MM2 predictions.

Many studies have also been carried out on internal rotations in cyclic or polycyclic compounds, the interconversion processes now simultaneously implying several bonds. MM calculations are quite valuable in this field, since they give information on the geometry of transient conformations not attainable by experiment.

5.5.2 Reactivity

For some years, increased interest arose towards the interpretation of chemical reactivity. Predicting the rearrangement or interconversion pathways, evaluating structural effects on reaction rate constants, etc. prompted numerous applications of molecular mechanics methods.

The principle of such approaches is straightforward for equilibrium studies, since it suffices to evaluate the thermodynamic parameters for both reactants and products. Molecular mechanics has recently been applied to two prototype Diels Alder reactions [99]: condensation of butadiene and ethylene and dimerization of butadiene. From the calculation of standard enthalpies of formation (for both products and reactants) and thermodynamic functions, an excellent agreement is obtained as to the evolvement of equilibrium constants on temperature. According to the authors, force field calculations can provide "a reliable alternative to laboratory methods" in terms of the thermodynamic equilibrium properties of gas-phase organic reactions.

The approach can also be extended to kinetic aspects. As stated by Allinger [2], if the calculations (presented here on stable starting states) can be performed (with suitable changes) on *transition states*, some information can be gained about the energetics of the process. As for conformational interconversions, this will be performed by exploring the multi-dimensional potential energy surface for given values of some geometrical parameters. So troughs corresponding to stable products and gorge, leading to the formation of products [119], can be described. A few examples are now discussed.

Solvolysis rate constants

The study of "true" chemical reactions, i.e. processes where bonds are broken and created, is somewhat more difficult than the study of rotational barriers,

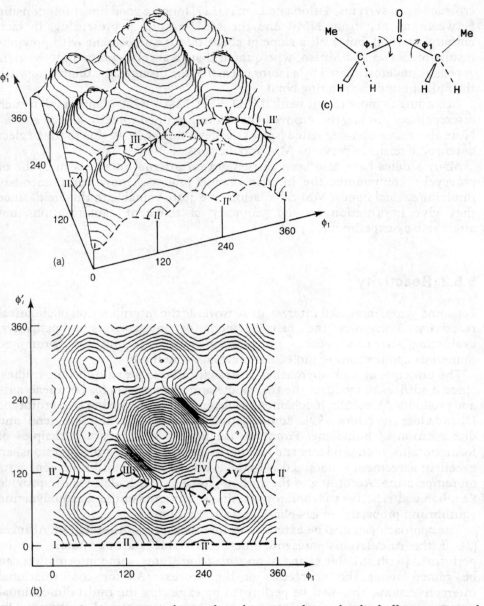

Figure 5.18 Steric energy surface and conformational map for diethylketone. Dotted lines represent interconversion pathways (from Cossé-Barbi [118]).

since we have to deal with transition states which do not correspond to usual molecules. Appropriate *models* therefore have to be sought.

However, if one compares a given reaction for a series of substrates, certain geometrical features of the transition state may be assumed constant, and the same is true for certain interaction mechanisms. This assumption allows for an easier description of the variations of the energetic factors within the series.

Solvolysis of tert-alkyl derivatives with a bridgehead-leaving group was the first reaction investigated. For such processes:

$$RX \rightarrow R^+ + X^- \rightarrow Products$$

The charge developed in the transition state is large, and presumably does not significantly vary from one compound to another one, so that the reaction is "sterically controlled". Rates of reaction can therefore be fairly well represented by the difference in strain energy between the reactants and the transition state (carbenium ion R^+). To calculate this difference in strain energy, Schleyer *et al.* [120] use the corresponding hydrocarbon as a model of the starting halogenated derivative (the difference in the interactions suffered by a hydrogen or a halogen within the molecules being reasonably assumed as constant for the whole series). More complex is the representation of the carbenium ion R^+. In pioneering work, Gleicher and Schleyer [120] calibrated the force field parameters from a learning set of five rate constants on a 12 log unit scale. They used them further for the prediction, through linear correlation, of an 18 log unit reactivity scale.

Subsequent work refined and completed the force field parameterization to represent carbenium ions, especially the Engler cation force field[1] or Muller parameters [121]. These, defined by successive adjustments to fit experimental data, lead to a good prediction of solvolysis rates (including families with varied leaving groups) on a large reactivity scale. Quite noteworthy, because of the large scope covered, is the observation that solvation or ion pairing effects do not disturb predictions.

These studies have recently been extended by Muller *et al.* [122], who investigated solvolytic rates of varied classes of compounds (chlorides, nitrobenzoates, alcohols). The authors showed that solvolytic rate constants of tertiary carbon substrates (including both bridgehead and non-bridgehead derivatives) can be rationalized by the computed steric energy difference between the incipient carbenium ion R^+ and the corresponding alcohol ROH.

Radical thermolysis of alkanes constitutes another example of the use of MM calculations to predict rate constants: the transition state is assumed to be neighbour to the radical formed by bond breaking, and is approximated by the corresponding alkane (see, however, Lomas [4] and the references quoted for the choice of a good model of transition states).

In some examples, reaction pathways have been determined by investigation of the stability of possible intermediates. So, in multistep carbenium ion rearrangements, the strategy was to calculate the energy of all the possible ions resulting from alkyl shifts in intermediates, and look for the reaction pathway of lowest energy (Figure 5.19). The method appeared very useful to determine stabilomers of polycyclic or cage hydrocarbons, and proposed likely pathways for their isomerization. Among the other reactions investigated, we can quote catalytic hydrogenation of cyclic hydrocarbons: the reaction pathway depends

---

[1] See references 10b, 65 and 69 in Osawa and Musso [3].

upon opening either the longer and more strained bond in the substrate or that which relieves more strain.

Numerous recent developments using more refined representations of the transition state have appeared in the last few years. They involve, among others, hydroboration, nitrile oxide cycloadditions, Diels Alder reactions, Claisen and Cope rearrangements, hydrogen transfer, lactonization, nucleophilic addition to carbonyls or Michael additions, radical additions to alkenes and boron enolate aldol reactions. Details can be found in the very well documented review by Esterowicz and Houk [123].

As basic hypotheses, it can be assumed that for a given type of reaction, breaking and forming bonds are fixed at some nearly constant lengths, leading to relatively constant transition state (TS) geometries, in agreement with quantum calculations, and that for such partially bonded systems, energy may be treated as an energy minimum (in fact, transition states correspond to an energy minimum in respect to all degrees of freedom except that corresponding to the reaction coordinate).

Simpler approximation uses the assumption that force constants are similar to those determined for neighbouring (true) molecules used as models. As to the geometry of the transition state, it can be derived from quantum mechanics. In the *rigid TS* model, atoms involved in bonding changes are given fixed positions and the attached groups are geometry optimized. A more refined model, the *flexible TS* model, optimizes all atom positions but requires the development of new (and often numerous) force field parameters. In

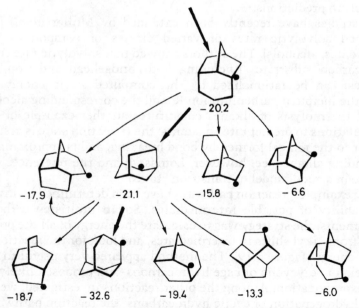

Figure 5.19 Carbenium ion rearrangement. The most probable pathway (bold arrows) is estimated from enthalpies of formation calculated for possible intermediates resulting from alkyl shifts and looking for the pathway of lowest energy (adapted from Osawa and Musso with permission [3]).

another approach, TS is considered as corresponding to a saddle point of the potential energy surface. Parameters can be derived by interpolation between reactants and products. An approximate TS structure may also be determined from intersection of the potential surfaces of reactants and products [124].

Modelling a complete reaction path with the capability of describing any desired point of the reaction coordinate, and not only the transition state, is a still more complex problem, requiring a modified force field for accurate prediction in regions of highly stretched bonds. This was recently tackled by Peyman and Beckhaus [125]. For dimerization of radicals, they proposed a *variable* force field where parameters characterizing the atoms involved in the bond-forming step are allowed to continuously vary according to the hybridization change.

These applications mainly concern *a posteriori* studies to rationalize experimental results, but MM can be also useful for *de novo* design: "What to make and how to make it" [126]. Predictions based on computed stabilities or the relative strain energy of intermediates can help in eliminating non-productive syntheses and give some information about the pathway to follow if several synthetic routes are possible (various examples can be found elsewhere [126]). We only quote here the work of Still *et al.* [127]. They prepared non-macrocyclic host molecules (podands) able to bind alkali metal cations thanks to a pre-organization. Once synthesized, these compounds show the expected activity.

As stated by Lipkovitz and Peterson [126], thanks to the development of more refined force fields and the broadcasting of user-friendly programs, "application of empirical force fields in organic synthesis looks promising."

5.5.3. Molecular associations

Although initially proposed for organic derivatives, and originally devoted to the study of intramolecular forces, molecular mechanics has recently been extended to molecular complexes, and there is particularly now a substantial number of studies involving molecular systems associated with metal centres [89, and the references quoted therein].

This interest relies on the essential role of metal ions for the maintenance of living systems through various mechanisms: control of the transmission of nervous impulses (sodium), secondary messengers (calcium), enzymatic redox processes (iron, copper). These studies suggest that useful information about molecular recognition and its geometrical and energetical features can be gained in this way.

Parameters necessary to describe these ion-molecule interactions are generally lacking in the usual force fields, particularly regarding non-bonded interactions for the ion and atoms with partial charges. The first task is obviously to define them according to appropriate protocols [128]. Fitting experimental or quantum results on simple model systems (such as $M^+\cdots O(CH_3)_2$ for crown complexation, for instance) constitutes a possible way.

The YETI program, oriented towards drug design applications on small molecule-protein complexes, includes directional potential functions for hydrogen bonding, salt linkages and metal ligand interactions. These directional functions appear to give a more realistic picture of short range interactions with only a small increase in the number of variables to treat [88]. Inclusion of intermolecular interactions with the surrounding molecules have been also considered, in order to take into account lattice effects [104] and provides better comparisons with X-ray data.

Host-guest complexation has been extensively studied. Kollman *et al.* [129, 130] discussed alkali cation complexes of 18-crown-6 and anisole spherands in terms of structural flexibility, ligand specificity and macrocyclic effect (Figure 5.20).

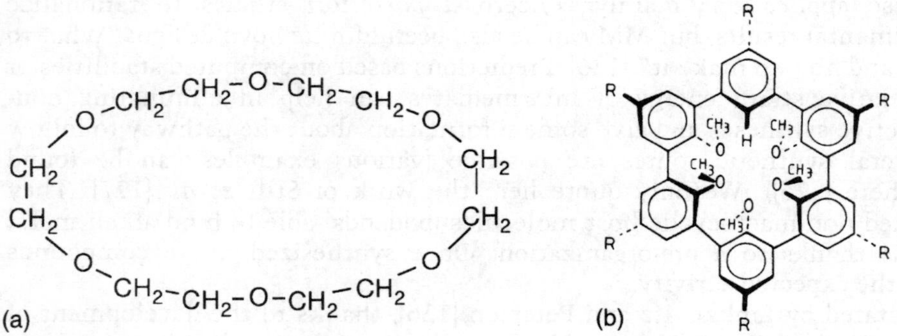

Figure 5.20 18-Crown-6 (a) and a typical anisole spherand molecule (b) (from Kollman *et al.* with permission [129]).

For the crown, several structures of comparable energy are found, in an order depending upon the dielectric constant, in agreement with crystallographic or NMR observations. However, in its complexed state, the crown exhibits a structural flexibility which allows it to adopt different conformations appropriate to its environment: the D_{3d} structure (favoured for the K$^+$ complex) would lead to a too large "hole" for the Na cation (the complex prefers then a C_i structure), whereas the larger Cs ion is moved out of the cavity (Figure 5.21 and 5.22).

Although the K$^+$/crown complex is intrinsically less stable than the Na$^+$/crown complex, it has a more negative formation energy in aqueous solutions. This is due to the fact that the difference in hydration energies of Na$^+$ and K$^+$ is calculated larger in magnitude than the intrinsic difference of complexation energies. The selectivity observed therefore results from a balance between the energies of crown-cation interaction and cation solvation.

Calculation on cation/crown/water shows that the crown shields cations from H$_2$O (decreasing their affinity for water) and so facilitates ion transport through hydrophobic environments (membranes, etc.). At last the enhanced affinity of crown compared to an open homologous chain is attributed to the greater stability of conformations other than those that can effectively interact

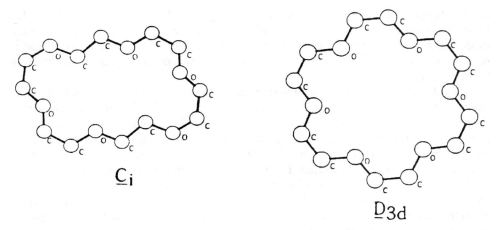

Figure 5.21 C_i and D_{3d} structures of 18-crown-6, viewed approximately perpendicularly to the mean molecular plane (from Damewood *et al.* with permission [128]).

with the cation for the open-chain derivatives and a better preorganization of cyclic systems for binding a guest ("macrocyclic effect") [129, 130]. Similar conclusions come from the study of anisole spherands, where flexibility is almost non-existent and which exhibit a still greater specificity.

Owing to these exceptional complexation capabilities, the interaction of crown ethers with neutral molecules (e.g. nitromethane, acetonitrile) was recently investigated [128]. In these systems, the overwhelming coulombic forces existing in charged complexes are absent and molecular complexation is controlled by weaker interactions (van der Waals, H-bonding or dipole-dipole forces). Acting as a guide, the comparison between experimental complexation enthalpies and the computed complexation energies (difference in energy between the complex and the two isolated partners) assesses the reliability of the approach. Among the more salient conclusions, an intrinsic positive cooperativity is calculated for the formation of the 2–1 complexes in solution,

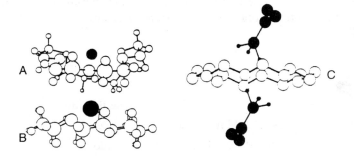

Figure 5.22 Complexes of crown with Na ions (A), Cs ions (B) and the 2–1 complex with nitromethane (C). The guest ions or molecules are shadowed (from Damewood *et al.* [128] and Wipff *et al.* with permission [130]).

i.e. the formation of the 1–1 complex preorganizes the host so that it is in the required conformation for 2–1 complexation. Binding energies are therefore more favourable for this process than for 1–1 complexation. The reverse trend, sometimes observed, results from the importance of entropic and solvation terms in such systems.

The influence of host-guest preorganization and complementarity have also been tackled recently by molecular mechanics and molecular dynamics on the example of cation complexes of a cyclic urea-anisole spherand [131]. The "size-match selectivity" concept was extensively studied by Hancock [132]: "a metal ion will form its most stable complex with the member of a series of macrocycles where the match between size of metal ion and macrocyclic cavity is closest." The effects of the chelate ring size seem to outweigh the effects of the macrocycle size regarding complex stability. Such studies confirm the potential power of molecular mechanics approach to gain some insight about host-guest interactions.

Several recent studies report the use of MM calculations to transition metal complexes. These coordination compounds were mainly investigated with the Boyd program [90] because desired coordination numbers can be handled. However, it does not take into account electrostatic interactions. This term is explicitly considered in MM2 but the program has to be modified to cope with coordination numbers higher than 4 [133]. The stability of hexacoordinate complexes of cobalt (III) and nickel (II) with aliphatic amines is related to the strain energy accompanying complex formation from a hypothetical "standard" state [see also 134,135].

Whereas in the preceding cases the usual procedures can be carried out with adequate parameterization for the metal environment, complexation with lanthanides, leading to heptacoordinate complexes, constitutes a more difficult problem. Natural bond angles (ligand-metal-ligand) are not easy to define, and the geometry is largely dependent upon ligand-ligand interactions. In the MM2 metal-extended force field (MM2MX), bending 1–3 interactions about a metal atom centre are omitted and replaced by van der Waals-type interactions between the corresponding atoms [89]. This method, which emphasizes steric interactions about the ligating atoms, was employed to determine the minimum energy conformations. It successfully reproduces the essential features of heptacoordinate lanthanide complexes (Figure 5.23).

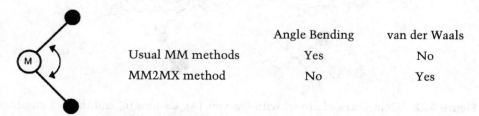

	Angle Bending	van der Waals
Usual MM methods	Yes	No
MM2MX method	No	Yes

Figure 5.23 Usual MM and MM2MX method comparison.

Iron complexation

Lack or excess of iron in plants or humans causes severe diseases; hence the
interest for complexing reagents able to control the concentration level of this
element and able to carry it in living systems. MM was used to investigate the
stability of new Fe(III) chelates containing two carboxy α-catechol moieties
separated by a spacer of several methylene units [136, 137]. After deriving
appropriate parameters for the ferric ion, energy minimization with the EMO
program [138] indicated that the more stable chelate is obtained for four
methylene subunits (Figure 5.24).

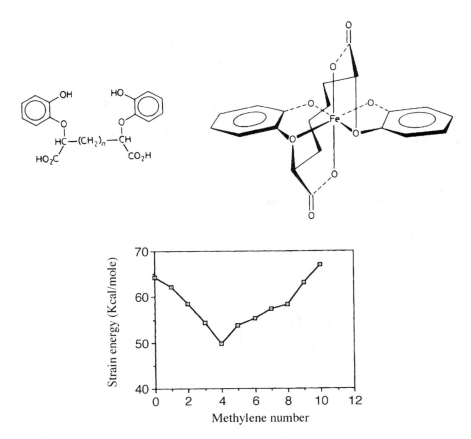

Figure 5.24 Variation of the strain energy for the chelate vs. the methylene number
(from Bouraoui *et al.* with permission [136, 137]).

5.6 TRENDS AND PROSPECTS

The high precision of MM is largely supported by comparison with other
(quantum) approaches as well as experimental data (X-ray crystallography).

Table 5.2   Engler *et al.* force field for saturated hydrocarbons.

Bond Stretching:
$$E(r) = 0.5 \, k \, (r - r_0)^2$$

	$10^{-2} \, k_r$	r_0
C–H	6.63	1.100
C–C	6.33	1.520

Angle Bending:
$$E(\theta) = 0.5 \, k_\theta \, (\theta - \theta_0)^2 \, [1 - 0.0096 \, |\, \theta - \theta_0 \,|\,] \qquad \text{for } |\, \theta - \theta_0 \,| < 25°$$

R	R'	R''	θ_0	$10^2 \, k_\theta$
C	C	C	109.5	2.50
C	C	H	110.1	2.50
C	H	H	110.4	2.50
H	C	C	109.2	1.75
H	C	H	109.0	1.75
H	H	H	109.5	1.75
H	C	C	109.1	1.45
H	C	H	109.2	1.45

Torsion:
$$E(\varnothing) = 0.5 \, k_\varnothing \, (1 + \cos 3\varnothing)$$

	$k_\varnothing$
H–C–C–H	0.69
H–C–C–C	0.69
C–C–C–C	0.45

Non-bonded:
$$E(\text{nb}) = \frac{\varepsilon}{1 - 6/a} \left[\left(\frac{6}{a} e^{a(1 - r/rm)} \right) - (r/rm)^6 \right]$$

	a	$10^2 \varepsilon$	rm
H $\cdots$ H	12.0	4.00	3.20
C $\cdots$ H	12.0	2.99	3.35
C $\cdots$ C	12.0	9.50	3.85

	CH_3	CH_2	CH	C
General increments heats of formation	–10.82	–5.88	–2.82	–0.82
Strain free increments (to be subtracted)	–10.05	–5.13	–2.16	–0.30

E Kcal/mol. distances in A°, angles in °

This type of method should continue to play an essential role in structural analysis, owing to the increasing interest devoted to large biomolecules (e.g. proteins) for which it is necessary to have at one's disposal a calculation method that is fast and easy to implement.

Refinement of the force field (for varied functional groups) and the definition of new parameters for atoms not previously considered, are among the developments logically expected.

Another way in which to speed up calculations should be the definition of transferable structural primitives, in place of the data derived from X-rays (which are sometimes disturbed by packing effects). For complex systems, this set of primitives should allow us to quickly guess more realistic geometries to start the optimization processes. Expert systems and databases are just beginning to be introduced in this field.

APPENDIX: ENGLER'S FORCE FIELD FOR SATURATED HYDROCARBONS

In Table 5.2 we reproduce the force field of Engler *et al.* [61]. Although more refined force fields have been proposed since, we retain it for illustration because of its simplicity (no interaction terms, etc.), and its good performance for most of the usual cases. For some indications about commonly available packages, see elsewhere [3, 4, 138–140], and for the way in which a calculation is carried out [140].

REFERENCES

1. D.H. Andrews *Phys. Rev.*, **36:** 1930; 544–554.
2. N.L. Allinger *Advances in Physical Organic Chemistry*, V. Gold and J. Bethell (eds.), Academic Press, London **13:** 1976; 1–82.
3. E. Osawa and H. Musso *Angew. Chem. Int. Ed. Engl.*, **22:** 1983; 1–12.
4. J. Lomas *L'Actualité Chimique*, 1986; 7–22.
5. S. Wilson *Chemistry by Computer*, Plenum Press, New York, 1986; 85–110.
6. C. Altona and D.H. Faber *Top. Curr. Chem.*, **45:** 1974; 1–38.
7. O. Ermer *Struct. Bonding*, **27:** 1976; 161–211.
8. S.R. Niketic and K. Rasmussen *The Consistent Force Field. A Documentation*, Springer-verlag, Berlin, 1977.
9. O. Ermer *Aspecte von Kraftfeldrechnungen*, Wolfgang Baur, Munich, 1981.
10. U. Burkert and N.L. Allinger *Molecular Mechanics* American Chemical Society, Washington, DC, 1982.
11. N.L. Allinger *J. Am. Chem. Soc.*, **99:** 1977; 8127–8134.
12. N.L. Allinger and J.H. Lii *J. Comput. Chem.*, **8:** 1987; 1146–1153.
13. N.L. Allinger, H.J. Geise, W. Pyckhout, L.A. Paquette and J.C. Gallucci *J. Am. Chem. Soc.*, **111:** 1989; 1106–1114.
14. N.L. Allinger, Y.H. Yuh and J.H. Lii *J. Am. Chem. Soc.*, **111:** 1989; 8551–8565.

15. J.H. Lii and N.L. Allinger *J. Am. Chem. Soc.*, **111:** 1989; 8566–8575.
16. J.H. Lii and N.L. Allinger *J. Am. Chem. Soc.*, **111:** 1989; 8576–8582.
17. A. Warshel and S. Lifson *J. Chem. Phys.*, **53:** 1970; 582–594.
17a. S. Lifson and A. Warshel *ibid*, **49:** 1968; 5116–5129.
18. N.L. Allinger, Molecular mechanics, *International School of Crystallography: Static and Dynamic Implications of Precise Structural Information*, Erice, Italy, 1985; 149–164.
19. W. Klyne and V. Prelog *Experientia*, **16:** 1960; 521.
20. R.A. Buckingham *Proc. Roy. Soc. London A*, **168:** 1938; 264–283.
21. T.L. Hill *J. Chem. Phys.*, **16:** 1948; 399–404.
22. D.E. Williams *J. Chem. Phys.*, **43:** 1965; 4424–4426.
23. U. Burkert *Tetrahedron*, **33:** 1977; 2237–2242.
24. N.L. Allinger and D.Y. Chung *J. Am. Chem. Soc.*, **98:** 1976; 6798–6803.
25. D.H. Wertz and N.L. Allinger *Tetrahedron*, **30:** 1974; 1579–1586.
26. S. Fitzwater and L.S. Bartell *J. Am. Chem. Soc.*, **98:** 1976; 5107–5115.
27. R.H. Boyd *J. Am. Chem. Soc.*, **97:** 1975; 5353–5357.
28. A. Bondi *J. Phys. Chem.*, **68:** 1964; 441–451.
29. S.J. Wiener, P.A. Kollman, D.A. Case, U.C. Singh, C. Ghio, G. Alagona, S. Profeta and P.K. Weiner *J. Am. Chem. Soc.*, **106:** 1984; 765–784.
30. J.M. Leonard and W.P. Ashman *J. Comput. Chem.*, **11:** 1990; 952–957.
31. A.J. Hopfinger and R.A. Pearlstein *J. Comput. Chem.*, **5:** 1984; 486–499.
32. J.L.M. Dillen *J. Comput. Chem.*, **13:** 1992; 257–267.
32a. J.L.M. Dillen *J. Comput. Chem.*, **11:** 1990; 1125–1138.
33. K. Palmo, L.O. Pietila and S. Krimm *J. Comput. Chem.*, **12:** 1991; 385–390.
33a. K. Palmo, L.O. Pietila and S. Krimm *J. Comput. Chem.*, **13:** 1992; 1142–1150.
34. U.C. Singh and P.A. Kollman *J. Comput. Chem.*, **5:** 1984; 129–145.
35. J. Mullay *J. Comput. Chem.*, **9:** 1988; 399–405.
36. W.J. Mortier, S.K. Ghosh and S. Shankar *J. Am. Chem. Soc.*, **108:** 1986; 4315–4320.
37. J. Mullay *J. Am. Chem. Soc.*, **108:** 1986; 1770–1776.
38. W.J. Mortier, K. Van Genechten and J. Gasteiger *J. Am. Chem. Soc.*, **107:** 1985; 829–835.
39. J. Gasteiger and M. Marsili *Tetrahedron*, **36:** 1980; 3219–3288.
40. R.J. Abraham and P. Smith *J. Comput. Chem.*, **9:** 1988; 288–297.
41. L.G. Hammarstrom, T. Liljefors and J. Gasteiger *J. Comput. Chem.*, **9:** 1988; 424–440.
42. R.J. Abraham and G.H. Grant *J. Comput. Chem.*, **9:** 1988; 244–256.
43. L. Dosen-Micovic, D. Jeremic and N.L. Allinger *J. Am. Chem. Soc.*, **105:** 1983; 1716–1722 and 1723–1733.
44. N.L. Allinger, S.H.M. Chang, D.H. Glaser and H. Honig *Isr. J. Chem.*, **20:** 1980; 51–56.
45. R.F. McGuire, F.A. Momany and H.A. Scheraga *J. Phys. Chem.*, **76:** 1972; 375–393.
46. H.A. Scheraga *Adv. Phys. Org. Chem.*, **6:** 1968; 103–184.
47. R.A. Scott and H.A. Scheraga *J. Chem. Phys.*, **45:** 1966; 2091–2101.
48. E.R. Lippincott and R. Schroeder *J. Chem. Phys.*, **23:** 1955; 1099–1106.
49. R. Schroeder and E.R. Lippincott *J. Phys. Chem.*, **61:** 1957; 921–928.
50. N.L. Allinger, R.A. Kok and M.R. Imam *J. Comput. Chem.*, **9:** 1988; 591–595.
51. M.K. Gilson and B. Honig *J. Comput-Aided Mol. Des.*, **5:** 1991; 5–20.
52. A. Warshel and A. Lappicirella *J. Am. Chem. Soc.*, **103:** 1981; 4664–4673.
53. N.L. Allinger and J.T. Sprague *J. Am. Chem. Soc.*, **95:** 1973; 3893–3907.
54. J.T. Sprague, J.C. Tai, Y.H. Yuh and N.L. Allinger *J. Comput. Chem.*, **8:** 1987; 581–603.
55. J.C. Tai, J.H. Lii and N.L. Allinger *J. Comput. Chem.*, **10:** 1989; 635–647.
56. M. Froimowitz *J. Comput. Chem.*, **12:** 1991; 1129–1133.
57. D.L. DeTar, S. Binzet and P. Darba *J. Org. Chem.*, **50:** 1985; 2826–2836.
58. W.H. Press, B.P. Flannery, S.A. Teutolsky and W.T. Vetterling *Numerical Recipes*, Cambridge University Press, Cambridge (USA), 1989; p 274–334.

59. J. Labanowski, I. Motoc, C.B. Naylor, D. Mayer and R.A. Dammkoehler *Quant. Struct.-Act. Relat.*, **5:** 1986; 138–152.
60. K.B. Wiberg *J. Am. Chem. Soc.*, **87:** 1965; 1070–1078.
61. E.M. Engler, J.D. Andose and P. von R. Schleyer *J. Am. Chem. Soc.*, **95:** 1973; 8005–8025.
62. J.F. Stanton and D.E. Bernholdt *J. Comput. Chem.*, **11:** 1990, 58–63.
63. K. Zimmermann *J. Comput. Chem.*, **12:** 1991, 310–319.
64. H.A. Scheraga *Peptides. Proceedings of the Fifth American Peptides Symposium*, M. Goodman and J. Meienhofer eds. J. Wiley. New York, 1977.
65. M. Kontoyianni, A.J. Hoffman and J.P. Bowen *J. Comput. Chem.*, **13:** 1992; 57–65.
66. E.L. Stewart and J.P. Bowen *J. Comput. Chem.*, **13:** 1992; 1125–1137.
67. L. Carballeira, R.A. Mosquera and M.A. Rios *J. Comput. Chem.*, **9:** 1988; 851–860.
68. L. Carballeira, R.A. Mosquera and M.A. Rios *J. Comput. Chem.*, **10:** 1989; 911–920.
69. S.A. Vasquez, M.A. Rios and L. Carballeira *J. Comput. Chem.*, **12:** 1991; 872–879.
70. B. Fernandez, M.A. Rios and L. Carballeira *J. Comput. Chem.*, **12:** 1991; 78–90.
71. J. Broeker, R.W. Hoffmann and K.N. Houk *J. Am. Chem. Soc.*, **113:** 1991; 5006–5017.
72. S. Grigoras and T.H. Lane *J. Comput. Chem.*, **9:** 1988; 25–39.
73. R.J. Abraham and G.H. Grant *J. Comput. Chem.*, **9:** 1988; 709–718.
74. S. Profeta Jr, R.J. Unwalla and F.K. Cartledge *J. Comput. Chem.*, **10:** 1989; 99–103.
75. J.B. Nicholas, A.J. Hopfinger, F.R. Trouw and L.E. Iton *J. Am. Chem. Soc.*, **113:** 1991; 4792–4800.
76. T. Liljefors, J.C. Tai, S. Li and N.L. Allinger *J. Comput. Chem.*, **8:** 1987; 1051–1056.
77. C.J.M. Huige, A.M.F. Hezemans and K. Rasmussen *J. Comput. Chem.*, **8:** 1987; 204–225.
78. M. Saunders and R.M. Jarret *J. Comput. Chem.*, **7:** 1986; 578–588.
79. P.K. Weiner and P.A. Kollman *J. Comput. Chem.*, **2:** 1981; 287–303.
80. W.L. Jorgensen and J. Tirado-Rives *J. Am. Chem. Soc.*, **110:** 1988; 1657–1666.
81. J. Pranata, S.G. Wierschke and W.L. Jorgensen *J. Am. Chem. Soc.*, **113:** 1991; 2810–2819.
81a. W.L. Jorgensen, J. Chandrasekhar, J.D. Madura, R.W. Impex and M.L. Klein *J. Chem. Phys.*, **79:** 1983; 926.
82. F.A. Momany, R.F. McGuire, A.W. Burgess and H.A. Scheraga *J. Phys. Chem.*, **79:** 1975; 2361–2381.
83. B.R. Gelin and M. Karplus *Biochemistry*, **18:** 1979; 1256–1268.
84. B.R. Brooks, R.E. Bruccoleri, B.D. Olafson, D.J. States, S. Swaminathan and M. Karplus *J. Comput. Chem.*, **4:** 1983; 187–217.
85. S. Lifson, A.T. Hagler and P. Dauber *J. Am. Chem. Soc.*, **101:** 1979; 5111–5121.
86. A.T. Hagler, E. Huler and S. Lifson *J. Am. Chem. Soc.*, **96:** 1974; 5319–5327.
87. O. Ermer and S. Lifson *J. Am. Chem. Soc.*, **95:** 1973; 4121–4132.
88. A. Vedani *J. Comput. Chem.*, **9:** 1988; 269–280.
89. D.M. Ferguson and D.J. Raber *J. Comput. Chem.*, **11:** 1990; 1061–1071.
90. R.H. Boyd *J. Chem. Phys.*, **49:** 1968; 2574–2583.
91. J.-H. Lii and N.L. Allinger *J. Comput. Chem.*, **13:** 1992; 1138–1141.
92. N.L. Allinger, M. Rahman and J.H. Lii *J. Am. Chem. Soc.*, **112:** 1990; 8293–8307.
93. L.R. Schmitz and N.L. Allinger *J. Am. Chem. Soc.*, **112:** 1990; 8307–8315.
94. N.L. Allinger, K. Chen, M. Rahman and A. Pathiaseril *J. Am. Chem. Soc.*, **113:** 1991; 4505–4517.
95. J.H. Lii and N.L. Allinger *J. Comput. Chem.*, **12:** 1991; 186–199.
96. H. Chuman, F.A. Momany and L. Schäfer *Int. J. Peptide Protein Res.*, **24:** 1984; 233–248.
97. V. Sasisekharan in *Conformation of Biological Molecules and Polymers*, E. Bergman, B. Pullman eds. Jerusalem, 1973; 247–260.
98. W. Olson and P.J. Flory *Biopolymers*, **11:** 1972; 1–23.
98a. W. Olson and P.J. Flory *Biopolymers*, **11:** 1972; 25–26.
98b. W. Olson and P.J. Flory *Biopolymers*, **11:** 1972; 57–66.

99. T.G. Lenz and J.D. Vaughan *J. Comput. Chem.*, **11**: 1990, 351–360.
100. P. Cieplak and P.A. Kollmann *J. Comput. Chem.*, **12**: 1991; 1232–1236.
101. D.L. Veenstra, D.M. Ferguson and P.A. Kollman *J. Comput. Chem.*, **13**: 1992; 971–978.
102. G. Alagona, C. Ghio and C. Pratesi *J. Comput. Chem.*, **12**: 1991; 934–942.
103. M. Clark, R.D. Cramer and N. Van Opdenbosch *J. Comput. Chem.*, **10**: 1989; 982–1012.
104. J.H. Lii, S. Gallion, C. Bender, H. Wikström, N.L. Allinger, K.M. Flurchick and M.M. Teeter *J. Comput. Chem.*, **10**: 1989; 503–513.
105. K. Gundertofte, J. Palm, I. Pettersson and A. Stamvik *J. Comput. Chem.*, **12**: 1991; 200–208.
106. B.H. Arnold *Intuitive Concepts in Elementary Topology*, Prentice-Hall, Englewood Cliffs NJ 1962; cited by N.J. Turro, *Angew. Chem. Int. Ed. Engl.*, **25**: 1986; 882–901.
107. J.E. Dubois, J.P. Doucet and A. Panaye *Bull. Soc. Chim. Belg.*, **98**: 1989; 31–43.
108. J.E. Dubois, J.A. MacPhee and A. Panaye *Tetrahedron*, **36**: 1980; 919–928.
109. A. Panaye, J.A. MacPhee and J.E. Dubois *Tetrahedron Lett.*, **21**: 1980; 3485–3488.
110. J.P. Doucet, A. Panaye, S.G. Yuan and J.E. Dubois *J. Chim. Phys.*, **82**: 1985; 607–611.
111. J.A. Marshall and H. Faubl *J. Am. Chem. Soc.*, **89**: 1967; 5965–5966.
112. D.J. Martella, M. Jones, P. von R. Schleyer and W.F. Maier *J. Am. Chem. Soc.*, **101**: 1979; 7634–7637.
113. W.F. Maier and P. von R. Schleyer *J. Am. Chem. Soc.*, **103**: 1981; 1891–1900.
114. C. Jaime *J. Comput. Chem.*, **11**: 1990; 411–415.
115. B. van de Graaf and J.M.A. Baas *J. Comput. Chem.*, **5**: 1984; 314–321.
116. E. Osawa *J. Comput. Chem.*, **3**: 1982; 400–406.
117. B. Tiffon and J. Lomas *Org. Magn. Reson.*, **22**: 1984; 29–33.
118. A. Cossé-Barbi personal communication.
119. K. Müller *Ang. Chem. Int. Ed. Engl.* **19**: 1980; 1–13.
120. G.J. Gleicher and P. von R. Schleyer *J. Am. Chem. Soc.* **89**: 1967; 582–593.
121. P. Muller, J. Blan and J. Mareda *Chimica*, **38**: 1984; 389.
122. P. Muller and J. Mareda *J. Comput. Chem.*, **10**: 1989; 863–868.
123. J.E. Eksterowicz and K.N. Houk *Chem. Rev.*, **93**: 1993; 2439–2461.
124. F. Jensen *J. Am. Chem. Soc.*, **114**: 1992; 1596–1603.
125. A. Peyman and H.D. Beckhaus *J. Comput. Chem.*, **13**: 1992; 541–550.
126. K.B. Lipkowitz and M.A. Peterson *Chem. Rev.*, **93**: 1993; 2463–2486.
127. T. Iimori, W.C. Still, A.L. Rheingold and D.L. Staley *J. Am. Chem. Soc.*, **111**: 1989; 3439–3440.
128. J.R. Damewood Jr, W.P. Anderson and J.J. Urban *J. Comput. Chem.*, **9**: 1988; 111–124.
129. P.A. Kollman, G. Wipff and U.C. Singh *J. Am. Chem. Soc.*, **107**: 1985; 2212–2219.
130. G. Wipff, P.K. Weiner and P.A. Kollman *J. Am. Chem. Soc.*, **104**: 1982; 3249–3258.
131. P.V. Maye and C.A. Venanzi *J. Comput. Chem.*, **12**: 1991; 994–1007.
132. R.D. Hancock *Acc. Chem. Res.*, **23**: 1990; 253–257.
133. Y. Yoshikawa *J. Comput. Chem.*, **11**: 1990; 326–335.
134. W.K. Lim, N.W. Alcock and D.H. Busch *J. Am. Chem. Soc.*, **113**: 1991; 7603–7608.
135. K.R. Adam, M. Antolovitch, L.G. Bridgen and L.F. Lindoy *J. Am. Chem. Soc.*, **113**: 1991; 3346–3351.
136. A. Bouraoui, M. Fathallah, B. Blaive, R. Gallo and F. M'Henni *Modelling of Molecular Structures and Properties.*, J.L. Rivail Ed., Studies in Physical and Theoretical Chemistry, Vol. **71**: Elsevier., 1990; p. 381–393.
137. A. Bouraoui, M. Fathallah, B. Blaive, R. Gallo and F. M'Henni *J. Chem. Soc. Perk. 2*: 1990; 1211–1214.
138. C. Roussel, B. Blaive, R. Gallo, J. Metzger and J. Sandstöm *Org. Magn. Reson.* **14**: 1980; 166–170.
139. D.B. Boyd and K.B. Lipkowitz *J. Chem. Ed.*, **59**: 1982; 269–274.
140. T. Clark *Handbook of Computational Chemistry*, J. Wiley, New York, 1985; p 12–92.

6 *Monte Carlo and molecular dynamics simulations*

For the sake of clarity, it is important to open this chapter by marking the difference between what physical chemists use to define molecular modelling and simulation. Indeed, there is a significant distinction to make between these two concepts, and a clear understanding of simulation techniques should begin by differentiating them from general modelling tools. Roughly speaking, a model is nothing else than a simplified representation of a system or of a process so as to better understand it. Taking the example of a weather forecast, data collected from ground-based measurements, upper-air readings and satellites provide the necessary information to build a model of the atmosphere at a given time. In molecular sciences, however, most models are microscopic, that is they have been obtained from computations performed on microparticles, i.e. on a single molecule or on a system comprising a few molecules. To this end, previous chapters have shown that both quantum chemistry and classical mechanics methodologies may be used. But in any case, the results lead to simplified representations of the real microscopic world. Molecular modelling may be defined, therefore, *as the construction and application of such microscopic models to rationalize molecular structure, function and interaction.* Of course, molecular modelling allows us to calculate both measurable and unmeasurable properties, and the former may be compared with experiments. Then, according to van Gunsteren and Berendsen [1], "the comparison validates or invalidates the model that is used. In the former case, the model may be used to study relationships between model parameters and assumptions or to predict unknown or unmeasurable quantities." For instance, the structural features of the crambin protein have recently been predicted using the MM2 model (see Chapter 5) [2].

On the other hand, molecular simulation may be defined as *the determination of the macroscopic properties of a system using the microscopic model which has been constructed to describe the main interactions between the particles of which it is made.* Simulation techniques are based on the laws of statistical mechanics, which give us the theoretical bases to make the connection between microscopic modelling and macroscopic behaviour [1]. Using these techniques, it is thus possible today to study the thermodynamic, structural and transport properties of ensembles of atoms or molecules, namely the bulk properties at finite temperatures of solids, liquids and gases [3]. Generally speaking, two kinds of molecular simulations, also known as computer experiments, may be performed: (i) the purely stochastic Monte Carlo method which randomly samples the configurational space of the system, leading, in a probabilistic mode, to static properties such as the lowest energy structure of a liquid; and (ii) the deterministic molecular dynamics method, producing trajectories in the configurational space and leading to both static and dynamic properties such as the distribution of kinetic energy or self-diffusion coefficients. Coming back to crambin, a molecular dynamics simulation of this protein would allow us to study the flexibility of the backbone or the structure of low-lying local minima and the energy barriers separating them. Finally, to vividly illustrate in more general terms the difference between modelling and simulation, we may turn to the example of a weather forecast outlined above. When starting from the static model of the atmosphere constructed from data collection, the calculation of the evolution of the model as a function of time (i.e. a time-dependent simulation) will allow us to predict temperature, wind velocities, humidity, cloud formation, etc., at a given place. Weather forecasting is therefore the result of a simulation of the behaviour of a model as a function of time.

6.1 MONTE CARLO SIMULATIONS

Monte Carlo methods are well known in applied mathematics as techniques based on the random sampling of large sets of numbers, which automatically accounts for the origin of their name. In particular, the Monte Carlo approach is well suited to multidimensional integrals not amenable to numerical integration and where the boundaries are complicated, the integrand is not strongly peaked in very small regions, and relatively low accuracy is acceptable [4]. The basic principle of Monte Carlo integration consists of evaluating the function at a random sample of points and estimating the integral based on that random sample (Figure 6.1). The problem is that the Monte Carlo method is slow to converge as the error is of the order of $n^{-0.5}$, where n is the number of points in the random sample.

In Monte Carlo simulations of molecular systems, sequences of configurations are generated with a given probability distribution, and thermodynamic properties of the system, such as free energies of solvation, are calculated as averages over these configurations. A good description of the basic features of the Monte Carlo method as applied to molecular simulations has been reported by Haile [5]. It may be summarized as follows. Let us assume that we have a system made of a fixed number of particles N occupying a fixed volume V at a constant temperature T. Suppose we would like to calculate the equilibrium value $\bar{A}$ of a thermodynamic quantity A, for example the potential energy function of the system, where $A = A(\mathbf{r}^N)$, $\mathbf{r}^N$ figuring the set of position vectors of the particles $(\mathbf{r}_1, \mathbf{r}_2, ..., \mathbf{r}_N)$. In this case, $\bar{A}$ represents the

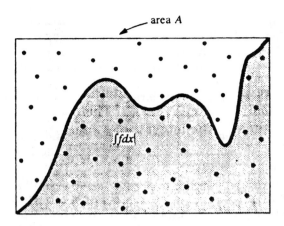

Figure 6.1 Monte Carlo integration. Random points are chosen within the area A. The integral of the function f is estimated as the area of A multiplied by the fraction of random points that fall below the curve f. (from Flannery *et al.* with permission [2]).

configurational part of the internal energy of the system [3]. According to statistical mechanics, $\bar{A}$ is given by the average:

$$\bar{A} = \frac{1}{Z}\int \ldots \int A(\mathbf{r}^N)e^{-\beta U(\mathbf{r}^N)}d\mathbf{r}_1 \ldots d\mathbf{r}_N \tag{6.1}$$

where $\beta = 1/kT$, k being the Boltzmann constant, and Z the configurational integral:

$$Z = \int \ldots \int e^{-\beta U(\mathbf{r}^N)}\,d\mathbf{r}_1 \ldots d\mathbf{r}_N \tag{6.2}$$

and $U(\mathbf{r}^N)$ is the intermolecular potential function, which is usually taken as the sum of isolated pair interactions $u(r_{ij})$ (neglecting three- and higher-body-terms):

$$U(r^N) = \sum_i \sum_{j<i} u(r_{ij}) \tag{6.3}$$

where r_{ij} is the distance between i and j particles.

Due to the 3N-dimensional character (don't forget that $d\mathbf{r}_i = dx_i\,dy_i\,dz_i$) of the integrals (6.1) and (6.2), it is not possible to evaluate them using standard numerical techniques such as the Gaussian quadrature or Simpson's rule. It is therefore necessary to resort to the Monte Carlo method.

A naive implementation of the Monte Carlo method to our problem would consist of randomly moving the N particles within the cell representative of volume V and of building in this way a statistically meaningful configurational sample. However, such a procedure would be inefficient because the Boltzmann factor $e^{-\beta U(r^N)}$ would be very small for a large number of configurations r^N, thus leading to many of them bringing no contribution to the integral. An elegant solution to this problem has been suggested by Metropolis *et al.* [6] in the form of what is known as the *importance sampling scheme.*

In a Metropolis Monte Carlo simulation of our system, the following steps are performed:

1. Assign initial positions r_i to the N particles and calculate the system energy U using (6.3), the $u(r_{ij})$ pair potentials being derived from the various components of a usual force field (see Chapter 5).
2. Construct a new configuration by arbitrarily choosing one particle and moving it by a random displacement from position r to r'. Calculate the new potential energy U' corresponding to this configuration.
3. If $U' < U$, accept the new configuration (i.e. allow the move to occur) and continue the process (i.e. proceed to step 2).
4. If $U' > U$, the new configuration is accepted with a probability proportional to $e^{-\beta \Delta U}$, where $\Delta U = U' - U$, that is select a random number i in the interval $(0, 1)$. Then:
    — if $e^{-\beta \Delta U} > i$, accept the new configuration and continue the process (i.e. proceed to step 2);
    — if $e^{-\beta \Delta U} < i$, reject the new configuration (i.e. do not allow the move to occur), count the old configuration as a new one and proceed to step 2.

For each configuration accepted, the integrand of equation (6.1) is evaluated and its value accumulated in the running sum $\overline{A}$. Several million configurations are generally needed to obtain statistically meaningful averages. If m is the total number of configurations considered, one therefore has as the result of the Monte Carlo evaluation of integral (6.1):

$$\overline{A} = \frac{1}{m} \sum_{i=1}^{m} A_i \tag{6.4}$$

with $A_i = A(r_i^N)$, r_i^N being the set of position vectors after the ith move.

It has been found that some properties converge more rapidly than others in Monte Carlo simulations [7]. For example, heat capacities require in general a much larger ensemble sampling than internal energies. Usually, the step size Δ, defined as:

$$r_i'^{x,y,z} = r_i^{x,y,z} + \Delta \cdot \xi^{x,y,z} \tag{6.5}$$

where $\xi^{x,y,z}$ is a vector constructed from a set of three random numbers taken in the interval $(0, 1)$, is chosen so that approximately 50% of the attempted moves are accepted, but this is undoubtedly a critical parameter: large Δ values may lead to poor acceptance scores, whereas small Δ values severely restrict the sampling of configuration space, which results in slow convergence of the properties. This explains why, for large systems exhibiting many degrees of freedom such as proteins, Monte Carlo methods are generally less efficient that molecular dynamics techniques. Consequently, Monte Carlo methods have been generally applied to systems where they are more effective, such as liquids or systems in solution, and we shall review here some typical studies in this field.

Actually, a prerequisite for performing reliable Monte Carlo simulations lies in the determination of accurate intermolecular pair potential functions $u(r_{ij})$ (equation (6.3)). To this end, a first approach consists in deducing them from a fitting of the parameters of simple potential functions to accurate *ab initio* potential energy surfaces calculated for two-particle or bimolecular interactions. An alternative is to fit the parameters involved by the $u(r_{ij})$ functions to experimental data deduced from X-ray studies, lattice dynamics investigations, infrared and NMR spectroscopic measurements, etc. [1]. In any case, the choice of a strategy in using or determining the best $u(r_{ij})$ functions depends upon the system investigated and on the quality of the force fields available, such as MM2, MM3, AMBER, CHARMM, etc. (for this latter point, the reader is referred to Chapter 5). For teaching purposes, a very simple Monte Carlo simulation of a liquid may be performed assuming it is made of hard spheres, i.e. of hard bodies interacting through a potential energy function of the form [5]:

$$u(r_{ij}) = \begin{cases} \infty & r_{ij} \leq \sigma \\ 0 & r_{ij} > \sigma \end{cases} \tag{6.5}$$

where σ is the diameter of the sphere. However, this potential is too simple to lead to a reliable simulation of real liquids.

One of the systems which have been mostly investigated using Monte Carlo simulations is undoubtedly liquid water. This is due to the importance of water itself, and of aqueous solutions both in chemistry and life sciences. The first Monte Carlo simulation of the properties of water was performed in 1969 by Barker and Watts [8]. These authors used an analytical, orientation-dependent additive pair potential energy function derived by Rowlinson [9], and they considered 64 water molecules in a cube with periodic boundary conditions at the experimental density at 25°C. In any molecular simulation, it is necessary to first equilibrate the system so that it becomes independent of initial conditions. In other words, the initial stage of the simulation, be it of the Monte Carlo or molecular dynamics type, consists of allowing the system to reach the equilibrium after a sufficient number of steps, and to "forget", so to speak, how it was prepared in the initial conditions [5]. In the case of the Monte Carlo simulation of water performed by Barker and Watts, equilibration was reached after sampling of 120 000 configurations. Then the simulation itself was carried out by generating 110 000 additional conformations, leading to satisfactory results as far as both the thermodynamic energy (referred to separate molecules): −8.36 kcal/mol (experimental value: −8.12 kcal/mol) and the specific heat: 20.5 cal/deg·mol (experimental value: 18 cal/deg·mol) are concerned.

An interesting feature of molecular simulations is that at each step the positions of all the particles are known. One may thus deduce a "local structure" of the system, i.e. the organization of the particles around one another. In mathematical terms, one uses distribution functions to describe how the particles are distributed in the volume. The most popular of these expressions in the radial distribution function $g(r)$, which is proportional to the probability of finding two particles, i.e. in liquid water, two H_2O molecules, separated by distance $r \pm \Delta r$ [5]. The function $g(r)$ may be seen as a measure of the extent to which the structure of the liquid deviates from a totally random distribution [10]. As it is possible to deduce $g(r)$ from X-ray or neutron-scattering experiments, the comparison with the corresponding function obtained from the simulation is important as it allows the user to estimate the validity of his calculations. Such a typical radial distribution curve obtained from neutron scattering experiments for liquid argon [11] is shown in Figure 6.2. It can be seen that the probability of finding an argon atom at a distance of about 4 Å from a given central atom is maximum, whereas some additional features are visible at larger distances (7, 10, 13 and 17 Å). All the maxima correspond to the various shells of neighbours surrounding a given atom in the liquid. Five successive shells may therefore be distinguished before $g(r)$ tends to unity, at large distances, which is characteristic of a random distribution. By integrating $g(r)$ along the r variable, the number of neighbours in each successive shell may be obtained.

Turning back to the early simulation of liquid water by Barker and Watts [8], the radial distribution function they report is in reasonable agreement with experiment (Figure 6.3). The number of neighbours calculated within 3.5 Å is about 6.4, which is to be compared with the experimental value 5.1. This is apparently due to the fact that the pair potential function used overestimates the bonding character of the water dimer. It is therefore natural that several

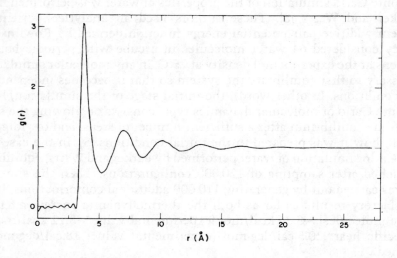

Figure 6.2 Radial distribution function obtained for liquid argon at 85 K from neutron-scattering experiments (from Yarnell *et al.* with permission [11]).

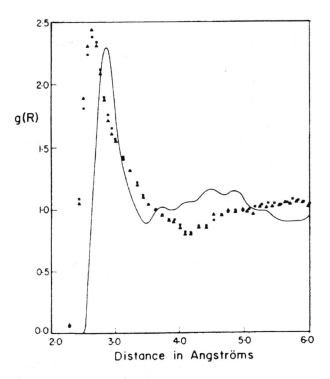

Figure 6.3 Radial distribution function of water at 25°C. Solid line, experiment; ▲: calculated from 54 000 configurations; ■: calculated from 110 000 configurations (from Barker *et al.* with permission [8]).

improved potentials have been reported after this very first investigation. If we concentrate on those which have been used in Monte Carlo simulations, significant progress was made in 1976 by Lie *et al.* [12], who employed a high-quality water-water interaction pair potential obtained from *ab initio* configuration interaction calculations.

The Monte Carlo simulation of Lie *et al.* [12] was performed on a cube of 343 water molecules with periodic boundary conditions at the experimental density at 25°C. After equilibration, the number of configurations used in the sample set was 600 000. In view of the quality of the interaction potential, it is not surprising that the O–O radial distribution function calculated by Lie *et al.* is in very good agreement with experiment (Figure 6.4). A similar result has also been obtained for O–H and H–H radial distribution functions, though some discrepancies between theory and experiment are observed for intensities. From integration of their O–O radial distribution curve, Lie *et al.* deduce an average number of nearest neighbours of 5 around a given H$_2$O molecule, which is in satisfactory agreement with the experimental value of 5.1. Thermodynamic properties, however, exhibit some discrepancies with respect to experimental values, which allows Lie *et al.* to conclude that three-

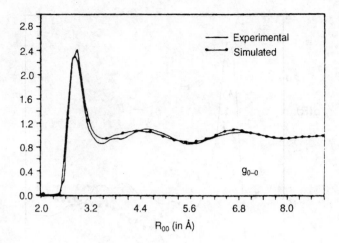

Figure 6.4   Comparison between simulated and experimental O–O radial distribution functions of liquid water (from Lie *et al.* with permission [12]).

body (and possibly four-body) interactions should be included in the interaction potential [12].

Many further Monte Carlo simulations of liquid water have appeared since the pioneering studies mentioned above were reported (for reviews, see elsewhere [13, 14]). Using more and more refined pair potentials, most of them show an even better agreement with experiment than the study of Lie *et al.* [12], while, in addition, allowing us to perform simulations at various temperatures and pressures. On the other hand, Clementi has derived three-body and four-body interactions potentials which provide a sound basis to perform high quality simulations [15]. In particular, a new potential obtained by fitting *ab initio* quantum chemical interaction energies, calculated at a high level of theory, has been shown by Corongiu and Clementi to accurately reproduce many structural and dynamical data of water from molecular dynamics calculations [16]. On the whole, one may assert without any overstatement that both the microscopic structure and the thermodynamical properties of liquid water are accessible today within the experimental error margins through Monte Carlo or molecular dynamics simulations.

After liquid water, the next steps in Monte Carlo applications are obviously to study aqueous solutions of simple solutes such as CH_4 [17] or Ar [18], and of more complex systems such as dimethyl phosphate anion [19]. The calculations of Swaminathan *et al.* [17] on a dilute aqueous solution of methane are interesting as the solute is a prototype of a nonpolar molecule dissolved in liquid water, and such a study is of great value in clarifying the role of water in maintaining the 3D structural integrity of biological molecules in solution. Without going into too many details, let us mention that this Monte Carlo simulation has been performed for a system made of one methane and 124 water molecules at 25°C at a density of $1\,g/cm^3$, using several analytical functions representative of *ab initio* calculations carried out at both SCF and MP2 levels. Integration of the radial distribution function for the centre of mass of water molecules with respect to the centre of mass of methane leads to an average water coordination number of 19.35, which is in satisfactory agreement with previous simulations [20]. This study in addition brings useful information as to the structural features of the system: the local solution environment of methane is likely to be a "distorted defective continuum pentagonal clathrate structure" [17] (Figure 6.5).

Finally, Swaminathan *et al.* show that, when quantifying the structural perturbations in the solvent, an increased four coordination is found, which reveals a stronger binding among H_2O molecules in the methane-water system as compared to bulk liquid [17]. This feature is characteristic of the hydrophobic interactions taking place in aqueous solutions of nonpolar residues.

Another illuminating application of Monte Carlo simulations in chemistry has been performed by Alagona *et al.* [19] for the dimethyl phosphate anion

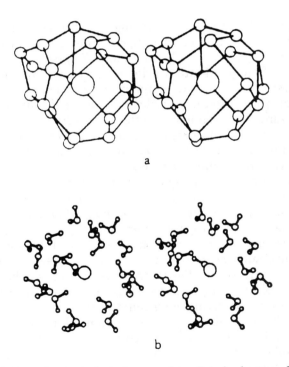

a

b

Figure 6.5 Stereographic view of methane and its first hydration shell taken from a structure with high statistical weight. (a) Disposition of the centres of mass of water molecules about methane with the quasiclathrate cage delineated; (b) disposition of water molecules about methane in the structure. Relative sizes of molecules scaled down for greater legibility; methane represented as a sphere (from Swaminathane *et al.* with permission [17]).

(DMP) $(CH_3O)_2PO_2^-$ in water. This molecule, which is a fragment prototype of biologically important systems, is interesting because it exhibits solvation sites which are very different in character: hydrophobic (CH_3), polar (ester O) and ionic (terminal O). In addition, it has two torsional degrees of freedom ϕ_1 and ϕ_2 (Figure 6.6), which represent one of the main sources of flexibility of nucleic acids. It was therefore of particular interest to investigate the solvation process of DMP and, in particular, the conformational features of solute-solvent interactions. Alagona *et al.* used in their simulation a system made of DMP and 216 water molecules and intermolecular pair potential functions consisting of electrostatic and van der Waals terms [19].

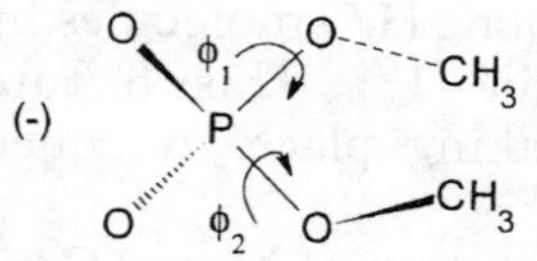

Figure 6.6   The ϕ_1 and ϕ_2 torsional angles of DMP molecule.

In view of the difficulty of performing an adequate sampling of (ϕ_1, ϕ_2) conformational probabilities in DMP, the two most stable forms of DMP in gas phase, namely $g+,g+$ $(\phi_1 = \phi_2 = 75°)$ and g,t $(\phi_1 = 75°, \phi_2 = 180°)$, were chosen to carry out two different Monte Carlo simulations. In addition to many statistical analyses of the results, simulation techniques were used to study the motions of the system particles by interpolating between various structures. A ligand displacement reaction is observed between 500 000 and 750 000 steps in the g,t simulation, in the sense that one water molecule coordinated to a terminal oxygen moves away to make place for an incoming second shell solvent molecule (Figure 6.7).

The coordination number of each of the terminal oxygen atoms is close to three, which is in agreement with previous calculations. As to the phosphate ester oxygen, water typically forms slightly more than one (weaker) hydrogen bond with it. Finally, the hydrophobic methyl groups lead, as expected, to large gaps in the water structure around them. Even though this study suffers from some deficiencies, such as the limited "umbrella" sampling of the (ϕ_1, ϕ_2) potential energy surface of DMP, the results show that Monte Carlo simulations are able to lead to a good description of the various hydrogen bonding interactions occurring between the different groups of DMP and water, provided that a flexible enough pair potential function is used.

After Monte Carlo applications devoted to simulations of water or aqueous solutions of simple solutes, let us turn to a study of the effects of hydration on the course of a chemical reaction. In a series of pioneering investigations, Jorgensen *et al.* have indeed performed Monte Carlo simulations of solvated reacting systems such as the S_N2 reaction between chloride ion and methyl chloride [21, 22], and the nucleophilic addition of hydroxide ion to formaldehyde [23]. We shall limit ourselves here to summarize the results of the first of these two investigations.

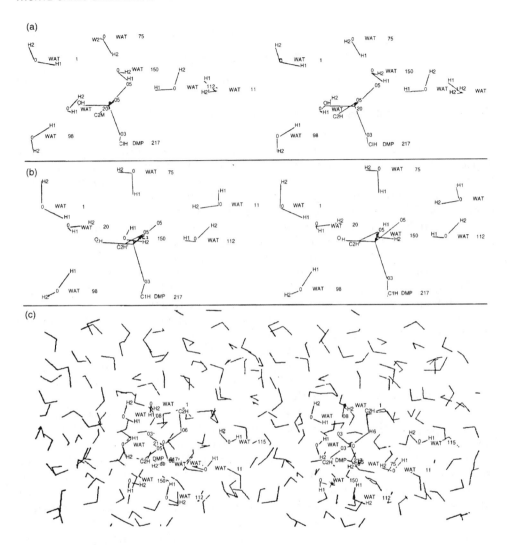

Figure 6.7 (a) Stereoscopic view of the water molecules lying near the anionic oxygens of g,t DMP after 5×10^5 steps (top); (b) same as (a) after 7.5×10^5 steps (centre); (c) same as (b) with a different viewpoint and all the water molecules included (bottom). After 5×10^5 steps, water 150 is coordinated to the anionic O_2 (a) and 2.5×10^5 steps further (b), water 11 has taken its place near that oxygen (from Alagona *et al.* with permission [19]).

Bimolecular S_N2 reactions such as $Cl^- + CH_3Cl' \rightarrow CH_3Cl + Cl'^-$ are prototypes of mechanisms which are strongly influenced by the medium, to the extent that solvation may affect even the qualitative nature of the reaction [22]. For example, the S_N2 reaction depicted in Figure 6.8 is known to exhibit a double well potential energy surface in the gas phase, whereas the corresponding

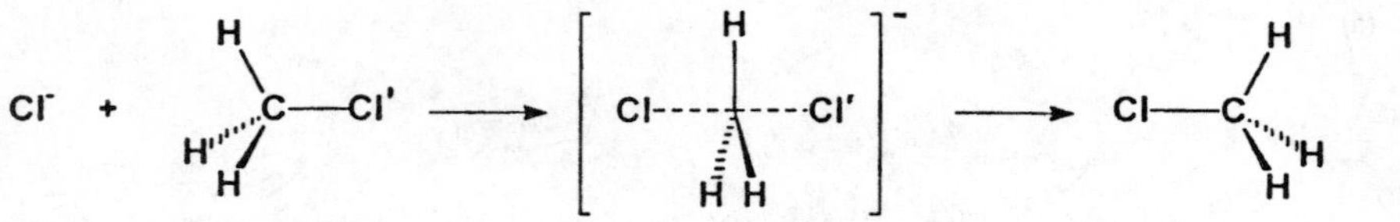

Figure 6.8   Schematic representation of the $Cl^- + CH_3Cl' \rightarrow CH_3Cl + Cl'^-$ S_N2 reaction.

profile is unimodal in solution [21]. In other words, two symmetric minima are expected to be found in the gas phase on the energy profile expressed as a function of the reaction coordinate $r_C = r_{CCl'} \cdot r_{CCl}$, corresponding to the electrostatic favourable interactions characterizing these ion-dipole complexes. These minima are separated by a symmetrical transition state ($r_C = 0$) which is at the origin of the activation energy or intrinsic barrier required for the S_N2 reaction to occur (see below). On the other hand, when the reaction proceeds in aqueous solution, desolvation of the ion somehow compensates the ion-dipole attraction and the minima are expected to disappear, thus leading to a unimodal energy profile. Simultaneously, the activation energy should significantly increase due to a weaker solvation of the transition state with respect to intermediate structures, which may be attributed to a larger delocalization of charge in the transition state.

Chandrasekhar *et al.* have therefore performed a three-step investigation of this prototype S_N2 mechanism: (i) determination of the gas-phase reaction surface using *ab initio* calculations; (ii) development of potential functions to describe solute-solvent interactions; and (iii) Monte Carlo simulations of the reaction in aqueous solution [22]. Whereas the *ab initio* calculations of the gas-phase energy profile have been performed at the SCF level using the 6–31G* basis set, step (ii) has been carried out with pairwise additive potential functions made of electrostatic and 6–12 van der Waals terms [22]. To reduce statistical errors, the Monte Carlo simulation of the S_N2 reaction profile in solution has been performed by determining the solvent-averaged potential mean force, which represents the relative free energy of the system as a function of r_C. To this end, a system made of the solute cluster and 250 water molecules at 25°C and 1 atom was constructed. The results, presented in Figure 6.9, are totally consistent with experiment.

Indeed, for the gas-phase energy profile, a complexation energy of −10.3 kcal/mol and an activation barrier of 13.9 kcal/mol have been calculated, whereas the corresponding experimental values are −8.6 [24] and 11.6 ± 1.8 kcal/mol [25] respectively. As expected, the calculated potential of mean force representative of the reaction in solution is markedly different from the gas-phase profile. Part of the difference is, of course, due to the fact that the former is a free energy curve while the latter describes the change in internal energy for a vibrationless system at 0 K, but the trend is essentially correct: the Monte Carlo simulation strongly suggests that S_N2 energy profiles are unimodal in aqueous solution with a large increase in activation energy due to solvation. Actually, the theoretical free energy of activation (26.3 ± 0.5 kcal/mol) is in quantitative agreement with the experimental value of

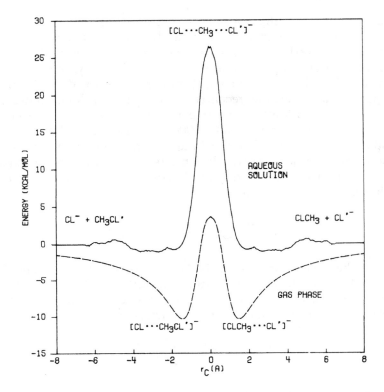

Figure 6.9 Calculated internal energies in the gas phase (dashed curve) and in solution (solid curve) for the system [Cl'CH₃Cl]⁻ as a function of the reaction coordinate r_C. (from Chandrasekhar *et al.* with permission [227]).

26.6 kcal/mol [26]. The enhanced barrier in solution may undoubtedly be attributed to a smaller solute-solvent attraction for the charge delocalized transition state than for the separate reactants. As for the other Monte Carlo simulations reviewed here, the aqueous solution calculations of the S_N2 mechanism similarly lead to a coherent description of the structural features of the various shells of water molecules around the solute cluster along the minimum energy reaction path [22]. This investigation therefore represents a good example of the wealth of information which may be deduced from Monte Carlo simulations of complex chemical processes in solution.

6.2 MOLECULAR DYNAMICS SIMULATIONS

Molecular dynamics (MD) simulations date back to 1957, with the pioneering study of a simple fluid made of two-dimensional hard disks performed by Alder and Wainwright [27, 28]. Originally designed to investigate relaxation phenomena and transport properties in liquids, MD simulations rapidly emerged as a powerful tool to calculate structural and thermodynamic properties of complex liquids, molten salts, crystals, polymers and proteins in solution [1, 3, 5, 7, 10, 15]. Actually, this development is due to a large extent to the spectacular progresses recently witnessed in both computer hardware and software, which allow us today to perform realistic simulations of systems as large as an enzyme surrounded by 3000 water molecules [29]!

As opposed to the Monte Carlo technique, a molecular dynamics simulation is a deterministic procedure which consists of sampling the configurational space by simultaneous integration of Newton's classical equations of motion for all the atoms i of the system:

$$m_i \frac{d^2 \mathbf{r}_i(t)}{dt^2} = \mathbf{F}_i(t) \quad i = 1, \ldots, N \tag{6.6}$$

where m_i is the mass of atom i, $\mathbf{r}_i(t)$ is the position of i at time t and $\mathbf{F}_i(t)$ is the force exerted on i by the other $N-1$ atoms at time t.

For each atom i, the force $\mathbf{F}_i$ is calculated at each time t as the negative gradient of the intermolecular potential function (6.3):

$$\mathbf{F}_i = -\frac{dU(\mathbf{r}_1, \mathbf{r}_2, \ldots, \mathbf{r}_N)}{d\mathbf{r}_i} \tag{6.7}$$

which implies that U is a differentiable function of the atomic coordinates $\mathbf{r}_i$ [1].

We see that, as in Monte Carlo simulations, the knowledge of the interaction potential U is an essential element in a MD study. However, instead of a random displacement of the atoms and of a Boltzmann sampling of the configurational space, the atoms move, in MD calculations, according to the laws of Newtonian mechanics, and the configurational space is sampled as a function of the time evolution of the system. Indeed, as equations (6.6) are integrated numerically by using small time steps Δt of the order of 1 fs (1 fs = 1 femtosecond = 10^{-15} s), the results of the calculation are the trajectories of the N atoms, i.e. a set of N $\mathbf{r}_i$ values ($i = 1, \ldots, N$) obtained for a set of n t_k values such as $t_k = t_{k-1} + \Delta t$ ($k = 1, \ldots, n$). As compared with Monte Carlo simulations, where properties were calculated as ensemble averages of functions depending on the particle coordinates only (see equation (6.4)), the expectation value of property A is calculated in MD as the time average:

$$\bar{A} = \frac{1}{\tau} \int_0^\tau A(t)dt \cong \frac{1}{\tau} \sum_{i=1}^n A(t_i)\Delta t \tag{6.8}$$

where τ is the total simulation time $\tau = n \cdot \Delta t$.

In principle, when the system is at equilibrium, the property value $\bar{A}$ (equation (6.8)) does not depend upon the initial time t_0. In addition, the knowledge of $r_i(t)$ at any time step allows one to calculate $v_i(t) = dr_i(t)/dt$, i.e. the momentum $m_i v_i(t)$ of each atom. This explains why, in contrast with Monte Carlo calculations, MD techniques offer the possibility of performing averages of dynamic properties, in addition to static ones, such as thermal transport coefficients. Finally, note that if we assume the ergodic hypothesis to be valid, the time average value of property A calculated from MD (equation (6.8)) is the same as the ensemble average derived from Monte Carlo (equation (6.4)), provided adequate samplings of the configurational space have been achieved.

Numerous algorithms have been proposed to integrate the MD equations of motion (6.6) [5, 7]. The most popular one is the finite-difference method suggested by Verlet [30], which computes atomic position vectors r_i at time $t + \Delta t$ from the forces and positions at previous times:

$$r_i(t + \Delta t) = 2r_i(t) - r_i(t - \Delta t) + \frac{F_i(t)}{m_i}(\Delta t)^2 + O\left[(\Delta t)^4\right] \qquad (6.9)$$

where the final term is the truncation error varying as $(\Delta t)^4$, which means that the calculation of successive atomic positions according to the Verlet algorithm is exact up to third order. Note that velocities are not present in the Verlet formula; they are generally estimated using the half-step equation [7]:

$$v_i(t) = \frac{r_i(t + \Delta t) - r_i(t - \Delta t)}{2\Delta t} + O\left[(\Delta t)^3\right] \qquad (6.10)$$

The Verlet algorithm as expressed by equation (6.9) has proved to be stable provided a sufficiently small time step, of the order of 1–10 fs, is used [10]. The method is relatively fast as it requires only one evaluation of the force by step, which is by far the most time consuming task. On the other hand, the calculation of velocities (equation (6.10)) involves the subtraction of two numbers of comparable magnitude and one has to be careful so as to avoid rounding errors. Examination of equation (6.9) shows that the Verlet algorithm is a two-step method as it estimates $r_i(t + \Delta t)$ from the current position $r_i(t)$ and the previous one $r_i(t - \Delta t)$. Hence, it is not self-starting: initial positions $r_i(0)$ and velocities $v_i(0)$ are not sufficient to begin a calculation, and one needs to know $r_i(-\Delta t)$ [5]. In general, the choice of the initial configuration is made by assigning the initial values $r_i(0)$ to some lattice structure or to values taken from a previous simulation. The initial velocities are chosen from a Maxwell–Boltzmann distribution at the appropriate temperature. Altogether, the Verlet algorithm presents the advantages being simple and reasonably fast. Its drawback is that in its original form it considers velocities as less important than positions, which is in conflict with the fact that phase-space trajectories are equally dependent upon positions and velocities. Among the other numerical integration methods of the MD equations (6.6), let us mention the Beeman [31] and leapfrog [32] algorithms, which have been derived from the Verlet scheme, and predictor-corrector techniques such as those devised by Rahman [33] and Gear [34].

Starting from the velocities as given from equation (6.10), one may calculate the temperature T of the system at any time t by using the relation:

$$T(t) = \frac{1}{(3N - c)k} \sum_{i=1}^{N} m_i v_i(t)^2$$

(6.11)

where $3N - c$ is the number of degrees of freedom, c being the number of constraints, and k is the Boltzmann constant. When sampling the configuration space, the advantage of MD over Monte Carlo techniques lies in the fact that the kinetic energy of the system (equation (6.11)) allows it to surmount energy barriers of the order of kT per degree of freedom. It is therefore possible to artificially raise the temperature to search larger portions of the configuration space or, alternatively, to cool down the system to reach minima on the Born–Oppenheimer energy surface, that is to remove kinetic energy from the system, which is known as *simulated annealing* [35]. Indeed, in MD simulations, monitoring the two components of the total kinetic and potential energy allows the user to explore in detail the most interesting portions of the configuration space [1].

As an example, Plate IV summarizes the results of an MD calculation performed for a single molecule of *n*-butane (CH_3–CH_2–CH_2–CH_3) so as to carry out a conformational search of the global energy minimum using simulated annealing. The *n*-butane molecule was chosen as it is a standard hydrocarbon exhibiting several conformational isomers with different C–C–C–C torsion angles (ω, Figure 6.10). The anti conformer ($\omega = 180°$) is known to be the most stable, with an energy roughly 1 kcal/mol lower than the two gauche forms ($\omega = 60°$ and $240°$), whereas the barrier height for the molecule to rotate from the gauche to the anti form is 3 kcal/mol. The purpose of the MD study of *n*-butane was to determine the capability of the simulated annealing technique to rearrange the molecular conformation from the gauche to the anti form. Examination of Plate IV shows that this is indeed the case: the simulation starts at 0 K with butane in a gauche conformation ($\omega = 60°$) and the temperature is raised rapidly up to 300 K in 0.1 ps (1 ps = 1 picosecond = 10^{-12} s), the time step being 0.0005 ps. Then, after a short equilibration period of 0.4 ps, the system is cooled down to 0 K in 3 ps. It is seen that, shortly after the beginning of the simulation, the torsion angle ω changes abruptly to a value fluctuating around 180°, i.e. the butane molecule has overcome the energy barrier separating the gauche and anti forms to end up in the latter conformation. This example illustrates both the efficiency of the simulated annealing technique and the importance of graphics to visualize conformational changes.

To summarize, a MD simulation therefore allows one to generate a trajectory of the system's constituents until the desired time interval has been spanned, relevant trajectories of course being obtained after the necessary equilibration phase has been completed. In general, the simulation may be performed over 10^5–10^6 steps of 1 fs, that is on intervals of 10^2–10^3 ps, depending upon the system investigated. Whereas many interesting motions of chemical systems will be achieved within this time scale, slow

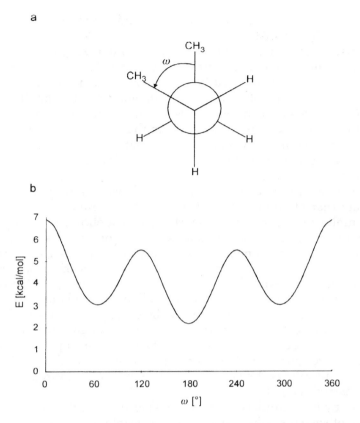

Figure 6.10 (a) Newman projection of the molecule of *n*-butane CH_3–CH_2–CH_2–CH_3 with definition of the torsion angle ω; (b) potential energy curve calculated for *n*-butane as a function of ω using the MM+ force field of the HyperChem package [36].

processes such as the folding of a protein are still beyond the reach of such simulations.

It should be noted, however, that simulation times are strongly dependent upon both the system investigated and the potential function used. For example, the large number of degrees of freedom in biomolecules makes the evaluation of potential energy and of its derivatives a computationally intensive task, which means that the simulation of such systems is generally limited to trajectories of 10–100 ps. This might well not be sufficient to reliably calculate average properties of the system, especially for those exhibiting longer relaxation times. Possibilities to lengthen the time scale of MD simulations are therefore the subject of intense investigation. Among them, let us mention the freezing of degrees of freedom which do not play an important role in the process simulated, the so-called *stochastification of degrees of freedom* using, for example, the Langevin technique, the activation barrier crossing using the method of umbrella sampling, etc. [1]. A detailed presentation of these techniques is beyond the scope of this book, and the reader should refer to specialized articles and textbooks for an in-depth presentation [1, 5, 7, 37, 38].

In contrast with inhomogeneous systems such as proteins, for simple homogeneous ones, such as a box of water molecules, simulations of only a few picoseconds are generally sufficient to calculate accurate structural and dynamic properties [37, 39]. However, in addition to the length of the simulation, the size of the system investigated (i.e. the number of constituent particles) may also play an important role in the convergence rate of computed properties. For example, Lybrand quotes that "simulations performed on systems with too few particles, e.g. too few solvent molecules to properly solvate a solute, may give misleading results" [7].

A further problem arises in simulations of systems of (necessarily) finite size, namely that of boundary conditions. When simulating a molecule or a system of molecules in the gas phase, the simplest choice is obviously that of vacuum boundary conditions. For solids, liquids or systems in solution, one generally uses periodic boundary conditions to minimize edge or wall effects. In this case, a unit cell is constructed which contains the proper number of particles taken in the simulation, surrounded by identical image cells in all directions, i.e. six. Then, the particles in all image cells are constrained to experience the same forces and to follow the same trajectories as the particles of the central unit cell. When a particle moves through a wall of the central cell, it enters the cell with identical velocity at the opposite side at the translated image position. This process, which is called mirror image convention, thus allows the simulation of a continuous system using a finite number of particles [1, 7].

A great many MD simulations have been performed in the last 20 years on a broad range of systems, which makes it impossible to systematically describe here even a small fraction of them. Rather, we shall summarize in the following pages the main features of some of the most important and spectacular such investigations.

As previously reported in the case of Monte Carlo calculations, liquid water has also extensively studied by MD simulations [10, 14, 15, 38, 39]. One of the first reference studies is undoubtedly that of Stillinger and Rahman, who used a two-body Lennard–Jones potential augmented by point charge electrostatic contributions representing the hydrogen bonding interactions [39]. In their calculations, Stillinger and Rahman simulate liquid water by using a cubic box of 216 H_2O molecules subject to periodic boundary conditions and a time step of 2×10^{-16} s, such a short increment being due to the strength and directional character of the hydrogen bonds. Whereas the maxima of the O–O radial distribution functions are still predicted to be too strong by these calculations, the overall agreement with experiment is very satisfactory as far as thermodynamic properties are concerned. These results have been slightly improved by more recent calculations, such as those of Neumann [40, 41], who concentrated on the dielectric constant of water and its frequency dependence, and those of Clementi *et al.* [15, 16, 42, 43], who investigated the dispersion of sound velocity and the behaviour of water in a 15 kbar range of pressure and a 1000 K range of temperature. In particular, the latter authors have shown the importance of introducing three-body and four-body contributions to their interaction potentials [44].

It is indeed seen in Figure 6.11 that the introduction of three-body forces leads to a significant improvement in the structural properties of liquid water. However, Fois *et al.* have recently pointed out that effective two-body potentials deduced from fittings to experimental data lead to better results that those derived from *ab initio* potential energy surfaces, even after inclusion of many-body terms, as far as the calculation of structure factors of liquid water by Monte Carlo simulations is concerned [14].

As for Monte Carlo simulations, the next step after liquid water in MD calculations is to investigate ions in aqueous solutions, and the subject has been reviewed recently by Marcus [45]. In addition to structural problems,

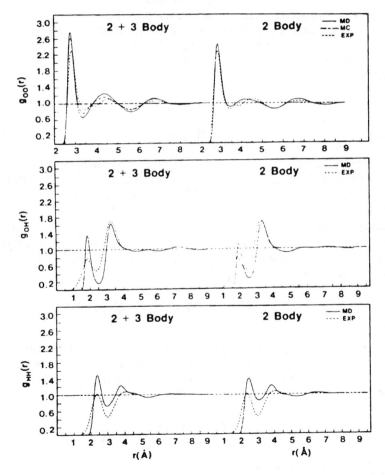

Figure 6.11 O–O, O–H and H–H radial distribution functions obtained from MD simulations using two-body and three-body potentials together with experimental results (from Wojcik *et al.* with permission [44]).

such as the number and position of water molecules in the first (and subsequent) solvation shells, MD methods may also lead to the calculation of enthalpies of formation of, e.g. $X(H_2O)_n$, $X = Li^+$, Na^+, K^+, Rb^+, Cs^+, F^-, Cl^-, Br^-, I^-, $n = 1$–6 [46], which, in general, compare very well with experiment. Note, however, that this result has been obtained using a so-called *polarizable electropole model* for water, i.e. by introducing explicitly an induction term into the water-water and ion-water potentials. As to the structure of the systems, Lin and Jordan have shown that, with the exception of Li^+, the larger the ion the fewer water molecules it coordinates in its first solvation shell, which correlates with the tendency of the systems to form unsymmetrical microclusters at 0 K [46].

Another important feature of MD simulations lies in their ability to calculate free energies, or more conveniently free energy differences between related systems. Indeed, free energies of molecular systems represent key thermodynamic properties which may be used to describe their tendencies to associate and to react by taking into account both enthalpy and entropy contributions [47]. The Gibbs free energy G, for example, is a state function whose variation upon a system change at constant temperature and pression is expressed as:

$$\Delta G = \Delta H - T\Delta S \tag{6.12}$$

where ΔH and ΔS are corresponding enthalpy and entropy variations respectively, T being the temperature.

Using statistical mechanics, G and consequently ΔG, may be calculated from the partition function, which involves a sum over all the Boltzmann-weighted energy levels of the system. This is quite impractical to implement in Monte Carlo or MD simulations, because of the difficulty in sampling adequately the configuration space [48]. Indeed, the sampling of configuration space should be enormous so as to incorporate the low-energy configurations which contribute the most in the direct calculation of the Gibbs free energy according to the formula:

$$G = -kT\ln\left\langle e^{-\beta U(\mathbf{r}^N)} \right\rangle \tag{6.13}$$

where $U(\mathbf{r}^N)$ is defined by equation (6.3) and $\langle \ \rangle$ refers to an ensemble average over the configuration space of all the possible states of the system.

However, free energy differences between related systems A and B characterized by potential functions $U_A(\mathbf{r}^N)$ and $U_B(\mathbf{r}^N)$ can be calculated as:

$$\Delta G = G_B - G_A = -kT\ln\left\langle e^{-\beta \Delta U(\mathbf{r}^N)} \right\rangle_A \tag{6.14}$$

where $\Delta U(\mathbf{r}^N) = U_B(\mathbf{r}^N) - U_A(\mathbf{r}^N)$.

The derivation of equation (6.14) is due to Zwanzig [49], and its meaning is the following: an ensemble of configurations is generated for state A and for each of them the expression $e^{-\beta \Delta U(\mathbf{r}^N)}$ is evaluated; then the result must be averaged over the whole ensemble of configurations sampled for A. ΔG may be conveniently evaluated using equation (6.14), which is the basic formula of *free energy perturbation calculations*, as it is much easier to sample adequately a configuration space involving the average of the $e^{-\beta \Delta U}$ expression rather than $e^{-\beta U}$ as implied by equation (6.13).

As an application of free energy perturbation calculations, let us turn to an investigation of the hydration of superoxide (O_2^-) by using a 216 molecules water box [50]. The free energy, enthalpy and entropy of hydration of O_2^- may be evaluated using the thermodynamic cycle:

$$
\begin{array}{ccc}
 & 4 & \\
H_2O(g) & \rightarrow & O_2^-(g) \\
3 \downarrow & & \downarrow \ 1 \\
H_2O(w) & \rightarrow & O_2^-(w) \\
 & 2 &
\end{array}
$$

where g stands for gas phase and w for aqueous solution. According to the thermodynamic cycle procedure, one has [51]:

$$\Delta G_1 - \Delta G_3 = \Delta G_2 - \Delta G_4 \qquad (6.15)$$

where ΔG_1, the Gibbs free energy of hydration of O_2^-, may be deduced from ΔG_3, the free energy of hydration of H_2O, and ΔG_2 and ΔG_4 which correspond to unphysical "transmutation" processes resulting from transformation of H_2O in O_2^- in aqueous solution and water, respectively. Whereas ΔG_3 can be obtained from experiment, the difference $\Delta G_2 - \Delta G_4$ can be evaluated from the thermodynamic perturbation or integration procedure frequently used in MD simulations [52].

The theoretical basis of equation (6.15) may be understood by the fact that, the free energy being a thermodynamic state function, any closed path which changes the state of a system back to itself leads to a zero net change in free energy. This means that the free energy change ΔG_1 associated with the hydration process of O_2^- may be evaluated via path 1 itself, or by using any other path that has the same initial and final states, namely 4–3–2. Though some care has to be taken to obtain reliable estimates of free energy change using the thermodynamic cycle [1, 29, 37, 53], this procedure offers, in the case of the hydration of O_2^-, an elegant alternative to the direct simulation of process 1, which is virtually impossible as it would involve the reversible removal of many water molecules from their hydration shells around the substrate. It was suggested first by Tembe and McCammon in 1984 [54], and then successfully applied to many important systems where the calculation of the free energy of ligand binding or the free energy of solvation was essential [53].

Turning back to the free energies of solvation of O_2^- as calculated by Shen *et al.* from simulations using the thermodynamic cycle procedure, they range from −87.6 to −90.8 kcal/mol, depending upon the computation parameters used, which compares very well with the experimental values lying in the −78 to −86 kcal/mol interval [50]. Finally, examination of the radial distribution functions shows that the average number of water molecules in the first hydration shell of O_2^- is 7.7.

In addition to the simulation of liquids and small systems in solution, MD calculations offer a sound basis for the description of structural and vibrational properties of solids such as zeolites [55]. These materials constitute a well-defined class of crystalline naturally occurring aluminosilicates, with three-dimensional structures arising from a framework of SiO_4^{4-} and AlO_4^{5-} coordination tetrahedra linked by their corners [56]. Their unique feature lies in their open structure, which exhibits channels and cages which are responsible for their shape-selectivity (at the molecular level) properties [57]. For example, the theta-1 zeolite is known to be a uni-dimensional medium pore (with a radius of 5.5 Å) aluminosilicate with 10 T-rings (T = Si or Al) [57] (Figure 6.12).

The channels and cavities present in zeolites can accommodate cations, water molecules or any reactant that fits within the available space. For example, the activity of the hydrogen forms of zeolites in acid-catalyzed

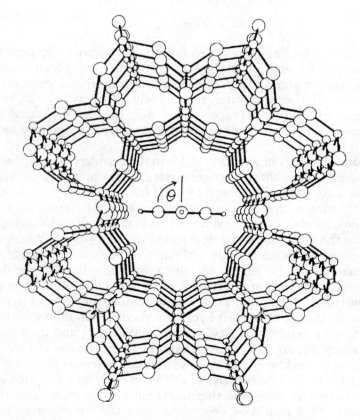

Figure 6.12   Channel axis projection of the theta-1 zeolite with a benzene molecule in the centre of the cage (from Catlow *et al.* with permission [59]).

reactions originates from the presence of protons balancing the additional negative charge of the AlO_4^{5-} tetrahedra [58]. It has recently been found that MD simulations, in addition to being able to adequately describe the framework motions of such materials [55], are ideally suited to investigate proton transfer and diffusion mechanisms, which are at the origin of their shape-selective catalytic properties [59]. In the latter work, the authors have investigated the behaviour of sorbed CH_4 and C_2H_4 in the ZSM-5 zeolite by incorporating both framework and sorbed species motions. The potential used has a quite elaborate form, and it has been obtained from a fitting to experimental data. The simulations have been performed using a box containing 576 framework atoms in addition to the sorbed species. Examination of the trajectories generated over time intervals of 30–120 ps shows that sorbed species may be trapped in a particular site for rather long periods of time (typically 5 ps), which was expected in view of the regular array of cavities and channels present in the ZSM-5 zeolite (Figure 6.13).

In addition to the trajectories of sorbed species, these calculations have led to diffusion coefficients for methane and ethylene in the ZSM-5 structure, which are in good agreement with experimental values [59]. Similar MD investigations have been reported on the self-diffusion of water [60] and benzene [61] in various types of zeolites.

Another very useful application of MD simulations has emerged in the last

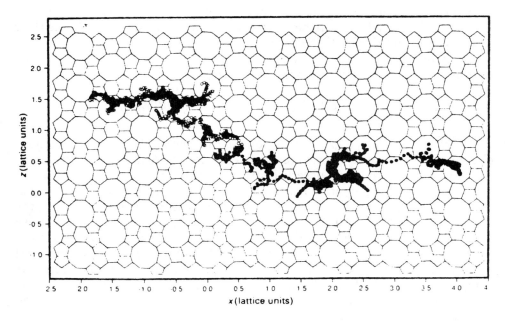

Figure 6.13 Simulated motions of the carbon atoms of two methane molecules in the ZSM-5 zeolite at 600 K. The two carbon atoms are depicted ○ and ●. The positions are plotted every 100 iterations; 1 lattice unit = 20.07 Å (from Catlow *et al.* with permission [59]).

few years, namely the determination of 3D molecular geometries on the basis of X-ray or neutron diffraction or 2D-NMR data [1,62–64]. The principle of this procedure is to determine a 3D structure for the system investigated, generally a protein or a nucleic acid, which (i) satisfies experimental data such as distance geometry or torsion angle constraints in the case of NMR, or a set of observed structure factor amplitudes in the case of diffraction results, and (ii) minimizes the potential energy function U as given by equation (6.3). In other words, one uses an effective potential $U_{tot}(\mathbf{r}^N)$ to run the MD global minimization search, such that:

$$U_{tot}(\mathbf{r}^N) = U(\mathbf{r}^N) + U_{cstr}(\mathbf{r}^N)$$

(6.16)

where $U(\mathbf{r}^N)$ is the potential energy function of the system (equation (6.3)) and $U_{cstr}(\mathbf{r}^N)$ is an adequate penalty or constraint potential function, the value of which increases the more an actual structure violates the experimental data. For example, the nuclear Overhauser effect (NOE) distance information deduced from 2D-NMR experiments may be incorporated into the MD simulation as stated by equation (6.16) by introducing a $U_{cstr}(\mathbf{r}^N)$ term of the form:

$$U_{cstr}(\mathbf{r}^N) = K_{NOE} \sum_i (d_i - d_{iNOE})^2$$

(6.17)

where the summation runs over all NOE constraints of the system, K_{NOE} being a pseudo force constant chosen to make NOE forces equivalent to those of $U(\mathbf{r}^N)$, d_i is the actual distance between the ith pair of protons, and d_{iNOE} is their distance as deduced from the NOE experiment.

Similarly, the penalty function introduced to perform a MD simulation leading to refine diffraction data is of the form:

$$U_{cstr}(\mathbf{r}^N) = K_{diff} \sum_{hkl} [F_{calc}(hkl) - F_{obs}(hkl)]^2$$

(6.18)

where hkl are the reciprocal lattice points of the crystal, K_{diff} is again a constant chosen so as to balance the two terms in equation (6.16), and $F_{calc}(hkl)$ and $F_{obs}(hkl)$ are the calculated and observed structure factor amplitudes, respectively.

During the process of refinement, MD trajectories are generally calculated at elevated temperatures to accelerate conformational sampling. Periodically, the system is cooled down using the simulated annealing procedure [35] to remove kinetic energy and to permit the trajectories to settle into local minimum energy conformations [7]. The technique of MD simulation with experimental constraints is very powerful, and it has proved to be an indispensable complement of both 2D-NMR and X-ray diffraction experiments performed on macromolecules. As such, it has been applied both to determine solution structures of many proteins, such as the complete folding of crambin, and to refine X-ray structures of macromolecules, such as the enzyme aspartate aminotransferase [64].

Turning finally to conventional MD simulations of proteins, which is nowadays a very popular tool in biomolecular sciences, we note that the tremendous progresses in computer hardware and software have made it possible to investigate the structure of globular proteins *in vacuo*, in solution and in hydrated crystals [7, 65, 66]. The latter calculations are particularly worth mentioning, as it is now a well established fact that solvent environment has a profound influence on protein structure and dynamics. Using the procedure of the thermodynamic cycle, the effect of side chain mutation can be simulated with a reasonable degree of reliability. Finally, promising modelings of the active sites of proteins may be achieved, which allows biochemists to investigate the process of ligand binding. It is probably not an overstatement to predict that MD simulations of macromolecules will become on indispensable partner of experiment in a broad range of applications such as drug design, protein engineering and molecular recognition.

REFERENCES

1. W.F. Van Gunsteren and H.J.C. Berendsen *Angew. Chem. Int. Ed. Engl.*, **29:** 1990; 992.
2. J.H. Lii, S. Gallion, C. Bender, H. Wikström, N.L. Ailinger, K.M. Flurchick and M.M. Teeter *J. Comput. Chem.*, **10:** 1989; 503.
3. S. Wilson *Chemistry by Computer. An Overview of the Applications of Computers to Chemistry*, Plenum, New York, 1986.
4. W.H. Press, B.P. Flannery, S.A. Teukolsky and W.T. Vetterling *Numerical Recipes. The Art of Scientific Computing*, Cambridge University Press, Cambridge, 1986.
5. J.M. Haile *Molecular Dynamics Simulation. Elementary Methods*, Wiley, New York, 1992.
6. N. Metropolis, A.W. Rosenbluth, M.N. Rosenbluth, A.H. Teller and E. Teller *J. Chem. Phys.*, **21:** 1953; 1087.
7. T.P. Lybrand in *Reviews in Computational Chemistry, vol. 1*, K.B. Lipkowitz and D.B. Boyd, (eds), VCH, New York, 1990; p. 295.
8. J.A. Barker and R.O. Watts *Chem. Phys. Lett.*, **3:** 1969; 144.
9. J.S. Rowlinson *Trans. Faraday Soc.*, **47:** 1951; 120.
10. D.M. Hirst *A Computational Approach to Chemistry*, Blackwell, Oxford, 1990.
11. J.L. Yarnell, M.J. Katz, R.G. Wenzel and S.H. Koenig *Phys. Rev. A*, **7:** 1973; 2130.
12. G.C. Lie, E. Clementi and M. Yoshimine *J. Chem. Phys.*, **64:** 1976; 2314.
13. W.L. Jorgensen and J.D. Madura *Mol. Phys.*, **56:** 1985; 1381.
14. E.S. Fois, A. Gamba, G. Morosi, A. Ponti, P. Demontis and G.B. Suffritti *Gazz. Chim. Ital.*, **120:** 1990; 591.
15. E. Clementi *J. Phys. Chem.*, **89:** 1985; 4426.
16. G. Corongiu and E. Clementi *J. Chem. Phys.*, **98:** 1993; 2241.
17. S. Swaminathan, S.W. Harrison and D.L. Beveridge *J. Am. Chem. Soc.*, **100:** 1978; 5705.
18. G. Alagona and A. Tani *J. Chem. Phys.*, **72:** 1980; 580.
19. G. Alagona, C. Ghio and P.A. Kollman *J. Am. Chem. Soc.*, **107:** 1985; 2229.
20. J.C. Owicki and H.A. Scheraga *J. Am. Chem. Soc.*, **99:** 1977; 7413.
21. J. Chandrasekhar, S.F. Smith and W.L. Jorgensen *J. Am. Chem. Soc.*, **106:** 1984; 3049.
22. J. Chandrasekhar, S.F. Smith and W.L. Jorgensen *J. Am. Chem. Soc.*, **107:** 1985; 154.
23. J.D. Madura and W.L. Jorgensen *J. Am. Chem. Soc.*, **108:** 1986; 2517.
24. R.C. Dougherty and J.D. Roberts *Org. Mass. Spectry*, **8:** 1974; 77.

25. S.S. Shaik and A. Pross *J. Am. Chem. Soc.*, **104:** 1982; 2708.
26. S.S. Shaik *J. Am. Chem. Soc.*, **106:** 1984; 1227.
27. B.J. Alder and T.E. Wainwright *J. Chem. Phys.*, **27:** 1957; 1208.
28. B.J. Alder and T.E. Wainwright *J. Chem. Phys.*, **31:** 1959; 459.
29. J.M. Goodfellow *Chem. Britain*, **26:** 1990; 1066.
30. L. Verlet *Phys. Rev.*, **159:** 1967; 98.
31. D. Beeman *J. Comput. Phys.*, **20:** 1976; 130.
32. R.W. Hockney and J.W. Eastwood *Computer Simulation Using Particles*, McGraw Hill, New York, 1981.
33. A. Rahman *Phys. Rev.*, **136:** 1964; 405.
34. C.W. Gear *Numerical Initial Value Problems in Ordinary Differential Equations*, Prentice-Hall, Englewood Cliffs, NJ, 1971.
35. S. Kirkpatrick, C.D. Gelatti Jr and M.P. Vecchi *Science*, **220:** 1983; 671.
36. HyperChem 3.0, Autodesk Inc., 2320 Marinship Way, Sausalito, CA 94965, USA.
37. M. Karplus and G.A. Petsko *Nature*, **347:** 1990; 631.
38. M.P. Allen and D.J. Tildesley *Computer Simulation of Liquids*, Clarendon, Oxford, 1987.
39. F.H. Stillinger and A. Rahman *J. Chem. Phys.*, **60:** 1974; 1545.
40. M. Neumann *J. Chem. Phys.*, **82:** 1985; 5663.
41. M. Neumann *J. Chem. Phys.*, **85:** 1986; 1567.
42. M. Wojcik and E. Clementi *J. Chem. Phys.*, **85:** 1986; 6085.
43. M.W. Evans, G.C. Lie and E. Clementi *J. Chem. Phys.*, **88:** 1988; 5157.
44. M. Wojcik and E. Clementi *J. Chem. Phys.*, **84:** 1986; 5970.
45. Y. Marcus *Chem. Rev.*, **88:** 1988; 1475
46. S. Lin and P.C. Jordan *J. Chem. Phys.*, **89:** 1988; 7492.
47. P. Kollman *Chem. Rev.*, **93:** 1993; 2395.
48. P.M. King in *Computer Simulation of Biomolecular Systems, vol. 2*, W.F. Van Gunsteren, P.K. Weiner and A.J. Wilkinson, (eds.), Escom, Leiden, 1993; p. 267.
49. R. Zwanzig *J. Chem. Phys.*, **22:** 1954; 1420.
50. J. Shen, C.F. Wong and J.A. McCammon *J. Comput. Chem.*, **11:** 1990; 1003.
51. T.P. Straatsma and J.A. McCammon *Annu. Rev. Phys. Chem.*, **43:** 1992; 407.
52. J.A. McCammon and S.C. Harvey *Dynamics of Proteins and Nucleic Acids*, Cambridge University Press, Cambridge, 1987.
53. D.L. Beveridge and F.M. Dicapua *Ann. Rev. Biophys. Chem.*, **18:** 1989; 431.
54. B.L. Tembe and J.A. McCammon *Comput. Chem.*, **8:** 1984; 281.
55. P. Demontis. G.B. Suffritti, S. Quartieri, E.S. Fois and A. Gamba *J. Phys. Chem.*, **92:** 1988; 867.
56. J. Weber, P.Y. Morgantini, P. Fluekiger and A. Goursot *Visual Computer*, **7:** 1991; 158.
57. A.K. Nowak, A.K. Cheetham, S.D. Pickett and S. Ramdas *Mol. Simulation*, **1:** 1987; 67.
58. A. Goursot, F. Fajula, F. Figueras, C. Daul and J. Weber *Helv. Chim. Acta*, **73:** 1990; 112.
59. C.R.A. Catlow, C.M. Freeman, B. Vessal, S.M. Tomlinson and M. Leslie *J. Chem. Soc. Faraday, Trans.*, **87:** 1991; 1947.
60. L. Leherte, J.M. André, E.G. Derouane and D.P. Vercauteren *J. Chem. Soc. Faraday Trans.*, **87:** 1991; 1959.
61. P. Demontis, S. Yashonath and M.L. Klein *J. Phys. Chem.*, **93:** 1989; 5016.
62. R. Kaptain, R. Boelens, R.M. Scheek and W.F. Van Gunsteren *Biochemistry*, **27:** 1988; 5389.
63. A.T. Brünger, J. Kuriyan and M. Karplus *Science*, **235:** 1987; 458.
64. A.T. Brünger and M. Karplus *Acc. Chem. Res.*, **24:** 1991; 54.
65. M. Levitt and R. Sharon *Proc. Natl. Acad. Sci. USA*, **85:** 1988; 7557.
66. D.S. Hartsough and K.M. Merz *J. Am. Chem. Soc.*, **115:** 1993; 6529.

7 Exploring the conformational space: distance geometry and model builders

As previously stated, the three-dimensional geometry is an important property for understanding or predicting the behaviour of molecular systems. It is the necessary starting point for the derivation of structural features (surface, volume, etc.), the estimation of steric requirements or the calculation of electronic properties. Geometry also plays a key role in the study of shape complementarity involved in host-guest interaction processes. In many drug design applications, the "active analogue" approach largely uses comparisons of molecular skeletons (possibly supplemented by consideration of more refined electronic or structural indices) to deduce possible pharmacophores (a particular spatial arrangement of atoms common to all active molecules, which is recognized by a single receptor). Therefore, this pharmacophore search directs or helps the synthesis of new drugs.

Apart from direct experimental determinations (X-ray crystallography, for instance), molecular geometry can be attained by database retrieval either directly for structures already stored in the database or by assembling relevant substructures (see Chapter 4). On the other hand, in Chapter 5, we emphasized the capabilities of computational approaches to derive possible molecular geometries. Methods such as molecular mechanics are of prime interest to complement X-ray data, and constitute a privileged avenue to geometrical determinations not (or not easily) available from experimentation: hypothetical structures not yet synthesized, low-energy conformers differing from those existing in the crystal state, and so on.

In fact, these energetics aspects are often complex, relying on the "multiple minimum" problem. First, for systems with numerous degrees of freedom, the hypersurface of potential energy (in the space of the parameters defining the system) may have a substantial number of local minima, and determining the global minimum is not always an easy task. Furthermore, finding that minimum energy conformation may be not sufficient. In flexible molecules, several conformations may be significantly populated in given conditions and the observed physicochemical properties correspond to an average over this conformational mixture. The receptor-bound conformation of a drug or the geometry of a complexed species may be different from that of the free molecule if specific stabilization processes (H-bonds, etc.) appear in the bound (or complexed) state. It is therefore essential for such analysis to sample the conformational space up to several kcal/mol above the global energy minimum.

For example, various spectroscopic data demonstrate that in the gas phase 1,3 dichloropropane exists as a mixture of two (or more) conformers: gauche-gauche (GG) and anti-gauche (AG), the former being the most stable [1] (Figure 7.1). As another example, we pointed out in Chapter 5 that 18-crown-6 complexed with K^+ ions adopts a D_{3d} conformation whereas the complex with Na^+ retains the C_i symmetry found for the free crown.

Computational conformation analysis comes up against two main problems:

1. Exploration of the conformational space, if exhaustive, may become a formidable task. Indeed, we saw that conventional minimization programs run only downhill from the initial geometry proposed without going through potential barriers. Some caution is necessary not to stop on a local minimum or keep hanging on singular points of the potential energy surface. In simpler cases, conformational analysis may be carried out by

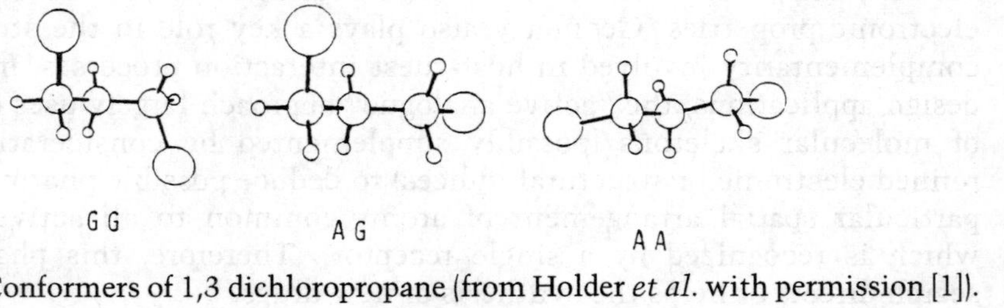

GG   AG   AA

Figure 7.1   Conformers of 1,3 dichloropropane (from Holder *et al.* with permission [1]).

varying only one or two geometrical parameters (usually dihedral angles), leading to usual conformational maps or interconversion profiles (such as those presented in Chapter 5). Such *systematic search* methods (or so-called *grid search* methods to remember the stepwise variations of independent geometrical parameters) look appealing, since they lead to an exhaustive generation of possible conformers. However, for systems with numerous degrees of freedom, a systematic study, to be sure that any interesting conformation has not been forgotten, would rapidly become a formidable and quite unrealistic task. For instance, assuming steps of 20°, examination of a system with six degrees of freedom would require us to consider 34 million conformations. Similarly, Lipton and Still noted that for a simple protein composed of 50 amino acids, analysis of the peptide backbone, with 120° torsional angle resolution, would keep a supercomputer, able to energy minimize one structure per second, fully occupied for more than 10^{40} years [2]. So, appropriate filtering methods have to be developed to eliminate unreasonable portions of the conformational space before more refined investigations are carried out. Steric interactions and ring closure conditions are among the most frequently used criteria.

2. Geometry determination or optimization (either by quantum or molecular mechanics methods) requires heavy numerical calculations and suffers some drawbacks:

   - an initial starting geometry is required,
   - empirical force field methods, interesting for large systems in view of their speed, fall down for interactions (torsional, angle bending) not previously parameterized.

In contrast to these approaches requiring heavy computational or experimental efforts, chemist in many cases can "manually" build very realistic models in assembling in the proper order predefined elementary units and using the general rules of conformational analysis. For instance, it is likely that a cyclohexyl ring will adopt a chair conformation (in the absence of other constraints), an alkyl chain prefers an extended arrangement, and so on. In such processes, the experts' knowledge, from various modelling sets, gives fairly correct and reasonably fast predictions. The interest of such developments is still enhanced in view of the increasing demand in drug design applications. Indeed, in this field the chemist frequently has to draw numerous 3D structures to get some insight about active molecules and propose new drugs. So, not surprisingly, there is increased interest in non-numerical model builders that quickly give realistic 3D geometries.

Relevant also in determination of possible molecular geometries, the "distance geometry" approach works primarily on interatomic distances and allows for the treatment of NOE enhancements in NMR (interesting experimental information, since it comes from solutions and not crystal-phase samples).

We will first briefly present this approach, since it also provides possible solutions to the more general problem of exploring the conformational space.

7.1 DISTANCE GEOMETRY

As stated by Crippen [3, 4] "Distance geometry refers to the study of geometric problems with an emphasis on distance between points".

The method utilizes a matrix of all pairwise atomic distances in a molecule to generate a set of Cartesian coordinates consistent with this distance matrix. Distances come from experiments (NOE measurements) or are derived from standard geometrical features.

The first aim of distance geometry seems to have been the search for a "robust" treatment of conformational calculations. It was then oriented towards a new approach of binding in drug design. The largely widespread use of NOE and 2D NMR to derive geometrical features of proteins will probably prompt various and numerous applications of distance geometry. Indeed, this approach relies on the concept of *upper* and *lower distance bounds*, so it seems particularly well-suited to the treatment of NOE factors, since these measurements generally give distance ranges rather than well-defined values.

The method is applicable to conformational search on small or medium size molecules [5, 6], but also appears very efficient to determine conformations of macromolecules in solution [7, 8] (EMBED algorithm [9], DIS-GEO program [10], DISMAN program [11]). An illuminating example in that field comes from a recent study of a 74-residue polypeptide (α-amylase inhibitor Hoe-467A) where distance geometry, starting with 401 NOE distance constraints, gives results quite consistent with an independent X-ray study [12, 13] as to the proposed backbone conformations.

In usual conformational investigations, a molecular system is defined by some geometrical variables: Cartesian or internal coordinates. Although a system of n atoms possesses $3n-6$ degrees of freedom, a current trend is to reduce the number of these variables (for instance, focusing interest on dihedral angles, and assuming standard values for bond lengths and valency angles). However, even with such a reduced set of coordinates, the treatment may suffer some instabilities. Small individual errors on successive dihedral angles may result in sizeable overall deviations for the end atoms. A slight angular variation can substantially modify the position of a remote atom via a "lever arm" effect. Clearly, working on pairwise distances between all couples of atoms leads to a more "robust" approach, i.e. a treatment which can accommodate some uncertainties and avoid propagation of errors. However, for n atoms there are now $n(n-1)/2$ distances and a large amount of redundancy is then introduced. Efficient algorithms are therefore mandatory to treat these numerous data.

7.1.1 Embedding

The general problem that distance geometry addresses is "embedding": starting from a matrix of all pairwise atomic distances, given a list of geometrical constraints, it consists of finding one or more sets of atomic coordinates for a molecule. These constraints can come from experiment (upper distances can be assigned to pairs of correlated protons in NMR experiments). The reasonable assumptions of nearly standard values for bond lengths or valency angles (at least for non heavily strained molecules), chirality, coplanarity of phenyl rings, etc. also impose numerous *a priori* constraints, so-called "holonomic constraints". The criterion that distances between non-bonded atoms cannot be less than the sum of the van der Waals radii defines a third type of constraint.

How can atom locations and coordinates be assessed starting from a set of interatomic distances? Remaining in a 2D space for the sake of simplicity, let us recall that a triangle is perfectly defined by its three edge lengths (Figure 7.2).

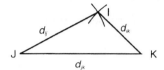

Figure 7.2 In the 2D space, the three distances d_{ij}, d_{ik}, d_{jk} fix the relative location of points i, j, k (if chirality is not considered).

So, starting from two reference points (three in a 3D space), one can successively locate the various atoms, one after the other, provided interatomic distances (and chirality in the 3D space) are known. However, some experimental uncertainties may occur, and there is no reason to privilege certain atoms with respect to the others. Furthermore, a construction taking into account simultaneously all the interatomic distances is likely to give an approach more able to cope with experimental errors and detect inconsistencies. In fact, any distance set does not always give an acceptable solution. Remaining in the 2D space, the triangle inequality tells us that for three points, any distance must lie between the sum and the difference of the other two.

Given three atoms (i, j, k), for any pair (i, j):

$$\left| d_{jk} - d_{ik} \right| \le d_{ij} \le d_{jk} + d_{ik}$$

Indeed, the Crippen algorithm is largely based upon that triangle inequality. A similar condition also exists for four points in the 3D space (i.e. given three non-colinear points and a fourth one, its distances to others must lie between the "planar cis or trans arrangements") (Figure 7.3). However, this condition seems more difficult to program, and the triangle inequality remains the most frequently used criterion.

The Crippen method lies on a distance matrix $(n*n)$, n being the number of atoms in the molecule. For actual distances, this matrix would be symmetrical, with diagonal elements equal to zero. Owing to experimental uncertainties and constraints, this distance matrix is here presented as being constituted of two parts. The upper triangle collects upper (maximal) bounds

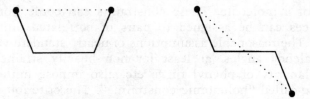

Figure 7.3   Upper and lower bounds for distance d_{ij}.

of the interatomic distances. In the lower triangle are indicated lower bounds for every distance. If some estimates are missing one can use a large number for upper bounds and zero or the sum of the van der Waals radii for the lower bounds (the Crippen distance matrix is shown below):

$$
\begin{vmatrix}
0 & \text{upper bounds} \\
& 0 \\
& & 0 \\
& & & 0 \\
& & & & 0 \\
& & & & & 0 \\
& & & & & & 0 \\
& \text{lower bounds} & & & & & & 0
\end{vmatrix}
$$

The following steps are then performed:

(a) Using the triangle inequality, distance bounds are smoothed (lowering some upper values and raising some lower values).
(b) For each distance, independent random values are chosen within the allowed range between upper and lower limits.
(c) This trial set of distances is converted to a "metric matrix" G, with centre of mass as the origin. G is a matrix of vector dot products, the elements of which being defined as:

$$g_{ij} = \mathbf{u}_i \cdot \mathbf{u}_j$$

where $\mathbf{u}_i$ is the vector from the origin to atom i. These $\mathbf{u}_i$ are not known when the calculation is in progress, but the g_{ij} can be easily evaluated from the interpoint distance matrix without reference to coordinates.

$$2g_{ij} = (d_{io})^2 + (d_{jo})^2 - (d_{ij})^2$$

with:

$$d_{io}^2 = n^{-1} \sum_{j=1}^{n} d_{ij}^2 - n^{-2} \sum_{j=2}^{n} \sum_{k=1}^{j-1} d_{jk}^2$$

(n = number of atoms).

The problem will now be to project these data into the 3D space and define coordinates such that the trial distances are reproduced as closely as possible. Such a reduction to the 3D space is carried out by searching for the first three principal axes of the initial data set, that is the eigenvectors corresponding to the three eigenvalues of the largest absolute magnitude.

(d) From this new three dimensional metric matrix, trial coordinates v_{rm} (for atom r, $m = 1$ to 3) are calculated. They are a good approximate solution, but there is no guarantee that they completely satisfy all the original constraints. They therefore have to be refined by minimizing an error function based on penalty terms for each violation of the constraints:

$$v_{rm} = (\lambda_m)^{0.5} w_{rm}$$

$m = 1$ to 3; w_{rm} = element of a three column matrix corresponding to the eigenvectors associated to the three largest eigenvalues (λ_m) of the original metric matrix.

This process leads to a randomly chosen conformation which more or less satisfies all the input constraints.

(e) go back to step (b) to generate another acceptable conformation.

Remark

Distances are not sensitive to mirror inversion about one chiral centre, so enantiomers cannot be distinguished by distance geometry. In principle, diastereoisomers would be distinguishable, but owing to the rather large tolerance on most distances, the problem is rather intricate. It is easier to introduce a "chirality violation" penalty term.

Given four points representing atoms surrounding an asymmetric carbon, chirality on this centre can be defined by the signed volume of the tetrahedron formed by the four points:

$$f = (\mathbf{V}_1 - \mathbf{V}_4).[(\mathbf{V}_2 - \mathbf{V}_4) \times (\mathbf{V}_3 - \mathbf{V}_4)]$$

the $\mathbf{V}_i$ vector defining the position of point i. If the numbering 1...4 obeys an ascending order according to the Cahn, Ingold, Prelog system:

$f > 0$ for an S configuration,
$f < 0$ for an R configuration.

7.1.2 Applications to geometrical or conformational problems

As previously indicated, distance geometry was successfully applied to propose reasonable conformations for either small molecules suffering strong geometrical constraints [5, 6] or large biomolecules [7, 10]. Embedding seems particularly useful for determining the conformation of small proteins and oligonucleic acids in solution thanks to NOE information. As an illuminating example, one can note the proposed spatial structure of protein basic pancreatic trypsin inhibitor. Using 508 proton-proton contacts shorter than

4 Å (extracted from a simulated ideal NOESY experiment), distance geometry reproduces crystal structure within 1.3 Å rms coordinate difference for the backbone atoms [10]. See also [14].

Crippen *et al.* [14, 14a] recently proposed an extension of the method for molecules where bond lengths and valence angles can be assumed as fixed. These holonomic constraints are enforced separately from experimental constraints by being introduced into the formulation of the problem. If bond lengths and angles are assumed fixed, the various degrees of freedom can be expressed by dihedral angle values only. They are therefore easily discussed in terms of spatial orientation of unit vectors defining the position of the bonds in local reference systems. The method is based on a *linearized representation* of the molecule as a tree graph (nodes = atoms and edges = bonds). Starting from a root atom, the position of every atom is defined thanks to local coordinate systems recursively set up after crossing each rotatable bond. Among the advantages of the method are a sizable decrease of the number of variables for molecules having rigid groups and a more accurate access to local geometries for rigid groups (phenyl chiral centres, etc.) (Figure 7.4).

Distance geometry was also used as part of the model builder proposed by Wegner and Smith [15] (see p. 222). In the "ensemble approach" of Sheridan *et al.* [16], the distance geometry method of Crippen has been modified to simultaneously treat two or several molecules as a single "ensemble". This extension can generate coordinates for the set of molecules in their "active conformation" (with their essential groups superimposed) in one step, and therefore gives a way in which to find a common pharmacophore.

Another important application of Crippen distance geometry is receptor modelling, particularly regarding the binding of small ligand molecules to sites of proteins or macromolecules. This point will be developed further in Chapter 12, devoted to molecular recognition.

As a concluding remark it may be noted that representing molecular structure as a set of distances sometimes offers definite advantages for features more easily expressed in terms of local distances than in dihedral angles or coordinates (ring puckering is an illuminating example). By contrast, one can easily specify, via dihedral angle constraints, a cis or trans arrangement: this is more difficult with distance bounds. Another advantage is that distance

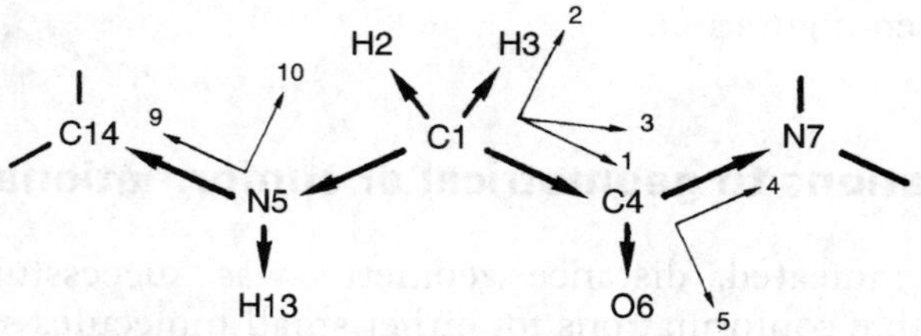

Figure 7.4 Linearized representation. Propagation of local coordinate systems (light arrows). Heavy arrows (along bonds) indicate the traversal through the graph from the origin C1. At C1, axis 1 is parallel to bond C1–C4, axis 2 is located at the plane C4–C1–H3 perpendicular to axis 1. Two axes are sufficient at C4, assuming coplanarity of the peptide fragment (from Crippen with permission [14a]).

geometry does not depend upon the number of rotatable bonds, a point which seriously hampers systematic conformational search for complex systems, as will be shown below.

When sampling the conformational space and choosing distances obeying the triangle inequality, it may be that distances are selected in a somewhat correlated manner leading to a biased sampling. Using, for example, torsion-angle sampling in place of 1,4 atomic distances and removing distance correlation enlarge the searchable conformational space and increase efficiency in the generation of possible conformers [17].

However, although distance geometry is an attractive and efficient way in which to convert distances into coordinates, it corresponds to a random exploration of the conformational space, within the constraints of the distance bounds but without any energy consideration. Proposed solutions therefore have to be more closely examined as to their stability. Furthermore, geometries deduced from random selection of distances within allowed ranges usually correspond to non-perfectly relaxed conformations. These therefore have to be optimized in a subsequent step by methods such as molecular mechanics or dynamics performing a local exploration of the conformational space around the solutions generated. Constrained molecular dynamics is frequently used for this purpose. It starts with some initial conformation (determined by distance geometry) suffering only minor constraint violations. Then molecular dynamics simulation is carried out, using as potential function a weighted sum of the usual empirical energy function plus the penalty function. Thermal motions send the molecule over small energy barriers towards a conformation that better satisfies the geometrical constraints and has a relatively good internal energy [7].

In addition, owing to the Monte Carlo nature of the sampling, one is never sure not to have missed any interesting solution.

7.1.3 Relaxing dimensionality and energy embedding

Distance geometry provides solutions consistent with geometrical constraints but does not consider energetic factors. To take them into account, when solving the multiple minimum problem, variants of the distance geometry approach were presented by Crippen with "energy embedding", whereas Scheraga *et al.* proposed an approach "relaxing dimensionality" [18–20a]. Basically, the latter method starts with a very low energy "conformation" located in a high-dimensional space where there are fewer local minima. Then it gradually contracts the dimensionality, perturbing energy as little as possible, so as to reach a low-energy structure in the usual three-dimensional space.

In a high-dimensional space (typically $n-1$ for n constraints on atoms), one can easily find a lower bound of the molecular energy, minimizing separately all the independent contributing terms (Lennard–Jones, hydrogen-bonding, torsion), the sum of which represents the molecular energy. Of course, the interatomic distances so obtained do not correspond to any realizable 3D structure. The game is therefore to restrict the dimensionality of the representation space down to a feasible 3D structure.

The method has been presented as approaching the global energy minimum "from below" (since energy increases as more distances are constrained), rather than "from above", as in usual minimization schemes, so that the likelihood of ending up with the good solution is increased.

As compared to neighbour distance geometry approaches, where the problem is also formulated in the distance space, it can be noted that in the Scheraga method, distances are used as the primary variables, without the need to come back to Cartesian coordinates in the minimization step. Results on model systems and the pentapeptide met-enkephalin look quite attractive, although the method requires rather high computer storage capabilities.

As to distance geometry with energy embedding, Crippen indicated that the reduction of dimensionality from an hyperspace R^{n-1} (for an n atom system) to the current R^3 space can be efficiently carried out as a constrained-optimization problem and solved by augmented Lagrangians [20b].

7.2 EXPLORING THE CONFORMATIONAL SPACE

Even assuming that bond lengths and valence angles are maintained at standard values, variations of dihedral angles induce a tremendously large number of internal degrees of freedom, even for medium-size molecules. Basically, for N dihedral angles and M possible values for each (for instance steps of ca. 30°), there are M^N conformations to be investigated to be sure not to overlook any valid structure (in fact, the actual number of structures will be less, if one discards high-energy conformers and takes into account ring closure conditions) [2].

Another problem is that the various energy minimization processes do not go through potential energy barriers. In other words, they only move downhill from the trial starting structure towards the nearest minimum, which may unfortunately be only a local minimum. Furthermore, one is generally interested not only in the best conformer but rather in the several best conformers of low energy (particularly, in a pharmacophore search, for selecting one common geometrical arrangement within the set of best conformers for different active molecules).

To overcome these difficulties, various approaches have been proposed to explore more efficiently (but safely) the conformational space. The common approach is to generate a set of possible trial conformations which are subsequently submitted to molecular mechanics energy minimization refining their geometry to nearby minimum energy conformers. Collecting the resulting conformers and discarding duplicates yields the required set of low-energy conformers.

Special attention was devoted to cycles since conformational searching in such systems is generally more intricate than for open chain compounds. It is not always easy to discriminate between the various conformers and to determine how many independent torsion angles must be varied to go from one conformer to another one. Furthermore, as stated by Gajewski *et al.* [21], it seems a difficult task to monitor by algorithms dihedral angle changes in cyclic structures. Particularly, great caution must be exercised with the usual bond-rotation algorithm, used for acyclic systems, which rotates all the atoms related to one extremity of the rotatable bond. A convenient solution, as proposed by Lipton and Still [2] and Chang *et al.* [22], is to temporarily open the ring and create a pseudoacyclic molecule which is processed as an acyclic structure but with additional constraints enforcing ring closure for the final structure selection. These ring closure constraints concern the distance between the two atoms forming the ring closure bond, the two adjacent bond angles and the three corresponding dihedral angles (Figure 7.5). A chirality test is also provided.

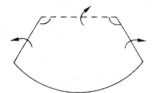

Figure 7.5   Closure parameters (Lipton and Still [2]).

7.2.1 Subunit optimizations

To limit the computational task, Scheraga *et al.* [23, 24] proposed a combination approach. Preliminary calculations are made on smaller subunits, then these subunits are combined to yield low-energy candidates for the whole structure. So, constraints *internal* to fragments are already solved, and in the final stage one has only to cope with interactions *between* fragments near their boundaries or with some long range interactions. Note that the same idea underlies the expert system AIMB (Analogy in Model Building; see below).

The approach was extensively used for energy-based structure prediction of proteins, by assembling peptide units. This point is presented in Chapter 13 (see section 13.4.2). Let us here only focus attention on some useful tricks:

- Use of nondegenerate minima, to limit the exponentially growing number of minima to be retained at each step of the building process. They are structures corresponding to the same backbone conformation but differing in side chain orientations.
- Modification of the force field, so that the energy function remains finite even in case of atomic overlap. Otherwise, for the very high energies resulting from such an overlap, minimization is computationally infeasible, although a small change could suffice to remove such singularities.

- Choice of a correct cut-off energy, since solvent stabilization may favour some conformers. This is empirically evaluated with incremental terms according to hydrogen-bonds with the surrounding medium.

7.2.2 Tree search methods

The algorithm proposed by Lipton and Still [2] offers a number of advantages: it works on internal coordinates and allows for sampling over all accessible regions of the conformational space using a tree search method. The approach begins with a geometry-optimized structure, and then performs stepwise torsional rotations about all rotatable bonds, retaining only conformations passing geometrical tests designed to reject high-energy structures.

Ignoring bond lengths and angle variations, the search is restricted to the torsional conformational space (dimension ca. N–3). The process can be viewed as a systematic grid search in a hyperspace where axes (associated with the various torsional angles) are divided into segments according to the variation range of the torsional angles for each step.

The search is performed in a sequential unidirectional fashion, starting at one end of the structure and travelling along the framework (for cyclic systems, rings are temporarily opened, then additional constraints will be placed to ensure ring closure). Ordered variations of torsional angles allow for representing the conformer generation process as a tree structure. On the associated graph, the leaves correspond to each of the possible conformations and each edge stands for a torsional rotation. At each node of the graph certain atoms become fixed, since they cannot move in the following rotations down in the tree. Tests for interatomic contacts applied to these newly fixed atoms allow for eliminating non-acceptable structures and pruning entire branches of the generation tree in a single operation. Contacts are looked for 1,5 interactions and more remote ones with cutoff distances of ca. 1.5Å between atoms (2Å for united atom structures where small adjustments of length, valence or dihedral angles may relieve strain). For cyclic molecules additional ring closure constraints are considered

The authors suggest that for medium size organic molecules, a 60° dihedral angle resolution seems convenient (smaller steps providing no new conformers). Tests on alkanes and cycloalkanes not only retrieve all known energy minima, but also propose new solutions. The algorithm is presented as optimized for both speed and memory requirement, so that global conformational searches for molecules up to 10^9 conformational possibilities in cyclic systems (10^6 in acyclics) are claimed feasible with reasonable CPU cost. However, it was quoted that the excision of branches of the tree depends upon the user's ability. Furthermore, rejection of branches before energy minimization may eliminate, on the basis of distances, structures that would appear acceptable after geometry optimization is carried out [25] (Figure 7.6).

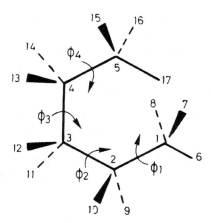

Figure 7.6 If, after examining Φ_1, the current rotation is that modifying Φ_2, atoms 1–3 and 6–10 are fixed. In the following rotation, Φ_3 will be modified: atoms 2–4 and 11–12 will remain unmoved, the others moving. So, in the current step only atoms 4, 11 and 12 need to be calculated (Lipton and Still [2]).

7.2.3 Stochastic approach

The stochastic approach (Monte Carlo method) is well suited for large molecules with complex interconversion pathways. As previously indicated, the common drawback of energy minimization algorithms is that these methods always work downhill and do not cross energy barriers. To be sure not to miss any possible conformers, one has to try numerous starting geometries. Monte Carlo methods, until now mainly used for liquid simulations, have been extended to tackle these conformational problems.

Typically, in the simplest stochastic search, from a given starting structure, random variations on selected coordinates (randomly updated) are drawn, and the internal energy of the resulting structure is calculated and energy-optimized. This new conformer is saved if its energy is lower than a preset value (for example, the energy of the previous minima saved) or, better, if the difference between its energy and the best one previously found is less than a fixed threshold:

$$E < E_{min} \quad \text{or} \quad E - E_{min} < \Delta E$$

(and if this conformer has not been already found). Then the cycle is repeated.

This is the principle of the Monte Carlo Multiple Minimum (MCMM) method of Chang et al., where "a single starting geometry is used repeatedly to accumulate its progeny" [22]. Alternatively, in the stochastic minimizer, one starts with the last generated conformer and randomly modifies its geometry. The same criteria as above are used to save or discard the new structure generated. This process corresponds to a "random walk along the potential energy surface in a continual loop" [25], updating the starting structure for minimization whenever a new extremum is reached in order to find all minimum energy conformations. As there is now some relation between the conformers successively created, the search is better.

In alternative versions, the previously saved structures are used the same number of times, or that structure which has been used the least is chosen [22] as the current starting point for updating. As stated by Saunders, "in such methods, one is not randomly exploring the conformational hyperspace, but is stepping, by random kicks followed by minimization from one local minimum to another one." Such random kicks are feasible both on Cartesian or internal coordinates [27, 28]. The number of parameters (or coordinates) to be varied at each Monte Carlo step to attain more efficiency is discussed by Chang *et al.* [22].

A random walk means that multiple runs are necessary to develop the confidence that the global minimum is found. So, when can the conformational mapping be stopped? The search is automatically terminated after a user-defined number of unsuccessful attempts (completed search steps in which no new acceptable structure, within a predetermined range, is located [25, 26]. This termination number is currently set at about a few hundred [27].

Conformational search may be directed towards the low-energy regions of the conformational space by using multiple low-energy conformers as initial structures and modifying only a fraction of the variable torsion angles. However, random search efficiency degrades rapidly as more conformers have been previously found. As a variant of MCMM, the SUMM (Systematic Unbounded Multiple Minimum search) approach of Goodman and Still proposes a new internal coordinate traversal of the conformational space [29]. Generation of conformers begins at low resolution (120° in torsion-angle space) sampling the whole conformational space, so that torsionally remote conformers are created early on in the search. Then, resolution is doubled and new points are sampled. This continuous process where the extent of the search need not be specified at the outset (in contrast with systematic searching where the number of steps must be fixed in advance), appears more efficient at finding all low-energy conformers for medium and large ring molecules.

In the RIPS (Random Incremental Pulse Search) method [28], starting with an input geometry, random changes (± 1Å) are performed on the coordinates of each atom (or a specified subset of atoms) and the perturbed geometry is then submitted to the usual (downhill) minimizers to find a new conformer. The process is then repeated again from that conformer to find another low-energy structure. To decide whether two conformations may (or may not) be considered as identical, two criteria are proposed: examining the conformational energy (which must be the same within a tolerance range of ca. 0.01 kcal/mol), or the values of the dihedral angles. Since full molecular mechanics minimizations would be time consuming, an initial screening rejects conformers lying too high in energy. A similar crude filtering test (examining severe non-bonded contacts or bad ring closure conditions) is also incorporated in the MCMM algorithm of Chang *et al.* [22].

To avoid hanging on a saddle point or to specify the minimum in a flat area of the potential energy surface, a refinement step introduces small random distortions (ca. 0.05 Å), that cause the collapse of transition state geometries and lead to true local minima [28].

A recent application derived from Monte Carlo/energy minimization techniques [22] concerns investigation of the conformational space of inhibitors bound to the active site of an enzyme (thermolysin, a bacterial enzyme, in the example treated) [30]. Accurate prediction of the bound conformers of enzyme inhibitors is a challenging problem for the modelling of new enzyme inhibitors, including prediction of their binding activity in advance of testing. Starting from a refined crystal structure, the inhibitor was removed and docked in random trial conformations that are energy minimized (solvation energies were estimated on the basis of the exposed solvent accessible surface area of the inhibitor). Several trials were analysed for exploring the low-energy domains of the potential energy hypersurface, and the process can be thereafter repeated to model the binding of other, structurally different, inhibitors. According to the authors, the methodology seems promising for the evaluation of potential new synthetic inhibitors in cases where the structures of the enzyme and a bound inhibitor are known.

As for all stochastic methods, there is no definite criterion to assess that all the acceptable conformations are found. However, from various studies, mainly on cycloalkanes (a challenging problem, since several low-energy forms with complex interconversion pathways are expected) [2, 22, 25–28], it appears that all the known conformers are easily retrieved, and that new acceptable solutions can be found, for instance for cyclononane. Of course, in that case, the newly discovered low-energy structure – 2.2 kcal/mol above the global minimum – would not intervene for more than 0.5% of the conformational mixture at room temperature. This percentage is low, but cannot be excluded in a detailed conformational analysis. Furthermore, it may change for other parent compounds.

The RIPS approach was recently used to scan the potential energy surfaces of C5–C8 cyclic alkanes to examine the consistency of molecular mechanics with semi-empirical or *ab initio* quantum methods. Good agreement is found for geometries, but relative energies derived from semi-empirical methods often deviated from those obtained by the other two methods [31].

It has also been said that Monte Carlo methods seem less effective when many covalent bonds are present (displacements "along" bonds, which correspond to larger energy changes, being disfavored vs. "perpendicular" ones, so that the generation of acceptable forms is somewhat biased) [7]. Scaling variables, so that the energy function is isotropic near its minimum, have been proposed to relieve this drawback (for dynamic aspects, see also Kawai *et al.* [33]).

Metropolis method

The Metropolis approach is similar to the stochastic minimizer, but introduces some probabilistic criteria to accept the new conformer generated.

For escaping potential wells of local minima more easily, one sometimes now accepts changes that increase energy with the hope that this will allow for going through a pass towards the global minimum well. In other words, a stochastic search is now performed such that if the resulting energy change ΔE is negative, the new structure is always accepted. If the energy change ΔE is positive, the structure is not automatically rejected, but it can be accepted, with a probability

$$\exp(-\Delta E / k_b T)$$

where k_b is the Boltzmann constant. The parameter T plays a role similar to that of temperature in Boltzmann's law, and is therefore called "temperature" by analogy [34].

This acceptance test is performed by picking a random number (between 0 and 1) and comparing it to $\exp-(\Delta E/k_B T)$: if this random number is lower, the change is accepted and yields a new geometry that can be used as a new starting point.

Higher temperatures accept larger steps, and so allow for large motions exploring the gross features of the conformational landscape. This makes it possible to escape local potential wells. On the contrary, lower temperatures provide more stringent conformational selection, the system remaining confined in smaller regions.

Various applications have dealt with the conformational search for polypeptides, in torsion angle space (with bond lengths and angles fixed). After a random change of one random torsion angle, the new structure is compared to the preceding one using the Metropolis criterion [32–33a]. Alternatively (and this seems to be a better solution), in the torsion-based method of Li and Scheraga to locate the global energy minimum [35], as in von Freyberg and Braun's study of met-enkephalin [36], the randomly generated conformation is first energy-minimized before the Metropolis acceptance test.

Interestingly, these authors proposed to jump to higher temperatures to avoid getting trapped in a local minimum (a situation possibly encountered in a constant temperature scheme) and then come back to a lower temperature to more thoroughly explore the vicinity of low-energy conformations. This process relies somewhat on the simulated annealing methods that will be detailed below.

For met-enkephalin (76 atoms), this approach found 1881 low energy solutions corresponding to 77 non-identical different conformations. Among them, 74 conformers, in an energy band of 2.5 kcal/mol, have backbones similar to that of the global minimum, with a gap of about 0.9 kcal/mol between this minimum and the next highest energy species.

Annealing

"Simulated annealing" [37] is widely used in optimization problems where one has to minimize an "objective (or cost) function" without being caught in a local minima. It can be viewed as a derivative of the Metropolis method in which both the energy and the temperature dependence of the Boltzmann distribution guide the search for the global minimum [38, 39]. We present this approach here for conformational search, by minimization of the steric energy to find the best conformer. However, other applications also appear in the field of molecular modelling: simulated annealing was proposed for the automated docking of substrates to proteins [40] and for the recognition of a common substructure in two molecules [41, 41a].

In conformational search, annealing is primarily designed for finding the global minimum [39] and samples low-energy conformations more sharply, whereas systematic search, for example, which can in principle find all minima, uniformly samples the whole conformational space

The name "simulated annealing" is derived by analogy with the process of crystallization: slowly cooling a melted sample produces the most stable (crystalline) arrangement, whereas a rapid freezing may lead to a metastable form (glass) corresponding to a local minimum. The main difference to the Metropolis method is that simulated annealing is monitored by a cooling schedule. The process is broken up into a number of cycles, each at a constant temperature and composed of a large number of individual steps corresponding to a stochastic search with a Metropolis criterion. Each cycle can be considered as reproducing the Boltzmann distribution of the conformational states (to be submitted to further minimization). Then the temperature is lowered and the process is iterated. By progressively lowering the temperature one explores conformations of decreasing cost in smaller and smaller regions of the conformational space, and finally the system falls into the global energy minimum.

Simulated annealing was, for instance, applied by Wilson and Cui [38, 39] to polypeptides. A first example treats individual amino acid dipeptide models: a bond is randomly chosen. It is rotated by a random amount between $-90°$ and $+90°$, and the energy is calculated to see if this solution can be accepted. Typically, 250 steps were performed for each of the 30 temperatures investigated, generating about 3500 structures per run. For dipeptide models for which the global energy minimum was already known, simulated annealing found the right solution. The method was then extended to polyalanine structures:

$$N-Ac(Ala)_n NHCH_3, \quad \text{with } n \text{ up to } 80$$

and the efficiency of the method thoroughly studied. As the cooling schedule is the original feature of annealing, an interesting point is the comparison of annealing with the common Metropolis algorithm at a constant temperature. Whereas annealing finds the global minimum, the fixed temperature Metropolis method does not always converge to it: at higher temperatures the molecule walks in the high energy regions too much, whereas at low temperatures it may remain trapped into a local minimum. According to the authors, a critical function of annealing would be to cool slowly over a phase transition. Another point of interest is that the random walk at each temperature should be long enough to reach a "steady state" distribution in accordance with the Boltzmann law. It appears that the percentage of runs finding the global minimum is determined not only by the number of local minima, but also the complexity of the energy surface. For polyalanine with four rotors, 250 steps are sufficient to find the minimum for 70% of the runs. The percentage falls to only 20% for 10 rotors, but in that case it increases to 70% if 1000 steps are carried out at each temperature (assuming three local minima per bond, there are now 3.49×10^9 minima).

A recent study carried out on tetrahydroionone [39] points out that the rigid geometry approximation (a commonly used optimization over only dihedral angles with constant bond lengths and angles) corresponds to a walk on a dihedral energy surface which may not be identical to the (real) total energy surface, so that the true minimum may be missed. This drawback can be avoided if a subsequent local minimization is performed on the solutions given by annealing in the rigid geometry approximation.

7.2.4 Minimization under constraints

In most of the previous examples, energy optimization was performed by common molecular mechanics or related methods, and additional constraints (severe non-bonded repulsions, ring closure conditions) were sometimes used to reject unreasonable structures. In some situations, experimental data (mainly from NMR with Overhauser enhancements or coupling constants) can not only help to reduce the search size but also guide the energy minimization process. As quoted by Scarsdale *et al.* [42], "such constraints are in no sense real energies" but they can be incorporated as pseudo-energy terms in potential energy calculations. They introduce penalty contributions tending to destabilize conformations where the distance between two protons is different to that assumed from an NOE measurement or a dihedral angle is not consistent with an observed coupling constant. As an example of application, we recall the prediction of the major solution conformation of tyrocidine A, a cyclic decapeptide, using 18 NOE measurements of interproton distances [43]. An empirical penalty function was developed in which any conformation which deviates by more than 10% from a defined NOE distance is penalized by approximately 2 kcal/mol:

Penalty $E(NOE)_{kcal/mol} = K*(R-R_0)^2/R_0^3$ with $K = 800$; R_0 = target distance

$$
\begin{array}{cc}
\text{D-Phe}\,4 - \text{Leu}\,3 - \text{Orn}\,2 - \text{Val}\,1 - \text{Tyr}\,10 & \\
| \qquad\qquad\qquad\qquad\qquad\qquad | & \text{A} \\
\text{Pro}\,5 - \text{Phe}\,6 -_{\text{D-}}\text{Phe}\,7 - \text{Asn}\,8 - \text{Gln}\,9 &
\end{array}
$$

$$\text{Ac} - \text{Ala} - \text{Ala} -_{\text{D-}}\text{Ala} - \text{Pro} - \text{Ala} -_{\text{D-}}\text{Ala} - \text{NHMe} \qquad\qquad \text{B}$$

$$\text{Ac} - \text{Ala} - \text{Ala} - \text{Ala} - \text{Ala} - \text{NHMe} \qquad\qquad\qquad\qquad \text{C}$$

The investigated decapeptide (A) is split into two open chain fragments (B and C) with common Ac and NHMe groups. For the sake of simplicity, amino acids are replaced by alanines (except for proline due to its cyclic character).

A first step consists in splitting the molecule into two polyalanine-type fragments with common terminal groups. Then, conformers of individual amino acids were generated using molecular mechanics and six intra-residue NOE constraints to restrict the allowed areas of conformational space in Ramachandran (ϕ, Ψ) maps. For the two open-fragment models, candidate structures were generated with every acceptable conformation of the backbone Φ, Ψ angles (discarding conformers with overlapping atoms) and energy-minimized, taking into account an NOE penalty function. This leads respectively to 23 and 71 possible fragments, which were classified by factor analysis. From the groupings observed in the space of the major factor axes, 5 and 4 representative conformers can be selected for the two halves of the backbone model. Merging these representative fragments, two possible target cyclic structures are then generated looking for the best superimposition of the terminal groups of the fragments and energy minimization with constraints. In the last step, side chains were added and the whole molecule minimized without constraints to give the final structure. According to the authors, this computer-built model is likely to represent the actual conformation in solution, since the proposed structure seems consistent with the hydrogen-bonding pattern and shows close agreement with the known conformation of identical sequences of residues in other peptides (Figure 7.7).

A similar approach, incorporating NOE constraints in molecular mechanics, was also carried out by Scarsdale *et al.* [42] to determine the solution conformation of glycolipids. The authors also developed a model allowing for the existence of interconverting conformers in NOE measurements.

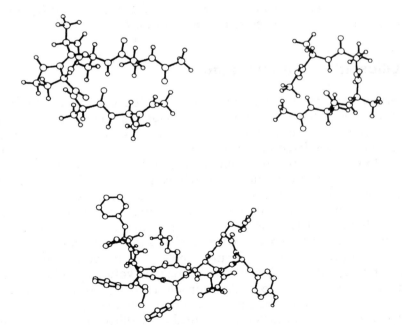

Figure 7.7 The selected open models (those giving the best end group superimposition) and the final structure obtained (from Tonge *et al.* with permission [43]).

7.2.5 Molecular dynamics

Molecular dynamics can also be used to derive low energy conformers. Basically, whereas X-ray data give a static, time-averaged picture of atomic locations, molecular dynamics simulates their instantaneous motions [7, 44–46]. Atom coordinates change with time, depending upon the kinetic energy terms and forces exerted by surrounding atoms (stretching bonds, bending angles). Starting with a low-energy structure, the method defines the evolution of the system by integration over time of Newton's second law of motion, according to the formalism of classical mechanics, and defines atom trajectories

Over short periods of time, motions can look erratic, but over longer periods slower coherent collective motions can be distinguished, giving some insight about preferred local fluctuations of selected groups. From these trajectories, low energy structures can be periodically sampled giving, in favourable cases, some insight about conformational changes [7]. However, owing to the high frequency of bond stretches, the time steps must be very short (typically about 1 fs), so that actual methodological and computational limitations restrict trajectory lengths to about a few hundred ps [7]. Conformational changes generally occur on longer time periods (particularly for large biomolecules) so molecular dynamics is not ideal for studying substantial barriers or sampling large regions of the conformational space. It appears to be more useful for local exploration.

Typically, one starts by generating about 100 structures and minimizing them. The results are used as starting points for dynamics simulation over a nanosecond at temperatures near 600°K. Selected configurations (for instance, every 100th) are then optimized to a local minimum.

7.2.6 Ellipsoid algorithm method

This method of constrained optimization, related slightly to the distance geometry approach, has been proposed as robust and avoiding local minima [47–49]. An interesting point is that the algorithm uses only one constraint per iteration, and makes large and discontinuous steps at the beginning of the process, allowing a start from random conformations. We suppose (as an example) that the conformational search can be carried out choosing as variables only n dihedral angles (to limit the dimensionality). Each conformer is assigned a point in an n dimension conformational hyperspace. From a randomly chosen initial conformation (a point in this conformational hyperspace), the algorithm starts with an ellipsoid which contains the entire conformational space of the molecule. Then, the algorithm will generate a sequence of ellipsoids with constantly decreasing volumes and containing at least one solution: the centre of the ellipsoid being the solution obtained at that iteration. Each iteration is based only on one inequality constraint, which is violated at the centre of the ellipsoid until a feasible conformation is found, that is a conformation in which no constraints are violated. Only at this point does the algorithm make an iteration based on an objective function (it may, for instance, be the conformational energy depending upon the dihedral angles). Constraints generally consist of bounds on distances (derived, for instance, from NMR measurements).

At each iteration, the new ellipsoid generated is defined as containing the entire half of the old ellipsoid on the side of the negative gradient for the constraint investigated (since this part must contain the optimum solution). This process results in an iterative decrease of the volume of the ellipsoid by a constant "shrink" factor, depending upon n, leading asymptotically to the minimum (when constraints are no longer violated and the objective function converges). The coordinates of the centre of the resulting ellipsoid define the dihedral angles for a conformation.

The method has been illustrated on Ala dipeptide. From a grid of 36 starting structures mapping a two-dimensional conformational space, 33 of them converged to the global energy minimum. Other test cases include energy refinement of met-enkephalin (a five aminoacid, 76 atom peptide determined by 20 dihedral angles) or an 11 residue segment [Arg-17-Met-27] of the polypeptide glucagon. The ellipsoid algorithm was also applied to the ionophore 18-crown-6, in conjunction with molecular dynamics refinement [48]. This approach found the conformations previously determined and new low-energy forms.

The ellipsoid algorithm was further extended to docking problems: determination of the three-dimensional structure of enzyme-ligand complexes (if at least one of the two partners of the association is known, e.g. from X-ray studies) [49].

7.2.7 Generic shape algorithm

To obtain start structures of macrocyclic compounds to be submitted to energy minimization, Gerber *et al.* [50, 51] proposed a "shape-guided conformation generation" retaining only the raw shape of the rings. "Generic shapes" are obtained through an approximate two-dimensional Fourier representation of real ring shapes (derived from existing crystallographic data), where only the two more significant coefficients, one radial and one axial, are retained. Give these "generic shapes" randomly scattered increments generates sets of different ring conformations. These representations violate small-scale requirements (correct bond lengths, angles), but it may be assumed that the little discrepancies they present will be quickly removed in the minimization stage. The process is stopped after a (user-defined) number of steps giving no new additional structure.

7.2.8 Efficiency and reliability

Flexible cycloalkanes constitute a privileged structural population to test the predictions of conformational search methods. For instance for cyclononane, four low-energy forms have been characterized from manual or algorithmic dihedral drivers (Figures 7.8, 7.9). The RIPS method [25, 26] found these low energy forms and also proposed three additional structures within 11 kcal/mol of the absolute energy minimum: the Twist-Chair-Twist-Chair (TCTC) form lying at only 2.3 kcal/mol above it. Note that two of these new forms have also been found by the tree search of Lipton and Still [2].

More recently, this problem was revisited by Saunders [52] using his stochastic search method and the MM2 or MM3 force field. An eighth conformer, of C_2 symmetry but with a methylene group folded into the interior of the ring and largely higher in energy than the seven found previously, was identified "as a definite minimum" on the potential energy surface. In the same study, Saunders also proposed to characterize the similarity between pairs of conformers by their "conformational distance": rms deviation of the dihedral angles between the two structures (taking into account a possibly different atom numbering). He also proposed a process using smaller and smaller kicks to determine the minimal geometrical perturbations necessary to induce conformational interconversion. This allows for determining the attraction basin of conformers (the part of the structure space where a starting geometry will refine to the conformer in question) and the most likely interconversion pathways.

To characterize the efficiency and completeness of various approaches, it was interesting to compare them on the same example. This was the aim of the work of Bohm *et al.* on a nine-membered lactam [53]. On the one hand, it may be expected that systematic search in principle generates all possible solutions, but is highly demanding of CPU time and storage location. On the other hand,

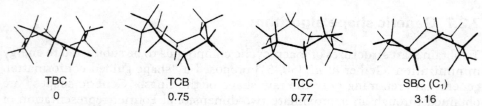

TBC   TCB   TCC   SBC (C_1)
0   0.75   0.77   3.16

Figure 7.8   The well established conformations of cyclonane. T = twist, B = boat, C = chair, S = skewed. Numbers give the relative MM2 energies (kcal/mol) above the minimum (TBC) (adapted from Ferguson *et al.* with permission [25]).

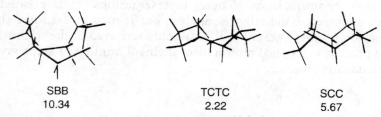

SBB   TCTC   SCC
10.34   2.22   5.67

Figure 7.9   The new conformations of cyclononane derived from RIPS. T = twist, B = boat, C = chair, S = skewed. Numbers give the relative MM2 energies (kcal/mol) above the minimum (TBC). Conformations TCTC and SCC were also proposed by the tree search of Lipton and Still ([2]) (adapted from Ferguson *et al.* with permission [25]).

molecular dynamics or Monte Carlo methods cannot guarantee the completeness of the search. From tests where each method considers about 1000 start structures, it appears that molecular dynamics, Monte Carlo methods or the generic shape algorithm are equally as efficient, finding most of the 39 local minima up to 60 kJ/mol above the global minimum. Generation from crystal data issued from the Cambridge database found a more limited set (23 local minima in place of 39), but had the advantage of an experimental origin. On the contrary, systematic search (working on a comparable set of start structures) appears "extremely inefficient", with only five solutions proposed (presumably because the number of starting structures was too low for a complete exploration of the conformational space) (Figure 7.10).

A comparison of diverse methods was also carried out on the example of *n*-octane [39]. Molecular dynamics stands apart, since it is more effective at local searches. Simulated annealing appears as the fastest (more rapid than the Monte Carlo method of Chang *et al.* [22] or the stochastic search of Saunders [27]. Systematic search with 60° steps is very low but, if the resolution is decreased down to 120° (which still appears enough for straight chains) it competes quite well with other methods. Unfortunately, distance geometry was not tested.

These conclusions are consistent with the comments of Howard and Kollman [7], who concluded that "systematic search, helped by expert systems, may be the most convenient approach for small or moderately sized systems, whereas larger molecules may be more amenable to distance geometry or molecular dynamics coupled with experimental distance constraints".

Energetic aspects also deserve some comments. So, Bohm *et al.* [53], in their comparative study, noted that some differences appear as to the ordering of the local minima, depending upon the force field or the quantum methods used, and recommended some caution for structural discussions based on energy differences. Furthermore, as stressed by Howard and Kollman [7], an important (but often neglected, and still difficult to treat) aspect is the role of the surrounding medium. For charged or highly polar species, interaction forces in aqueous or biological environments are expected to largely differ from those existing in the gas phase or in the crystal, and their influence has to be taken into account. Explicit inclusion of solvent molecules or appropriate approximations to simulate their effects are necessary (but of course costly) processes to get a better insight about the relative energies of the active conformers in living media.

Another comparative study was recently carried out on highly flexible cycloheptadecane, including systematic and random searches in both internal

Figure 7.10

(torsion angle) and external (Cartesian) coordinates, molecular dynamics and distance geometry. Within 3 kcal/mol of the global minimum, 262 conformations were discovered, and it can be assumed that the conformer of minimum energy is only 8% in the conformational mixture at room temperature. None of the methods found all the low-energy conformations in a single search, but all, except distance geometry, located the same global minimum. In this study, distance geometry and molecular dynamics seem to be less effective in finding all low energy conformations in comparable CPU times [54].

In a recent study [55], distance geometry and molecular dynamics were compared in their ability to search the conformational potential energy surface of β cyclodextrin. Possible structures were first generated by distance geometry and then used as starting points for molecular dynamics simulation. Distance geometry appears able to find structures of a lower energy than those obtained by minimization of X-ray or neutron diffraction structures. Molecular dynamics simulations find structures still lower in energy than those derived from distance geometry. However, owing to the low temperature adopted (298K), they can only explore regions around the distance geometry-starting structure whereas distance geometry traverses much larger regions of the conformational space. Molecular dynamics therefore appears as a way in which to locally refine the distance geometry structures. For neighbouring applications, see elsewhere [56, 57].

7.3 MODEL BUILDERS

When constructing compounds, at least in the simplest cases, a chemist can guess well enough, thanks to the general rules of conformational analysis, low-energy conformers close to the solution given by an energy minimization program. It can be hoped that mimicking the chemists reasoning by automated exploitation of the usual knowledge would significantly speed up conformational analysis.

In this field of molecular architecture, the use of computer's power to derive a user-friendly interface, providing the chemist with reasonable conformations close enough to the true local minimum energy, is of course a highly attractive prospect. An estimated geometry not too far from the real one saves steps (and CPU time) in energy minimization processes. It lessens the risk of stopping on false or local minima. Furthermore, in some fields (e.g. surface and volume calculations) it may give reasonably acceptable solutions, without the need for more refined treatments.

7.3.1 From 2D to 3D representations

Such model builders may also appear to bridge the gap between 2D and 3D representations, i.e. producing a reasonable 3D representation from a topological description (connection table augmented with some stereochemical descriptors). Indeed, numerous computer-aided structural treatments (structure/activity or structure/property relationships, structural elucidation, spectral simulation, etc.) largely rely on 2D representations of the molecular framework. In other words, they consider only the molecular topology (as do planar structural formulae), possibly augmented by flags specifying stereochemical features (configuration of chiral centres, cis or trans orientation, etc.). On the other hand, in molecular modelling, conformational refinements or the derivation of electronic features start from a 3D representation of the structure as the input of downstream simulation programs.

Early attempts seemed mainly devoted to giving a guess geometry before starting minimization programs, whereas the aim of more recent works is to propose reasonable solutions able to be used directly, thanks to reasoning relying on the artificial intelligence methodology. The expert system approach, in so far as it mimics the chemist's reasoning, has been used in systems such as WIZARD relying on predicate calculus. By other respects, CONCORD uses both rules and tables to derive bond lengths and angles, with the help of a simplified energy-minimizer. The model builder AIMB proposes analogical reasoning from the knowledge already captured by crystallography.

Among the basic features, it can be noted that at a first level of approximation (except for severe steric constraints), bond lengths and valency angles may be considered as fixed at standard values (according to atom type or hybridization), and the problem is mainly to determine dihedral (torsional) angles. Relieving bad van der Waals contacts and ensuring ring closure are therefore the main points to examine. A major difficulty arises from the flexibility of acyclic chains or large ring systems able to have several conformations of comparable energy, whereas smaller rings exhibit more rigidity, possibly with largely distorted geometries for the more severely strained small-ring structures. We shall now briefly present the main features (in our opinion) of these recently proposed model builders (for a detailed review, see elsewhere [58]). As quoted by the authors, the main requirements these automatic model builders must satisfy are: robustness, the ability to process large files with a high conversion rate, speed in automated mode, and of course, high quality models (including stereochemistry) for varied chemical types.

7.3.2 Earlier attempts

The MBLB [59] program of Gordon and Pople uses a connection table, the geometrical information being conveyed implicitly in the ordering of entries within the connection table. The model is able to reproduce user-defined rotameric states around bonds. However, it cannot cope with structures arbitrarily created by a structure generator. Non-sequential ordering within the connection table may also lead to problems.

Other methods start from a planar representation (such as those obtained on the screen of a graphic terminal) with symbols to specify up and down orientations with respect to the plane of the screen. This gives approximate Cartesian coordinates sufficient to start geometry optimizations [60].

It has been also proposed (SCRIPT program) to use a library of templates (i.e. molecular fragments) with a set of rules governing the assembly of these fragments into complete molecules [61]. This makes possible systematic and automatic generation of complex cyclic structures (including fused, spiro or bridgehead systems). Similarly, the SCA system (Systematic Conformational Analysis) [62, 63] uses a list of allowed conformations for single rings and determines torsion constraints from a set of rules.

7.3.3 Application of distance geometry

The approach of Wenger and Smith [15] appears as a variation of the distance geometry of Crippen *et al.* The input is formed by a connection table augmented by stereochemical indicators, parity bits (0 or 1) specifying the configuration of chiral centres, cis or trans orientation, etc.

The next step is to set distance ranges in the distance matrix of the Crippen approach. The connection table provides information about atom connectivity, but none on distances. However, some (approximate) values can be derived for 1,2 and 1,3 distances and limits on possible 1,4 distances:

- 1,2 distances are set equal to standard bond lengths.
- 1,3 distances (between 2 atoms, i,k bonded to a same atom j) are easily determined from 1,2 distances and the valency angle (θ):

$$d^2{}_{ik} = d^2{}_{ij} + d^2{}_{jk} - 2d_{ij}d_{jk}\cos\theta$$

(For 1,2 and 1,3 distances, a common value is used as the lower and upper bounds in a Crippen distance matrix).

- Minimal and maximal 1,4 distances are then determined. So, for an sp^3 moiety they respectively correspond to dihedral angles of 0° and 180°.
- Other distances are set to 2.0 Å (minimal value) to avoid overlap between non-bonded atoms, and for upper values to $10\,n^{1/3}$ Å (where n is the number of atoms in the molecule).

The program can also deal with the constraints existing in small cycles, and uses additional geometrical information provided by the user. Given the Crippen distance matrix, the next step searches for a set of Cartesian coordinates of the atoms consistent with these interatomic distance ranges. This is solved in the usual way detailed previously (see p 202) by computing a metric matrix (with distances referring to the centre of mass), determining its first three greatest eigenvalues and the corresponding eigenvectors, and then refining the coordinates to satisfy configuration and distance constraints.

Although very attractive, this approach suffers some limitations (already indicated for distance geometry). The computation process begins with the placement of atoms at random locations (consistent with the distance ranges). So, it only leads to a random sampling of the conformational space, rather than to a systematic exploration. Furthermore, no energetic criteria are considered, so that structures satisfying the distance constraints, but chemically "unreasonable", can be proposed and these have to be examined by usual energy minimizers.

7.3.4 WIZARD

The WIZARD system [64], and the more recent COBRA[1] [65, 66] program, apply expert system techniques to propose preliminary estimates of likely low-energy conformers. Although limited in its first version to saturated acyclic hydrocarbons, it is now able to treat more complex structures up to 200 atoms and accept varied file formats. Rather than directly proposing an exact solution, the aim of WIZARD at the beginning was mainly to suggest a tentative conformation within the energy well of a local minimum, that will be subsequently refined by energy minimization.

Let us first emphasize some common components of expert systems, and examine them through WIZARD implementation. The main parts of any expert system are (Figure 7.11):

- A *knowledge database* which gathers a body of knowledge extracted from a human expert.

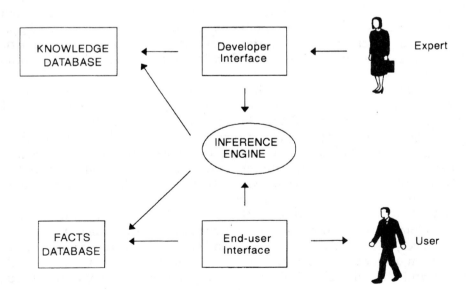

Figure 7.11   WIZARD expert system schematic.

[1]COBRA is available from Oxford Molecular Ltd, Oxford, UK.

- A set of tools for querying and utilizing this knowledge. This is performed by an *inference engine*, which uses the knowledge to derive rules. These rules will be applied to the study of the problem under investigation.
- A *database of facts* which gathers data about the problem to solve.

Among the characteristics of expert systems, one can note that, in contrast to exploiting a database, the knowledge can be applied to situations not previously encountered. Thanks to these new problems, this knowledge can be updated in a way similar to that of human experts.

Knowledge is stored as rules, which often take the form:

IF such condition is fulfilled
THEN such consequence results.

Each rule draws one conclusion about a subject, and is dependent upon one or more terms (either observed or inferred). For instance:

"the bond angle at an atom can be inferred to be increased
IF it is between two bulky groups (observed) and
IF these groups show steric interaction (inferred)" [64]

To fire an action the system uses backward chaining through the rules until all the inferred terms are reduced to observational terms: according to whether the response is yes or no, it decides upon the appropriate conclusions.

The way in which the knowledge is encoded significantly influences the efficiency of such expert systems. With procedural approaches, the knowledge is directly expressed in a programming language such as Fortran. In a declarative fashion, the knowledge is stored as declaration relationships between objects. Then an interpreter in an advanced artificial intelligence (AI) language (such as Prolog) ensures diverse tasks: maintaining and updating the database, performing backward chaining, monitoring input and output, etc.

Although procedural approaches deal more rapidly with numerical problems, the declarative method clearly separates the tools and the knowledge which they address. So, one can focus on the expression of knowledge without worrying about procedural details. The ability of the AI languages to manipulate high level concepts makes the reasoning easier. It allows an efficient way to update the knowledge database and solve conflicts between rules. For these reasons, WIZARD uses both Prolog and Fortran (for some subroutines and subsequent MM2 calculations).

In WIZARD, the rules forming the knowledge base stem from a double origin. Some have been extracted from chemists. For instance (Figure 7.12): "Place everything on a diamond lattice and see if it fits without problems. If it does, then quit, otherwise continue..."; or (pentane rule): "a C5 chain with a $g+/g-$ pair (dihedral angle $= \pm 60°$) is much less stable than a conformer lacking such a pair." [64]

Others are derived from examination of about 100 model molecules analysed by the MM2 molecular mechanics program. About 200 rules were used, in which fewer than 50 are specific of conformational analysis.

Figure 7.12 Mapping the carbon framework onto the diamond lattice. Butane anti (top), gauche (right) and cyclooctane (left) (from Randic *et al.* with permission [67]).

Predicate calculus formalizes arguments and inference by manipulating symbolic formulae. Reasoning is independent of the domain.

After the data have been read in from a 2D representation, the program identifies "conformational units" (patterns such as phenyl, cyclohexane, carboxyl) and generates a "unit graph", an abstract representation of the structure to be built. Nodes correspond to conformational units and edges specify the mode of junction (acyclic, fused or bridged rings). Every unit is then assigned one (or a few) subconformation, taken from a library where a series of templates are stored [65, 66]. The basic assumption is that each unit will exhibit the same behaviour in large molecules as in a small, isolated molecule, unless strain deforms it. A "suggested" structure is built by assembling these local subconformations. "Suggested" structures are criticized before being accepted, since subconformations only reflect local interactions and do not deal with long range effects: tests concern energetics, junction between units, bad van der Waals contacts, violation of pentane rule, etc. An estimation of the strain energy is obtained by simple summation of a few selected interaction terms (constituting a very crude force field). So, for a simple butane bond, the gauche arrangement is set approximately 0.8 kcal less stable than the anti conformation (Figure 7.13). If the suggestion is rejected another "assemblage" is tried.

With the COBRA program, a total search of the conformational space can be performed by examining all possible combinations of units. To find the lowest energy structures more efficiently, the search is directed using a tree representation of the search space and evaluating a cost path at each step of the assembling process without explicit use of minimization methods (in contrast to the subunit optimization of Scheraga *et al.* [23, 24], where energy minimization is invoked at each step).

From the examples given (first on acyclic alkanes, then on functionalized

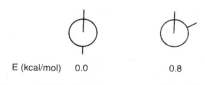

Figure 7.13 Strain energy estimation.

compounds), it appears than WIZARD does suggest conformers quite close to the solutions obtained from complete MM2 optimization, and so significantly speeds up the exploration of the conformational space.

The same methodology is used to build strained structures. If a problem occurs in one conformation, a similar process of suggestions and criticisms is performed on deformations of templates, and rules are activated to relieve strain. The treatment of highly branched molecules, however, would require a very large set of special case rules without the certainty that the system is complete. So, they were replaced by a few suggestions for strain resolution, ordered according to their probable hierarchy in energy (dihedral twist first, then angle bond opening, and finally bond lengthening). Hydrogen-bonding can be also considered. These rules are sequentially examined and criticized: if they lead to conflicting decisions, one goes to the following rule to perform the resolution of the conformer. For instance, in di *tert*-But methane, the alteration of dihedral angles appears insufficient to relieve van der Waals repulsions, and increasing the central angle tBu–C–tBu is invoked (Figure 7.14).

Figure 7.14 Alternative alteration of dihedral or valence angles.

For more refined applications, the combination of WIZARD and MM2 was claimed to explore the conformational space of a saturated acyclic hydrocarbon more rapidly than by using the torsion angle driver of MM2 alone, and with a greater confidence of complete coverage than manual generation followed by MM2 minimization (up to seven degrees of freedom can be efficiently handled). Thanks to its high level symbolic reasoning, it also allows for finding otherwise ignored minima, and avoids some poor suggestions. The system was recently extended [68] combining model-building and distance geometry, the latter method performing a conformational search if templates are lacking (Figure 7.15).

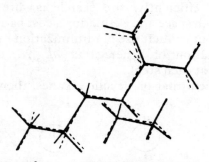

Figure 7.15 Comparison between WIZARD's suggestion (dashed lines) and MM2 optimization (solid lines); non-strained conformer of isopropylhexane (from Dolata and Carter with permission [64]).

7.3.5 CONCORD

CONCORD aims to generate approximate 3D structures from a connection table representation of molecules [69, 70]:[2]

CONnection table → COoRDinates

For any given molecule, CONCORD gives a default geometry assumed to be of low energy, and usually proposes shapes not too far from exact minimization (by molecular mechanics or quantum methods). The basic organic elements are treated (H, C, Si, N, P, O, S and the halogens) with a maximum atomic connectivity of 4. The upper limit for acyclic, mono-, poly- or heterocyclic structures is about 200 non-hydrogen atoms. CONCORD is a hybrid system which combines the expert system approach and a pseudo-molecular mechanics approach. Many decisions about bond and torsion angles are derived from rules, whereas an approximate optimization procedure deals with complex cases, as for example, less common cycles. Basically, the majority of the effort is spent on the more rigid fragments while little time is wasted on flexible chains and their environment-dependent conformations.

At the origin, CONCORD treats as input the compounds stored in the MACCS database[2], using SMILES (Simplified Molecular Interpretative Line Entry System), a linear notation system based on graph theory and encoding molecules as alphanumerical strings [71–75]. More recently, additional input formats have also become accepted. Similarly, at output a variety of formats are provided for interfacing with molecular or quantum mechanics programs, or various databases.

Some limitations have been pointed out:

- Only one conformer is generated. This may be a drawback for molecules that are able to exist in various low energy states, and for which the reacting conformation is not in all cases the most stable or the crystal one. However, it was stressed that owing to the difficulty of exploring the conformational space with complete safety on one hand, and on the other hand storing many conformers, "searches need to be suggestive and not exhaustive" [69].
- Until now, only organic structures have been treated, although extension to other elements of the periodic classification is planned.
- The rules encoded are very general, and may lead in some cases to unrealistic structures that must be rejected (however, some warning or error messages are included in the program).

Recently, Hendrickson *et al.* [76] compared geometries proposed by CONCORD with experimental ones from the Cambridge Crystallographic Database. The conversion rate was about 69%. For 90 structures where both types of data were available, it appears that CONCORD gave quite satisfactory geometries for rigid systems, but failed for flexible structures with several rotatable bonds.

---

[2]CONCORD distributed by TRIPOS Associates, St Louis, MO. 1987.
[3]MACCS, The Molecular ACCess System. Molecular Design Limited, San Leandro CA, USA.

7.3.6 CORINA

The CORINA model builder [77, 78] automatically generates 3D structures from connection tables. For flexible molecules, it proposes one conformation (that assumed to be the lowest energy), but a series of different conformations can be also created. The basic units are monocentric configurations (an atom with the stemming-bond directions) which are gradually put together, respecting appropriate bond lengths and torsion angles. Deviations in standard geometrical parameters are allowed for cyclic compounds or strained systems. A pseudo force field (ensuring rough molecular mechanics calculations) helps in refining cyclic structures. (A detailed description can be found elsewhere [58].) From the examples presented, CORINA predictions show good agreement with experimental crystallographic structures (from the Cambridge Database) and the program seems particularly interesting for macrocyclic, polymacrocyclic molecules or organometallic compounds.

The reliability of CORINA predictions was tested in comparison with 639 X-ray data from the Cambridge Database. CORINA was able to process the whole set with a very high conversion rate; 89% of ring atoms and 42% of all non-hydrogen atoms were located well, with an accuracy comparable to X-ray data, showing the efficiency of the approach. It must be noted that "for more than one third of the structure, the X-ray geometry was reproduced, including also the flexible parts of the molecules" [58].

7.3.7 Analogy In Model Building: AIMB

Another strategy was adopted with the AIMB project [79, 80]. The aim was to provide an automated model builder using *analogy* to propose 3D models through symbolic reasoning. Rather than using rules, as the previous systems, AIMB relies on existing data which implicitly contain knowledge on building molecules. The process avoids heavy minimization programs and works like an expert system. It therefore looks not too dissimilar to the usual behaviour of the chemist when he manually builds molecular structures.

In this field of model building, the analogy concept may be expressed as "if two compounds have a similar structure, they may have a similar geometry".

Like an expert system, AIMB stores all its knowledge about molecular geometry in a knowledge base, which gathers the conformations (3D coordinates) of typical molecules captured from the Cambridge X-ray Crystallographic Database.

Given a target molecule to be constructed, the program looks for analogous structures in the database. If there are none, the target is divided into smaller parts and each of these subcomponents is in turn examined as a new target to be sought for in the crystal file, and so on until analogy is found or the program stops if no analogous solution exists in the knowledge base.

Underlying this approach is the basic assumption that recognizing subunits provides partial solutions where the interactions within the subunit are already minimized. So, the global problem is now reduced to cope with interactions between subunits (that is, interactions near their frontier). To provide an easier treatment of these interactions, each subcomponent is given some information about the remainder of its structural environment thanks to dummy atoms (atoms α to atoms involved in solving the subunit identification). These dummy atoms, which to some extent specify the direction of the hanging free valences, are also very useful for the alignment steps when making the new bonds assembling subunits into the structure.

Some of the main components of the AIMB system are now briefly presented and these processes summarized in Figure 7.16 a–d [79].

The *knowledge base* is given a hierarchical organization: three classes distinguish ring assemblies, rings and chains. In each class the components are differentiated according first to their size, then on the basis of atom and bond types.

After perceiving the target structure (thanks to a graphical interface converting the chemist's 2D input-diagram into a connection table with symbolic stereochemistry), the *analyser* identifies rings, chains, aromaticity and stereochemistry. It determines if close analogous solutions (or the target itself) are contained in the knowledge base, and it selects the most closely related ones. If no analogous solutions are found, the analyser breaks down the target into subunits. Then each of them is treated as a new problem.

The *analogy finder* looks for analogies to each subproblem in the knowledge base. An *evaluator* scores each match to determine the best solution

Target

Ring 1   Chain 1   Chain 2   Chain 3   Ring 2

Fragment 1   Fragment 2   Fragment 3   Fragment 4

Sub-problems

(a)

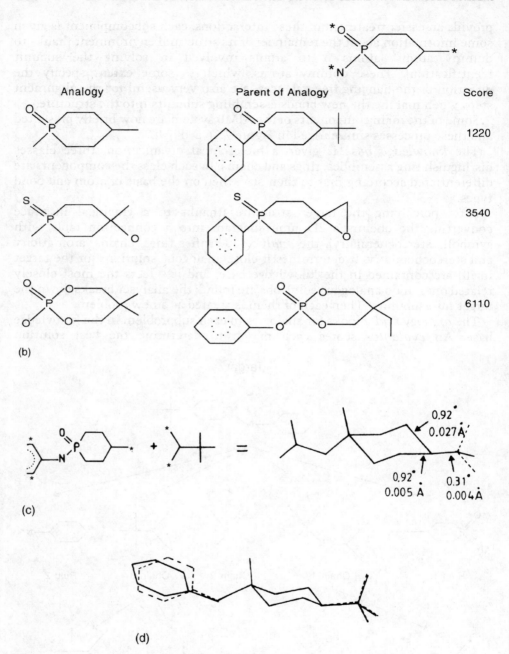

Figure 7.16 Steps in the AIMB process. (a) Division into subproblems when no compound analogy is directly found for the target; (b) analogy scores for ring subproblem 1 (higher value, lower analogy) (dummy atoms are represented by stars); (c) assembling fragments using dummy atoms for alignment (fusion misfits between dummy and real atoms are indicated); (d) comparison of AIMB model (dashed lines) with crystal structure (solid lines) (from Wipke and Hahn with permission [80]).

depending upon the atom and bond attributes such as type, valence, hybridization, stereochemistry, bond type, etc.

The *model assembler* fuses the solved pieces, using dummy atoms for alignment, and displays the structure. An *explanation module* gives supporting data for the proposed model. Display of the parent compounds of the subunits retained, and literature references and data about the constraints that were (or were not) met are also provided.

The results presented by the authors, even at an initial prototype stage, suggest that the approach is very efficient (for a detailed discussion of the performance of the method, see Wipke and Hahn [80]. Note that AIMB leads to solutions fairly close to those obtained by true energy minimizations, and works at a rate allowing for interactive manipulations. Increasing the knowledge base not only gives better, more realistic models, but also, albeit more surprisingly, increases speed: better analogies are found sooner, since fewer constraints need to be relaxed before analogues are found. Furthermore, since the program evaluates more than one analogy per subproblem, it is able to generate alternate or best models, so it gives some capabilities to conformational search.

As to the drawbacks mentioned, prediction would be poor if only a few distant analogues were found. Another problem stems from long range interactions. Although dummy atoms generally allow for conveniently assembling subunits, in a few cases long range interactions may cause some trouble. However, this situation seems quite infrequent: such effects are generally implicitly taken into account and encoded in the experimental crystal structures, so they rarely heavily intervene when assembling subunits.

A combined distance geometry and joining approach also relying on a knowledge base was proposed by Ai and Wei [81]. In this knowledge base, a record is a molecular fragmentary structure or a conformational unit that includes the bond lengths, angles and torsion angles of a centre atom in a specific environment. So the knowledge base is small and easily handled.

7.3.8 Generation of databases of calculated structures

Increased efforts in drug design have recently concentrated on the search for new leads from which potentially active drugs may be derived, a field where geometrical information is obviously of prime importance. This has prompted various manufacturers or users to build databases gathering "reasonable" 3D structures inferred from 2D structural formulae. Such databases are therefore complementary to the Cambridge Crystallographic Database gathering experimentally determined geometries.

In this quest for new leads, an important preliminary step is the search for pharmacophoric patterns, that is, a group of atoms considered as necessary for a molecule to be recognized by a receptor and therefore possess the desired activity. This research area, which closely relies on the concept of "molecular similarity" is presented in Chapters 11 and 12. The field is now rapidly evolving thanks to the development of varied 3D search techniques. New tools include 3D similarity searching, representation and searching of flexible molecules, and the design of new molecules fitting 3D constraints.

However, beyond pharmacophore search (which remains up to now the main concern), current applications of 3D databases are now also concerned with chemical documentation, modelling, synthesis, etc. and reach new structural areas such as macromolecules or proteins [92].

Two main avenues may be distinguished:

- introduction of 3D information and property values in the large files of the Chemical Abstract Service [82], or
- the extension to 3D structures of the large existing commercial or proprietary databases associating 2D structural information with biological or pharmaceutical activity [83–85].

The generation of these large 3D databases was extensively carried out thanks to various model builders, such as CONCORD [69], AIMB [79, 80], WIZARD or COBRA [64–66] as detailed in the preceding paragraphs.

CONCORD, for example, was used on a very large scale to generate 3D databases. Rusinko *et al.* [69] indicated that the Lederle Laboratories (American Cyanamid Company) have built a database of ca. 224,000 structures. Similarly, using CONCORD, Henry *et al.* [83] proposed two 3D databases, MACCS-II DrugData Report–3D (MDDR–3D) and the Fine Chemical Directory 3–D (FCD 3–D), built from existing 2D databases at Molecular Design Limited. According to the authors, in their 1990 releases, MDDR contains 20,637 compounds (release 90–2) and FCD 66,000 (release 90–1). The THOR database of Martin *et al.* gathers ca. 70,000 structures translated from the MACCS database using the MEDCHEM program (MACCS acronym for Molecular ACCess System, the structure database management system from MOLECULAR DESIGN LIMITED) [72–74].

For years, the Chemical Abstract Service (CAS) has collected very large databases encoding about 11 million compounds and about 100,000 Markusch (generic) structures. Using CONCORD, a 3D database for about 4.5 million substances was generated [82]. To evaluate new techniques or strategies of exploitation, two subsets (60,000 compounds) have been extracted, one general, the other one limited to rigid substances. For 6000 structures, besides 3D coordinates, varied characteristic properties or indices are now available: elctronic and energetic information from the semi-empirical molecular orbital MOPAC package (HOMO, LUMO, maximum, minimum and mean values of electronic populations), hydrophobicity, van der Waals surface area and volume, and topological indices such as flexibility etc.

In other respects, the Chapman and Hall Dictionary of Drugs, the source of information on over 6000 drugs and their derivatives, was similarly converted to a 3D database built from the Chem-X software [84][4], according to a method similar to AIMB of Wipke and Hahn [79].

---

[4]Chem-X: molecular modelling software developed and distributed by Chemical Design Ltd, Oxford, UK.

Obviously, beyond the efficiency of the model builder used to generate a 3D database rapidly and with high conversion ratios, (as discussed some pages above), the interest of such 3D databases largely depends on some key points:

- reliability of the generated geometries
- availability of efficient database management systems and search strategies.

As to *management systems*, the MACCS system, for example, used for storing and searching chemical structures as 2D entities was extended in MACCS 3D to include additional files containing the 3D information (linked by a structure/model correspondence to the 2D files) [85]. An obvious advantage is the capability of merging data from varied fields (spectroscopy, computational structural chemistry) into a corporate database allowing scientists from different disciplines to access the information more efficiently. This point was discussed by Güner *et al.* [93]. The possibility of accepting the diverse description languages used for automated treatment of 2D structural formulae or various formats for connection tables is also quite useful.

Special strategies had to be defined for searching three-dimensional substructures (or pharmacophores) [86, 87]. For example, speed can be attained by dividing searches into two parts: a fast prescreen uses an inverted key system, of the form:

$$\text{atom type 1} \quad \text{distance} \quad \text{atom type 2}$$

where the atom type includes five fields (element, neighbours, π electrons, hydrogens, formal charge). After the key search (finding structures containing pairs of atoms with the correct type at roughly the correct distances), an atom by atom geometric search is performed, relying on the Ullman algorithm (isomorphous subgraph algorithm, see p 345). Angle or dihedral constraints can be handled. Excluded volumes (no atoms of the substructure at given positions) can also be taken into account. The examples presented showed that test pharmacophores (already proposed in the literature) can be easily found. Interestingly, the system can also identify classes of compounds structurally different from those used for the derivation of the pharmacophore but still containing it. In some cases, the search can even indicate bond "frameworks" in which the important atoms are connected in novel ways or held rigidly in the desired geometry. Such answers may constitute a possible origin of worthwhile suggestions for new directions in drug synthesis [88]. See also Chapter 12 for more details.

Conformational flexibility must also be taken into account, since it is a problem of prime importance in the handling of 3D structures. Indeed, the active conformation of a drug (that bound to its receptor) is not necessarily the lowest energy conformation (in solution) nor the crystal geometry. This was established for instance for acetylcholine in its active conformation, binding the nicotinic receptor. Flexibility can be introduced both in the query (combining rigid and flexible parts for allowing more degrees of freedom) and in the search (performing conformational analysis or "flexible fitting"), or in the database, storing several conformations for the same structure [85, 89, 92, 94].

This latter solution, retaining for each compound some of the lowest energy conformers, would in fact require much disk space. Modules were therefore developed allowing work with a database storing only one conformation per compound but able to build and search for various low energy conformations for each compound without explicitly storing them, as with the ChemDBS-3D module of the Chem-X commercial software [89–90, 95]. Using rule-based conformational searching, the algorithm generates additional low energy geometries, allowing only certain values of torsional angles for rotatable bonds. Searching is carried out with keys evaluated for each possible conformation generated. Generation of low-energy conformers is also possible with the ALADDIN program [96], for example in order to test hypotheses about a pharmacophore model. Structural descriptions may also include some flexibility [97].

Several works discuss the reliability of the structures generated in these databases. A recent paper [90] explores the ability of various packages to generate the structure of small molecules in their bound conformations and compare them with those in ligand complex crystal entries in the Brookhaven Protein Data Bank. The programs reviewed include CONCORD, COBRA, ChemDBS-3D, developed by Chemical Design Ltd. [89, 90] and CONVERTER[5]. This package developed by Biosym. Technologies, relies on a distance geometry method. The aim was to examine whether such programs can generate the active conformation of a drug (that which bounds the receptor), a challenging problem due to conformational flexibility. It is to be noted that CONCORD produces a single low-energy conformation whereas the other programs can propose more than one low energy structure wherever possible. Furthermore COBRA was devoted to conformational analysis rather than rapid 2D–3D conversion of large files. According to the authors, ChemDBS-3D is limited by the size of the structures and achieves only a 62% conversion rate, but then gives structures very close to experiment. CONCORD performs well, with a high conversion rate, and generally good results. More problems arise with COBRA for that type of application. CONVERTER, with 100% conversion and good proposals, is claimed to be the best. Rather puzzling is the fact that the PDB (Brookhaven) and the CSD (Cambridge) structures are not always in agreement. This may reflect conformational changes occurring during binding, the influence of conformational flexibility, or perhaps a consequence of the accuracy of the coordinates of the ligands in the PDB [90].

Computer-generated structures obtained from ChemModel, the model builder proposed by Chem-X, were also compared to X-ray crystallographic data [91]. In 57% of the cases, computed structures and experimental data are in good agreement, whereas CONCORD gives only 38% satisfactory results. According to its authors, the superior performance of ChemModel is due to the generation of multiple structures covering the entire conformational space.

In conclusion, it must also be recalled that searching 3D databases is far more demanding on computational resources than 2D database searching [90].

---

[5]CONVERTER is a product of Biosym Technologies Inc, San Diego, CA. 1992.

Indeed, in a pharmacophore search, the query generally comprises few atoms, distance ranges are rather large, and proprietary databases may contain several analogous compounds. This generally results in numerous answers unless efficient strategies are introduced. Parallel computing seems therefore an attractive alternative [90]. However, not surprisingly, a 3D search is not by definition more efficient than a 2D search: if a good conformer is not considered, the solution can be missed [95].

Although the field is rapidly evolving, 3D searching has up to now been a complex and cumbersome procedure. To obtain a molecule containing a given pharmacophore, rather than searching for a structure already stored in a database, the recent approach of "*de novo* design" tries to build from atoms or fragments molecules meeting the imposed criteria or constraints. The efficiency is now limited by the number of available structural primitives and not by the extent of the database. Such applications will be developed in Chapter 12.

REFERENCES

1. A.J. Holder and D.L. Wertz *J. Comput. Chem.*, **9**: 1988; 684–688.
2. M. Lipton and W.C. Still *J. Comput. Chem.*, **9**: 1988; 343–355.
3. G.M. Crippen *Distance Geometry and Conformational Calculations*, Research Studies Press (D. Bawden, Ed.), Wiley, New York 1981.
4. G.M. Crippen *J. Math. Chem.*, **6**: 1991; 307–324.
5. G.M. Crippen *J. Comput. Phys.*, **24**: 1977; 96–107.
6. P.K. Weiner, S. Profeta Jr, G. Wipff, T.F. Havel, I.D. Kuntz, R. Langridge and P.A. Kollman *Tetrahedron*, **39**: 1983; 1113–1121.
7. A.E. Howard and P.A. Kollman *J. Med. Chem.*, **31**: 1988; 1669–1675.
8. I.D. Kuntz *Protein Eng.*, **1**: 1987; 147–148.
9. G.M. Crippen and T.F. Havel *Distance Geometry and Molecular Conformation*, D. Bawden (Ed.), Chemometrics Research Studies Series, Research Studies Press, Wiley, New York, 1988.
10. T.F. Havel and K. Wüthrich *Bull. Math. Biol.*, **46**: 1984; 673–698.
11. M. Billeter, T. Schaumann, W. Braun and K. Wüthrich *Biopolymers*, **29**: 1990; 695–706.
12. A.D. Kline, W. Braun and K. Wüthrich *J. Mol. Biol.*, **189**: 1986; 377–382.
13. J.W. Pflugrath, G. Wiegand, R. Huber, and L. Vertesy *J. Mol. Biol.*, **189**: 1986; 383–386.
14. H.I. Mosberg, K. Sobczyk-Kojiro, P. Subramanian, G.M. Crippen, K. Ramalingam and R.W. Woodward *J. Am. Chem. Soc.*, **112**: 1990; 822–829.
14a. G.M. Crippen *J. Comput. Chem.*, **10**: 1989; 896–902.
15. J.C. Wenger and D.H. Smith *J. Chem. Inf. Comput. Sci.*, **22**: 1982; 29–34.
16. R.P. Sheridan, R. Nilakantan, J.S. Dixon and R. Venkataraghavan *J. Med. Chem.*, **29**: 1986; 899–906.
17. C.E. Peishoff and J.S. Dixon *J. Comput. Chem.*, **13**: 1992; 565–569.
18. E.O. Purisima and H.A. Scheraga *Proc. Natl. Acad. Sci. USA*, **83**: 1986; 2782–2786.
19. E.O. Purisima and H.A. Scheraga *J. Mol. Biol.*, **196**: 1987; 697–709.
20. G.M. Crippen *J. Comput. Chem.*, **3**: 1982, 471–476.
20a. G.M. Crippen *J. Phys. Chem.*, **91**: 1987; 6341–6343.
20b. G.M. Crippen and P.K. Ponnuswamy *J. Comput. Chem.*, **8**: 1987; 972–981.

21. J.J. Gajewski, K.E. Gilbert and J. McKelvey in *Advances in Molecular Modeling*, D. Liotta (Ed), JAI Press Inc, London, 1990; 65–92.
22. G. Chang, W.C. Guida and W.C. Still *J. Am. Chem. Soc.*, **111**: 1989; 4379–4386.
23. M.R. Pincus, R.D. Klausner and H.A. Scheraga *Proc. Natl. Acad. Sci. USA*, **79**: 1982; 5107–5110.
24. K.D. Gibson and H.A. Scheraga *J. Comput. Chem.*, **8**: 1987; 826–834.
25. D.M. Ferguson, W.A. Glauser and D.J. Raber *J. Comput. Chem.*, **10**: 1989; 903–910.
26. D.M. Ferguson and D.J. Raber *J. Comput. Chem.*, **11**: 1990; 1061–1071.
27. M. Saunders *J. Comput. Chem.*, **10**: 1989; 203–208.
27a. M. Saunders *J. Am. Chem. Soc.*, **109**: 1987; 3150–3152.
28. D.M. Ferguson and D.J. Raber *J. Am. Chem. Soc.*, **111**: 1989; 4371–4378.
29. J.M. Goodman and W.C. Still *J. Comput. Chem.*, **12**: 1991; 1110–1117.
30. W.C. Guida, R.S. Bohacek and M.D. Erion *J. Comput. Chem.*, **13**: 1992; 214–228.
31. D.M. Fergusson, I.R. Gould, W.A. Glauser, S. Schroeder and P.A. Kollman *J. Comput. Chem.*, **13**: 1992; 525–532.
32. T. Noguti and N. Go *Biopolymers*, **24**: 1985, 527–546.
33. H. Kawai, T. Kikuchi and Y. Okamoto *Prot. Eng.*, **3**: 1989; 85–94.
33a. J. Skolnick and A. Kolinski *J. Mol. Biol.*, **212**: 1989; 787–817.
34. N. Metropolis, M.N. Rosenbluth, A.W. Rosenbluth, E. Teller and A. Teller *J. Chem. Phys.*, **21**: 1953; 1087–1092.
35. Z. Li and H.A. Scheraga *Proc. Natl. Acad. Sci. USA*, **84**: 1987; 6611–6615.
36. B. von Freyberg and W. Braun *J. Comput. Chem.*, **12**: 1991; 1065–1076.
37. S. Kirkpatrick, C.D. Gelatt Jr and M.P. Vecchi *Science*, **220**: 1983; 671–680.
38. S.R. Wilson and W. Cui *Biopolymers*, **29**: 1990; 225–235.
39. S.R. Wilson, W. Cui, J.W. Moskowitz and K.E. Schmidt *J. Comput. Chem.*, **12**: 1991; 342–349.
40. D.S. Goodsell and A.J. Olson *Proteins, Structure, Function and Genetics*, **8**: 1990; 195–202.
41. M.T. Barakat and P.M. Dean *J. Comput.-Aided Mol. Des.*, **4**: 1990; 295–316.
41a. M.T. Barakat and P.M. Dean *J. Comput. Aided Mol. Des.*, **4**: 1990; 317–330.
42. J.N. Scarsdale, P. Ram, J.H. Prestegard and R.K. Yu, *J. Comput. Chem.*, **9**: 1988; 133–147.
43. A.P. Tonge, P. Murray-Rust, W.A. Gibbons and L.K. McLachlan *J. Comput. Chem.*, **9**: 1988; 522–538.
44. N.C. Cohen, J.M. Blaney, C. Humblet, P. Gund and D.C. Barry *J. Med. Chem.*, **33**: 1990; 883–894.
45. J.A. McCammon and S.C. Harvey *Dynamic of Proteins and Nucleic Acids*, Cambridge University Press. New York, 1987.
46. J.A. McCammon and M. Karplus *Acc. Chem. Res.*, **16**: 1983; 187–193.
47. M. Billeter, T.F. Havel and K. Wüthrich *J. Comput. Chem.*, **8**: 1987; 132–141.
48. M. Billeter, A.E. Howard, I.D. Kuntz and P.A. Kollman *J. Am. Chem. Soc.*, **110**: 1988; 8385–8391.
49. M. Billeter, T.F. Havel and I.D. Kuntz *Biopolymers*, **26**: 1987; 777–793.
50. P.R. Gerber, K. Gubernator and K. Mueller *Helv. Chim. Acta*, **71**: 1988; 1429–1441.
51. K. Mueller, H.J. Ammann, D.M. Doran, P.R. Gerber, K. Gubernator and G. Schrepfer *Bull. Soc. Chim. Belg.*, **97**: 1988; 655–667.
52. M. Saunders *J. Comput. Chem.*, **12**: 1991; 645–663.
53. H.J. Böhm, G. Klebe, T. Lorenz, T. Mietzner and L. Siggel *J. Comput. Chem.*, **11**: 1990; 1021–1028.
54. M. Saunders, K.N. Houk, Y.D. Wu, W.C. Still, M. Lipton, G. Chang and W.C. Guida *J. Am. Chem. Soc.*, **112**: 1990; 1419–1427.
55. D.A. Wertz, C.X. Shi and C.A. Venanzi *J. Comput. Chem.*, **13**: 1992; 41–56.
56. J. de Vlieg, R.M. Scheek, W.F. van Gunsteren, H.J.C. Berendsen, R. Kaptain and J. Thomason *Proteins, Structure Function and Genetics*, **3**: 1988; 209–218.
57. H. Pepermans, D. Tourwe, G. van Binst, R. Boelens, R.M. Scheek; W.F. van Gunsteren and R. Kaptain *Biopolymers*, **27**: 1988; 323–338.

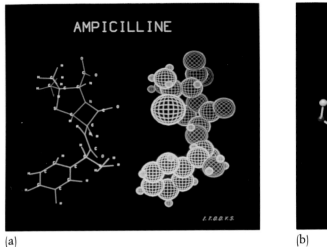

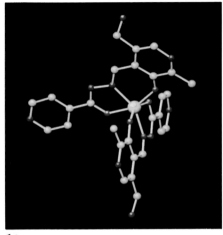

(a) (b)

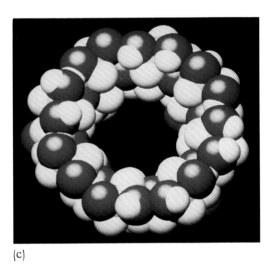

(c)

Plate I Usual molecular models: (a) wire frame and space filling models of ampicilline represented on a calligraphic system (note the small radius chosen for hydrogen sphere to emphasize the shape of the framework of heavy atoms); (b) ball and stick model: crystal structure of the octahedral complex between iron and pyridoxal-isonicotinoyl-hydrazone Fe(PIH)$_2$. The metal ion is coordinated to two molecules of PIH located in nearly perpendicular planes; (c) space filling model of cyclodextrine.

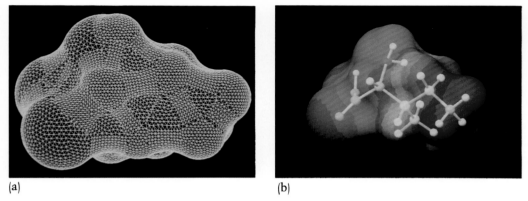

(a)　　　　　　　　　　　　　　　　　　(b)

Plate II Hydrophobic potential for iso-leucine. From Connolly dot surface,
(a) triangulation gives a set of planar facets. Parts corresponding to the re-entrant
surface patches are clearly visible. (b) Molecular surface colour-coded according to the
hydrophobic potential, represented together with the structural skeleton; blue areas
correspond to hydrophilic regions near the carboxyl and amino groups.

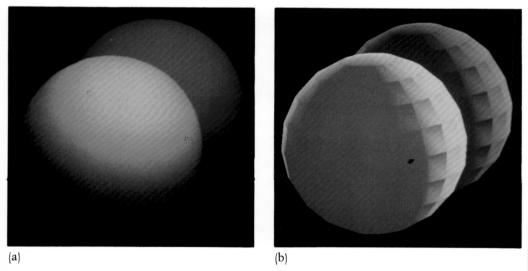

(a)　　　　　　　　　　　　　　　　　　(b)

Plate III Building atomic spheres by rotation of a spherical lune. Note the smooth
appearance obtained with a high number of facets (a), whereas with a low number
(b), facets still appear, as on a golf ball, with shading distortion at their junction.

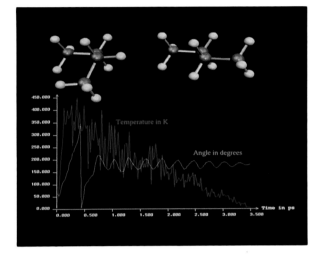

Plate IV Summary of the results of an MD simulation performed for *n*-butane using the simulated annealing technique and the MM+ force field as implemented in the HyperChem program [36]. Top left: initial *gauche* structure; top right: final *anti* form. The diagram in the lower part of the figure displays the evolution of temperature and torsion angle ω as a function of time.

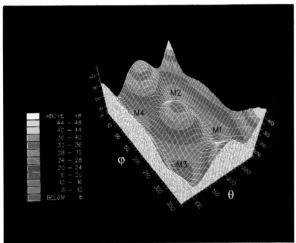

Plate V Potential energy surface of the 1,2 difluorohydrazine HFN–NFH molecule calculated at the Hartree Fock (3–21G basis set) level as a function of two variables: φ and θ, the FNNF and FNNH torsion angles, respectively. Energy values, in kcal/mol, refer to differences with respect to the absolute minimum M3.

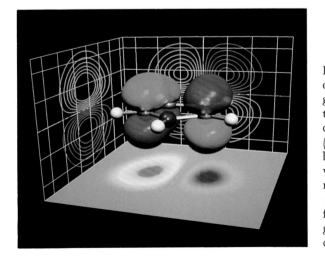

Plate VI Solid model of the HOMO of pyrrole represented as red and green isosurfaces (at ± 0.05 au) together with a ball and stick model of the structure of the compound (the nitrogen atom is depited by the blue ball). The horizontal and vertical planes contain coloured area maps and contour levels (from ± 0.0006 au, by steps of ± 0.0001 from outside to inside), respectively, generated from the MO values in the corresponding planes.

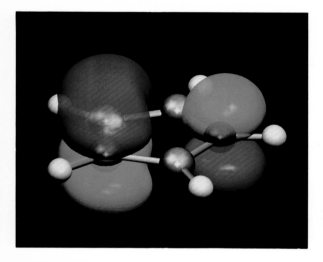

Plate VII Isovalue surfaces of the HOMO–1 of pyrrole represented as solid models in the same conditions as those of Plate VI.

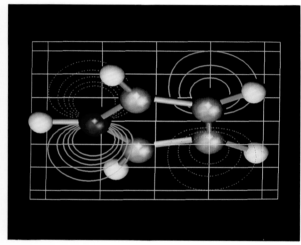

Plate VIII Selected contour levels of the HOMO–1 of pyrrole represented in the vertical mirror plane of the molecule and superimposed to the structural model. Dashed contours correspond to negative values. Contour values (from outside to inside): ± 0.05, ± 0.075, ± 0.10; ± 0.125; ± 0.15 au.

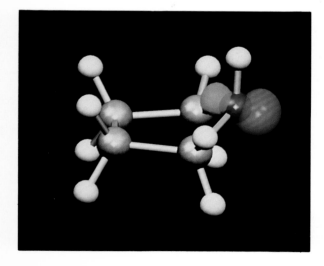

Plate IX Isovalue surfaces (at ± 0.2 au) of the HOMO of pyrrolidine represented as solid models.

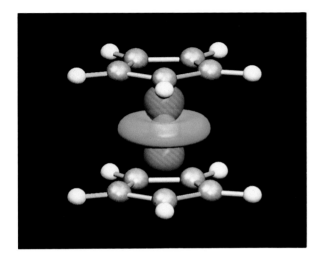

Plate X Isovalue surfaces (at ± 0.07 au) of the $5a_{1g}$ MO of ferrocene represented as solid models.

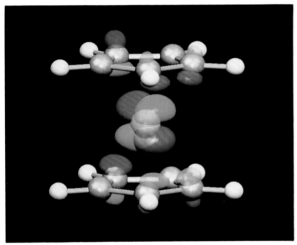

Plate XI Isovalue surfaces (at ± 0.07 au) of the $4e_{1g}$ MO of ferrocene represented as solid models.

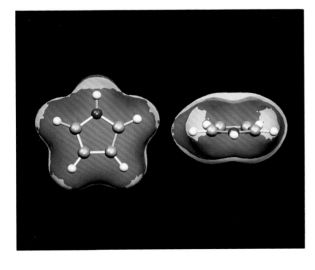

Plate XII Two different views of the superimposed Connolly envelope (red) and isovalue (at 0.002 au) molecular electron density surface (yellow) of pyrrole.

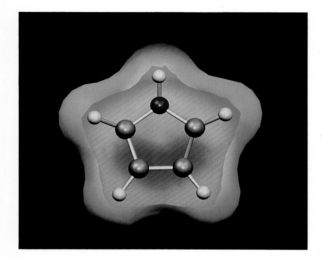

Plate XIII Solid model of the molecular surface of pyrrole coloured according the MEP value (ab initio SCF calculation using the 3–21G basis set). The lowest (negative) values correspond to the red zone, with a minimum at −38 kcal/mol.

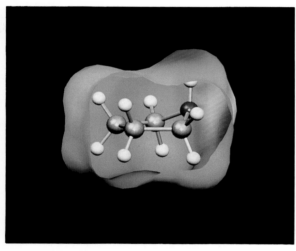

Plate XIV Solid model of the molecular surface of pyrrolidine coloured according the MEP value *(ab initio* calculation using the 3–21G basis set). The lowest (negative) values correspond to the red zone, with a minimum at −72 kcal/mol.

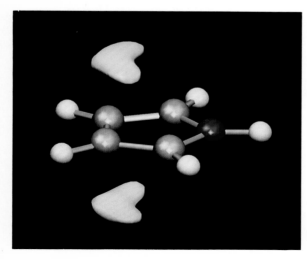

Plate XV Isoenergy surfaces (at −35 kcal/mol) of the MEP of pyrrole (calculated as in Plate XIII).

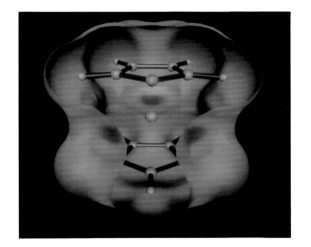

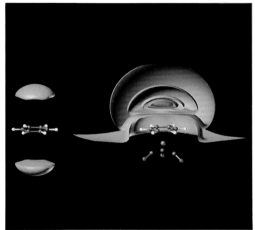

Plate XVI Molecular surface of ferrocene [Fe(C₅H₅)2] coloured according to E_{int} (equation 10.9) for electrophilic attack. Protonation sites are easily seen as the red zones around the metal atom. Reprinted with permission from J. Weber and P.Y. Morgantini, *EPFL Supercomputing Review*, 1990; **14**; © 1990 Ecole Polytechnique Fédérale de Lausanne.

Plate XVII Structural models of benzene (left) and benzene-chromium(CO)₃ (right) represented together with solid models of E_{int} isovalue surfaces calculated for nucleophilic attack. Colour coding of the surfaces: blue, –0.15; purple, –5.0; green, –10.0; yellow, –15.0; brown, –17.0 and red, –18.0 kcal/mol. Reprinted with permission from G. Bernardinelli, *et al. Chimia*, **46**: 1992; 126; © 1992 Schweizerischer Chemiker-Verband.

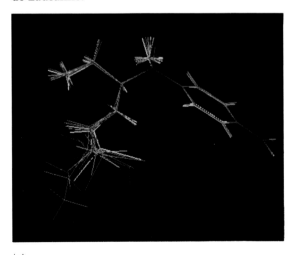

(a)

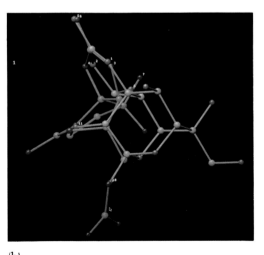

(b)

Plate XVIII Molecular superimpositions. (a) Superimposition of a family of chloramphenicol congeners (antibacterial derivatives) with all phenyl rings coincident (wire frame representation), (b) superposition of ball and stick models of saxitoxin (green) and tetrodotoxin (yellow) for the best fit of atom positions.

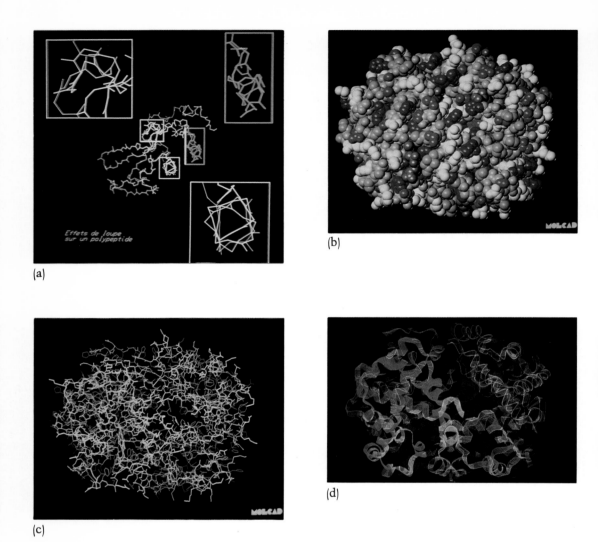

(a)

(b)

(c)

(d)

Plate XIX Representation of macromolecules. (a) Simplified representation of lyzozyme (a small protein of 129 amino acids). Only the main chain is displayed. The lowest inset (pink) corresponds to an helix seen along its axis. (b) Space filling model of human haemoglobin. (c) Stick model of human haemoglobin.The structure is coloured according to the various residues. (d) Ribbon model of human haemoglobin. The subchains are displayed in different colours. (b–d: from J. Webber *et al.*, reference 131 of Chapter 13 with permission.)

58. J. Sadowski and J. Gasteiger *Chem. Rev.*, **93**: 1993; 2567–2581.
59. M.S. Gordon and J.A. Pople, MBLD Programm, QCPE, **10**: 1975; 135.
60. W.T. Wipke in *Computer Representation and Manipulation of Chemical Information*. W.T. Wipke, S.R. Heller, R.J. Feldmann and F. Hyde (Eds). Wiley, New York, 1974; 153.
61. N.C. Cohen, P. Colin and G. Lemoine *Tetrahedron*, **37**: 1981; 1711–1721.
62. P.J. De Clercq *J. Org. Chem.*, **46**: 1981; 667–675.
63. J. Hoflack and P.J. De Clerc *Tetrahedron*, **44**: 1988; 6667–6676.
64. D.P. Dolata and R.E. Carter *J. Chem. Inf. Comput. Sci.*, **27**: 1987; 36–47.
65. A.R. Leach, K. Prout and D.P. Dolata *J. Comput. Chem.*, **11**: 1990; 680–693.
66. A.R. Leach and K. Prout *J. Comput. Chem.*, **11**: 1990; 1193–1205.
67. M. Randic, B. Jerman-Blazic and N. Trinajstic *Computers Chem.*, **14**: 1990; 237–246.
68. A.R. Leach and A.S. Smellie *J. Chem. Inf. Comput. Sci.*, **32**: 1992; 379–385.
69. A. Rusinko III., R.P. Sheridan, R. Nilakantan, K.S. Haraki, N. Bauman and R. Venkataraghavan *J. Chem. Inf. Comput. Sci.*, **29**: 1989; 251–255.
70. A. Rusinko III, J.M. Skell, R. Balducci and R.S. Pearlman, CONCORD. University of Texas. Austin; distributed by Tripos Associates, St Louis MO, 1987.
71. D. Weininger *J. Chem. Inf. Comput. Sci.*, **28**: 1988; 31–36.
72. D. Weininger, A. Weininger and J.L. Weininger *J. Chem. Inf. Comput. Sci.*, **29**: 1989; 97–101.
73. D. Weininger and A. Weininger, THOR thesaurus oriented chemical database, version 3.54, Daylight Chemical Information Systems. Claremont CA 91711, 1989.
74. D. Weininger, A. Weininger and A.J. Leo *MedChem Software Manual release 3.52 Medicinal Chemistry Project*. Pomona College, Claremont CA., 1987.
75. Y.C. Martin, E.B. Danaher, C.S. May and D. Weininger *J. Comput.-Aided Mol. Des.* **2**: 1988; 15–29.
76. M.A. Hendrickson, M.C. Nicklaus, G.W.A. Milne and D. Zaharevitz *J. Chem. Inf. Comput. Sci.*, **33**: 1993; 155–163.
77. J. Gasteiger, C. Rudolph and J. Sadowski *Tetrahedron Computer Methodology.*, **3**: 1990; 537–547.
78. J. Sadowski, C. Rudolph and J. Gasteiger *Anal. Chim. Acta*, **265**: 1992; 233–241.
79. W.T. Wipke and M.A. Hahn *Chemical structures*, W.A. Warr Ed. Springer Verlag, Berlin, 1988; 267–268.
80. W.T. Wipke and M.A. Hahn *Tetrahedron Computer Methodology*, **1**: 1988; 141–167.
81. Z. Ai, Y. Wei *J. Chem. Inf. Comput. Sci.*, **33**: 1993; 635–638.
82. W. Fisanick, K.P. Cross and A. Rusinko III *Tetrahedron Computer Methodology*, **3**: 1990; 635–652.
83. D.R. Henry, P.J. McHale, B.D. Christie and D. Hillman *Tetrahedron Computer Methodology* **3**: 1990; 531–536.
84. K. Davies and R. Upton *Tetrahedron Computer Methodology* **3**: 1990; 665–671.
85. B.D. Christie, D.R. Henry, W.T. Wipke and T.E. Moock *Tetrahedron Computer Methodology*, **3**: 1990; 653–664.
86. R.P. Sheridan, R. Nilakantan, A. Rusinko III, N. Bauman, K.S. Haraki and R. Venkataraghavan *J. Chem. Inf. Comput. Sci.*, **29**: 1989; 255–260.
87. A.T. Brint and P. Willett *J. Mol. Graphics*, **5**: 1987; 49–56.
88. R.P. Sheridan, A. Rusinko III, R. Nilakantan and R. Venkataraghavan *Proc. Natl. Acad. Sci. USA*, **86**: 1989; 8165–8169.
89. N.W. Murrall and E.K. Davies *J. Chem. Inf. Comput. Sci.*, **30**: 1990; 312–316.
90. E.M. Ricketts, J. Bradshaw, M. Hann, F. Hayes, N. Tanna and D.M. Ricketts *J. Chem. Inf. Comput. Sci.*, **33**: 1993; 905–925.
91. M.C. Nicklaus, G.W.A. Milne and D. Zaharevitz *J. Chem. Inf. Comput. Sci.*, **33**: 1993; 639–646.
92. Y.C. Martin and P. Willett *Tetrahedron Computer Methodology*, **3**: 1990; 527–530.

93. O.F. Güner, D.W. Hughes and L.M. Dumont *J. Chem. Inf. Comput. Sci.*, **31**: 1991; 408–414.

94. O.F Güner, D.R. Henry and R.S. Pearlman *J. Chem. Inf. Comput Sci.*, **32**: 1992; 101–109.

95. K.S. Haraki, R.P. Sheridan, R. Venkataraghavan, D.A. Dunn and R. McCulloch *Tetrahedron Computer Methodology*, **3**: 1990; 565–573.

96. J. H. Van Drie, D. Weininger and Y.C. Martin *J. Comput.-Aided Mol. Des.*, **3**: 1989; 225–251.

97. D.E. Clark, P. Willett and P.W. Kenny *J. Mol. Graphics*, **10**: 1992, 194–204.

8 Molecular surfaces and volumes

The surface and volume of a molecule represent important parameters involved in various research areas: intermolecular interactions, drug design and protein folding [1]. Molecular volume intervenes in many computational approaches of drug design [2] since steric fit is an important feature in the "lock-and-key" or "hand-and-glove" models of molecular interactions involved in pharmacophore recognition. It also appears as one of the fundamental properties of macromolecules [3, 4], directly related to their physicochemical characteristics. Density, for instance, is a useful parameter in the study of the tertiary structure of proteins: its local variations in the interior regions and packing defects have been related to conformational fluctuations, folding or hydrogen exchange [5].

The external surface represents the interface with the surrounding medium. It gives a first image of the areas able to bind ligands or other macromolecules, interact with the solvent, etc. This view may be refined by the concept of "structured surfaces" [1], i.e. surfaces encoded (or "coloured") with the value of a property such as electrostatic potential or hydrophobicity to obtain a better insight as to the chemical complementary needed for corresponding areas of interacting systems. Many research papers deal with molecular surfaces or volumes, dedicated either to their quantitative evaluation or representation on a graphical display. Although these goals may appear, at first glance, quite distinct, they often rely on the same approaches and are therefore closely related. So here we will discuss the two aspects of evaluation and visualization at the same time.

Two general approaches have been developed: *numerical* and *analytical*. The analytical methods give an exact answer from a set of equations but require somewhat involved geometrical computations. Numerical algorithms subdivide the object into a large number of small elementary similar subunits, for example elementary cubes or "voxels" (volume elements), the union of which reproduces the actual object. The method is only approximate, but can be efficiently programmed and is very convenient for the logical operations (Boolean operations) involved in the comparison of surfaces or volumes for several neighbouring molecules.

8.1 DEFINITION OF MOLECULAR SURFACES AND VOLUMES

Various definitions of the external surface of a molecule have been proposed: for a brief historical summary see Richards [6]. Figure 8.1 shows the various molecular surfaces.

8.1.1 Van der Waals surface

Elementary models generally represent a molecule as an assembly of rigid hard spheres featuring the atoms. Their radii are chosen equal to the atomic van der Waals radii (associated with the closest approach distance for non-bonded atoms). The union of these atomic spheres is accepted as a common representation of the molecular body, and is known as the van der Waals contour (or volume) [7, 8]. For large biomolecules, one can choose not to represent the hydrogen atoms but only groups built from every heavy atom and its attached hydrogens ("united atoms"), these groups being still represented as spheres [3].

The van der Waals contour gives a good estimate of the molecular shape for

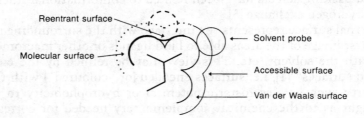

Figure 8.1   Various molecular surfaces.

small molecules. However, for larger biomolecules, an important part of this surface is embedded (buried) in the "interior", and therefore less able to interact with an incoming ligand or a solvent molecule. This leads to the concept of "solvent-excluded volume", gathering its own van der Waals volume and the interstitial volume (empty but too small to accommodate a solvent molecule). The extent of this "solvent-excluded" part is, of course, difficult to estimate, since it is not directly obtainable by experiment. However, with reasonable assumptions, Connolly estimated the van der Waals volume of the small protein crambin (i.e., 4245 Å^3) to represent only 82% of the solvent-excluded volume, assuming a radius of 1.5 Å for a spherical probe representing the solvent. The percentage even decreases to 76% for a 3.0 Å probe radius [5]. Some comments about the influence of the selected probe radius will be given later.

8.1.2 Solvent accessible surface

Another definition is therefore necessary to characterize the external boundary area that can be occupied by the solvent without penetrating the molecule. Lee and Richards [9] first introduced the *solvent-accessible surface*. A small probe sphere simulating the solvent molecule is rolled onto the molecular shape and the solvent-accessible surface is defined as the locus of the centre of this probe sphere. In fact, it corresponds to a contour pushed out from the van der Waals surface by a distance equal to the radius of the rolling probe sphere (usually 1.4 Å to mimic a water molecule).

8.1.3 Accessibility

For each atomic centre one can define its own *accessibility*: that percentage of the accessible area compared to the total area of the associated atomic sphere (with a radius equal to the atomic van der Waals radius plus the probe radius). Several computer algorithms implement this definition [see the references in [10]. For instance, Lee and Richards [9] used sections by a set of parallel planes with a given spacing: surface elements are computed as the product of the non-overlapping arcs of circles (corresponding to the section of the atomic spheres) by the spacing of the planes. Another model spreads a set of 92 points on the atomic surface and counts the remaining ones after eliminating those located in the intersection volumes [11].

Results obtained for some proteins (ribonuclease S, lysozyme and myoglobin) indicate that non-polar atoms (C, S) occupy approximately 40–50% of the accessible area (Figure 8.2). Changing from a hypothetical extended chain to the native folded conformation leads to an average change in accessibility by a factor of about 3, although individual values for the various residues may be significantly different [9].

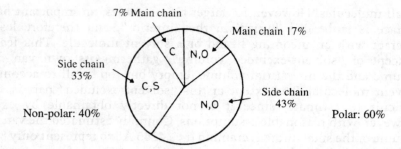

Figure 8.2 Relative accessibility of polar and non-polar atoms for lysozyme (from Lee and Richards with permission [9]).

This concept of accessibility was later modified by Lavery *et al.* [12]:

1. To better evaluate the effect of the macromolecular environment on a given subunit (base, sugar, phosphate), an "intrinsic accessibility" was defined. It takes into account the fact that when atoms are engaged in a molecule, their accessibility decreases (due to steric hindrance with other atoms of the molecule). So, the accessibility of a given atom in a subunit of a macromolecule must be compared to its intrinsic accessibility in the subunit (considered alone) rather than referred to a free atom.

2. One has to differentiate the way in which the solvent approaches the macromolecule. For water, for instance, the contact can correspond to the hydrogen or the oxygen atoms. Two indices have been proposed: one, the "accessible area", measures the area of the receptor protein within which the attacking molecule may be placed, in at least one of its possible configurations, without intersection with the macromolecule envelope. The second, "fractional configurational flexibility", relates to the fraction of the possible configurations of the attacking molecule placed on the accessible area, which does not suffer intersections with the macromolecule. It is to be noted that these accessibility parameters are related to the van der Waals surface (and not the accessible one). So, some artefacts in the definition of Richards are suppressed (for certain atoms, accessibility would increase if the radius of the probe gets larger). Furthermore, to avoid lengthy calculations, the simple spherical model of water can be maintained, provided that different radii are chosen (1.2 Å if the atom contacting the receptor envelope is a hydrogen, 1.4 Å if it is an oxygen).

Some results [12] as to the role of the environment for a macromolecule have been presented for fragments of a B–DNA model formed from five base pairs (poly(dG–dC) or poly(dA–dT)). Passing from isolated nucleic acids to a single helical strand and then to double helical strands induces a successive reduction in the accessible area of all the base atoms. However, the extent of this reduction largely changes from site to site. For example, for isolated guanine, O_6, N_7 and N_3 have the largest accessibility in the free base. In the single helix, N_3 becomes practically inaccessible, and in the double helix, N_7 is the more accessible. The effect of sequence variations appears, in many cases, non-negligible and specific. Comparing, for instance, sequences **a** (GC–GC–GC–GC–GC), **b** (AT–AT–GC–AT–AT) and **c** (TA–TA–GC–TA–TA), it comes that $O_6(G)$ is nearly as accessible in the three sequences, whereas $N_3(G)$, totally inaccessible in sequence **a**, becomes accessible in **b** or **c**.

When interpreting the reactivity of a given site within a macromolecule, this criterion of accessibility may play an important role, in complement to electronic factors (expressed, for instance, by the molecular electrostatic potential). For guanine, in DNA motifs, N_3, in spite of a high value of the MEP, is only slightly reactive toward various alkylating agents because of a low accessibility (Figure 8.3).

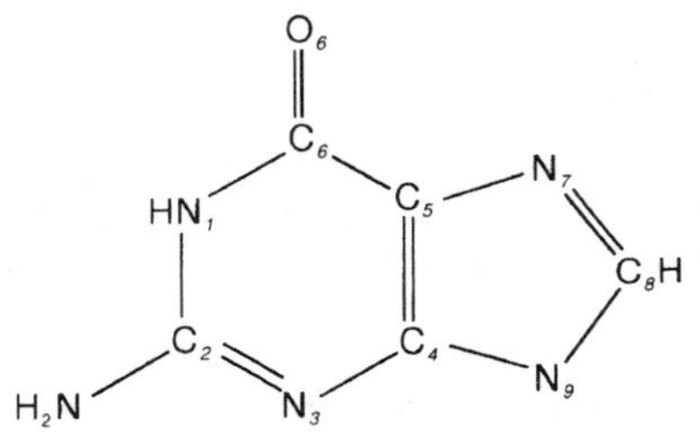

Figure 8.3   Guanine.

8.1.4 Molecular surface: Richards' surface

Later, the *molecular surface* ("Richards' surface") was defined as the *contact surface* (part of the van der Waals surface accessible to the probe sphere) and the *reentrant surface* (inward facing surface of the probe when it is simultaneously in contact with two or more atoms) [3]. Patches of the convex van der Waals surface are joined by concave parts of the probe spheres "blocked at the opening of the narrow molecular fissures". The join between faces is smooth, in contrast to the van der Waals surface, which presents "sharp crevices" at the intersection of the atomic spheres. This surface envelope corresponds to the *solvent-excluded volume* [3].

8.2 ANALYTICAL EVALUATIONS OF SURFACES OR VOLUMES

Connolly presented an analytical method for calculating the Richards' molecular surface [10]. To generate the various parts of the surface, the method uses, as a working tool, a probe sphere rolling over the molecule, tangential to each atom, each pair or triplet of neighbouring atoms. In this respect it thus looks like an outgrowth of an earlier numerical evaluation (the "dot algorithm", see below). Depending upon the probe-sphere situation, different

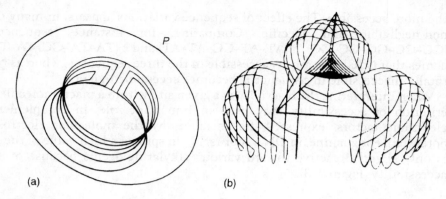

Figure 8.4 Parts of the Connolly model. (a) Representation of the reentrant surface associated with a probe sphere in contact with two atoms (part of a torus). Circle P corresponds to the trajectory of the centre of the probe sphere; (b) placement of a probe sphere tangential to three atom spheres. The inward facing part of the probe sphere is shown as a curved triangle in the centre of the figure (from Connolly with permission [10]).

parts of the surface are generated forming a connected network covering the molecule (Figure 8.4).

These pieces, looking like the faces of some curved polyhedron, are either convex or concave parts of spheres or saddle shapes (parts of a torus) between neighbouring atoms, and join at circular arcs. They are defined by the surface they lie on and their boundary contours. The corresponding geometrical expressions are detailed by Connolly [10]. The Connolly algorithm begins from the probe sphere tangential to three atoms (therefore with no more translational degrees of freedom). This fixes the end points of the trajectory of the probe. Then, these trajectories limit the area of a rolling probe with two degrees of freedom. This approach is convenient not only for evaluating the molecular surfaces (thanks to the analytical formulae given by Connolly [10], but also for displaying these surfaces on a video screen (see below).

The analytical approach has been further extended to the calculation of the van der Waals and solvent-excluded volumes [5] of a molecule considered as an assembly of rigid hard spheres, the starting point being the analytical representation of the molecular envelope (Figure 8.5). Thanks to an analytical partition, the molecular volumes are treated as formed of simpler shapes. Internal polyhedra are derived from the solvent accessible area or van der Waals contours. They are calculated as truncated triangular prisms. Their bases are the flat triangles built from the centres of triples of atoms forming a concave triangle. The volume outside the polyhedron but inside the surface is decomposed in a set of disjointed pieces: convex pieces (volume between the centre and the convex face of an atom), saddle pieces (formed of conical pieces and a part of the hole of a torus) and concave pieces (triangular pyramid minus a piece of probe sphere). Corrections are introduced for cusps (when the molecular surface intersects itself).

The method of evaluation is exact for the van der Waals volumes, whereas

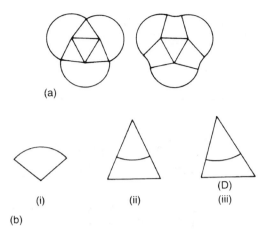

(a)

(D)

(i)   (ii)   (iii)

(b)

Figure 8.5 The Connolly approach of molecular surfaces and volumes (2D schematization). (a) Analytical partition of solvent-excluded volume (right) and van der Waals volume (left), and (b) constituent pieces. Interior polyhedra are not represented (from Connolly with permission [5]). (i) convex piece, spherical; (ii) concave piece, triangular pyramide minus a part of a probe sphere; (iii) saddle piece, revolution axis: interatomic line segment D (from Connolly [5] with permission).

for the solvent-excluded volumes, the error (due to cusps) is estimated at about 0.01%. Connolly stresses, however, that the extent to which physicochemical properties can be represented by a static assembly of rigid hard spheres, with no account of potential flexibility, would also have to be questioned.

Some applications have been presented. For instance a comparison of the volumes of two bovine pancreatic trypsin structures (the diisopropyl-fluorophosphate-inhibited (DIP) and the benzamidine-inhibited ones) was carried out. The volumes differ by no more than 240 Å^3 for a total van der Waals volume of about 21 000 Å^3. The high accuracy in the evaluation makes the method attractive for comparing 3D structures of macromolecules in slightly differing conformations (due to oxidation state, presence of ligands, crystal form, etc.). Other potential applications suggested may be volume measurements for packing defects in protein interiors, ligand-binding pockets on protein surfaces or void volumes at subunit interfaces [10].

As previously indicated, the analytical approach of Connolly can be used for surface representation. Each face defining the surface is converted into a set of concentric curved polygons which can be drawn on a calligraphic device (Figure 8.6). A method for displaying an analytical molecular surface on a raster system has also been developed by Connolly [13, 13a].

The accessible surface is directly involved in solvation phenomena, and it may be interesting to examine its variations with conformations (and resulting effects on solvation energy) when carrying out energy minimization. This is seldom possible due to the heavy calculations needed. To get numerical expressions, quickly evaluated and able to be differentiated during mini-mization processes, Hasel *et al.* [14] proposed a pairwise function of radii and

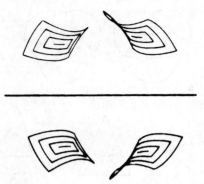

Figure 8.6 Reentrant surface construction. Accessible parts of the inward facing surface of the torus are represented as a disconnected set of saddle-shaped rectangles (from Connolly with permission [10]).

atomic coordinates. The basis of this is the approach of Wodak and Janin [15]. The approximate solvent-accessible surface area is obtained by summation of individual atomic accessible areas (A_i). For any atom i:

$$A_i = S_i \prod_j (1.0 - b_{ij}/S_i)$$

with:

$$S_i = 4\pi(r_i + r_s)^2$$
$$b_{ij} = \pi(r_i + r_s)(r_i + r_i + 2r_s - d)[1 + (r_j - r_i)/d]$$

S_i is the total accessible area of isolated atom i (radius r_i) for a given probe (r_s), and b_{ij} the accessible area removed from atom i by overlap with atom j (at a given separation d).

Although working well when "atoms" represent polyatomic fragments (protein residues), the method underestimates the area when multiple overlap is important, and the authors turned to a parameterization of the preceding equation:

$$A_i = S_i \prod_j (1.0 - p_i p_{ij} b_{ij}/S_i)$$

Atom parameter p_i takes into account hybridization and substitution of atom i, and connectivity parameter p_{ij} distinguishes bound atoms j from more distant ones, having less positional predictability. Parameters p_i and p_{ij} were optimized so as to reproduce the area calculated by an exact analytical method, over a learning set of about 270 largely varied molecules. The results indicate that the model works fairly well, although areas are systematically underestimated by approximately 8%. According to Ooi *et al.* [16], solvation energies can be related to solvent-exposed surface area. Indeed, from areas calculated with the model of Hasel *et al.*, solvation energies are satisfactorily reproduced, provided a scaling coefficient (about 1.27) is used. Analytical formulae are also proposed for the calculation of first and second derivatives, with a view to the possibility of incorporating crude but rapid solvation energy evaluation into molecular mechanics minimizations.

The more complex problem in the derivation of the molecular surface is related to the determination of spaces inaccessible to the solvent and limited by the reentrant surface. In the GEPOL model [17, 18] such spaces are filled with a new set of spheres created between the original atomic spheres (Figure 8.7). These new spheres are subsequently used in surface and volume calculations. The program examines all the pairs of spheres that it is possible to generate from the original atomic spheres. Next, it determines whether there is some space inaccessible to the solvent probe, and if so, it creates a new sphere between the pair. This is added to the original set (to update the set of sphere pairs), and the process is iterated until no new spheres can be created. According to the authors, the program favourably competes with the widely used Connolly's dot approach at both precision and time levels. For a fast calculation of molecular surface and volume, the approach has been completed [18] by a triangular tessellation process. To select parts of the spherical surfaces that form the molecular surface, each sphere is divided into 60 spherical triangles (by projection of a pentakisdodecahedron) used as starting points for the representation or evaluation of properties.

8.3 NUMERICAL METHODS

Various possibilities have been proposed to represent the molecular envelope. A first approach distributes points uniformly on the surface (or inside the

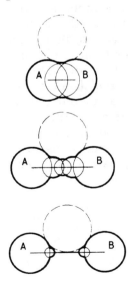

Figure 8.7   In the GEPOL model, the reentrant surface is approximated by inserting new spheres among the original atomic spheres (from Pascul-Ahuir and Silla with permission [17]).

molecular body if volume information is needed). It looks very attractive, since only a limited number of points is necessary for visual perception. Relevant to this approach is the dot algorithm of Connolly, widely used for the representation of structured surfaces [13, 13a].

Rather than distributing points on the surface to be investigated, several numerical methods prefer to sample the space through a 3D lattice of a given mesh size. These grid or cube methods can perform both surface and volume calculations or representations. Another advantage is that their data structure is able to easily reflect some neighbourhood relations between surface points. Although not commonly used in that field, octree techniques can give additional efficiency in the treatment [1] thanks to their hierarchical organization and recursive subdivision methods.

8.3.1 Dot method

Among the various algorithms presented for representation of the solvent-accessible area (or the van der Waals surface), one of the most popular is the dot algorithm of Connolly [13, 13a]. It relies on the same concepts as analytical surface calculations, but is limited to generating surface points without computing boundary arcs and faces. This type of representation eases the calculations and provides a simple graphical display. Other advantages are that this dot representation does not overload the drawing, and it allows for a simultaneous visualization of the atomic framework, which is very attractive for chemists.

To derive the surface dots, a probe sphere is placed tangential to each atom, each pair or each triplet of neighbouring atoms. When the probe is free from overlapping with other atoms, points on the contact circle (latitude circle) are selected to represent the molecular surface. The probe is then moved to generate another set of dots. The same is done for points of the reentrant surface or concave faces (contact with two or three atoms). The spacing of the points between and along the latitude circles is fixed so as to obtain a regular density of dots per unit area. The probe is moved in angular increments, so that the method is a numerical one. The algorithm also produces outward pointing normals (useful for hidden line removal or rendering treatments) and approximate elementary areas associated with the points (Figure 8.8).

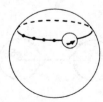

Figure 8.8 Dots are located along contact circles of a probe sphere rolling along the molecular shape.

Scattering dots onto the molecular surfaces provides a very convenient way in which to represent some physicochemical properties on the molecular envelope or on a homologous shape pushed away from it by a constant quantity (for instance, the radius of a solvent molecule). The property (molecular electrostatic potential, hydrophobicity potential, for instance) is evaluated on the Connolly's dot positions, and these are colour-coded according to the values obtained, giving "structured surfaces" ("4D representations") (see Plates IIb and XII).

Related to dot algorithms, the USURF method [19] was presented as providing a significant saving in computer time. In a first step, dots are regularly scattered on the accessible surface. This surface is "regarded as a solvation layer where dots represent the position of probe spheres in contact with the molecule". Each dot on this envelope is assigned a parent atom. Then for each dot, probe hemispheres facing their parent atom are generated and surfaced with evenly distributed dots. Those penetrating other probes are eliminated. The remaining ones give a good approximation of the Richards' surface (Figure 8.9). This surface is composed only of concave patches, but in most cases, it does not have too rough an appearance, and the gain in computer time seems quite interesting.

8.3.2 Grid or cube methods

The papers of Greer and Bush [20] or Pavlov and Fedorov [21] constitute pioneering works on surface evaluation via numerical approaches. However, the recent study of Stouch and Jurs [22] provides a thorough discussion of such numerical (discrete) methods as to the methodology, accuracy, time expenditure and storage requirements. It is therefore used here as the basis of our presentation.

The molecule, defined by the 3D coordinates and the van der Waals radii of its constituent atoms, is embedded in a three-dimensional grid of a variable size and density of nodes. Nodes are encoded as interior ("in") or exterior ("out") to the molecular volume using a simple distance test: if a node-atom distance is less than or equal to the van der Waals radius of that atom, the node is "interior". If the distances of a given node to all atoms are greater than the corresponding radii, the node is "exterior". The node status ("in" or "out") can

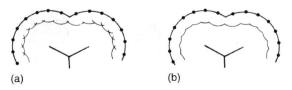

(a)   (b)

Figure 8.9 The USURF method. (a) Scattering dots on the accessible surface and drawing probe hemispheres; (b) surfacing hemispheres on each probe and removing overlapping parts leads to an approximation of the Richards' molecular surface (Moon and Howe [19]).

be encoded using only one bit. Furthermore, if one decides to explore the grid along a fixed pathway, for instance successive slabs along the z direction and for each slab traversal of rows of nodes along the y axis (first rows of the first slab from the lowest row to the highest, then rows of the second slab, etc.) then the whole grid can be described as a string of bits. Each node is mapped onto a bit in this string: the bit value characterizes the status of the node, and its location in the string defines the geometrical position of the node in the 3D grid (Figure 8.10).

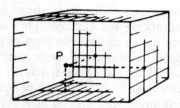

P(X,Y,Z) → (I,J,K) → Rank in the bit string

Figure 8.10 If the grid is traversed according to a given order, there is a one-to-one relationship between the location of a node in the 3D space and the rank of the corresponding record in the bit string describing the grid.

Such a data structure is very efficient in terms of both storage and handling of the structural information. A medium-sized molecule such as hydrocortisone $(C_{21}O_5H_{30})$ is reported to require only 13,000 bytes of storage (i.e. 1/70th of a normal 3½ in. PC floppy disk).

From the encoded nodes, volume calculation is straightforward, simply counting the "in" nodes, since each of them can be assumed to represent an elementary cubic volume element (voxel) within the shape (Figure 8.11).

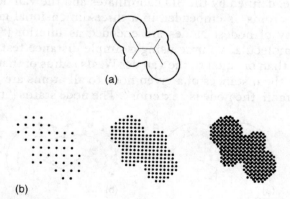

Figure 8.11 Higher node density gives a more realistic estimate of the molecular body. Cross-section of the van der Waals volume of trans 1,2 dichloroethylene in a plane parallel to the double bond, 1 Å above it. (a) Exact solution, (b) grid approximation with node densities of 1.5, 3 and 6 bla (bit per linear angstrom).

It seems obvious that the higher the node density, the higher the accuracy, but also the higher are the storage requirements and computer time; so an acceptable compromise has to be found. Two criteria have been looked for: a comparison of the so-calculated volumes with, firstly, experimental heats of vaporization of alkanes, and then, volume values derived from an exact analytical calculation [23]. A grid density of about 2 or 3 bits per linear angstrom (bla) seems very convenient: for the test examples reported, a rather crude grid density of 1 bla leads to results within 10% of the actual values, and the error is only 1.5% for a density of 3 bla (Figure 8.12).

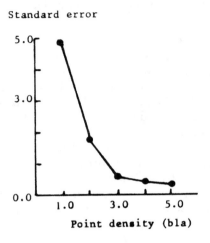

Figure 8.12   Decreasing the standard error on volume calculation with a higher point density (from Stouch and Jurs with permission [22]).

The surface may be approached retaining "in" nodes having at least one of its immediate neighbours with a different status (node "in" with one (to five) neighbour node(s) "out"). However, owing to the discrete nature of the sampling, the calculation is less precise. In fact, such "surface" nodes are not strictly located on the external envelope but more or less in its vicinity; therefore, each "surface" point does not represent an identical surface area. Satisfactory results have been obtained thanks to a weighting factor: for example, in the Stouch and Jurs treatment, an "in" node with 5 neighbouring "out" nodes is certainly very near the actual surface, and will be weighted by a factor of 5, whereas an "in" node with only one "out" neighbour is likely to be further apart, and will be weighted by 1 (Figure 8.13). A comparison with a rigorous analytical evaluation [23] shows that a density of 3 bla is still convenient, a higher density giving only a small benefit.

The influence of some elements in the interatomic clefts has been considered by Meyer [24], since points near the surface of two overlapping spheres may (or may not) be counted twice.

The same approach can be readily extended to the surface-accessible area.

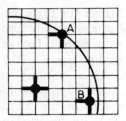

Figure 8.13 Weighting surface nodes according to the number of their "in" neighbours. In this 2D schematization, each node has four neighbours. Point A, with only two "in" neighbours, is more likely to be located in the vicinity of the surface than point B, with three "in" neighbours.

That is an important feature for large molecules. It contains its own volume of the molecule (i.e. the van der Waals volume) and a contribution from roughly half the layer of the solvent probe around the surface of the molecule). The molecule is now considered as an assembly of overlapping spheres with radii equal to the sum of the van der Waals radius of the atom and the radius assumed for the solvent probe.

As to molecular surfaces (according to Richards' definition), which imply "reentrant" parts, they can be determined (as a first approximation) with a similar approach: starting from a node [25] or voxel [21] representation of the accessible area and then erasing parts (nodes or voxels) in the accessible volume that are located within a solvent radius from the points defining the accessible surface contour (Figure 8.14).

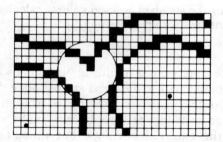

Figure 8.14 Shadowed voxels represent van der Waals (inner contour) and accessible (outer) surfaces. Erasing appropriate voxels leads to an approximate Richards' surface. Dots correspond to atom centres (from Pavlov and Fedorov with permission [21]).

Very close to the previous approaches are the methods of Higo and Go [26] or Karfunkel and Eyraud [1], combining the grid technique with the octree data structure, for the computation and representation of surfaces and volumes. The latter method even allows us to treat not only individual species but also "hypermolecules", a set of overlapping molecules (representing, for instance, several effectors binding to the same site of an unknown receptor) mapped on the same grid. The methods rely on the decomposition of the space containing the molecules into cubic regions, organized in an octree structure. The molecule is embedded in a box formed of elementary cuboids (voxels). According to their position with respect to the constituent atomic spheres, they are classified as "exterior" or "interior", or "surface" if they cross the surface. There is no need to further consider octants completely inside or outside the molecular surface. For Higo and Go [26], surface octants are considered as contributing one half to the molecular volume. Surface octants are subsequently decomposed into eight smaller octants tested to determine whether they are interior, exterior or surface. The approximation of course becomes better as the size of these voxels decreases, the process being repeated again until the resolution desired is obtained.

An original feature in the approach of Karfunkel and Eyraud [1] is the consideration of a double layer of points, either on the outer surface of the molecular body or just below these outer points. So, the volume may be represented via the surface without saving grid points inside the system. In Higo and Go's program [26], surface-voxel encoding can be approximately, but rapidly, performed by asking whether a minimum sphere containing the voxel is interior, surface or exterior to the atomic spheres (these minimum spheres have a radius $r\text{min} = (3^{0.5}/2)a$, where a is the length of the voxel side). This is answered by a simple test on the distance (D) between the centres of the voxel and the atomic sphere: comparing D with the sum or difference of the radii of the atomic sphere (R) and the minimum sphere (r_{min}): $R + r_{\text{min}}$ and $|R - r_{\text{min}}|$ (Figure 8.15).

The accuracy of this approach has been carefully checked both on simulation models (artificial "molecules" built from an assembly of spheres) and a small protein (ca. 751 atoms) with various orientations with respect to the embedding box. Starting from initial voxel sides of 2.0 Å with five levels of subdivision (down to 0.125 Å) appears to give quite satisfactory results with a calculation that is carried out rapidly (Figure 8.16) [26].

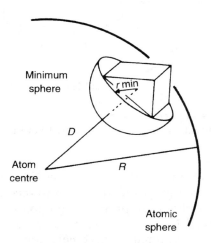

Figure 8.15 The distance test is carried out with a minimum sphere containing the voxel (Higo and Go [26]).

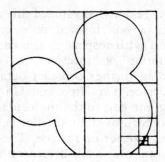

Figure 8.16 Using the octree data structure, only "surface" regions need to be further subdivided.

Stouch and Jurs [22] emphasized the fact that the status of each node (internal or external) can be encoded on one bit. To increase speed, Bohacek and Guida [27] proposed updating the string description of the voxel grid thanks to bit-encoded templates representing atoms. To encode voxels, previous methods evaluated the distances of each node of the grid to atomic centres. Here, each atom is represented as a string of bits locally indicating the voxels occupied around its centre. To build the set of occupied voxels featuring the molecular body, for each atom, given its location, this "local" string is transferred at the correct position of the global bit array (with an offset representing the appropriate atomic position). Some points (in overlapping regions) are written twice, but this is of minor importance in view of the overall gain in computer time (Figure 8.17).

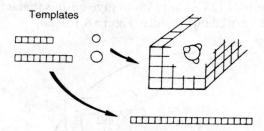

Figure 8.17 The bit string representing the molecule is built by locating at the appropriate position template strings corresponding to atoms.

For graphical display, rather than exploring the full bit array, a special algorithm (MAZE) [27] was proposed: it takes into account only those voxels at the surface boundary, without sampling interior points. The name is presented as derived from the well known rule for escaping a maze: "always go left". Once an "on" bit is encountered, surface cells are sought by systematically turning left into "on" cells until an "off" cell is reached. The previous cell was therefore a surface cell and its position is recorded; the process is then repeated. The results presented suggest that this method can treat, with good precision, molecules containing several hundred atoms in less than one minute of CPU time on a VAX-class computer.

8.3.3 Graphical display

As for grid or box methods, they also provide an easy representation of the molecular shape on a graphical display, simply plotting the "in" nodes on the "surface". Here also a grid density of about 2 or 3 bla seems sufficient to give a representation looking like the common CPK models. If details are necessary (to illustrate small local conformational changes, for example) the density can be increased in the relevant area only, so as to maintain a reasonable computational time.

If the sampling is not fine enough, only rough contours are defined (since one gets only nodes of the lattice in the vicinity of the actual surface). Refinements have therefore been proposed to derive more aesthetic images. In Pearl and Honegger's treatment [25], a positional correction is applied by moving the point along the vector connecting it to the host atom's centre, so that it lies exactly at the specified radius for that atom (Figure 8.18).

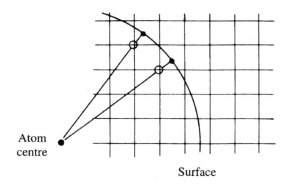

Figure 8.18   In the Pearl and Honegger method, the contour appearance is refined thanks to a homothetical correction [25]. Bold circles, actual surface points; hollow circles, relevant (approximate) grid nodes.

In the POLYMOD system of Dubois *et al.* [28], mainly devoted to the representation of electronic shapes, space structuralization by a grid is essential for locating the molecular information (Figure 8.20). "True" surface points (at the correct distance of the host atom) are sought on the edges of the 3D grid lattice by interpolation from the selected "surface" nodes (Figures 8.19 and 8.20).

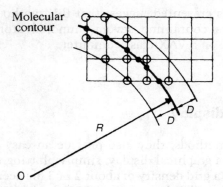

Figure 8.19 Refinement of the molecular contour in POLYMOD. Frontier nodes are selected within a slice $(R + D, R - D)$ (hollow circles). The intersections with the voxel edges (at the correct interatomic distance R) are determined by interpolation (black circles) (from Dubois *et al.* [28]).

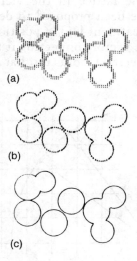

Figure 8.20 POLYMOD system. Successive steps in the display of molecular shapes (cross-section in a plane). (a) Extracting the frontier nodes (Dubois *et al.* [28]); (b) determining the intersection points; (c) drawing the contour line.

8.4 BOOLEAN OPERATIONS AND MOLECULAR COMPARISONS

Molecular shape analysis is closely related to recognition problems involved in drug design or more generally in the field of molecular interactions [22]. When comparing two or more molecular shapes, the data structures (strings of bits) obtained in grid methods look very efficient to quantify and visualize similarities between molecular shapes thanks to Boolean operations. *Union* allows for determining the total volume shared, *intersection* defines the parts common to all structures, and the *exclusive-or* operation selects only unique aspects of each structure (parts belonging to one molecule but not to the other) (Figure 8.21).

All these operations are easily performed on the strings representing the two molecules, using logical operators (AND, OR, etc. and their combinations) on a bit-by-bit basis. The graphical display of the results is easy, since the data structure is maintained (Figure 8.22). Such processes suppose, of course, as a prerequisite, that the molecules to be compared have previously been superimposed and mapped onto the same grid. The problem of comparing molecules with different orientations with respect to the referential relies on pattern recognition techniques or common substructure search, and will be discussed separately.

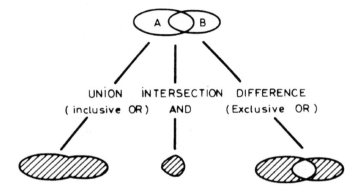

Figure 8.21 Elementary Boolean operations of union, intersection and symmetrical difference.

8.5 TOWARDS QUANTITATIVE RELATIONSHIPS

Although molecular surface or volume are not directly attainable by experimentation, some quantitative correlations with physically observable quantities have been proposed. So, on a set of about 100 simple molecules (alkanes, cycloalkanes and chlorinated or brominated derivatives), Meyer points out linear relationships between the (calculated) van der Waals volume and the *molar refraction* or the *b constant* (covolume) in the van der Waals equation of state [29–30a][1]. Similarly, the heats of vaporization of *n*-alkanes are linear in the van der Waals volumes [31]. More recently, a correlation was proposed between molecular areas (calculated by the GEPOL program) and the logarithm of the aqueous solubility for 67 molecules of ethers, esters, alcohols and ketones [18].

[1]The molecular volume derived from the density measured in the liquid phase (not solution) $V_{in} = M/N d$ is an "inner volume", characterizing the bulk of the substance. It comprises the van der Waals volume plus a free volume amounting sometimes to about 50% of the total).

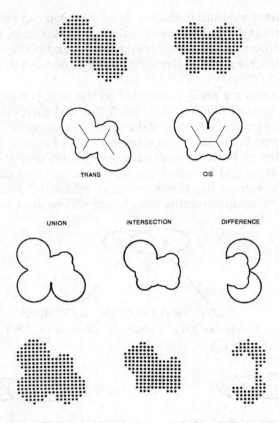

Figure 8.22 Boolean operations on molecular shapes using grid representation. The example corresponds to cis and trans dichloroethylene. The drawing (exact contour and node approximation) represents the sections of the molecular shapes by a plane parallel to the double bond, 1 Å above it.

Van der Waals volumes are rather insensitive to conformational features, but surfaces vary much more. Within sets of closely related species they correlate with the free volume that encapsulates the molecule in the bulk, so that consideration of both the van der Waals volume and surface gives some insight into the molecular shape. In a subsequent study, Meyer [32] proposed extending these concepts of volume and surface to substituent groups to deal with steric effects on reaction rates. Two descriptors are considered: the shape is characterized by the substituent's *surface to volume ratio* (G), bulk being approached by the *volume of the portion of the group lying within 0.3 nm of the reaction centre* (V_a). In the quantitative treatment of steric effects, this step appears as an update of the approaches from Charton [33–33b] or Verloop *et al.* [34], who suggested taking into account the actual shape of the substituent rather than evaluating its size by only one parameter (as the steric constant of Taft [35]). For alkyl substituents, rate constants for varied reactions, including acid hydrolysis (a reference process for defining steric effects) correlate with G

and V_a. However, the fit is less satisfactory when these new parameters are compared to the usual steric constants (Taft's E_s).

These quantitative relationships assess the physical meaning of the so-defined surface and volume calculations. In return, one can reasonably use such calculations to obtain some insight into molecular interaction processes. In a recent paper, Meyer et al. investigated the adsorption of alkanoic acids on silica. Cross-sectional areas of adsorbates, derived from a fractal analysis of adsorption data, were compared to values computed from atomic coordinates. In this way, it was established that these molecules are adsorbed parallel to the surface, a rather surprising result since soap-like molecules are generally adsorbed perpendicularly [36].

Hydration is certainly an important factor for the definition of the favoured conformers of polar molecules in biological media. However, until now, its treatment usually requires complex computational tasks which cannot be easily incorporated in programs aiming at rapid conformational analysis. For the inclusion of hydration effects in empirical conformational energy calculations of polypeptides, Ooi et al. [16] proposed breaking down the free energy of hydration into additive contributions from the various intervening groups. These contributions are assumed proportional to the accessible surface area of the groups, and have been scaled by fitting experimental free energy of solvation for small test compounds.

Relationships between the molecular surface area and physicochemical properties have been refined by explicitly taking into account the nature of the groups or atoms, and more precisely, the charge information. So, for the evaluation of polar intermolecular interactions, various authors proposed separating hydrocarbon or non-polar portions (with isotropic hydrophobic interactions) from functional or polar moieties (which are largely hydrated or solvated): the properties investigated were, for example, aqueous solubility, partition coefficients and hydrophobicity [37–42]. In a more precise treatment, Stanton and Jurs [37] define a set of 25 Charged Partial Surface Area (CPSA) descriptors combining molecular surface area and charge information (partial atomic charges being calculated from an empirical method [43, 44]). These CPSA descriptors, combined with topological ones, were included in a multiparametric regression treatment to correlate chromatographic retention-times, boiling points or surface tension values.

Surface area was also invoked to calculate the physical properties of pure organic substances such as critical temperature, critical volume and related properties for molecules in the range 40–500 u.a. in the Molecular Surface Interaction approach of Grigoras [45]. However, since in the critical state molecules are in contact with one another, van der Waals radii are no longer convenient to delineate the molecular body, and new atomic radii, considered as adjustable parameters, must be defined.

Some information about crystal state can also be gained through volume or surface calculations, as shown by Gavezzotti [46, 46a]. In this work, the discrete sampling is carried out on local polar coordinates rather than on an orthogonal lattice. From each atom centre a probe vector originates with a length equal to the selected radius of that atom, and its direction is given systematic variations by steps of ca. 2–3° in longitude or latitude. Surface points inside another atom sphere are counted as interior (N_i), other points being encoded as lying on the surface (N_o). The *free atomic surface* for each atom is:

$$4 \pi R^2 (N_o / (N_i + N_o))$$

Summation of atomic free surfaces leads to the *molecular free surface* S_m. Rather unexpectedly, a linear relation has been observed between these free surfaces and the number of valence electrons in the molecule. In fact, although for isolated atoms van der Waals spheres may be considered as an envelope of the outer electrons, in a molecule the relation between the actual electron distribution and a surface defined as the union of atomic spheres, taken as rigid bodies, is not obvious. From this observation, of entirely empirical origin, a set of group increments can be defined, the summation of which leads to a rapid evaluation of the molecular free surfaces S_m.

Deviations of calculated molecular and atomic free surfaces with respect to group increments, which correspond to standard situations, contain some information about molecular conformations and local strain. Steric crowding, for example, reduces free surface area. For moderately polar organic compounds, S_m is related to the crystal packing energy. Atomic free surfaces indicate how strain or crowding are distributed, and then specify the amount of cohesive energy provided by each atom. This can lead to interesting discussions towards a better understanding of crystal growth in relation to the close packing principle (in an ideal close-packed crystal, all atoms are exposed equally and have the same energetic relevance, with a maximum of intermolecular contacts in their coordination sphere).

8.5.1 Topological model

Quite different to the previous approaches (except, in some ways, Gavezotti's studies), working on real 3D coordinates, Govers and de Voogt [47] proposed calculating the van der Waals volume from molecular topology only (indication of atoms and bonds between them) by summation of fragmental volumes. This model relies, to some extent, on the previous additive schemes of Bondi [48] and Moriguchi [49] and the observations of Kier and Hall [50] that the van der Waals volume of these fragments correlates with their connectivity indices.

As a basis for the model, indices denote the number of non-hydrogen valence electrons. Fragments are limited to one (heavy) atom and its corresponding hydrogens ("united atoms"). Their volumes are considered as a sum of orbital volumes for valence electrons. One distinguishes sp³ CH orbitals, other σ orbitals, π orbitals and lone pairs. Methane is the reference. Scaling factors allow for a quantum level dependence (according to the principal quantum number value).

For atomic fragments:

$$V = F_n V_o - G_n [(\sigma - h)(V_h - V_\sigma) + \pi(V_h - V_\pi) + p(V_h - V_p)]$$

where $\sigma - h$, π, p denote the number of non-hydrogen σ electrons, π electrons and lone pairs in the fragment, and V_o is the reference volume, that of methane $V_o = 4 V_b$ $(V_h - V_\sigma)$, $(V_h - V_\pi)$, $(V_h - V_p)$ represent the difference between the volume of the combined hybrid orbitals and substitute orbitals $(1s + ... sp^*)$. Factors F and G express the dependence upon quantum numbers. However, for second row elements of the periodic system $(n = 2)$, $F_n = G_n = 1$.

Other similar formulae can be derived to get expressions more akin to that of Kier and Hall [50]:

$$V = V_o - n^{-2}(\delta + \delta^v) A V^*$$

$\delta = \sigma - h$, molecular connectivity; $\delta^v = \sigma - h + \pi + 2p$; A and V^* representing an auxiliary constant and a "hypothetical volume". In this approximation, $F_n = 1$ and $G_n = n^{-2}$.

Tests performed on a reference set of 33 fragments, where previously calculated volumes can be compared to the proposed model, establish the consistency of the previous expressions, also confirmed in the study of polycyclic aromatics. The model has been subsequently used to derive 28 additional fragmental-volumes not directly available, greatly broadening its structural scope.

8.6 CONCLUDING REMARKS: ROUGHNESS AND FRACTAL SURFACES

The accessible and Richards' molecular surfaces involve a probe (mimicking the solvent molecule) rolling on to the van der Waals contour. What might be the influence of the radius of that probe sphere on the shape or extent of these surfaces? From another point of view [51], at the molecular level, what role can the surface irregularity or roughness play in the interactions occurring in molecular recognition?

At first glance, it appears that a large probe cannot enter sharp crevices at the surface of the molecule, and would thus characterize only a global shape. On the other hand, a very small probe can roll over the entire van der Waals contour and penetrate grooves (remember that if the probe radius becomes zero, the accessible surface tends to the van der Waals body).

These points are more easily quantified using the concept of fractal dimension of a surface [51]. The fractal dimension (D) can be derived from the ratio of the variations of the log of the surface area vs. those of the log of the probe size used:

$$2 - D = d(\log A) / d(\log R)$$

where D = fractal dimension, A = molecular surface and R = probe radius.

As to the meaning of the fractal dimension [52], recall that the concept of fractal dimension relies to some extent on the notion of homothety, and the elementary relationship:

$$\log (a^d) = d * \log a$$

D can be expressed as the ratio:

$$D = (\log Q) / (\log k)$$

where Q is the ratio of the measures of the objects and k the homothety ratio.

So, if a segment l is submitted to a homothety (centre at origin) of ratio a, the measure of the new segment obtained is $a * l$. Then:

$$Q = al / l = a \quad \text{and} \quad D = \log a / \log a = 1$$

Similarly, for a square (in a plane) or a cube (both of edge l) the area or the volume become $(al)^2$ or $(al)^3$, corresponding to ratios $Q = a^2$ or a^3 and fractal dimensions of 2 (planar figure) or 3 (volume). Turning again to molecular surfaces, the fractal dimension characterizes the degree of irregularity: for smooth surfaces $D\#2$, whereas for a nearly completely space-filling surface $D\#3$.

Such calculations have been applied to proteins: D is evaluated from the slope of a plot of $\log A$ (calculated, for instance, from Connolly's method) vs. $\log R$.

It was shown that D approaches 2 for both large and very small probes, whereas significant variations (up to $D = 2.4$) are observed with probe radii between 1–3.5 Å, revealing a high degree of irregularity at the atomic scale. Direct experimental determination of the fractal dimension of the polypeptide backbone agrees with this conclusion [53]. Interestingly, as quoted by Lewis and Rees [51], water molecules and side chains fall into this size range, suggesting that local variations of the fractal dimension should reflect specific interactions. For some molecules (e.g. lysozyme, superoxide dismutase) a display of these fractal dimensions as spherical projections allows for examining the variations of D over protein surfaces. Regions able to form thigh complexes with other proteins (interfaces) appear to be more irregular than on average, suggesting that an irregular surface leads to more stabilizing contacts. In contrast, active sites, interacting transiently with ligands without forming stable complexes, seem to be located in smooth regions (Figure 8.23).

In the examples investigated, D is unrelated to residue mobility or exposed surface area, although regions accessible to large probes are more probably associated with smooth regions than with residues on irregular parts. So, fractal dimension may constitute a new characteristic of the molecular surfaces. This recent concept of local irregularity would perhaps provide an original approach to the origin of specificity in molecular recognition processes (Figures 8.24, 8.25).

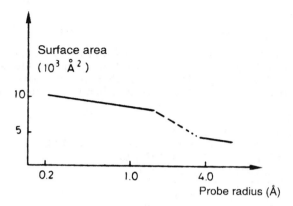

Figure 8.23 Variation of the molecular surface area vs. probe radius for lysozyme (log scale) (redrawn from Lewis and Rees [51] with permission).

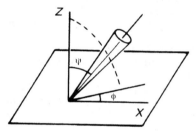

Figure 8.24 Angular coordinates for determining local accessibility (redrawn from Lewis and Rees [51] with permission).

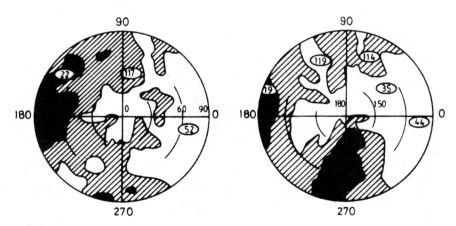

Figure 8.25 Spherical projection of the fractal surface of lysozyme. For direction (ϕ, ψ), the area is calculated for atoms within a cone $(\pm 30°)$. ψ is measured from the crystallographic z axis (some values are indicated along the light arcs of internal circles), and ϕ from the x axis in the equatorial plane (some values are indicated on the external circle). The location of some residues is shown. Shadowed regions correspond to $D > 2.3$ and darker regions to $D > 2.5$ (redrawn from Lewis and Rees [51] with permission).

REFERENCES

1. H.R. Karfunkel and V. Eyraud *J. Comput. Chem.*, **10**: 1989; 628–634.
2. G.R. Marshall, C.D. Barry, H.E. Bosshard, R.A. Dammkoehler and D.A. Dunn in *Computer Assisted Drug Design*, ACS Symposium Series 112, E.C. Olson and R.E. Christoffersen (Eds.), American Chemical Society, Washington, DC, 1979.
3. F.M. Richards *Ann. Rev. Biophys. Bioeng.*, **6**: 1977; 151–176.
4. H.R. Karfunkel *Match*, **19**: 1986; 67–87.
5. M.L. Connolly *J. Am. Chem. Soc.*, **107**: 1985; 1118–1124.
6. W.G. Richards *J. Mol. Graphics*, **4**: 1986; 186.
7. N.L. Max, *J. Mol. Graphics*, **2**: 1984; 8–13.
8. N.L. Max *J. Med. Sys.*, **6**: 1982; 485–499.
9. B. Lee and F.M. Richards *J. Mol. Biol.*, **55**: 1971; 379–400.
10. M.L. Connolly *J. Appl. Cryst.*, **16**: 1983; 548–558.
11. A. Shrake and J.A. Rupley *J. Mol. Biol.*, **79**: 1973; 351–371.
12. R. Lavery, A. Pullman and B. Pullman *Intern. J. Quantum Chem.*, **XX**: 1981; 49–62.
13. M.L. Connolly *Science*, **211**: 1981, 661–666.
13a. M.L. Connolly *Science*, **221**: 1983; 709–713.
14. W. Hasel, T.F. Hendrickson and W.C. Still *Tetrahedron Computer Methodology*, **1**: 1988; 103–116.
15. S.J. Wodak and J. Janin *Proc. Natl. Acad. Sci. USA*, **77**: 1980, 1736–1740.
16. T. Ooi, M. Oobatake, G. Nemethy and H.A. Scheraga *Proc. Natl. Acad. Sci. USA*, **84**: 1987; 3086–3090.
17. J. L. Pascual-Ahuir and E. Silla *J. Comput. Chem.*, **11**: 1990; 1047–1060.
18. E. Silla, I. Tunon and J.L. Pascual-Ahuir *J. Comput. Chem.*, **12**: 1991; 1077–1088.
19. J.B. Moon and W.J. Howe *J. Mol. Graphics*, **7**: 1989; 109–112.
20. J. Greer and B.L. Bush *Proc. Natl. Acad. Sci. USA*, **75**: 1978; 303–307.
21. M.Y. Pavlov and B.A. Fedorov *Biopolymers*, **22**: 1983; 1507–1522.
22. T.R. Stouch and P.C. Jurs *J. Chem. Inf. Comput. Sci.*, **26**: 1986; 4–12.
23. R.S. Pearlman, in *Physical Chemical Properties of Drugs*, S.H. Yalkowsky, A.A. Sinkula and S.C. Valvani (Ed.) M. Dekker, New York, 1980; 321–347.
24. A.Y. Meyer *J. Comput. Chem.*, **9**: 1988, 18–24.
25. L.H. Pearl and A. Honegger *J. Mol. Graphics*, **1**: 1983; 9–12.
26. J. Higo and N. Go *J. Comput. Chem.*, **10**: 1989; 376–379.
27. R.S. Bohacek and W.C. Guida *J. Mol. Graphics*, **7**: 1989; 113–117.
28. J.E. Dubois, S.Y. Yue and J.P. Doucet *The Visual Computer*, **2**: 1986; 367–378.
29. A.Y. Meyer *J. Chem. Soc. Perkin Trans. II*, 1985; 1161–1169.
30. A.Y. Meyer *J. Mol. Struct.*, **124**: 1985; 93–106.
30a. A.Y. Meyer *J. Comput. Chem.*, **7**: 1986, 144–152.
31. M. Marsili, P. Floersheim and A.S. Dreiding *Comput. Chem.*, **7**: 1983; 175–181.
32. A.Y. Meyer *J. Chem. Soc. Perkin Trans.* **2**: 1986; 1567–1572.
33. M. Charton *Prog. Phys. Org. Chem.*, **8**: 1971; 235–317.
33a. M. Charton *J. Am. Chem. Soc.*, **97**: 1975; 1552–1556.
33b. M. Charton *J. Am. Chem. Soc.*, **91**: 1969; 615–618.
34. A. Verloop, W. Hoogenstraaten and J. Tipker, in E.J. Ariens (Ed.): *Drug Design, Vol. VII*, Academic Press, New York, 1976; 165.
35. R.W. Taft *J. Am. Chem. Soc.*, **74**: 1952; 2729–2734.
36. A.Y. Meyer, D. Farin and D. Avnir *J. Am. Chem. Soc.*, **108**: 1986; 7897–7905.
37. D.T. Stanton and P.C. Jurs *Anal. Chem.*, **62**: 1990; 2323–2329.
38. P. Camilleri, S.A. Watts and J.A. Boraston *J. Chem. Soc. Perkin Trans.*, **2**: 1988; 1699–1707.
39. G.L. Amidon, S.H. Yalkowsky, S.T. Anik and S.C. Valvani *J. Phys. Chem.*, **79**: 1975; 2239–2246.
40. W.J. Dunn III, M.G. Koehler and S. Grigoras *J. Med. Chem.*, **30**: 1987; 1121–1126.

41. M.G. Koehler, S. Grigoras and W.J. Dunn III *Quant. Struct.-Act. Relat.*, **7**: 1988; 150–159.
42. K. Iwase, K. Komatsu, S. Hirono, S. Nakagawa and I. Moriguchi *Chem. Pharm. Bull.*, **33**: 1985, 2114–2121.
43. R.J. Abraham and P.E. Smith *J. Comput. Chem.*, **9**: 1988; 288–297.
44. S.L. Dixon and P.C. Jurs *J. Comput. Chem.*, **13**: 1992; 492–504.
45. S. Grigoras *J. Comput. Chem.*, **11**: 1990; 493–510.
46. A. Gavezzotti *J. Am. Chem. Soc.*, **105**: 1983; 5220–5225.
46a. A. Gavezzotti *J. Am. Chem. Soc.*, **107**: 1985; 962–967.
47. H. Govers and P. de Voogt *Quant. Struct. Act. Relat.*, **8**: 1989; 11–16.
48. A. Bondi *J. Phys. Chem.*, **68**: 1964; 441–451.
49. I. Moriguchi, Y. Kanada and K. Komatsu *Chem. Pharm. Bull.*, **24**: 1976; 1799–1806.
50. L.B. Kier and L.H. Hall *J. Pharm. Sci.*, **70**: 1981; 583–589.
51. M. Lewis and D.C. Rees *Science*, **230**: 1985; 1163–1165.
52. J.B. Touchard *Images Numériques*, Cedic/Nathan (Ed.), 1987, p. 145.
53. J.P. Allen, J.T. Colvin, D.G. Stinson, C.P. Flynn and H.J. Stapleton *Biophys. J.*, **38**: 1982; 299–310.

9 *Key features of quantum chemistry methods used in CAMD*

The previous chapters have shown how strongly the basic features of computer-aided molecular design (CAMD) are connected to the generation and graphical representation of molecular architectures. Indeed, it is essential for a chemist to rapidly determine what is the spatial arrangement of atoms in a given molecular system towards an inceptive evaluation of its chemical properties. It is therefore indisputable that the very first and most popular application of molecular graphics consists in building or retrieving, visualizing and manipulating computerized models of chemical compounds. However, molecular graphics and, by extension, CAMD would be of limited interest without the possibility to also evaluate and display the physico-chemical properties of these compounds, such as their molecular orbitals, electron densities, electrostatic potentials, reactivity indices, etc. Indeed, stereo-chemistry is only one component of the multi-faceted aspects of molecular structure, taken in a broad sense, and a modelling with subsequent graphical representation of properties related to electronic structure is an indispensable extension of the basic steps of CAMD. To this end, the recourse to quantum chemistry is an inescapable reality.

Quantum chemistry may be defined as the application of quantum mechanics to atomic and molecular systems, which means that the behaviour of the microscopic particles they are made of, namely nuclei and electrons, will be described according to quantum theory. It is well known that quantum mechanics is based on the Schrödinger equation, the solution of which consists of the wavefunction of the system of particles. In principle, this wavefunction contains all the information needed to describe the properties of the system of interest. This had led to the famous statement made by Dirac: "The underlying physical laws necessary for the mathematical theory of a large part of physics and the whole of chemistry are thus completely known, and the difficulty is only that the exact application of these laws leads to equations much too complicated to be soluble" [1]. Not only for its historical value, it is also very interesting to report the continuation of Dirac's fundamental article: "It therefore becomes desirable that approximate practical methods of applying quantum mechanics should be developed, which can lead to an explanation of the main features of complex atomic systems without too much computation" [1]. Through the advent of computers and of large-scale computing, this very lucid prediction of Dirac has now become a reality: whereas it is true that an *exact* solution of the Schrödinger equation is still an impossible mathematical problem today for systems of chemical interest, acceptable approximate solutions may be routinely obtained for molecules containing up to several hundreds of electrons. The main difficulty which now faces the computational chemist is thus no longer related to a struggle for an *exact* solution – an impossible task anyway – but rather to a clear understanding of the approximations inherent in a given model, and to an unbiased estimation of their influence on the results. In view of the great many quantum chemical models proposed so far and available as standard computer packages, this is by no means an easy task.

Many well documented textbooks dealing with both methodological and applied aspects of quantum chemistry have been published recently [2–13]. This means that we are not going to present the subject in detail, but rather summarize the main features and performances of the most commonly used quantum chemical models in CAMD, whose hierarchy is depicted in Figure 9.1.

Although it can be a little arbitrary to compare the respective merits of quantum chemical models on the basis of the approximations they rest on – one would definitely prefer a comparison based on the performance of the methods – the advantage of the scheme shown in Figure 9.1 lies in a clear-cut presentation of the most common tools of quantum chemistry. A detailed discussion of all the approximations inherent in the models reported in this scheme undoubtedly lies beyond the scope of the present monograph. However, before examining the key features of these methods and the role they play in CAMD applications, it is worth stressing some important points which are connected with the scheme:

1. All the models reported here, ranging from simple Hückel theory to sophisticated MCSCF methodology, have their own merit and field of applicability: there is no universal model that is equally valid for small and large compounds, whether organic or inorganic, and for the prediction of properties so different as conformation, electric or magnetic behaviour, reactivity, etc.
2. Traditional models of quantum chemistry are located on the left part of the scheme: they range from the popular semi-empirical schemes such as EH or AM1 to the sophisticated post-SCF schemes such as MCSCF or MBPT;

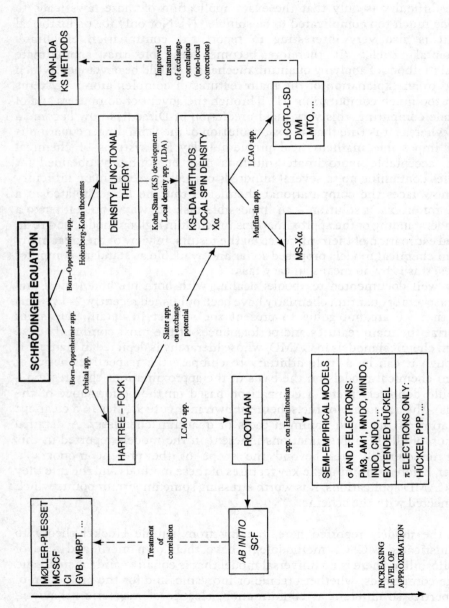

Figure 9.1 Scheme 1.

on the other hand, the more recent density functional models are found on the right part of the scheme. Broadly speaking, the latter methods represent the models of choice to predict with good accuracy the properties of large compounds (up to 500–1000 electrons) due to the reasonable computational effort they require.

3. When starting a CAMD application necessitating the use of quantum chemistry, the choice of model is generally strongly dictated by the size of the system under study, the availability of adequate computer hardware/ software and human experience, etc., which in principle drastically reduces the possibilities offered by the scheme.

The plan of the present chapter is clearly depicted by the scheme in Figure 9.1: starting from a survey of the Schrödinger equation and the main problems connected with its exact solution, the approximation procedures commonly used will be reviewed and discussed on the basis of their practical applicability to CAMD problems. We shall therefore be concerned by three main groups of methods: *ab initio* Hartree–Fock and post-SCF procedures, semi-empirical models and density functional approaches. Finally, recent applications illustrating the capabilities and limitations of these methods will be presented.

9.1 THE TIME-INDEPENDENT SCHRÖDINGER EQUATION

According to the postulates of quantum mechanics, the energy and properties of a stationary state of a system of microscopic particles are obtained by solution of the Schrödinger equation [14]:

$$H\Psi = E\Psi \tag{9.1}$$

where, for a system made of n electrons and N nuclei, the Hamiltonian H is the five-component energy operator:

$$H = T_e + T_n + V_{en} + V_{ee} + V_{nn} \tag{9.2}$$

E being the energy of the state described by the wavefunction $\Psi(1,2,...n,$ $1,2,...N)$; $1,2,...n$ and $1,2,...N$ representing the spatial (Cartesian) and spin coordinates of electrons and nuclei respectively.

To the operators of equation (9.2) correspond the following physical signification and mathematical expressions:

T_e = kinetic energy of the electrons,
T_n = kinetic energy of the nuclei,
V_{en} = potential energy arising through Coulombic interactions between electrons and nuclei,
V_{ee} = potential energy arising through Coulombic interactions between electrons,

V_{nn} = potential energy arising through Coulombic interactions between nuclei.

$$T_e = -\frac{h^2}{8\pi^2 m} \sum_{i=1}^{n} \left[\frac{\partial^2}{\partial x_i^2} + \frac{\partial^2}{\partial y_i^2} + \frac{\partial^2}{\partial z_i^2} \right]$$

(9.3)

where h is Planck's constant and m the electron mass, the summation running over the electrons with spatial coordinates (x_i, y_i, z_i).

The operator $\frac{\partial^2}{\partial x_i^2} + \frac{\partial^2}{\partial y_i^2} + \frac{\partial^2}{\partial z_i^2}$ is called the *Laplacian*, and denoted by ∇_i^2.

$$T_n = -\frac{h^2}{8\pi^2} \sum_{v=1}^{N} \frac{1}{M_v} \nabla_v^2$$

(9.4)

where M_v is the mass of nucleus v, the summation running over the nuclei with spatial coordinates (X_v, Y_v, Z_v).

$$V_{en} = -\sum_{i=1}^{n} \sum_{v=1}^{N} \frac{Z_v e^2}{r_{iv}}$$

(9.5)

where Z_v is the atomic number of nucleus v, e the electron charge and r_{iv} the distance between electron i and nucleus v.

$$V_{ee} = \frac{1}{2} \sum_{i=1}^{n} \sum_{j \neq i}^{n} \frac{e^2}{r_{ij}}$$

(9.6)

where r_{ij} is the distance between electrons i and j.

$$V_{nn} = \frac{1}{2} \sum_{\mu=1}^{N} \sum_{v \neq \mu}^{N} \frac{Z_\mu Z_v e^2}{R_{\mu v}}$$

(9.7)

where $R_{\mu v}$ is the distance between nuclei μ and v.

Expression (9.1) is known as the *time-independent Schrödinger equation*, as H does not depend explicitly upon time. Indeed, time-dependent interactions, such as the effect of a variable electric or magnetic field, are omitted in H as expressed by equation (9.1), which is the eigenvalue equation for the energy of our system; it has several well behaved solutions, known as stationary states Ψ_k, to each of them corresponds a discrete energy value E_k. The state of lowest energy E_o is called the ground state of the system.

As the Hamiltonian of the Schrödinger equation (9.1) does not contain any operator acting on the spin of the particles, it commutes with the usual spin operators. In addition, solving equation (9.1) will yield the dependence of the wavefunctions Ψ_k on spatial coordinates only.

When working on systems containing more than two particles, i.e. when $n + N > 2$, which is of course the case for any system of chemical interest, the

Schrödinger equation (9.1) is not exactly solvable. This is the quantum equivalent of the well known three (or more) body problem in classical physics. Following Dirac's advice [1], it is "therefore desirable that approximate practical methods" of solving the Schrödinger equation are available and, fortunately, many of them can be found on the computational chemistry market.

The first approximation performed on equation (9.1) consists of the separation of nuclear and electronic motions, as suggested by Born and Oppenheimer [15]. Indeed, as the proton mass is larger than that of the electron by a factor of 1836, the ratio between the mean velocity of the electrons and that of the nuclei is so large that the electrons adapt almost instantaneously their motion to the small changes in configuration of nuclei. It is therefore reasonable to assume that the electrons move in a field generated by fixed nuclei, which allows us to write the total wavefunction $\Psi(r,R)$ as a product of an electronic wavefunction $\Psi_R^e(r)$ and a nuclear wavefunction $\Psi^n(R)$:

$$\Psi(r,R) = \Psi_R^e(r) \cdot \Psi^n(R) \tag{9.8}$$

where r and R are shorthand notations for electron and nuclear coordinates, respectively.

Born and Oppenheimer have then shown that the electronic wavefunction $\Psi_R^e(r)$, which depends parametrically upon the nuclear coordinates R, is the solution of the electronic Schrödinger equation:

$$H^e\,\Psi_R^e(r) = E_R^e\,\Psi_R^e(r) \tag{9.9}$$

where:

$$H^e = T_e + V_{en} + V_{ee} \tag{9.10}$$

and E_R^e is the electronic energy of our system, calculated for the nuclear configuration R.

On the other hand, the nuclear wavefunction $\Psi^n(R)$ is the solution of the nuclear Schrödinger equation:

$$[T_n + (E_R^e + V_{nn})1]\,\Psi^n(R) = E'\,\Psi^n(R) \tag{9.11}$$

where $E_R^e + V_{nn}$ is the potential-energy function for the motion of the nuclei (1 being the unit operator) and E' an approximation to the exact energy E. $E_R^e + V_{nn}$ is the quantum chemical equivalent to the strain energy defined in section 5.3 using empirical force fields: a knowledge of the main features of this function of $3N-6$ variables (for non-linear molecules), also known as the potential energy surface, is essential for conformational analysis purposes, as its minima correspond to the various stable geometries of the system, whereas its saddle points should be associated with transition states. A good example of a potential-energy surface is presented in Plate V, which illustrates the case of the 1,2-difluorohydrazine molecule [16].

Potential energy surfaces represented as functions of two variables, such as that of Plate V, are very useful for a quantitative description of the various conformers of a given molecule, allowing, for example, differentiation between local vs. global minima. In the case of the HFN–NFH molecule, three local minima are found (M1 and M2, which are equivalent, and M4) in addition to the global one, M3.

We may now come back to the solution of the Schrödinger equation of our system. Practically, the main result of the Born–Oppenheimer approximation is that, unless one is interested in infrared and Raman spectroscopies, it suffices to solve the electronic Schrödinger equation (9.9). However, for performing conformational analysis, this task should be repeated for many positions R of the nuclei in order to have a good knowledge of the potential energy surface. As in this chapter we are only concerned with electronic structure and molecular conformations, one may omit the lowerscripts and superscripts of equation (9.9), which becomes:

$$H\,\Psi(r) = E\,\Psi(r)$$

(9.12)

where H is the Hamiltonian given by equation (9.10).

Practically, equation (9.12) is not very much easier to solve rigorously than equation (9.1) because of the many-body problem it involves. The next step is therefore to make the so-called orbital approximation [17], which assumes that the many-electron wavefunction $\Psi(r) = \Psi(1,2,...n)$ may be written as a product of one-electron wavefunctions ψ_i called *spin orbitals*:

$$\Psi(1,2,...n) = \psi_1(1)\,\psi_2(2)...\psi_n(n)$$

(9.13)

This is undoubtedly a rough approximation, which amounts physically to assuming that the n electrons move independently of one another. The one-electron wavefunctions ψ_i of the so-called Hartree product constituting equation (9.13) are defined as the product of a spatial function $\phi_i(x,y,z)$ and a spin function $\eta(\xi)$:

$$\psi_i(x,y,z,\xi) = \phi_i(x,y,z)\eta(\xi)$$

(9.14)

where x,y,z are the coordinates of the electron and ξ its spin variable.

The spatial functions ϕ_i are called the *molecular orbitals* (MOs) of the system and as there are only two possible spin functions $\alpha(\xi)$ and $\beta(\xi)$ (corresponding to the popular spin-up and spin-down picture), spin orbitals are of the form $\phi_i(x,y,z)\alpha(\xi)$ or $\phi_i(x,y,z)\beta(\xi)$. This amounts to say that a given MO ϕ_i may accommodate two electrons: one with α spin corresponding to the spin orbital $\phi_i\alpha$, and one with β spin corresponding to the spin orbital $\phi_i\beta$.

A basic problem with equation (9.13), i.e. by approximating an n-electron wavefunction by a single product of spin orbitals, is that does not have the property of antisymmetry as required by the Pauli principle for such a function [18]. This principle implies indeed that any n-electron wavefunction must change sign if we interchange the coordinates of any pair of electrons:

$$\Psi(1,2,3,...n) = -\Psi(2,1,3,...n)$$

(9.15)

As a suitable n-electron wavefunction satisfying the Pauli principle, one therefore uses, instead of equation (9.13), a Slater determinant [19] built over the spin orbitals:

$$\Psi(1,2,\ldots n) = \frac{1}{\sqrt{n!}} \begin{vmatrix} \psi_1(1) & \psi_1(2) & \cdots & \psi_1(n) \\ \psi_2(1) & \psi_2(2) & \cdots & \psi_2(n) \\ \vdots & \vdots & \vdots & \vdots \\ \psi_n(1) & \psi_n(2) & \cdots & \psi_n(n) \end{vmatrix} \tag{9.16}$$

where $1/\sqrt{n!}$ is the normalizing factor, calculated assuming that the spin-orbitals are orthonormal, ensuring that:

$$\int \Psi^*(1,2,\ldots n)\, \Psi(1,2,\ldots n)\, d\tau_1 d\tau_2 \ldots d\tau_n = 1 \tag{9.17}$$

with the asterisk denoting complex conjugation and $d\tau = dx\,dy\,dz\,d\xi$.

The antisymmetry property required by equation (9.15) then follows directly for the wavefunction (9.16) from the properties of determinants (interchange of two columns changes the sign of a determinant). In addition, a determinant with two identical rows of zeros means that all the spin orbitals ψ_i must differ either by ϕ_i or by η. This leads to the well-known Pauli exclusion principle [18], i.e. two electrons cannot have the same set of quantum numbers, which is actually a consequence of the antisymmetry principle.

The problem is now to find the best possible MO ϕ_i to use in the determinantal wavefunction (9.16). It will be solved by the Hartree–Fock method, presented in the next section.

9.2 HARTREE–FOCK AND ROOTHAAN EQUATIONS: *AB INITIO* METHODS

For the sake of simplicity, we may now consider a so-called "closed-shell" system made of N nuclei and $2n$ electrons in such a configuration that all the occupied MOs accommodate two electrons, one with α spin and the other one with β spin. For such a system, the Slater determinant writes:

$$\Psi(1,2,\ldots 2n) = \frac{1}{\sqrt{(2n)!}} \begin{vmatrix} \phi_1(1)\alpha(1) & \phi_1(2)\alpha(2) & \cdots & \phi_1(2n)\alpha(2n) \\ \phi_1(1)\beta(1) & \phi_1(2)\beta(2) & \cdots & \phi_1(2n)\beta(2n) \\ \phi_2(1)\alpha(1) & \phi_2(2)\alpha(2) & \cdots & \phi_2(2n)\alpha(2n) \\ \vdots & \vdots & \vdots & \vdots \\ \phi_n(1)\beta(1) & \phi_n(2)\beta(2) & \cdots & \phi_n(2n)\beta(2n) \end{vmatrix} \tag{9.18}$$

The value of the normalizing factor of this determinant has been calculated assuming that the MO ϕ_i values are orthonormal, i.e.:

$$\int \phi_i(1)\phi_j(1)dv_1 = \delta_{ij} = \begin{cases} 1 & i = j \\ 0 & i \neq j \end{cases} \tag{9.19}$$

where dv_1 represents $dx_1 \, dy_1 \, dz_1$.

This is an important condition which ensures orthonormality of the spin-orbitals ψ_i, the spin functions α and β being orthonormal on their own.

The best MOs of our system will be given by using the variational principle, which is satisfied by any approximate solution $\Psi'(1,2,...2n)$ of the Schrödinger equation. This principle states that the energy associated with the normalized approximate wavefunction Ψ', expressed in the Dirac notation [20] as:

$$E' = <\Psi'|H|\Psi'> \tag{9.20}$$

where H is the Hamiltonian defined by equation (9.10), and is an upper limit to the ground state energy E_o yielded by the exact solution of equation (9.12):

$$E' \geq E_o \tag{9.21}$$

In the case of an approximate function of determinantal type such as equation (9.18), the energy E' will depend upon the nature of the one-electron wavefunctions ϕ_i, and the best MOs will be obtained by requiring that the energy (9.20) is minimum, subject to the constraint that the functions ϕ_i keep orthonormal. The two basic steps of the Hartree–Fock method [21, 22] are thus:

- the derivation of total electronic energy $E = <\Psi|H|\Psi>$ using as Ψ the Slater determinant (9.18) and H as the Hamiltonian (9.10),
- the minimization of E under the imposed constraint (9.19).

This leads to the well-known Hartree–Fock one-electron equations [2,5]:

$$F(1)\,\phi_i(1) = \varepsilon_i\,\phi_i(1) \tag{9.22}$$

where ε_i is the energy of orbital ϕ_i and F, the Fock operator, is given by:

$$F(1) = h(1) + \sum_{j=1}^{n}[2J_j(1) - K_j(1)] \tag{9.23}$$

1 representing the spatial coordinates of electron 1. The one-electron operators of equation (9.23) are defined as:

$$h(1) = -\frac{h^2}{8\pi^2 m}\nabla_1^2 - \sum_{v=1}^{N}\frac{Z_v e^2}{r_{1v}} \tag{9.24}$$

$$J_j(1)\,\phi_i(1) = \left[\int \phi_j(2)\frac{e^2}{r_{12}}\phi_j(2)dv_2\right]\phi_i(1) \tag{9.25}$$

$$K_j(1)\,\phi_i(1) = \left[\int \phi_j(2) \frac{e^2}{r_{12}} \phi_i(2) dv_2 \right] \phi_j(1) \tag{9.26}$$

$h(1)$ is therefore made of the kinetic and nuclear energy operators for electron 1, whereas $J_i(1)$ and $K_i(1)$ are the well-known Coulomb and exchange operators, respectively. These last two operators are responsible for interelectronic repulsion, but it is seen from equations (9.25) and (9.26) that electron 1 interacts with electron 2 located at an *average* position in its MO ϕ_i, not an *instantaneous* one as it should be; we will return to this important point later.

A careful examination of equations (9.22)–(9.26) allows one to conclude that solving the Hartree–Fock equations is not as simple as it would seem at first sight. Indeed, instead of being an eigenvalue (ε_i) – eigenvector (ϕ_i) system, equation (9.22) actually consists of coupled integro-differential equations where the ϕ_i functions, which act as eigenfunctions, are included in the J_i and K_i operators. This means that these equations must be solved using the iterative procedure depicted in the scheme of Figure 9.2 and known as the *self-consistent field* (SCF) method.

In the SCF method, one starts from an initial guess made of trial functions $\phi_i^{(0)}$, $i = 1,2,...n$, which are used to calculate the integrals involved in the Coulomb and exchange operators of $F^{(1)}$, the Fock operator of iteration 1. Then, the Hartree–Fock equations are solved, leading to a new set of MO $\phi_i^{(1)}$, which are subsequently used to construct the new Fock operator $F^{(2)}$. The process is repeated until convergence test parameter δ, which may be defined as the maximum difference between the wavefunction values obtained in consecutive iterations, is smaller than a given criterion. Generally speaking, there are two problems associated with the SCF procedure in quantum chemistry: (i) in many pathological cases, such as open-shell systems (configurations with one or more MOs accommodating a single electron), convergence may be difficult to reach; (ii) the system of SCF equations has to be solved m times, which considerably increases the computational effort.

For atoms, the Hartree–Fock equations may be solved rigorously by taking advantage of their spherical symmetry, which allows one to separate the variables. For molecules, however, this is no longer possible, and one generally uses the method proposed independently by Hall [23] and Roothaan [24], known as the *linear combination of atomic orbitals* (LCAO) approximation. In this method, the MO ϕ_i are expanded over a basis set of atomic orbitals $\{\chi_\mu\}$ of dimension m:

$$\phi_i(1) = \sum_{\mu=1}^{m} c_{\mu i} \chi_\mu(1) \tag{9.27}$$

and the problem is now to determine the coefficients $c_{\mu i}$ of our system for a given AO basis $\{\chi_\mu\}$. Using the variational method with the orthonormality constraint (9.19), Hall and Roothaan have shown that they are solutions of the equations:

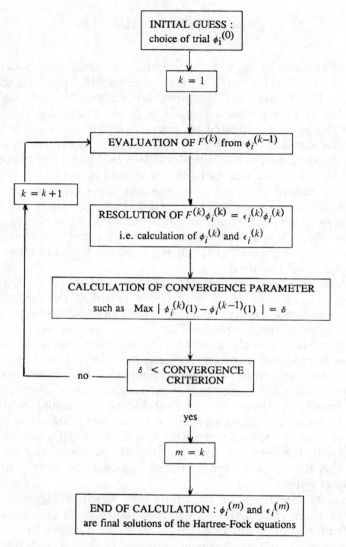

Figure 9.2   Scheme 2.

$$\sum_{v=1}^{m}\left(F_{\mu v}-\varepsilon_i S_{\mu v}\right)c_{vi}=0 \qquad\qquad \mu=1,2,\ldots m \qquad (9.28)$$

where:

$$F_{\mu v}=H_{\mu v}+G_{\mu v} \qquad (9.29)$$

with:

$$H_{\mu v}=\int \chi_{\mu}(1)\, h(1)\, \chi_v(1)\, dv_1 \qquad (9.30)$$

and:

$$G_{\mu\nu} = \sum_{\lambda\rho} P_{\lambda\rho}[<\mu\nu|\lambda\rho> - 1/2 <\mu\lambda|\nu\rho>] \qquad (9.31)$$

In expression (9.31), $P_{\lambda\rho}$ is the first-order density matrix defined as:

$$P_{\lambda\rho} = 2\sum_{i=1}^{n} c_{\lambda i} c_{\rho i} \qquad (9.32)$$

and $<\mu\nu|\lambda\rho>$ is a short-hand notation for the two-electron integral:

$$<\mu\nu|\lambda\rho> = \iint \chi_\mu(1)\chi_\nu(1) \frac{e^2}{r_{12}} \chi_\lambda(2)\chi_\rho(2)\, dv_1\, dv_2 \qquad (9.33)$$

In the Roothaan equations (9.28), ε_i is the one-electron energy of MO ϕ_i and $S_{\mu\nu}$ is the atomic overlap integral, defined as:

$$S_{\mu\nu} = \int \chi_\mu(1)\, \chi_\nu(1) dv_1 \qquad (9.34)$$

The relations (9.28) are the algebraic form of the Roothaan equations, whose matrix form is:

$$FC = SCE \qquad (9.35)$$

where F, C and S are $m\times m$ matrices with elements $F_{\mu\nu}$, $c_{\mu i}$ and $S_{\mu\nu}$ respectively, E being a diagonal matrix with elements ε_i.

Solving the Roothaan equations (9.35) leads to a set of MOs $\{\phi_i\}$ $i = 1,...m$; the ground state configuration is then obtained when the n lowest energy MOs each accommodate two electrons, the $m - n$ remaining ones being called *virtual MOs*.

Since it involves a great number of matrix operations, for which computers are very well suited, the matrix form (9.35) of Roothaan equations is solved preferentially in computational chemistry programs. However, when solving equation (9.35), one is faced with two well known problems:

1. As $F_{\mu\nu}$ depends upon the solutions through the density matrix $P_{\lambda\rho}$, the SCF procedure must be used (the scheme in Figure 9.2).
2. The number of $<\mu\nu|\lambda\rho>$ integrals to be evaluated is proportional to m^4, which places serious limitations to the size of the basis sets of practical use.

It is necessary here to make a fundamental distinction between the LCAO-MO methods, which attempt to solve the Roothaan equations (9.35): in the *ab initio* approach (i.e. literally, from first principles), these equations are solved for a given basis set without further approximations. On the contrary, the *semi-empirical* schemes introduce in the $F_{\mu\nu}$ matrix elements (equation (9.29)) both simplifying approximations and adjustable parameters, so as to lead to acceptable agreement with experimental results. The semi-empirical methods will be discussed later on in this chapter, and we turn now to the *ab initio* techniques.

The choice of an adequate AO basis set $\{\chi_\mu\}$ is an important problem in *ab initio* quantum chemistry. It is not possible to develop this question at length here, and the reader should refer to recent reviews, such as that published recently by Feller and Davidson [25]. Let us just mention that, for practical reasons, the atomic basis functions used in *ab initio* calculations are generally of the Gaussian type, i.e. they are characterized by a drop-off as $\exp(-\zeta r^2)$, where ζ (zeta) is a constant called the orbital exponent [8]. Atomic Gaussian functions, which were introduced by Boys [26], are generally defined in such a way as to correspond to the usual angular symmetries of hydrogen-like atomic orbitals [2], i.e. one may label them as 1s, 2s, $2p_x$, $2p_y$, $2p_z$, etc. However, as the radial behaviour of Gaussian functions leads to a poor representation of the actual electron density of atoms (no cusp at the origin, drop-off different from the $\exp(-\zeta r)$ dependence of hydrogen-like orbitals), one generally uses a (so-called contracted) linear combination of Gaussian functions, with fixed coefficients, as the expression of each basis function χ_μ.

It is clear that the maximum accuracy of the Roothaan method, the so-called Hartree–Fock limit, is achieved with a basis set of infinite dimension, i.e. $m = \infty$ in equation (9.27). As computers have not yet been designed to solve matrix equations of this dimension, the choice of basis sets in *ab initio* calculations is, as summarized by Hehre *et al.* [8], "a compromise between accuracy and efficiency." Roughly speaking, depending upon both the size of the molecule and the computational resources available, three different types of AO basis sets can be used in *ab initio* quantum chemistry:

1. *Minimal basis sets*, comprising one contracted Gaussian function (single zeta) for each occupied AO in the separated-atom limit ground state of the molecule under study. A typical example is the popular STO-3G basis set of Hehre *et al.* [27], where each AO of the minimal basis set is made of a contraction of three individual Gaussian functions. For the hydrogen fluoride (HF) system, the use of such a basis leads to a dimension (m) of 6: 1s, 2s, $2p_x$, $2p_y$ and $2p_z$ AOs on fluorine and 1s AO on hydrogen.
2. *Multiple-zeta basis sets*, which are characterized by the fact that two or more contracted Gaussian functions, instead of one, are used to describe the AOs of the minimal basis set. The double-zeta basis set, for example, comprises twice as many basis functions as the minimal one, the triple-zeta three times as many, etc. Such basis sets are much more flexible than minimal ones, and generally lead to a significant improvement in the description of the electronic structure of molecular systems. However, as core electrons are little affected by the formation of chemical bonds, in most cases it is not necessary to use a multiple-zeta description of their AOs and a good compromise is found in split-valence basis sets. In this case, a single contracted function corresponds to each core AO, whereas double-zeta functions are used for each valence AO. A good example is found in the 6–31G basis of Hehre *et al.* [28], where a contraction of six primitive Gaussians is used for core AOs, the valence AOs being described

by contracted functions made of three and one Gaussians, respectively. For the HF molecule in the 6-31G basis set, the value of m is 11: 1s, 2s, 2s′, $2p_x$, $2p_x'$, $2p_y$, $2p_y'$, $2p_z$, $2p_z'$ AOs on fluorine, 1s and 1s′ AOs on hydrogen.

3. *Polarized basis sets*, which differ from the previous ones by the addition of functions corresponding to AOs with higher angular quantum number l than that actually found in the ground state configurations of the constituent atoms. Such functions are known as *polarization functions* because they allow us to describe charge polarization effects resulting from the internal molecular electric fields. The 6–31G* basis set of Hariharan and Pople [29] is among the simplest polarized basis sets: it is constructed by the addition of six d-type Gaussian primitives (expressed in Cartesian coordinates, there are six second-order Gaussian functions: d_{xx}, d_{yy}, d_{zz}, d_{xy}, d_{xz} and d_{yz}) to the split-valence 6–31G description of each atom ranging from Li to Cl. For the HF molecule in the 6-31G* basis set, m amounts now to 17: 1s, 2s, 2s′, $2p_x$, $2p_x'$, $2p_y$, $2p_y'$, $2p_z$, $2p_z'$, $3d_{xx}$, $3d_{yy}$, $3d_{zz}$, $3d_{xy}$, $3d_{xz}$, $3d_{yz}$ AOs on fluorine, 1s and 1s′ AOs on hydrogen. It is seen that in view of the m^4 dependence of the number of two-electron integrals to be evaluated, the use of polarized basis sets involves a major computational effort for molecules containing more than 10–15 heavy (non-hydrogen) atoms.

It is not possible to conclude this section without mentioning the problem of electron correlation and the usual ways in which to solve it, at least partially, by resorting to post-Hartree–Fock treatments. We have previously pointed out that the Coulomb (equation (9.25)) and exchange (equation (9.26)) operators of the Hartree–Fock equations involve the *average*, not *instantaneous* as they should, repulsive interactions between electron pairs. Actually, this deficiency is due to the independent electron approximation, equation (9.13), and consequently to the single determinantal wavefunction, equation (9.16), postulated as the solution for the ground state configuration of the n electron system. Rigorously, the correlation energy E_c is defined as the difference between the true, non-relativistic energy E_o of the system and the energy E_{HF} of the Hartree–Fock wavefunction [30]:

$$E_c = E_o - E_{HF} \tag{9.36}$$

As the variational principle states that $E_o < E_{HF}$, it follows that E_c is always negative. Even though in practice E_c amounts typically to only 1% of E_{HF}, its absolute value is large due to the order of magnitude of E_{HF}. As an example, the Hartree–Fock energy of benzene is roughly −231 hartrees, which means that the correlation energy of this molecule is of the order of −1450 kcal/mol (1 hartree = 627.5 kcal/mol)! However, it is not always necessary to go beyond the Hartree–Fock treatment when using the *ab initio* model: geometries (bond distances in error by 0.02–0.03 Å and bond angles by 5°) and a large number of electronic properties (charge densities, electrostatic potentials, ionization energies of organic molecules, etc.) are generally reasonably well predicted at the Hartree–Fock (commonly denoted SCF) level, at least with an accuracy good enough for many CAMD applications. Clearly, post-Hartree–Fock treatments are much more demanding in terms of computer time than the SCF ones and, therefore, they should be reserved for cases of absolute necessity, such as the prediction of accurate spectroscopic properties or of chemical reaction paths involving bond breaking or making, etc. [3, 8, 9].

Broadly speaking, there are two categories of post-Hartree–Fock treatments, known as configuration interaction (CI) and multi-body perturbation methods respectively. In the CI approach, the molecular n-electron wavefunction is written as a sum of Slater determinants Ψ_I:

$$\Psi(1, 2, \ldots n) = \sum_I C_I \Psi_I(1, 2, \ldots n) \qquad (9.37)$$

where, in addition to the ground-state configuration (equation (9.16)), the Slater determinants Ψ_I correspond to excited states obtained by promoting electrons from occupied to virtual MOs ψ_i. The unknown coefficients C_I are determined variationally, i.e. by minimizing the energy E_{CI} calculated as:

$$E_{CI} = \langle \Psi | H | \Psi \rangle \qquad (9.38)$$

where Ψ is the multideterminantal wavefunction given by equation (9.37) and H the electronic Hamiltonian of equation (9.10). The selection of excited configurations to be introduced in the variational wavefunction (9.37) has a direct influence on the amount of correlation energy which can be recovered by this treatment. A standard choice, known as CISD (CI with singles and doubles), consists of introducing into the summation of equation (9.37) all the configurations differing from the ground-state by single and double excitations to virtual MOs, which allows us in principle to recover more than 90% of the correlation energy one would obtain by using the untruncated expansion (9.37) [31, 32]. However, due to uncompleteness of the one-electron (AO) basis set, the correlation energy recovered by a CISD calculation is usually much smaller than 90% of E_c because equation (9.36) assumes the use of the Hartree–Fock limit wavefunction to evaluate correlation effects (see the H_2O example presented below!). The main problems associated with CI calculations are their cost, which may be prohibitive (the number of possible excited configurations increases very rapidly with both the number of electrons and the quality of the basis set), and the fact that they are not size-consistent, i.e. correlation energies they lead to are not proportional, even roughly, to the size of the molecule.

Another problem connected with post-Hartree–Fock treatments of CI type is their slow convergence as a function of the number of configuration state functions (CSF), i.e. of excited state configurations taken in the calculation. This is illustrated by Figure 9.3, which displays the total energies of the H_2O molecule obtained by Lee *et al.* [32] from various *ab initio* SCF and post-SCF calculations. The dramatic increase of the number of CSFs as a function of the degree of excitation is clearly seen (i.e. when going from CISD to CISDTQ), together with the small energy lowering resulting from introduction of triply and quadruply excited states. It can be easily deduced from Figure 9.3 that the CISD calculation performed using the DZ and DZP basis sets allow recovery of only 34% and 53% respectively, of the correlation energy. Actually, a very

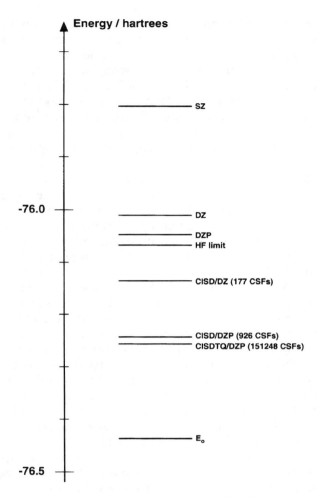

Figure 9.3 Total energy of the H_2O molecule calculated using different *ab initio* procedures [3, 32] and compared with the E_0 value of equation (9.36). SZ = single zeta, DZ = double zeta + polarization basis sets, CISD = CI including single and double excitations, CISDTQ = CISD including triple and quadruple excitations.

large basis set including several polarization functions is required to obtain, at the CISD level, a total energy for the H_2O molecule recovering 86% of the exact correlation energy, i.e. of the difference between E_0 and E_{HF} [33].

Multi-body perturbation methods provide an interesting alternative to post-Hartree–Fock treatments of the CI type. Indeed, as seen in the example reported above, convergence of CI expansions is slow and a very large number of configurations has to be used for an accurate evaluation of the correlation energy [34]. It was therefore of interest to devise more economical strategies to recover the largest fraction of E_c. In the Møller–Plesset (MP) [35] formulation of the multi-body perturbation theory (MBPT), which is the most popular version of MBPT, the Hartree–Fock operator is extracted from the Hamiltonian of equation (9.10) and the remaining part is treated as a perturbation [3, 8]. For a given AO basis set, the exact ground state wavefunction and energy can then be expressed using perturbation theory, i.e. by a perturbation expansion up to infinite order. Practically, the expansion is truncated to 2nd, 3rd or 4th order, which leads to the well known MP2, MP3 or MP4 models. It can be shown that the MP2 total energy is made of the Hartree–Fock energy for the ground state of our system plus a second-order contribution resulting from all the double excitations to the virtual MOs [8]. Again, only double excitations contribute to the MP3 total energy, whereas single, triple and quadruple ones lead to the MP4 correction.

Practically, MP methods are convenient as they require less computational effort than conventional CI. For example, the CPU time needed by a MP2 calculation is generally larger than a SCF one by a factor of 2.0–4.0 only. In addition, MP methods have the advantage of being size-consistent, and MP2 has proven to be very efficient for accurate structural predictions [8]. However, the MP methods present the limitations of: (i) being not variational (it is not unusual that the MP2 energy correction is larger in absolute value than the correlation energy); and (ii) not being able to calculate excited states.

As a conclusion, *ab initio* quantum chemical methods are undoubtedly useful tools in CAMD, provided the size of the target compounds enables one to perform "good quality" and relevant calculations (choice of a sufficiently large basis set and possibly post-SCF treatment). Referring to two popular application fields of CAMD, namely pharmacophores and new materials, frequently these conditions are not fulfilled, and the users have to turn to simpler and more approximate models. In addition, *ab initio* methods are too demanding in computational resources to be used for interactive (or pseudo-interactive) molecular graphics applications such as the building of structural models or the evaluation of reactivity indices. However, these models are indispensable and do bring an invaluable contribution when: results of chemical accuracy are needed for small molecules or model fragments, or new force field parameters have to be determined for molecular mechanics or dynamics investigations. Significant progresses have recently been achieved in post-Hartree–Fock treatments, resulting in performing methods such as coupled-cluster (CC), multi-configuration SCF (MCSCF), etc. For a good review, the reader should refer, for instance, to R.J. Bartlett and J.F. Stanton, in *Reviews in Computational Chemistry*, vol 5, K.B. Lipkowitz and D.B. Boyd, (eds.), VCH Publishers, New York, 1994; p 65. For the reasons mentioned in this Chapter, these methods are, broadly speaking, still beyond the scope of standard CAMD applications and they need no further treatment here.

9.3 SEMI-EMPIRICAL METHODS

For the obvious reason of the lack of adequate computing facilities required by accurate quantum chemical models, approximate semi-empirical methods have older roots in quantum chemistry than *ab initio* ones. Historically, the first of them is the qualitative π-electron *Hückel* MO (HMO) method, which traces back to the early 1930s [36], i.e. 20 years before Hall and Roothaan developed the LCAO-SCF model! Actually, the HMO model rests on a phenomenological Hamiltonian describing in a purely empirical way the interactions between π electrons of planar unsaturated molecules [37]. In spite of its limitations, the model is able to rationalize many properties of organic molecules, and it is not unusual, even nowadays, to see HMO calculations reported in prestigious journals [38, 39]!

As pointed out by Pople and Beveridge [40], semi-empirical methods may be conceived from two basically different points of view: (i) construction of a purely empirical Hamiltonian, as in the HMO and its successor, the extended Hückel (EH) MO [41] models, without further justification than reproducing experimental results; and (ii) introduction of physical approximations and empirical parameters as well into the F Hamiltonian matrix of the Roothaan equations (9.35). As both approaches have led to the development of successful methods, they will be briefly presented and their respective merits outlined.

Semi-empirical methods may be defined as approximate procedures which rely on a set of empirical parameters to calculate the wavefunctions of valence electrons only. As compared with *ab initio* methods, which solve rigorously the Hall–Roothaan equations in a given basis set, semi-empirical methods are therefore based on the use of approximate effective Hamiltonians acting on the valence space of the molecule, the inner shell electrons being treated with the nucleus as an unpolarizable core [42].

9.3.1 Extended Hückel method

In the Hückel theory [36], the MOs of unsaturated organic molecules are built from π AOs of the carbon skeleton, using a one-electron Hamiltonian (equation (9.30)) whose matrix elements are defined as:

$$H_{\mu\mu} = \alpha$$
$$H_{\mu\nu} = \beta \quad \text{if } \mu \text{ and } \nu \text{ belong to bound atoms} \qquad (9.39)$$
$$H_{\mu\nu} = 0 \quad \text{otherwise} \qquad (9.40)$$

where α and β, the Coulomb and resonance integrals respectively, are determined empirically by fitting the results to observed data. In addition, the electron-repulsion matrix elements $G_{\mu\nu}$ (equation (9.31)) are neglected, and the overlap integrals $S_{\mu\nu}$ (equation (9.34)) are taken such as $S_{\mu\nu} = \delta_{\mu\nu}$.

The resulting linear equations:

$$\sum_{\nu}^{m} (H_{\mu\nu} - \varepsilon_i \delta_{\mu\nu}) c_{\nu i} = 0 \quad \mu = 1, 2, \ldots m \qquad (9.41)$$

where m is the number of carbon atoms, are then solved using the HMO determinantal equation:

$$|H_{\mu\nu} - \varepsilon_i \delta_{\mu\nu}| = 0 \qquad (9.42)$$

In addition to being useful for understanding the gross features of chemical bonding, especially its topological character, the HMO method has some success in calculating properties such as oxidation-reduction potentials, ionization energies and hyperfine splittings of hydrocarbon radicals [37–39, 43]. However, being seriously limited to specific classes of molecules and properties, the method is not general enough to be an efficient tool in most CAMD applications. The next step, suggested by Hoffmann and known as the *extended Hückel method* [41], was therefore to extend the HMO formalism to all the valence electrons. The features of the EH method are the following:

1. The AO basis set $\{\chi_\mu\}$ is made of Slater-type functions [44], characterized by a drop-off as $\exp(-\zeta r)$, spanning the valence shell of all the atoms of the molecule. As an example, the AO basis set of an EH calculation performed on ferrocene, $Fe(C_5H_5)_2$, is made of the following functions:

   Fe : $3d_{z^2}$, $3d_{x^2-y^2}$, $3d_{xy}$, $3d_{xz}$, $3d_{yz}$, $4s$, $4p_x$, $4p_y$, $4p_z$
   each C atom : $2s$, $2p_x$, $2p_y$, $2p_z$
   each H atom : $1s$

   i.e. it is of dimension 59.
2. Whereas electron repulsion matrix elements $G_{\mu\nu}$ (equation (9.31)) are still neglected, the overlap integrals $S_{\mu\nu}$ (equation (9.34)) are all retained and calculated using the Slater basis orbitals.
3. The matrix elements of the one-electron Hamiltonian (equation (9.30)) are defined as:

$$H_{\mu\mu} = -I_\mu \qquad (9.43)$$

$$H_{\mu\nu} = 0.5K(H_{\mu\mu} + H_{\nu\nu})S_{\mu\nu} \qquad \mu \neq \nu \qquad (9.44)$$

where I_μ is the valence state ionization energy (VSIE) of orbital μ, deduced from atomic spectroscopic data, and K is the Wolfsberg–Helmholz constant [45], usually taken as 1.75. Contrary to the HMO model, all non-diagonal $H_{\mu\nu}$ matrix elements are retained and calculated using equation (9.44), whether or not μ and ν belong to bound atoms.

4. Using this approximated Hamiltonian, the matrix form of the EH equations:

$$HC = SCE \qquad (9.45)$$

is then solved in a way similar to the Roothaan equations (9.35). However, as H does not depend upon the solutions $c_{\mu i}$, solving the system (9.45) does not require the SCF procedure.

As simple as it is, the EH method has proven very successful for analysing and interpreting the ground state properties of organic, organometallic and inorganic compounds [46–48]. It is still very much in use today in many CAMD applications, as it generally leads to electronic structures enabling the user to assess and discuss the reactivity of the compounds under study. Another advantage of the EH method is its versatility, which allows it to be used for any kind of compounds comprising practically any element of the periodic table. Indeed, in addition to the K constant, the only parameters of the model are the exponents of the Slater functions and the VSIE values, both of which have been determined for all the atoms in their standard oxidation states.

However, the model suffers from two well known deficiencies, which have been partly overcome by various techniques:

1. For polar and ionic compounds, it generally leads to overestimated atomic charges, unless one uses charge-dependent VSIEs, as suggested by Basch *et al.* in their self-consistent charge and configuration (SCCC) procedure [49]. In particular, the use of the SCCC technique is indispensable for transition metal complexes. In these conditions, the EH method leads in general to a coherent description of the gross features of electronic structures, as exemplified by detailed comparisons with more elaborate models [50, 51].

2. The approximate expression (9.44), employed for non-diagonal matrix elements of EH Hamiltonian, accounts essentially for covalent bonding, and no provision is made for electrostatic interactions such as core-core repulsions. Whereas this deficiency has little influence on the calculated electronic structures, it is of dramatic importance for potential energy surfaces, with the result that it leads in many cases to atomic collapse, i.e. to continuously decreasing total energies when bond distances tend to vanish. To overcome this deficiency, Anderson and Hoffmann have shown that an approximate two-body repulsive energy can be added to the EH binding energy, resulting in a reasonable behaviour of the modified EH (MEH) model for the prediction of geometries [52]. More recently, Calzaferri *et al.* have shown that the MEH model can be improved by introducing a parametrized and distance-dependent Wolfsberg–Helmholz K constant (equation (9.44)) [53], and Table 9.1 presents some results that we have recently obtained for organometallics using this simple technique. It is seen that the calculated geometries compare reasonably

Table 9.1 Comparison between MEH and experimental bond distances (Å) of some transition metal carbonyl complexes.

Bond[a]	V(CO)$_6$	Cr(CO)$_6$	Fe(CO)$_5$	Ni(CO)$_4$
M–C$_{eq}$(calc)	1.890	1.856	1.834	1.805
M–C$_{eq}$(exp)	2.008	1.910	1.833	1.838
M–C$_{ax}$(calc)			1.806	
M–C$_{ax}$(exp)			1.810	
C–O$_{eq}$(calc)	1.149	1.146	1.150	1.154
C–O$_{eq}$(exp)	1.139	1.141	1.153	1.140
C–O$_{ax}$(calc)			1.140	
C–O$_{ax}$(exp)			1.153	
RRMS[b]	0.051	0.024	0.005	0.017

[a]eq = equatorial; ax = axial. The distinction between equatorial and axial ligands applies only to Fe(CO)$_5$.
[b]RRMS = relative root mean square of the error on calculated parameters.

well with the experimental values, which suggests that the EH model can be used for quantitative or semi-quantitative predictions of both potential energy surfaces and chemical reactivities in CAMD applications. In Chapter 10, we present and discuss a reaction potentials technique recently developed along these lines [54].

9.3.2 CNDO method

Historically, the Complete Neglect of Differential Overlap (CNDO) method was the first one attempting to solve in an approximate way the Roothaan–Hall equations [55, 56]. Indeed, contrasting with the HMO and EHMO models, it is the first member of the series of semi-empirical models referred to in the beginning of this section as being based on the introduction of physical approximations and empirical parameters into the F Hamiltonian matrix of the Roothaan equations (9.35).

As in all the semi-empirical techniques reviewed in this section, the AO basis set used in CNDO is built over Slater orbitals spanning the valence shell of the atoms present in the molecule. However, as opposed to the EH model, the CNDO method and its successors have in general not been parametrized for transition metal complexes, and consequently their AO basis set does not include d orbitals, a few cases excepted, where they may be used as polarization functions.

The sequence of approximations leading to the CNDO scheme will be briefly reviewed as it provides the necessary basis to delineate its limitations and to understand the reasons of the improvements brought about by its successors.

1. The zero-differential overlap (ZDO) approximation [57] is used:

$$\chi_\mu(1)\chi_\nu(1) = 0 \quad \text{if } \mu \neq \nu \tag{9.46}$$

for any position of electron 1. This entails, of course:

$$S_{\mu\nu} = \int \chi_\mu(1)\chi_\nu(1)\,dv_1 = \delta_{\mu\nu} \tag{9.47}$$

which in turn leads to a simplified form of the Roothaan equations (9.35):

$$F\,C = C\,E \tag{9.48}$$

In addition, the ZDO approximation leads to a drastic reduction in the number of two-electron integrals (equation (9.33)):

$$<\mu\nu\,|\,\lambda\rho> = <\mu\mu\,|\,\lambda\lambda>\,\delta_{\mu\nu}\,\delta_{\lambda\rho} \tag{9.49}$$

which means that all three- and four-centre two-electron integrals are neglected.

2. To keep rotational invariance, i.e. independence of the CNDO results upon the choice of coordinate system, all two centre two-electron integrals involving AOs on a given pair of atoms are taken as equal, i.e.:

$$< \mu\mu \mid \lambda\lambda > = \gamma_{AB} \quad \text{for any AO } \mu \text{ belonging to A} \qquad (9.50)$$
$$\text{and any AO } \lambda \text{ belonging to B}$$

Approximations (1) and (2) were concerned with two-electron matrix elements $G_{\mu\nu}$. Let us turn now to approximations (3) and (4), which deal with one-electron matrix elements $H_{\mu\nu}$.

3. The $H_{\mu\nu}$ matrix elements (equation (9.30)) may be written:

$$H_{\mu\nu} = < \mu \mid h \mid \nu > \qquad (9.51)$$

where h is the Hamiltonian (9.24). One therefore has:

$$H_{\mu\nu} = < \mu \left| -\frac{h^2}{8\pi^2 m} \nabla_1^2 - \sum_B V_B \right| \nu > \qquad (9.52)$$

the summation running over all the atoms of the molecule, V_B being defined as:

$$V_B = \frac{Z_B e^2}{r_{1B}} \qquad (9.53)$$

Assuming that the AO μ belongs to atom A, one may write:

$$H_{\mu\nu} = \underbrace{< \mu \left| -\frac{h^2}{8\pi^2 m} \nabla_1^2 - V_A \right| \nu >} - \sum_{B \neq A} < \mu \mid V_B \mid \nu > \qquad (9.54)$$

$$= \qquad U_{\mu\nu} - \sum_{B \neq A} < \mu \mid V_B \mid \nu > \qquad (9.55)$$

Similar to the treatment of $< \mu\mu \mid \lambda\lambda >$, approximation (3) then amounts to assuming:

$$< \mu \mid V_B \mid \nu > = V_{AB} \, \delta_{\mu\nu} \qquad (9.56)$$

which means that all three centre nuclear attraction integrals are neglected and that two centre ones are set equal for a given pair of atoms A and B.

4. Non-diagonal matrix elements $H_{\mu\nu}$ (equation (9.51)), with AOs μ and ν belonging to atoms A and B respectively, are approximated using an expression analogous to that of the EH model (equation (9.44)):

$$H_{\mu\nu} = \beta_{AB} \, S_{\mu\nu} \qquad (9.57)$$

where β_{AB} is a parameter depending only upon the nature of atoms A and B.

The effect of approximations (1)–(4) is to reduce the matrix elements of the Fock operator (equation (9.48)) to their CNDO expressions:

$$F_{\mu\mu} = U_{\mu\mu} + (P_{AA} - 1/2\, P_{\mu\mu})\gamma_{AA} + \sum_{B \neq A} (P_{BB}\, \gamma_{AB} - V_{AB}) \qquad \mu \text{ on atom A} \qquad (9.58)$$

$$F_{\mu\nu} = \beta_{AB}\, S_{\mu\nu} - 1/2\, P_{\mu\nu}\, \gamma_{AB} \qquad \begin{array}{l} \mu \text{ on atom A} \\ \nu \text{ on atom B} \end{array} \qquad (9.59)$$

where:

$$P_{AA} = \sum_{\mu \varepsilon A} P_{\mu\mu} \qquad (9.60)$$

with $P_{\mu\mu}$ given by equation (9.32).

In the CNDO method, the following parameters therefore have to be assigned values before calculations may be performed: $U_{\mu\mu}$, γ_{AB}, V_{AB} and β_{AB}. It is then relatively easy to solve the CNDO matrix equation (9.48), though the SCF procedure must be used in view of the dependence of the Hamiltonian matrix elements upon solutions $c_{\mu i}$ via the first order density matrix P.

For the calculation of potential energy surfaces (equation (9.11)), a standard core-core repulsive contribution V_{nn}, analogous to equation (9.7):

$$V_{nn} = \frac{1}{2} \sum_{A} \sum_{B \neq A} \frac{Z_A Z_B e^2}{R_{AB}} \qquad (9.61)$$

is added to the CNDO total electronic energy. In equation (9.61), the summations run on the cores of the molecule, each of them having a charge Z_A, and R_{AB} being the distance between A and B.

Several parametrizations (i.e. several coherent sets of values for the aforementioned parameters) have been proposed for the CNDO scheme. The most successful one is undoubtedly the so-called CNDO/2 version, which leads in general to satisfactory geometries and atomic charges [40, 58]. However, in view of the large number of two-electron integrals neglected and the rather small quantity of parameters to fit on experimental or *ab initio* data, the CNDO/2 method is not elaborate enough to be able to yield reliable results for a whole range of organic compounds. It has therefore been superseded in CAMD applications by its INDO- or NDDO-type successors.

9.3.3 INDO method

One of the main deficiences in the CNDO formalism lies in the impossibility of reproducing energy splittings between different spin states arising from the same electronic configuration. This is due to the ZDO approximation, according to which the one centre exchange integrals of $< \mu\nu \mid \mu\nu >$ type responsible for these energy splittings are neglected. The next step is therefore to retain monoatomic differential overlap in one centre two-electron integrals, which leads to the so-called Intermediate Neglect of Differential Overlap (INDO) method [59]. Actually, in a basis set made of pure s and p AOs (no hybrids), the only non-vanishing, one centre, two-electron integrals are $< \mu\mu \mid \mu\mu >$ and $< \mu\mu \mid vv >$, both of which were already included in the CNDO scheme, and $< \mu v \mid \mu v >$, which is the only new type of integral to be introduced in the INDO method. Consequently, the Fock matrix elements are written as follows in the INDO formalism:

$$F_{\mu\mu} = U_{\mu\mu} + \sum_{\lambda \varepsilon A} P_{\lambda\lambda} [< \mu\mu \mid \lambda\lambda > -1/2 < \mu\lambda \mid \mu\lambda >]$$

$$+ \sum_{B \neq A} (P_{BB} \gamma_{AB} - V_{AB}) \quad \mu \text{ on atom A} \tag{9.62}$$

$$F_{\mu v} = P_{\mu v} [1.5 < \mu v \mid \mu v > -0.5 < \mu\mu \mid vv >] \quad \mu \text{ and } v \text{ both on atom A} \tag{9.63}$$

whereas the two -centre non-diagonal $F_{\mu v}$ matrix element is the same as in CNDO.

Generally, for INDO one uses the same parametrization as that of CNDO/2 and, as the two methods are closely related, they lead to very similar results for closed-shell molecules [40]. For open-shell systems, however, which correspond to configurations exhibiting one or more singly occupied MOs, INDO results are significantly better due to the improved description of the different interactions taking place between electrons with the same or opposite spins.

It is worthwhile here briefly commenting on the different strategies which have been adopted to parametrize a given semi-empirical model such as INDO. In both CNDO/2 and INDO, the school of Pople has tried to optimize a set of parameters so as to reproduce the results of minimum basis set *ab initio* SCF calculations. Indeed, these semi-empirical methods attempt to solve in an approximate way the Hall–Roothaan equations, and it is natural, according to the school of Pople, to try to find optimum values of the parameters which would more or less compensate for the approximations made (i.e. for the neglect of a large number of one- and two-electron integrals), so that finally, for a given set of molecules, the semi-empirical and *ab initio* results will be as close one another as possible. This approach, however, has been criticized by Dewar, who argued that the parametrization should be primarily directed towards agreement with experimental data [60]. Indeed, as we have seen previously in this chapter, minimum basis set *ab initio* calculations present several deficiencies, the most serious one consisting of the neglect of correlation effects, and Dewar is probably right pointing out that a good semi-empirical model should try to overcome them by introducing part of the correlation into the effective Hamiltonian used [42]. Starting from these principles, Dewar and his group have developed, in the last 20 years, a whole range of semi-empirical models which are now very useful tools in CAMD applications to organic chemistry. In the last part of this section, we are going to review the most popular of them, namely MINDO/3, MNDO and AM1.

9.3.4 MINDO method

In his series of Modified Intermediate Neglect of Differential Overlap (MINDO) semi-empirical methods, among which the most successful one is MINDO/3 [61], Dewar has used the INDO Hamiltonian developed by Pople (equations (9.62) and (9.63)) and, at this level of approximation, has performed an extensive re-parametrization of the various terms in order to have at hand a general model applicable to the calculation of the largest possible number of properties. In addition, this effort also involved the extension of the parametrization to a large number of elements (without transition metals, still), so as to allow the study of large classes of organic compounds.

It is not our purpose here to describe in detail the modifications brought by Bingham *et al.* [61] in MINDO/3 to the various terms of the Hamiltonian matrix elements (equations (9.62) and (9.63)), especially as several good reviews have appeared recently [7, 34, 42, 62]. Let us just mention the following differences with respect to INDO:

1. Instead of being taken from atomic spectra, the $U_{\mu\mu}$ integrals are treated as adjustable parameters.
2. The two-electron integrals γ_{AB} are no longer calculated using s AOs, but approximated using the distance-dependent formula suggested previously by Dewar and Sabelli [63].
3. Instead of being an average of atomic values, the β_{AB} parameter is characteristic of the pair of atoms A–B.
4. The ζ exponents of the Slater AOs are all taken as adjustable parameters.
5. The core-core repulsive energy V_{nn} (equation 9.61) is made a function of electron-electron repulsion and "true" core-core repulsion with a parameter α_{AB} characteristic of the pair A–B.

All the parameters of the MINDO/3 method were optimized by a least-squares fit to experimental heats of formation and geometries of a selected set of organic molecules. To illustrate the computational effort involved by this fitting, let us just quote that the original parametrization of 10 atoms required to optimize 159 parameters in MINDO/3, as compared with 30 parameters in INDO! Some well known pitfalls apart [34, 62], the performance of the MINDO/3 model is good, even for properties or types of compounds not used in the parametrization. As an example, the calculated heats of formation of a series of closed-shell molecules containing C, O, N and H atoms show an average error of 11 kcal/mol [62]. In addition, the geometries of hydrocarbons are in good agreement with experiment, the errors on bond lengths being generally less than 0.02 Å and on bond angles less than 4° [34]. These results, together with the small computational effort as compared with *ab initio* calculations (generally by a factor of less than 1/1000), undoubtedly help in understanding the popularity of MINDO/3 in CAMD applications. However, some caution is required, broadly speaking, for types of compounds or properties where few, if any, MINDO/3 results are available. In these cases, patient comparisons and calibrations of the model are indispensable for assessing the validity of the approach and estimating an error bar on the calculated structures and properties.

9.3.5 MNDO method

In view of the very elaborate parametrization of MINDO/3, Dewar and Thiel soon realized that further improvement of the semi-empirical techniques could only be achieved by working at a reduced level of approximation [64], that is by starting from the Neglect of Diatomic Differential Overlap (NDDO) scheme proposed in 1965 by Pople *et al.* [55]. Indeed, in both CNDO and INDO models, the repulsion between lone-pair electrons located on adjacent atoms, such as in hydrazine, is not taken into account, and this important effect can probably not be entirely compensated for by a judicious choice of empirical parameters [62]. The next less approximate level of theory is therefore the NDDO scheme, in which the ZDO approximation is applied only to orbitals located on different atoms, i.e. equation (9.46) is modified in such a way:

$$\chi_\mu(1)\, \chi_\nu(1) = 0 \quad \text{if } \mu \text{ and } \nu \text{ belong to different atoms} \tag{9.64}$$

The NDDO approximation entails that integrals of $< \mu\nu \mid \lambda\rho >$ type are only neglected if μ and ν, or λ and ρ, belong to different atoms. One has to realize, however, that this improvement in the treatment of two-electron integrals induces in turn a significant increase of the computational effort, as it raises the number of two-electron two-centre integrals by a factor of 100 for each pair of non-hydrogen atoms in the molecule [62]. In the NDDO formalism, the Fock matrix elements write as follows:

$$F_{\mu\mu} = H_{\mu\mu} + \sum_{\lambda\in A} P_{\lambda\lambda}[< \mu\mu \mid \lambda\lambda > -1/2 < \mu\lambda \mid \mu\lambda >] +$$
$$+ \sum_{B} \sum_{\rho\in B} \sum_{\sigma\in B} P_{\rho\sigma} < \mu\mu \mid \rho\sigma > \quad \mu \text{ on atom A} \tag{9.65}$$

$$F_{\mu\nu} = H_{\mu\nu} + 1/2\, P_{\mu\nu}[3 < \mu\nu \mid \mu\nu > - < \mu\mu \mid \nu\nu >] +$$
$$+ \sum_{B} \sum_{\rho\in B} \sum_{\sigma\in B} P_{\rho\sigma} < \mu\nu \mid \rho\sigma > \quad \mu \text{ and } \nu \text{ both on atom A} \tag{9.66}$$

$$F_{\mu\nu} = H_{\mu\nu} - 1/2 \sum_{\lambda\in A} \sum_{\rho\in B} P_{\mu\nu} < \mu\lambda \mid \nu\rho > \quad \begin{array}{l}\mu \text{ on atom A} \\ \nu \text{ on atom B}\end{array} \tag{9.67}$$

the summation on B in equations (9.65) and (9.66) running over all the atoms different from A in the molecule.

As was the case for MINDO/3, Dewar and his group have worked out a very elaborate parametrization of the NDDO Hamiltonian (equations (9.65)–(9.67)), which gave rise to the so-called modified NDDO (MNDO) scheme. In this latter model, all the parameters already present in MINDO/3 are treated similarly, except for the α_{AB} and β_{AB} expressions which are taken as averages over atomic parameters instead of being characteristic of the various pairs AB. The additional two-electron two centre integrals are evaluated using classical expansions in terms of semi-empirical multipole-multipole interactions [64]. All the MNDO parameters were optimized for a given set of standard molecules so as to reproduce by the calculation experimental heats of formation, dipole moments, ionization energies and molecular geometries.

The MNDO model generally leads to good results for a vast number of organic molecules and their properties. When comparing with MINDO/3, the absolute values of the average errors of heats of formation, structural parameters, etc., are uniformly smaller by a factor of about 2 [65]. This has undoubtedly contributed to the popularity of the MNDO model, which has been, and still is, largely used in numerous CAMD applications. In addition to being a generally reliable semi-empirical model, MNDO has been the subject of extensive programming effort with introduction into the MOPAC [66] or AMPAC [67] packages of powerful geometry optimizers (gradient minimizations) and calculation of properties such as vibrational frequencies [61]. For small to medium size molecules, typically containing up to 10–15 atoms, it leads to CPU times for geometry optimizations of the order of 1 hour on modern workstations, which allows one to run these calculations as batch background jobs while performing standard modelling or graphics applications.

9.3.6 AM1 and PM3 methods

In spite of the important effort made towards an optimum parametrization of MNDO, some deficiencies remain, such as an overestimation of the repulsions between non-bonded atoms [62–68], which represents a serious limitation for the study of hydrogen-bonded systems. To overcome this problem, Dewar *et al.* have suggested including additional terms in the expression of core-core repulsive energy, while keeping all the MNDO parameters unchanged. This led to the so-called Austin Model 1 (AM1) method [68], which represents a significant improvement over MNDO, the mean absolute errors of practically all calculated properties being reduced and the model being able to treat adequately weakly bonded systems.

Using a slightly different approach from that of Dewar, Stewart has in parallel performed a complete re-parametrization of MNDO, with a core-core repulsion term similar to that of AM1, which he called Parametric Method Number 3 (PM3) [69]. To this end, he has used an automatic procedure to optimize the parameters [70], which is a promising tool for the development of new semi-empirical techniques. The results obtained using PM3 are still rather scarce, and it is difficult to make a clear assessment of its performance. However, taken as a whole, it should not be very different to that of AM1, and the reader should refer to recent reviews in order to find more detailed information [13, 62].

As a conclusion, let us mention that it is very fortunate that all the Dewar group methods are included in large packages such as MOPAC or AMPAC, as they enable the CAMD user to rapidly test the different models for a given series of molecules, for which experimental data are available, and then to select the best one for the application to be performed. It is therefore not an overstatement to assert that these packages today represent an important tool to rationalize the properties of organic molecules and, consequently, to carry out CAMD applications requiring their prediction at a quantitative level.

9.4 DENSITY FUNCTIONAL METHODS

Density functional methods are generally considered as a valuable alternative to the traditional *ab initio* quantum chemical model. Indeed, they are in principle also based on a parameter free theory, i.e. they attempt to find solutions "from first principles" to the SCF mean-field model of electronic structure, while treating the electron correlation problem differently from the post-Hartree–Fock techniques seen in section 9.2. Together with an approximate (local) expression of the exchange operator, this leads to a new set of one-electron equations, the solution of which involves a substantial reduction of computational effort as compared with Hartree–Fock. Consequently, density functional methods can be advantageously applied to large systems, such as clusters of transition metals or organometallic complexes.

As for the other quantum chemical models previously seen, several good textbooks and reviews of the density functional methodology have appeared recently [71–74], and we shall only summarize its main features. The basic idea of Density Functional Theory (DFT) is to use the electron density $\rho(\mathbf{r})$ as the variable of the system, instead of the electronic wavefunction $\Psi(1,2,...n)$. This choice can be done without loss of rigor, as it has been shown by Hohenberg and Kohn that the ground state energy of a multi-electron system is completely and uniquely determined by its density, although the explicit functional dependence of the energy on density is not entirely known [75]. However, in spite of this difficulty, the energy functional satisfies the variational principle, i.e. it is minimum for the true electron density of the system. Without any approximation, the electronic energy of an n-electron system is written, in atomic units:

$$E = -\frac{1}{2}\sum_i \int \psi_i(\mathbf{r}_1)\, \nabla_1^2\, \psi_i(\mathbf{r}_1)d\mathbf{r}_1 + \sum_A \int \frac{Z_A}{|\mathbf{r}_1 - \mathbf{R}_A|}\rho(\mathbf{r}_1)d\mathbf{r}_1$$

$$+\frac{1}{2}\int\int \frac{\rho(\mathbf{r}_1)\,\rho(\mathbf{r}_2)}{|\mathbf{r}_1 - \mathbf{r}_2|}\, d\mathbf{r}_1\, d\mathbf{r}_2 + E_{xc} \tag{9.68}$$

where the first term is the kinetic energy of a reference system of non-interacting electrons with the same total density $\rho(\mathbf{r}) = \sum_i n_i\, |\psi_i(\mathbf{r})|^2$ as the actual system of interacting electrons [74], ψ_i and n_i being spin-orbitals and their occupation numbers respectively. The second term represents the usual potential energy arising through electron-nuclei interactions, the summation running over the nuclei with charges Z_A located in R_A, whereas the third term is the classical Coulomb energy resulting from the interaction between the electron densities $\rho(\mathbf{r}_1)$ and $\rho(\mathbf{r}_2)$. The last term of expression (9.68) is more delicate: known as the *exchange-correlation energy*, it represents the energy contributions arising from the exchange interactions and correlation effects seen previously in this chapter, plus the difference in kinetic energy between interacting and non-interacting systems with density $\rho(\mathbf{r})$ [76]. Actually, the major problems of DFT are due to E_{xc}, as there is no exact formulation for this term in the case of our n-electron system and approximations will have to be sought.

Applying the variational principle to the energy given by equation (9.68), Kohn and Sham have subsequently reformulated the density functional theory by deriving a set of one-electron Hartree-like equations leading to the wavefunctions $\psi_i(\mathbf{r})$ involved in the calculation of $\rho(\mathbf{r})$ [77]. In their usual formulation, the Kohn–Sham (KS) equations are written as follows:

$$\left[-\frac{1}{2}\nabla_1^2 + \sum_A \frac{Z_A}{|\mathbf{r}_1 - \mathbf{R}_A|} + \int \frac{\rho(\mathbf{r}_2)}{|\mathbf{r}_1 - \mathbf{r}_2|}\, d\mathbf{r}_2 + V_{xc} \right] \psi_i(\mathbf{r}_1) = \varepsilon_i\, \psi_i(\mathbf{r}_1) \qquad (9.69)$$

where the expression in brackets is the effective one-electron Kohn–Sham Hamiltonian h_{KS}, the exchange-correlation potential V_{xc}, which contains the multi-electron effects, being defined as:

$$V_{xc}[\rho] = \frac{\partial E_{xc}[\rho]}{\partial \rho} \qquad (9.70)$$

Note that the Kohn–Sham Hamiltonian is a local operator $h_{KS}(1)$, fully determined in principle from the knowledge of the electron density [77]. This is the main difference with respect to the Hartree–Fock equations (9.22), which contain a non-local operator, namely the exchange part of the potential operator (equation (9.26)). In addition, the KS equations incorporate the correlation effects through V_{xc}, whereas they are lacking in the Hartree–Fock SCF scheme. Nevertheless, though the latter model cannot be considered as a special case of the KS equations, there are some similarities between Hartree–Fock and Kohn–Sham methods, as both lead to a set of one-electron equations allowing to describe an n-electron system.

In principle, the KS equations (9.69) would lead to an exact solution to our problem, provided the formulation of the exchange-correlation energy functional E_{xc} was known. However, in practice, approximate expressions of E_{xc} must be used, and the search of adequate functionals for this term is probably the greatest challenge of DFT [71]. The simplest model has been proposed by Kohn and Sham: if the system is such that its electron density is smooth (i.e. it exhibits little variations within the molecular volume), the local density approximation (LDA) may be introduced:

$$E_{xc}[\rho] = \int \rho\,(\mathbf{r})\,\varepsilon_{xc}\,(\rho)\,d\mathbf{r} \qquad (9.71)$$

where $\varepsilon_{xc}(\rho)$ is the exchange and correlation energy per particle of a uniform and homogeneous electron gas of density ρ. V_{xc} may then be easily deduced from this approximate expression of E_{xc} by using equation (9.70) and the KS equations can be solved. As for Hartree–Fock based methods, the SCF procedure must be used since the h_{KS} Hamiltonian depends explicitly upon the solution $\rho(\mathbf{r})$.

As shown by Parr and Yang [71], the exchange (ε_x) and correlation (ε_c) contributions to $\varepsilon_{xc}(\rho)$ can be separated as:

$$\varepsilon_{xc}\,(\rho) = \varepsilon_x(\rho) + \varepsilon_c(\rho) \qquad (9.72)$$

and the exchange part is usually taken from electron gas theory:

$$\varepsilon_x(\rho) = -\frac{3}{4}\left(\frac{3}{\pi}\right)^{1/3}\rho(\mathbf{r})^{1/3} \qquad (9.73)$$

If no correlation is introduced ($\varepsilon_c = 0$), the KS equations reduce to the well known Xα method proposed by Slater [78] as a simplification of the Hartree–Fock scheme with a local exchange operator:

$$\left[-\frac{1}{2}\nabla_1^2 + \sum_A \frac{Z_A}{|\mathbf{r}_1 - \mathbf{R}_A|} + \int \frac{\rho(\mathbf{r}_2)}{|\mathbf{r}_1 - \mathbf{r}_2|}d\mathbf{r}_2 - \frac{3}{2}\alpha\left\{\frac{3}{\pi}\rho(\mathbf{r})\right\}^{1/3}\right]\psi_i(\mathbf{r}_1) = \varepsilon_i\psi_i(\mathbf{r}_1) \quad (9.74)$$

where α is an adjustable parameter. Actually, the Xα formalism is the simplest DFT method based on the LDA approximation, and a large number of more sophisticated exchange-correlation potentials have been proposed by various authors (for a review, see Salahub [76]). Suffice it to mention here that, generally, the $\rho^{1/3}$ functional is retained for exchange, whereas formulations, based on quantum Monte Carlo calculations, have been proposed for the correlation contribution.

Broadly speaking, the LDA formalism, based on equation (9.71), is applicable to systems with slowly varying electron densities, a situation which is rarely encountered in atoms and molecules. However, experience has shown that DFT methods based on LDA give surprisingly good results for the electronic structure and related properties of a broad range of compounds, including clusters of transition metals [76] and organometallic complexes [74]. Recently, it has been pointed out that the LDA formalism can be further improved by addition of so-called *non-local gradient correction terms*, which lead to molecular spectroscopic properties in very good agreement with experiment [73].

Whatever the form of the V_{xc} exchange correlation potential used, various schemes have been proposed to solve the Kohn–Sham one-electron equations (9.69). The oldest method is undoubtedly the so-called *multiple scattering* (MS) or scattered wave one, where the molecular volume is partitioned into (spherical) atomic, interatomic and (spherical) extramolecular regions [79]. The KS potential is generally taken from the Xα formalism, i.e. without explicit inclusion of correlation, and approximated as a spherical or volume average in each region. In spite of these simplifications, the MS-Xα model has led to good predictions [80, 81], as exemplified by Table 9.2 which presents a comparison between experimental and MS-Xα ionization energies of cobaltocene. However, although it is still in use today for such calculations [83, 84], this model does generally lead to an inconsistent variation of total energy as a function of structural parameters, which makes it impractical for predicting the minima of potential energy surfaces. As the optimization of geometrical parameters is often considered to be the top priority in molecular modelling applications, the MS-Xα method can at best be used as a second

Table 9.2 Comparison between MS-Xα and experimental ionization energies (eV) of cobaltocene.

Level	Type[a]	MS-Xα	Experiment[b]
$4e_{1g}$	M 3d	4.87	5.55
$5a_{1g}$	M 3d	7.75	7.15
			7.65
$3e_{2g}$	M 3d	8.12	7.99
$4e_{1u}$	L 2pπ	8.34	8.72
$3e_{1g}$	L 2pπ	9.05	9.92
$4a_{2u}$	L 2pπ	11.08	
$2e_{2g}$	L 2pσ	11.86	
			12.34
$2e_{2u}$	L 2pσ	11.95	
			13.43
$3e_{1u}$	L 2pσ	12.15	
$4a_{1g}$	L 2pπ	12.16	
$2e_{1g}$	L 2pσ	12.19	
$3a_{2u}$	L 2pσ	15.22	
$3a_{1g}$	L 2pσ	15.43	16.98
$1e_{2u}$	L 2pσ	16.06	
$1e_{2g}$	L 2pσ	16.15	

[a]Predominant atomic contribution: M = metal, L = ligand.
[b]From Cauletti *et al.* [82].

step technique, once the geometry of the compound under study has been extracted from a structural data bank or optimized by another tool. It is therefore not surprising that the MS-Xα model has been supplanted by more elaborate DFT methods, which rest generally on the use of the popular LCAO approximation.

Two different DFT schemes have been proposed almost simultaneously to solve the KS equations using the LCAO approximation: the so-called Discrete Variational Method (DVM) [85], which uses a Slater-type orbital basis set, and the Linear Combination of Gaussian-Type Orbitals (LCGTO)-local spin density (LSD) technique suggested by Sambe and Felton [86]. Both techniques, which have been considerably refined and improved in the last few years [74], rely on the fit of the electron density to one centre auxiliary functions to achieve a faster calculation of the Coulomb (and exchange in the case of LCGTO-LSD) operator of h_{KS}. This leads roughly to a N^3 scaling of the computational effort involved by a LCAO-DFT calculation, where N is the size of the one-electron basis set, as compared with the N^4 and N^5 dependences of the SCF Hartree–Fock and CI schemes respectively. This explains the growing popularity of these LCAO-DFT schemes, as they allow one to perform calculations on large clusters (comprising typically 10–20 transition metal atoms) and organometallics (made of up to 50 atoms). In addition, the range of application of these techniques has been recently extended by the development of pseudopotentials, relativistic corrections and the calculation of analytical energy gradients for geometry optimization purposes [74]. As an example, Table 9.3 presents the results obtained for various properties of the carbon monoxide molecule using the deMon LCGTO-LSD program developed recently by St-Amant and Salahub [87]. It is easily seen that the overall performance of this model is of the same level as that of sophisticated

Table 9.3 Properties of the CO molecule calculated by various quantum chemical models.

Property	Method			
	LCGTO-LSD	HF-SCF	HF-CI	Experiment
d_{C-O} [Å]	1.136	1.114[a]	1.133[a]	1.128[b]
vibrational frequency [cm^{-1}]	2'160	2'438[a]	2'113[c]	2'170[b]
proton affinity[d] [kcal/mol]	140.9	134.8[e]	139.4[e]	141.9[f]
dipole moment [D]	0.16	−0.33[a]	0.12[a]	0.11[g]

[a] 6–31G* basis set (Hehre *et al.* [8]).
[b] Huber and Herzberg [88].
[c] MP2 result (Hehre *et al.* [8]).
[d] Zero-point vibrational corrections included.
[e] Jasien and Stevens [89].
[f] Lias *et al.* [90].
[g] Hehre *et al.* [8].

Hartree–Fock + CI calculations, which suggests that DFT-based techniques could be integrated as the tool of choice in future molecular modelling packages.

REFERENCES

1. P.A.M. Dirac *Proc. Roy. Soc. Lond. A*, **123**: 1929; 714.
2. R. Zahradnik and R. Polak *Elements of Quantum Chemistry*, Plenum, New York, 1980.
3. P. Carsky and M. Urban *Ab Initio Calculations. Methods and Applications in Chemistry: Lecture Notes in Chemistry, vol 16*, Springer, Berlin, 1980.
4. A. Szabo and N.S. Ostlund *Modern Quantum Chemistry*, MacMillan, New York, 1982.
5. P.W. Atkins *Molecular Quantum Mechanics (2nd ed)*, Oxford University Press, Oxford, 1983.
6. G.W. Richards and D.L. Cooper *Ab Initio Molecular Orbital Calculations for Chemists (2nd ed)*, Clarendon, Oxford, 1983.
7. T. Clark *A Handbook of Computational Chemistry*, Wiley, New York, 1985.
8. W.J. Hehre, L. Radom, P.v.R. Schleyer and J.A. Pople *Ab Initio Molecular Orbital Theory*, Wiley, New York, 1986.
9. C.E. Dykstra *Ab Initio Calculation of the Structures and Properties of Molecules, Studies in Physical and Theoretical Chemistry*, **58**, Elsevier, Amsterdam, 1988.
10. R.G. Parr and W. Yang *Density-Functional Theory of Atoms and Molecules*, Oxford University Press, New York, 1989.
11. J.L. Rivail *Eléments de Chimie Quantique à l'Usage des Chimistes*, InterEditions/Editions du CNRS, Paris, 1989.
12. B. Webster, *Chemical Bonding Theory*, Blackwell, Oxford, 1990.
13. J.J.P. Stewart MOPAC: A semiempirical molecular orbital program, *J. Comp. Aided Mol. Design*, **4:** 1990; 1.
14. E. Schrödinger *Ann. Physik*, **79:** 1926; 361.
15. M. Born and J.R. Oppenheimer *Ann. Physik*, **84:** 1927; 457.
16. R. Zahradnik, B. Schneider, P. Hobza, Z. Havlas and H. Huber *Can. J. Chem.*, **63:** 1985; 1639.
17. D.R. Hartree *Proc. Cambridge Phil. Soc.*, **24:** 1928; 426.
18. W. Pauli *Z. Physik*, **31:** 1925; 765.
19. J.C. Slater *Phys. Rev.*, **35:** 1930; 509.
20. F.L. Pilar *Elementary Quantum Chemistry*, McGraw-Hill, New York, 1968.
21. D.R. Hartree *Proc. Cambridge Phil. Soc.*, **24:** 1928; 89.
22. V. Fock *Z. Physik*, **61:** 1930; 126.
23. G.G. Hall *Proc. Roy. Soc. Lond. A*, **205:** 1951; 541.
24. C.C.J. Roothaan *Rev. Mod. Phys.*, **23:** 1951; 69.
25. D. Feller and E.R. Davidson in *Reviews in Computational Chemistry*, vol. 1 K.B. Lipkowitz and D.B. Boyd (eds.), VCH Publishers, New York, 1990, p. 1.
26. S.F. Boys *Proc. Roy. Soc. Lond. A*, **200:** 1950; 542.
27. W.J. Hehre, R.F. Stewart and J.A. Pople *J. Chem. Phys.*, **51:** 1969; 2657.
28. W.J. Hehre, R. Ditchfield and J.A. Pople *J. Chem. Phys.*, **56:** 1972; 2257.
29. P.C. Hariharan and J.A. Pople *Chem. Phys. Lett.*, **16:** 1972; 217.
30. P.O. Löwdin *Adv. Chem. Phys.*, **2:** 1959; 207.
31. C.W. Bauschlicher and P.R. Taylor *J. Chem. Phys.*, **85:** 1986; 2779.
32. T.J. Lee, R.B. Remington, Y. Yamaguchi and H.F. Schaefer *J. Chem. Phys.*, **89:** 1988; 408.

33. W. Meyer *Int. J. Quant. Chem., Symp.*, **5:** 1971; 341.
34. D.M. Hirst *A Computational Approach to Chemistry*, Blackwell, Oxford, 1990.
35. C. Møller and M.S. Plesset *Phys. Rev.*, **46:** 1934; 618.
36. E. Hückel *Z. Physik*, **70:** 1931; 204.
37. E. Heilbronner and H. Bock *The HMO Model and its Application*, Wiley, London, 1968.
38. E.I.V. Nagy-Felsobuki *J. Chem. Educ.*, **66:** 1989; 821.
39. D.E. Manopoulos, J.C. May and S.E. Down *Chem. Phys. Lett.*, **181:** 1991; 105.
40. J.A. Pople and D.L. Beveridge *Approximate Molecular Orbital Theory*, McGraw-Hill, New York, 1970.
41. R. Hoffmann *J. Chem., Phys.*, **39:** 1963; 1397.
42. W. Thiel *Tetrahedron*, **44:** 1988; 7393.
43. J.P. Lowe *Quantum Chemistry*, Academic Press, New York, 1978.
44. J.C. Slater *Phys. Rev.*, **36:** 1930; 57.
45. M. Wolfsberg and L. Helmholz *J. Chem. Phys.*, **20:** 1952; 837.
46. M. Elian and R. Hoffmann *Inorg. Chem.*, **14:** 1975; 1058.
47. M. Simonetta and A. Gavezzotti *Struct. Bonding*, **27:** 1976; 1.
48. R Hoffmann *Angew. Chem. Int. Ed. Engl.*, **21:** 1982; 711.
49. H. Basch, A. Viste and H.B. Gray *Theor. Chim. Acta*, **3:** 1965; 458.
50. J.H. Ammeter, H.B. Bürgi, J.C. Thiebeault and R. Hoffmann *J. Am. Chem. Soc.*, **100:** 1978; 3686.
51. A. Gavezzotti, G.F. Tantardini and H. Miessner *J. Phys. Chem.*, **92:** 1988; 872.
52. A.B. Anderson and R. Hoffmann *J. Chem. Phys.*, **60:** 1974; 4271.
53. G. Calzaferri, L. Forss and I. Kamber *J. Phys. Chem.*, **93:** 1989; 5366.
54. C. Daul, A. Goursot, P.Y. Morgantini and J. Weber *Int. J. Quant. Chem.*, **38:** 1990; 623.
55. J.A. Pople, D.P. Santry and G.A. Segal *J. Chem. Phys.*, **43:** 1965; S129.
56. J.A. Pople and G.A. Segal *J. Chem. Phys.*, **43:** 1965; S136.
57. R.G. Parr *J. Chem. Phys.*, **20:** 1952; 239.
58. J.A. Pople and G.A. Segal *J. Chem. Phys.*, **44:** 1966; 3289.
59. J.A. Pople, D.L. Beveridge and P.A. Dobosh *J. Chem. Phys.*, **47:** 1967; 2026.
60. M.J.S. Dewar *Science*, **187:** 1975; 1037.
61. R.C. Bingham, M.J.S. Dewar and D.H. Lo *J. Am. Chem. Soc.*, **97:** 1975; 1285.
62. J.J.P. Stewart in *Reviews in Computational Chemistry* vol. 1, K.B. Lipkowitz and D.B. Boyd, (eds.), VCH Publishers, New York, 1990, p. 45.
63. M.J.S. Dewar and N.L. Sabelli *J. Chem. Phys.*, **34:** 1961; 1232.
64. M.J.S. Dewar and W. Thiel *J. Am. Chem. Soc.*, **99:** 1977; 4859.
65. M.J.S. Dewar and W. Thiel *J. Am. Chem. Soc.*, **99:** 1977; 4907.
66. J.J.P. Stewart *QCPE*, 1983, *Prog.* 455.
67. M.J.S. Dewar and J.J.P. Stewart *QCPE*, 1986, *Prog.* 506.
68. M.J.S. Dewar, E.G. Zoebisch, E.F. Healy and J.J.P. Stewart *J. Am. Chem. Soc.*, **107:** 1985; 3902.
69. J.J.P. Stewart *J. Comput. Chem.*, **10:** 1989; 209.
70. J.J.P. Stewart *J. Comput. Chem.*, **10:** 1989; 221.
71. R.G. Parr and W. Yang *Density Functional Theory of Atoms and Molecules*, Oxford University Press, New York, 1989.
72. E.S. Kryachko and E. Ludena *Density Functional Theory of Many-Electron Systems*, Kluwer Academic Press, Dordrecht, 1990.
73. A.D. Becke, in *The Challenge of d and f Electrons*, D.R. Salahub and M. Zerner, (eds.), *ACS Symp. Series 394*, American Chemical Society, Washington, 1989; p. 165.
74. T. Ziegler *Chem. Rev.*, **91:** 1991; 651.
75. P. Hohenberg and W. Kohn *Phys. Rev.*, **B136:** 1964; 864.
76. D.R. Salahub, in *Ab Initio Methods in Quantum Chemistry*, K.P. Lawley, (ed.), *Adv. Chem. Phys. vol. 69*, Wiley, Chichester, 1987; p. 447.
77. W. Kohn and L.J. Sham *Phys. Rev.*, **A140:** 1965; 1133.

78. J.C. Slater *Phys. Rev.*, **81:** 1951; 385.
79. K.H. Johnson *Adv. Quant. Chem.*, **7:** 1973; 143.
80. J. Weber, A. Goursot, E. Penigault, J.H. Ammeter and J. Bachmann *J. Am. Chem. Soc.*, **104:** 1982; 1491.
81. D.A. Case *Ann. Rev. Phys. Chem.*, **33:** 1982; 151.
82. C. Cauletti, J.C. Green, M.R. Kelly, P. Powell, J. Van Tilborg, J. Robbins and J. Smart *J. Electron Spectrosc. Relat. Phenom.*, **19:** 1980; 327.
83. A. Grand, P.J. Krusic, L. Noodleman and R. Subra *J. Mol. Struct. Theochem.*, **226:** 1991; 251.
84. P.K. Ross and E.I. Solomon *J. Am. Chem. Soc.*, **113:** 1991; 3246.
85. E.J. Baerends, D.E. Ellis and P. Ros *Chem. Phys.*, **2:** 1973; 41.
86. H. Sambe and R.H. Felton *J. Chem. Phys.*, **61:** 1974; 3862.
87. A. St-Amant and D.R. Salahub *Chem. Phys. Lett.*, **169:** 1990; 387.
88. K.P. Huber and G. Herzberg *Molecular Spectra and Molecular Structure, Vol IV. Constants of Diatonic Molecules*, Van Nostrand, New York, 1979.
89. P.G. Jasien and W.J. Stevens *J. Phys. Chem.*, **83:** 1985; 2984.
90. S.G. Lias, J.L. Liebman and R.D. Levin *J. Phys. Chem. Ref. Data*, **13:** 1984; 695.

10 *Derivation and visualization of molecular properties*

In this chapter, we shall review the derivation and visualization of molecular properties other than conformations, that is, we will concentrate on electronic properties. It is indeed true that in most CAMD applications it is necessary to go beyond the structural model by clothing it using selected electronic properties which will help in rationalizing or predicting the chemical behaviour of the compound. To this end, various "property builders", as compared with the well known structural model builders, based mostly on the quantum chemical methods described in the previous chapter, must be used, and they will be reviewed here.

We have deliberately chosen to place the emphasis in this chapter on local, rather than global, properties which can be advantageously represented as a complement of molecular models. This indicates that we will present and discuss properties amenable to graphical representations only, excluding all the global properties, such as spectroscopic ones, which in any case are of little use in CAMD. It is therefore natural to begin with the molecular orbitals themselves, and to proceed with electron densities and reactivity indices, which leads to a very rational framework for this chapter.

10.1 MOLECULAR ORBITALS

Historically, molecular orbitals (MOs) were the first electronic property to be visualized on simple graphics hardware such as printers, plotters, etc., before being advantageously represented on the screen of PCs and workstations. Actually, MOs of selected compounds were first reported as reproductions of listings presenting the numerical values of their coefficients [1, 2] (Figure 10.1) before the first textbooks appeared with three-dimensional plots [3], in-plane contour levels [4] and wire-frame surfaces [5]. Simultaneously, in the 1970s, several programs for drawing MOs on a display terminal or plotter linked to it were made available, which significantly contributed to

Molecular Orbital Coefficients

		1 (B1U)--O	2 (AG)--O	3 (AG)--O	4 (B1U)--O	5 (B2U)--O
EIGENVALUES --		- 9.59890	- 9.59821	- 0.67947	- 0.52038	- 0.43192
1 1C 1S		0.69858	0.69965	- 0.18412	- 0.14427	0.00000
2	2S	0.04948	0.02989	0.45277	0.40764	0.00000
3	2PX	0.00000	0.00000	0.00000	0.00000	0.00000
4	2PY	0.00000	0.00000	0.00000	0.00000	0.41437
5	2PZ	- 0.00748	0.00443	- 0.09724	0.21152	0.00000
6 2C 1S		- 0.69858	0.69965	- 0.18412	0.14427	0.00000
7	2S	- 0.04948	0.02989	0.45277	- 0.40764	0.00000
8	2PX	0.00000	0.00000	0.00000	0.00000	0.00000
9	2PY	0.00000	0.00000	0.00000	0.00000	0.41437
10	2PZ	- 0.00748	- 0.00443	0.09724	0.21152	0.00000
11 3H 1S		- 0.00917	- 0.00864	0.12427	0.21439	0.23169
12 4H 1S		- 0.00917	- 0.00864	0.12427	0.21439	- 0.23169
13 5H 1S		0.00917	- 0.00864	0.12427	- 0.21439	0.23169
14 6H 1S		0.00917	- 0.00864	0.12427	- 0.21439	- 0.23169

		6 (AG)--O	7 (B3G)--O	8 (B3U)--O	9 (B2G)--V	10 (B2U)--V
EIGENVALUES --		- 0.35259	- 0.30779	- 0.21130	0.05830	0.41232
1 1C 1S		0.01886	0.00000	0.00000	0.00000	0.00000
2	2S	- 0.02781	0.00000	0.00000	0.00000	0.00000
3	2PX	0.00000	0.00000	0.63129	0.81903	0.00000
4	2PY	0.00000	0.41184	0.00000	0.00000	0.75123
5	2PZ	0.52378	0.00000	0.00000	0.00000	0.00000
6 2C 1S		0.01886	0.00000	0.00000	0.00000	0.00000
7	2S	- 0.02781	0.00000	0.00000	0.00000	0.00000
8	2PX	0.00000	0.00000	0.63129	- 0.81903	0.00000
9	2PY	0.00000	- 0.41184	0.00000	0.00000	0.75123
10	2PZ	- 0.52378	0.00000	0.00000	0.00000	0.00000
11 3H 1S		0.19345	0.33625	0.00000	0.00000	- 0.69260
12 4H 1S		0.19345	- 0.33625	0.00000	0.00000	0.69260
13 5H 1S		0.19345	- 0.33625	0.00000	0.00000	- 0.69260
14 6H 1S		0.19345	0.33625	0.00000	0.00000	0.69260

Figure 10.1   Listing of the coefficients of the MOs calculated for ethylene using the Gaussian package [3].

popularizing the use of MO methods among chemists [6–9] (Figure 10.2). Subsequently, with the tremendous progresses witnessed in both computer graphics hardware and software, sophisticated packages allowing us to build and manipulate solid models representative of the 3D features of MOs on graphics workstations have appeared, and they are too numerous to be quoted here *in extenso*. Let us mention, to take a single example, the spectacular representation of MOs as coloured isovalue surfaces treated as 3D models by the SPARTAN package [10], which is actually the successor of

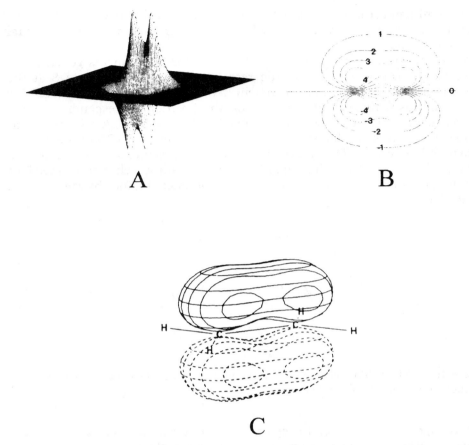

A   B

C

Figure 10.2   Three-dimensional plot (A), in-plane contour levels (B, the sign of the label indicates the sign of the orbital lobes) and wire-frame isovalue surface representation (C) of the $1b_{3u}$ HOMO of ethylene. From [6] (A and B) and [5] (C) (from Thalmann *et al.* [5] and Jorgensen *et al.* [6]).

the program used by Hout *et al.* to produce the models, unfortunately in black and white, of the MOs of numerous compounds in their pioneering textbook [11].

More than 20 years after the publication of the milestone theoretical treatise by Woodward and Hoffmann [12] about the conservation of orbital symmetry, MOs are more popular than ever to rationalize the broad features of chemical reactivity. According to these authors, it is indeed possible to describe the stereo- and regiochemical aspects of basic bonding processes in organic chemistry, such as pericyclic reactions, from an inspection of the shapes and symmetries of the frontier orbitals (FO) of the reacting species. As these concepts may be extended to a broad range of mechanisms in organic and organometallic chemistry, it is clear that the construction and manipulation, using an interactive graphics system, of any type of MOs, be they obtained from semi-empirical or *ab initio* quantum chemical models, enables the chemist to perform modelling which goes far beyond the structural representations.

A basic postulate of FO theory [13] is that, for mechanisms governed by orbital control [14], an electrophile reacts with a given substrate S at the atomic centre with the largest coefficient in the highest occupied MO (HOMO) of S, whereas a nucleophile will attack the position corresponding to the largest coefficient in the lowest unoccupied MO (LUMO). If we take pyrrole as an example, Fukui [13] found, using the HMO model, the HOMO coefficients shown in Figure 10.3. The values of these coefficients indicate that carbon C2 is most likely to undergo electrophilic attack, a result which is confirmed by a characterization of the halogenation reaction product obtained by Mazzara and Borgo [15].

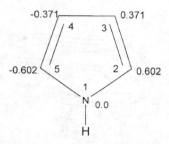

Figure 10.3   The HOMO coefficients of the pyrrole molecule as calculated using the Hückel molecular orbital method.

Although there are several exceptions to these simple rules, they are very useful to assess in a first approximation the regioselectivity of electrophilic and nucleophilic additions or substitutions to organic substrates. In this case, in view of the approximations inherent to this crude model, it is not necessary to use "accurate" *ab initio* MOs, and the representation (or examination of the corresponding coefficients) of semi-empirical MOs is sufficient to apply FO theory. However, the innocent reader should be warned of the limitations of this approach and, in cases where reliable information is needed, of the necessity to use a more elaborate model involving, for example, the calculation of reaction pathways. In addition, for inorganic and organometallic reaction mechanisms, the FO approach is of a more dubious quality, in view of the presence of several energy levels lying close to the HOMO and LUMO of the substrates, and the use of more refined procedures, even if they are based on simple semi-empirical MO models such as EH, is strongly advisable [16–18] (see section 10.4).

In this section, we illustrate the role pictorial representations of MOs can play in CAMD by using two examples pertaining to organic and organometallic chemistry. The first is concerned with the molecules of pyrrole **1** and pyrrolidine **2** which both are five-membered heterocycles (Figure 10.4).

However, the aromaticity of **1** involves participation of the lone-pair

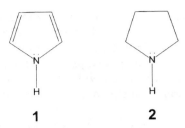

Figure 10.4 The molecules of pyrrole (**1**) and pyrrolidine (**2**).

electrons of nitrogen to the π-electron system, which is not the case for the saturated heterocycle **2**. As a consequence, **1** is much less basic than **2** as the lone pair of its nitrogen atom is much less available for electrophilic attack or, in the case of intermolecular associations, for hydrogen bonding [19]. Let us see now whether the electronic structures of **1** and **2**, in particular the composition and shape of their frontier orbitals, reflect their very different nucleophilic properties. The results reported for these compounds have been obtained using the *ab initio* SCF model with the 3–21G one-electron basis set [20].

   We first note that the HOMO of pyrrolidine lies at an energy of –9.52 eV, i.e. 1.63 eV lower than that of pyrrole, which already indicates that **2** is a better donor and a more basic species than **1**. Looking now at Plate VI, which displays the HOMO of **1**, it is seen that it is of π type without any contribution from nitrogen which, according to the FO model and using the criterion of the largest atomic contribution to the HOMO, suggests that protonation does not occur on this atom but on the carbon located in the α position. This is in good agreement with experimental evidence, pointing out that electrophilic substitution occurs predominantly at the α position in five-membered heterocycles [21], and also with theoretical studies of the protonation mechanism of pyrrole [22, 23]. This nice application of the FO model illustrates the potential of the method and how useful it can be to display MOs so as to simply rationalize some basic reaction mechanisms of organic chemistry.

   Looking at another MO of pyrrole, we find as HOMO–1 a second π MO made of the lone pair of nitrogen and of contributions from the β carbon atoms (Plate VII). As mentioned above, the lone pair orbital of nitrogen is therefore strongly involved in the π system of the molecule, and it does lie at a lower energy than the HOMO, thus being less available for interaction with an incoming electrophile. The display of contour levels of the HOMO–1 of pyrrole (Plate VIII) of course confirms the conclusions drawn from the solid model representation (Plate VII). They may, however, lead to a more accurate description of the characteristics of the orbitals in a given plane. It is, for example, immediately seen from inspection of Plate VIII that the HOMO–1 of pyrrole is more localized on nitrogen than on the π bond between β carbon atoms, as indicated by the presence of higher value contours in the π region of nitrogen.

Comparing now the reactivity of species **1** and **2** towards electrophiles by means of the FO model, it is seen in Plate IX that the HOMO of pyrrolidine is indeed made of the lone pair of nitrogen, which means that according to the FO formalism, this site should be attacked preferentially by an electron acceptor. As mentioned above, this correlates well with the nucleophilic properties of saturated heterocycles, and with the larger basicity of pyrrolidine as compared with pyrrole.

Our second example is devoted to the well known molecule of ferrocene, which is the most prominent member of the metallocene series (Figure 10.5). Of course, the point here is not to make a detailed description of the structural and reactive features of this fascinating compound; the interested reader will find this information in other textbooks [24–26]. Instead, we shall see how a pictorial representation of some MOs helps understanding some properties of ferrocene.

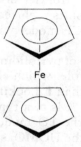

Figure 10.5   The ferrocene molecule.

An energy diagram of the MOs of ferrocene in the frontier region is represented in Figure 10.6. It has been obtained by performing LCGTO-LSD density functional calculations using the deMon program [27]. As usual for ferrocene, the MOs are labelled according to the D_{5d} point group, even though the compound is known from gas-phase electron diffraction data to exhibit a D_{5h} symmetry with eclipsed and planar ligand rings [28].

Ferrocene is a closed-shell molecule with the $(3e_{2g})^4(5a_{1g})^2(4e_{1g})^0$ $^1A_{1g}$ configuration. The predominantly 3d orbitals are, by order of increasing energies, $3e_{2g}$ (HOMO–1), $5a_{1g}$ (HOMO) and $4e_{1g}$ (LUMO), and it is seen in Figure 10.6 that they are energetically sandwiched between the occupied π and virtual π^* MOs of ligands. The iron atom is formally in the II oxidation state with a $3d^64s^0$ configuration, though covalency effects have been shown to significantly modify this simplistic description [29]. Indeed, some occupied MOs with predominant ligand π character such as $3e_{1g}$ exhibit large admixtures of metal 3d orbitals, which leads to a significant ligand-to-metal charge donation. As there is simultaneously much less metal-to-ligand back donation arising through the $3e_{2g}$ and $5a_{1g}$ MOs, it is not surprising to find in usual

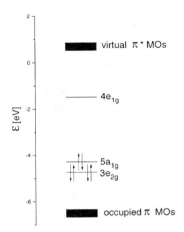

Figure 10.6 Energy level diagram of the MOs of ferrocene as deduced from LCGTO-LSD calculations.

quantum chemical calculations a substantial positive charge on metal, ranging roughly from +0.7 to +1.4 [29].

Returning to the MOs of ferrocene, $4e_{1u}$ exhibits a typical ligand π character with little admixture of metal 4p. However, suppress the metal 4p contribution to this MO is much smaller than the corresponding 3d admixture to $3e_{1g}$, which suggests, as expected, that the role played by the metal 4p AO in chemical bonding is almost negligible. The pictorial representation of the predominantly metal 3d orbitals illustrates perfectly their different nature in such complexes: $3e_{2g}$ is of $3d\delta$ type, whereas $5a_{1g}$ (Plate X) and $4e_{1g}$ (Plate XI) are of $3d\sigma$ and $3d\pi$ types respectively. The latter one is the LUMO with a strongly antibonding character, as it exhibits a large out-of-phase admixture of ligand π AOs.

When applying the FO model to ferrocene, i.e. deriving the most favourable sites for electrophilic and nucleophilic attacks from an inspection of the HOMO (Plate X) and LUMO (Plate XI), respectively, we conclude that both are located on metal as this atom has predominant contributions to both of these MOs. However, whereas this prediction is essentially correct for the addition of an electrophile, as both experimental [30] and theoretical [31] studies have shown that ferrocene protonates readily on metal in strong acids, the site of nucleophilic attack as calculated in the FO model does not agree with experiment. It is indeed known experimentally that the isoelectronic cobalticinium ion undergoes nucleophilic addition on the exo-face of a ligand ring [26], whereas the $4e_{1g}$ MO of this compound is very similar to that of ferrocene and still exhibits its largest coefficient on metal [32]. This is our first example of failure of the very simple reactivity index derived from the FO model and we are going to see some other ones in sections 10.3 and 10.4. This explains why some more sophisticated indices should be used, particularly when organometallic substrates are examined, and this will be a major topic of the present chapter.

10.2 ELECTRON DENSITIES

The electron density is undoubtedly the easiest property to derive from the one-electron wavefunctions calculated for a given compound. Formally, the electron density $\rho(\mathbf{r})$, i.e. the number of electrons per unit volume centred in $\mathbf{r}$, of a system approximated by a single Slater determinant (equation (9.16)) is defined as

$$\rho(\mathbf{r}) = \sum_i n_i |\psi_i(\mathbf{r})|^2 \tag{10.1}$$

where the summation runs over all the spin orbitals ψ_i of the system, each of them being characterized by an occupation number n_i, such as $n_i = 1$ for occupied spin orbitals and $n_i = 0$ for virtual ones.

In the case of a closed-shell system, equation (10.1) becomes:

$$\rho(\mathbf{r}) = \sum_i n_i |\phi_i(\mathbf{r})|^2 \tag{10.2}$$

where the ϕ_i are MOs with occupation numbers n_i equal to 2 (occupied) or 0 (virtual ones).

In cases where post-SCF calculations, such as a CI treatment, have been performed, it is no longer possible to define the electron density using equations (10.1) or (10.2). Instead, Löwdin has shown that a corresponding expression may be used [33]:

$$\rho(\mathbf{r}) = \sum_i n_i |\chi_i(\mathbf{r})|^2 \tag{10.3}$$

where the χ_i are the so-called natural orbitals, whose property is to diagonalize the one-electron density matrix, with occupation numbers n_i possessing any fractional value between 0 and 1.

In the case of most quantum chemical calculations, be they of the SCF or CI post-SCF type, it is therefore possible to rapidly evaluate the electron density at any point of the molecular volume. Generally, as a Hartree–Fock wavefunction leads to an electron density which is exact to the second order [34], an SCF calculation is sufficient to provide a reliable description of the electron distribution. The possibility of easily calculating $\rho(\mathbf{r})$ has prompted theoreticians to use this property to derive additional information on the nature of chemical bonds and, possibly, on molecular properties such as reactivity [35–38]. However, these latter properties are mostly related to the distribution of valence electrons and the main features of the distribution of total electron density within the molecular volume is strongly dominated by core electrons [39, 40]. This is clearly depicted in Figure 10.7, which displays contour levels of the total electron density in the molecular crystal of a substituted anthracene as obtained from high-resolution X-ray crystallography [40].

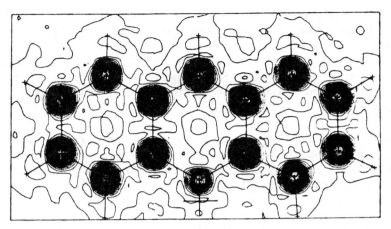

Figure 10.7 Total electron density in 9-tert-butylanthracene (from Angermund *et al.* with permission [40]).

Total electron densities are indeed characterized by high-value circular contour levels lying close to the atoms, indicating that atomic core electron densities remain practically undistorted and spherical during bond formation. This suggests that a more sensitive property, describing better the changes in electron distribution induced by interatomic bonding, can be obtained by subtracting from $\rho(\mathbf{r})$ the spherical electron densities of unperturbed atoms, and we shall return to this point below. However, the total electron density of a given compound can be advantageously used to describe and possibly quantify its shape. It is indeed possible to "measure" the size of a compound or to compare molecular shapes by building isovalue contours or surfaces around the molecular skeleton. To be meaningful, this procedure must be carried out by selecting a constant $\rho(\mathbf{r})$ value for the contours or surfaces calculated and represented, and generally the value 0.002 au is used, as it has been shown that this choice gives physically reasonable molecular dimensions [35, 41, 42]. Indeed, an isovalue surface at 0.002 au of the molecular electron density generally encompasses at least 95% of the electronic charge.

Practically, the graphical representation of $\rho(\mathbf{r})$ can be achieved in a similar way as that of the MOs we have seen in the previous section, namely as contour levels in selected planes, three-dimensional plots and solid models. Of course, this latter mode of representation requires all the attributes of standard 3D modelling i.e. color, shading, clipping, etc. The 3D solid representation of an isovalue $\rho(\mathbf{r})$ surface is in many respects very similar to the Connolly molecular envelope we have seen in Chapter 8. Indeed, as we have seen before that molecular electron densities are strongly dominated by spherical core electron distributions, it is not surprising that isodensity surfaces are very much like envelopes built from van der Waals radii. Of course, this striking resemblance is only observed when choosing an adequately small value of $\rho(\mathbf{r})$, such as 0.002 au.

To enable a better comparison between the 0.002 au isodensity surface and the Connolly envelope, Plate XII presents a superposition of these two surfaces in the case of the pyrrole molecule. It is seen that the two surfaces are indeed very similar in their shapes and sizes, the Connolly one being slightly more elongated in the molecular plane. Simultaneously, the isodensity surface has a larger size perpendicular to the molecular plane, which should be related to the significant π-electron density of pyrrole.

In any case, no attempt has been made to optimize the isodensity surface value so as to have a closer agreement between these two envelopes. Nevertheless, it is clear that both surfaces could be equally used to roughly quantify molecular shapes or to carry additional information such as a reactivity index (see sections 10.3 and 10.4).

Let us turn now to the use of electron densities to discuss the main features of chemical bonding. As mentioned above, this must be achieved using a more sensitive property than $\rho(\mathbf{r})$ itself: the electron density deformation (EDD) $\Delta\rho(\mathbf{r})$ defined as:

$$\Delta\rho(\mathbf{r}) = \rho(\mathbf{r}) - \sum_A \rho_A(\mathbf{r}) \tag{10.4}$$

where $\rho(\mathbf{r})$ is the molecular (total) electron density and $\rho_A(\mathbf{r})$ is the spherically averaged electron density of atom A, the summation running over all the atoms of the molecule.

The EDD, which was first proposed by Roux and Daudel [43], represents the deformation of the spherical charge distribution of the atoms resulting from the formation of chemical bonding, i.e. from molecule formation. As valence electrons are the more concerned and redistributed during bond formation, the EDD will largely reflect the subtle electronic effects accompanying the transformation of isolated spherically averaged atoms into a molecule. As a general rule, the EDD maps exhibit electron density accumulation near the midpoints of covalent bonds and in lone pair regions [44] (Figure 10.8).

An advantage of the EDD local property is that it can in several cases be determined experimentally from X-ray diffraction. The EDD maps thus obtained in the form of contour levels can therefore be used for detailed comparison with calculated EDDs, which allows calibration of the one-electron and N-electron basis sets of *ab initio* calculations.

For theoreticians, such tests are of great value, as they provide an immediate answer as to the quality of their calculations, without requiring one to evaluate elaborate spectroscopic properties. In CAMD applications, however, EDDs are of limited value as they can hardly be used as reactivity indices or as descriptors of properties useful in molecular design [45]. On the other hand, Bader has recently developed a model allowing one to correlate the Laplacian of the total molecular charge density with chemical reactivity [38], which leads essentially to similar results as those of the FO theory.

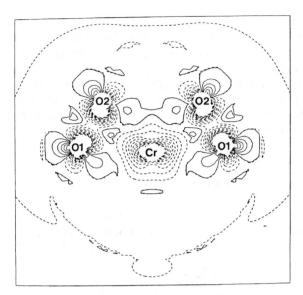

Figure 10.8 Contour levels of the EDD calculated for valence electrons for the $[Cr(O_2)_4]^{3-}$ compound in the (1,1,0) plane. Contour values differ by 0.10 e/Å^3; solid lines indicate positive contour values, dashed lines negative ones (from Roch *et al.* with permission [44]).

10.3 ELECTROSTATIC PROPERTIES

There is a long tradition in the use of electrostatic models in chemistry, which probably traces back to Faraday [46]. Actually, it is interesting to notice that such approximate methods have not been totally superseded by quantum theory. On the contrary, they are nowadays becoming more and more popular because of their simplicity and possible uses in CAMD.

Basically, the generalized Hellmann–Feymann theorem allows one to make the connection between classical electrostatics and quantum theory, stating essentially that "the forces on the nuclei in a molecule can be calculated by classical electrostatics provided one describes the electron distribution using the correct quantum chemical wavefunction" [47]. The important point is therefore that classical electrostatics may apply to derive molecular properties *provided* quantum chemical calculations are performed to derive a reasonably accurate electron distribution (see the previous section). The purpose of this section is to describe why and how electrostatics is used in this context.

Due to their long-range nature arising from their r^{-1} behaviour, electrostatic energies between interacting fragments play a crucial role in solving problems of molecular recognition such as drug-receptor, substrate-enzyme or ion-ionophore interactions. However, several difficulties are encountered when attempting to perform an accurate theoretical treatment of these interactions:

solvent effects may play an important role on them, the atomic charges and higher multipole moments derived from the electron density are not uniquely defined, and it is not easy to describe the polarization of charges when the fragments approach each other, etc. [48]. As these problems are complex in nature, which explains why intensive research work is still in progress in this field, we will discard them here and concentrate on two very popular approaches to the problem of electrostatic interactions, namely the molecular electrostatic potential and the electrostatic field.

10.3.1 Molecular electrostatic potentials

As first defined by Bonaccorsi *et al.* [49], the molecular electrostatic potential (MEP) $V(\mathbf{r})$ represents the value, at first order of perturbation, of the interaction energy between molecule M and a proton located in $\mathbf{r}$ (Figure 10.9):

$$V(\mathbf{r}) = \sum_A \frac{Z_A}{|\mathbf{r} - \mathbf{r}_A|} - \int \frac{\rho(\mathbf{r}')}{|\mathbf{r} - \mathbf{r}'|} d\mathbf{r}' \tag{10.5}$$

where the first term corresponds to nuclear repulsion, the summation running over all atoms of M, with nuclear charge Z_A located in $\mathbf{r}_A$. The second term originates from electronic attraction, $\rho(\mathbf{r}')$ being the electron density of M and the integration being carried out over the whole space.

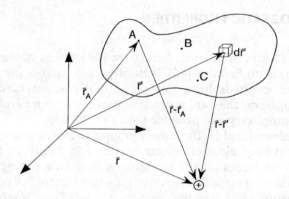

Figure 10.9　Representation of the various parameters involved in the electrostatic interaction energy between molecule M and a proton.

Using the definition of $\rho(\mathbf{r})$ as deduced from the one-electron approximation for a closed-shell system (equation (10.2)) and the LCAO approximation for the molecular orbitals $\phi_i(\mathbf{r})$ (equation (9.27)), it is straightforward to deduce the following expression for $V(\mathbf{r})$:

$$V(\mathbf{r}) = \sum_A \frac{Z_A}{|\mathbf{r} - \mathbf{r}_A|} - \sum_\mu \sum_\upsilon P_{\mu\upsilon} \int \frac{\chi_\mu(\mathbf{r}')\chi_\upsilon(\mathbf{r}')}{|\mathbf{r} - \mathbf{r}'|} d\mathbf{r}' \tag{10.6}$$

where $P_{\mu\nu}$ is the first-order density matrix defined by equation (9.32) and the χ_μ are the AO basis functions used.

The calculation of $V(\mathbf{r})$ in any LCAO model, be it semi-empirical or *ab initio*, thus requires, according to equation (10.6), the evaluation of one-electron integrals of nuclear-attraction type as those involved in equation (9.30). The computational machinery necessary for this evaluation is therefore immediately available in all *ab initio* and most semi-empirical programs [50]. For practical applications, however, we shall see below that the MEP $V(\mathbf{r})$ has to be evaluated repeatedly at selected points in 2D or 3D space, which may lead to a prohibitive computational effort for large molecules or when extended one-electron basis sets are used. In such cases, the molecular electron density may be replaced by a set of point charges q_i, generally located on the atoms and calculated from the wavefunction itself by means of a Mulliken population analysis [51] or, better, by a fitting procedure of the "exact" MEP evaluated at a limited number of grid points surrounding the molecule and located beyond the van der Waals surface [52]. This leads to the following approximate expression for $V(\mathbf{r})$:

$$V(\mathbf{r}) = \sum_i \frac{q_i}{|\mathbf{r} - \mathbf{r}_i|} \tag{10.7}$$

where $\mathbf{r}_i$ is the position of point charge q_i.

One generally imposes the constraint that the sum of the q_i must be equal to the molecular charge. It has been shown recently that MEP-derived charges, which may also be used advantageously to parametrize the electrostatic term in empirical force fields (see Chapter 5), are adequate to reproduce using equation (10.7) the gross features of "exact" MEPs calculated from equation (10.6) [53]. Actually, at rather large distances from the molecular skeleton (typically beyond van der Waals atomic spheres), it is possible to approximate the MEP in a very accurate way using an analytical, multicentre, multipole expansion which considerably reduces the computational time with respect to the quantum evaluation of the MEP using equation (10.6) [54]. To this end, the electron density calculated using the chosen quantum chemical model must first be expanded as a finite number of multicentre multipole contributions. This procedure is very efficient for large molecules where the expansion is made in terms of approximately transferrable fragments [55]. However, it has the drawback of being valid at large distances from the atoms, but in any case this is the region where intermolecular energies can be approximated to a good extent by the electrostatic component [56].

Interestingly, we note when inspecting equation (10.5) that the MEP at any point $\mathbf{r}$ is given by summing up two components of opposite signs: the positive nuclear one and the negative electronic one. The sign of $V(\mathbf{r})$ will then depend upon whether the nuclear or electronic interaction is dominant at position $\mathbf{r}$. As Weinstein *et al.* have shown that the MEP of an isolated, neutral atom is positive everywhere [57], we expect that negative regions of $V(\mathbf{r})$ will be found in molecules at regions of electron density accumulation accompanying the formation of chemical bonds, i.e. close to the lone pairs of heteroatoms or to multiple bonds. Roughly speaking, the negative regions of the MEPs should therefore be found in positive regions of the EDDs we have seen in the previous section [45], that is in electron-rich portions of the molecular volume resulting from build-up of electron density.

It is worth noting that the electrostatic potential is a real physical quantity which can be determined experimentally by X-ray diffraction or electron diffraction techniques [58], which in principle allows us to perform meaningful comparisons with their theoretically derived counterparts.

One of the main advantages of the MEP property is that it can be used as a reactivity index to predict the regions and sites of a substrate S which are the most reactive towards protonation or, more generally, towards an electrophilic attack. Indeed, an incoming electrophile R is likely to be initially attracted by S in the regions of negative $V(\mathbf{r})$ and the sites of addition will be those corresponding to the MEP minima, with the most favorable site for $S–R$ interaction corresponding to the global minimum. It has to be said, however, that the MEP is only the long range part of the total $S–R$ interaction energy and that additional components, such as the energy due polarization of electron distribution of S by the proton and that arising through $S–R$ charge transfer, become important when R penetrates the van der Waals atomic spheres of S (see the next section). In addition, the MEP reflects the exact $S–R$ electrostatic energy only when R is a proton, i.e. a hard electrophile. For other electrophiles and especially the softer ones, (frontier) orbital interactions should be taken into account [14]. Moreover, for this category of electrophiles R, the electrostatic component E_{es} of their interaction energy with S contains two additional terms, namely the attraction and repulsion of the electron density of R by the nuclei and electrons of S respectively, and the MEP as given by equation (10.5) is only an approximation of the true E_{es}.

We now turn to some examples of the usefulness of the MEP property in chemistry. Again, as previously stressed in this chapter, graphics play a very important role for a rapid and comprehensive examination of the features of any MEP: due to the local character of $V(\mathbf{r})$, this property must be calculated at many points in the 3D space surrounding the substrate, and it would be impossible to process this large amount of information without graphics. The following representations of the MEP are therefore most commonly used: contour levels in selected planes, historically the first procedure used [49]; coloured area maps, where areas between successive contour levels in a plane are coloured as a function of the MEP value [45]; colour-coded dots on the Connolly molecular surface [59]; mapping of the MEP value onto solid models of electron density isosurfaces [60] or of molecular surfaces [61]; and isoenergy surfaces represented as wire-frame [62] or solid models [45]. Let us examine and briefly comment upon these various representations.

The display of MEPs as contour levels in selected plane(s), as depicted in Figure 10.10, is very convenient for an accurate evaluation of the characteristics of $V(\mathbf{r})$ [63] but, of course, it leads only to a 2D representation and it could be misleading for complex molecules or when competing sites are present in even simple compounds. The same statement applies to coloured

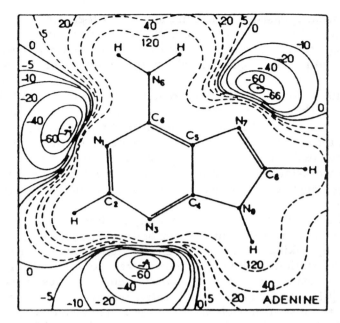

Figure 10.10 Contour levels (in kcal/mol) of the MEP of adenine; solid lines indicate positive contour values, dashed lines negative ones (from Bonnaccorsi *et al.* with permission [63]).

area maps, which are mainly used for a rapid evaluation of the main features of MEPs in selected planes.

However, being restricted to a representation of the MEP property in various planes, contour levels or coloured area maps are generally of limited value for a rapid estimation of the most favourable site for electrophilic attack, except for two cases:

- small molecules with only one basic centre, such as H_2O, NH_3 or pyrrolidine, for which one can anticipate in which plane the MEP minimum will lie;
- the modelling software used allows the instantaneous construction of coloured area maps in a given plane, and subsequently to translate interactively that plane within the whole molecular volume, with simultaneous reconstruction of the MEP maps, which leads to a fast localization of the global minimum; such a facility is implemented in various packages such as UNICHEM [64].

Significant progress was made in 1982 with the procedure suggested by Weiner *et al.* [59] and consisting of the representation of the MEP as colour-coded dots on the Connolly surface of the substrate. Indeed, since these surfaces were first depicted as dots defining the molecular envelope, the idea was to colour these dots according to the MEP value, using a standard colour code, so as to generate four-dimensional (three for structural features and one for the MEP) models. In this way, it is fairly easy to localize in a single step (i.e. using a single graphical representation) the most reactive site of the substrate towards electrophilic attack, except for compounds with several possible sites lying close to one another in MEP values and for which this kind of modelling might be ambiguous.

As Connolly dot surfaces are generally made of 2000–5000 pixels for compounds with up to 50 atoms, this representation does not in general require one to evaluate the MEP at a number of points which is significantly larger than for contour levels in a single plane. It is therefore very popular and implemented in practically all the molecular modelling packages running on workstations (or even on PCs), with a MEP calculated using the point charge approximation (equation (10.7)) in order to accelerate the calculations.

Instead of using the dot surface of Connolly to map MEP values, it is possible alternatively to generate solid models of the surface and to colour them accordingly by taking advantage of all the rendering facilities offered by graphics workstations: smooth shading, large colour scale available, the possibility of using several light sources, etc. The result is in general spectacular [61], although one should realize that it is much more difficult to manipulate these solid models in real time as compared with their colour-coded dot analogues. Examples are provided by Plates XIII and XIV, which display solid models of the Connolly surface of pyrrole and pyrrolidine respectively, coloured according to the MEP and properly clipped so as to also visualize the structural model. There is a general agreement among software developers as to the colour-coding range of the surfaces, from red and yellow to blue, extending smoothly over the numerical range of E_{int} from the most negative to zero to the most positive values, which means that the red zones correspond to preferred sites of electrophilic attack. It is seen in Plates XIII and XIV that whereas the nitrogen atom of pyrrolidine is indeed predicted by the MEP to be the most favourable site for electrophilic attack, which is in agreement with the experimental evidence, the β carbon atoms of pyrrole are the most reactive ones towards incoming electrophiles in this MEP approximation, and this is in contradiction with both structural and kinetic studies pointing to an α attack. However, in agreement with the relative basicities of pyrrole and pyrrolidine, the minimum value of the MEP on the molecular surface of pyrrolidine lies at a considerably lower energy than the corresponding value obtained for pyrrole. The fact that the regioselectivity of electrophilic attack on pyrrole is not correctly described indicates that the use of the MEP as a reactivity index for electrophilic addition reactions should be made with some care, as it is only a static property taking account of the ability of the undistorted substrate, with frozen electron density, to react with a proton [65]. In several cases, more elaborate models accounting for both structural deformations and electron density rearrangements should be used, and we will return to this point later on in this chapter.

Turning back to the different modes of representation of MEPs, we should mention that it has also been suggested to map their values on van der Waals envelopes [66] or electron density isosurfaces at 0.002 au [60]. The latter have recently been advocated as being better than other envelopes because they should better reflect unique molecular features such as bond formation, lone pairs, etc. [42]. However, in view of the very close matching between such surfaces and Connolly molecular envelopes, this argument is most probably of little value. Finally, MEPs are also displayed as isoenergy models in 3D space, either in the form of wire-frame [62] (Figure 10.11) or solid surfaces [43] (Plate XV).

This type of representation requires immersing the substrate into a 3D grid box and calculating the MEP at a large number of points, generally of the order of 10^5–10^6, which of course involves a significant computational effort, unless an approximate evaluation of the MEP is performed. However, it can be seen that whereas such models lead in the case of positive isoenergy surfaces to another type of molecular surface, to be compared with Connolly or electron isodensity envelopes, they are able to perfectly locate the MEP minimum in the molecular volume (Plate XV). This is undoubtedly useful for molecules with several MEP minima, as lowering progressively the energy of the isosurfaces leads inevitably to the local minima disappearing and the global one staying as the only isoenergy envelope.

Numerous applications of MEPs have appeared in the last ten years, and the most important ones have been reviewed by Politzer et al. [67, 68]. In particular, it is shown that MEPs are of invaluable help in rationalizing and predicting several important processes such as electrophilic addition or substitution reaction mechanisms, biological recognition, hydrogen bonding, etc. Interesting correlations have been found between MEP minima and chemical properties such as solvatochromic hydrogen-bond donor and

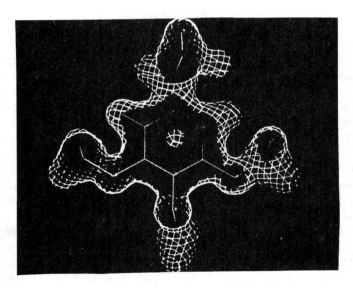

Figure 10.11   Wire-frame model isoenergy ($V = 100$ kcal/mol) surface of mescaline (from Doucet et al. with permission [62]).

acceptor parameters α and β, inductive and resonance substituent constants σ_1 and σ_2, pK_a values, etc.

It has also been suggested that the MEP may be used as a reactivity index for nucleophilic attack, in which case the reactant is attracted towards positive regions of this local property, provided one considers the maxima of MEPs calculated on surfaces sufficiently removed from the nuclei [42]. Indeed, the absolute maxima of the MEPs are located on the atoms because they correspond to maximum nuclear repulsion, and the use of a molecular surface is indispensable to prevent the incoming nucleophiles to collapse on to the nuclei. This amounts to assuming the presence of an infinite exchange-repulsion potential on the molecular surface (hard spheres approximation), and we shall return to this point in the next sections.

Undoubtedly, the MEP is an indispensable tool in CAMD applications because of its simplicity and versatility. This is particularly true for organic molecules and fragments where intermediate range effects such as charge transfer or polarization are generally negligible, which means that the reactivity of frozen substrates may be described to a good extent by their ability to attract or repel incoming species through long range electrostatic interactions.

10.3.2 Molecular electrostatic fields

Instead of the MEP $V(\mathbf{r})$ (equation (10.5)), several authors have suggested using the electrostatic field (EF) $\mathbf{E}$ to describe long-range electrostatic interactions, with $\mathbf{E}$ defined as:

$$\mathbf{E} = -\nabla V(\mathbf{r}) \tag{10.8}$$

where ∇ is the gradient with respect to $\mathbf{r}$, i.e. to the proton position [69–71]. By definition, the EF is a vector normal to the isoenergy MEP surface passing by point $\mathbf{r}$, which represents the force acting on a unit test positive charge located in $\mathbf{r}$. It may be therefore useful to calculate and display the flow of EF vectors around a molecule, especially in the region which is most likely to be involved in molecular recognition (Figure 10.12). Indeed, these vectors, whose length is of course proportional to the magnitude of EF, correspond to the orientation of an external point dipole experiencing the molecular field and, thus, they provide an approximate description of the docking of polar X–H bonds into hydrogen bonding positions [71]. As was the case for the MEPs, the EFs can be conveniently approximated using a multicentre multipolar expansion of the electron density, which considerably reduces the computer time involved in the calculation of the flow of EF vectors surrounding a given compound.

We have briefly mentioned in this section the limitations of the electrostatic approximation, in particular of the MEP, to predict reliably the preferred sites for electrophilic attack. To summarize them, the MEP is a static reactivity index which does not take account of the important changes in both the structure and charge distribution of the substrate which may arise through

Figure 10.12 Representation of the electrostatic field surrounding the molecule of pyrimidine (from Price *et al.* with permission [71]).

protonation [72]. In other words, the MEP and EF are better to describe kinetically controlled processes with an early transition state, that is mechanisms for which these changes are relatively unimportant. However, these indices are likely to be less reliable for thermodynamically controlled processes, and we shall review in the next section some other procedures which could be useful alternatives for such cases.

10.4 REACTIVITY INDICES

We have previously seen that the MEP may be used as a local reactivity index, i.e. as a property defined at any point of the molecular volume and whose value directly correlates with the propensity of the compound to react with an incoming species located at this point. The advantage of local reactivity indices lies in the fact that they may be easily represented on graphics workstations using the various techniques described in the previous section, which allows the chemist using such CAMD tools to rapidly localize the hot regions corresponding to favourable interaction energies and to roughly identify low-energy reaction pathways [73]. Of course, such models of chemical reactivity are approximate, and in no way can they be considered as substitutes for accurate reaction paths calculated using adequate quantum chemical techniques. However, taking account of the considerable computational effort which is generally involved by the latter calculations, especially when performed at the *ab initio* level, local reactivity indices are very useful tools in interactive, or pseudo-interactive, CAMD applications even though their results should be taken with a pinch of salt.

Beyond some force field models, which have been reviewed in Chapter 5, several approaches have been proposed in an attempt to introduce into a reactivity index further components of the intermolecular interaction energy than the MEP [18, 60, 72, 73]. Whereas some of them are based on the introduction of additional components (orbital interaction, polarization, etc.) using second order perturbation theory [72, 73], the procedure of Kahn *et al.* incorporates the exact treatment of the electrostatic term when the incoming reactant is a nucleophile (e.g. a hydride anion) [60]. Indeed, in this case, additional terms to equation (10.5) are required to evaluate rigorously the electrostatic energy between substrate and reactant, namely those involving the interaction between the electrons of the latter species with both the nuclei and electrons of the substrate. In any case, these developments have shown that in general a more refined treatment of intermolecular interactions than the single MEP is important as it improves the reliability of the reactivity index.

We would like to present here some results obtained recently by one of us using such an "augmented" reactivity index (with respect to the MEP) developed mainly for organometallic compounds [18]. Indeed, for the latter systems, the MEP component is generally not sufficient to describe intermediate to long range interactions between a substrate S, containing a transition metal atom, and a reactant R because of the importance of a charge-transfer component arising mainly through overlap between the d orbitals of transition metal-containing substrate and the orbitals of the reactant [74–77].

To describe the reactive properties of substrate S in the presence of reactant R, a local reactivity index made of the S–R intermolecular interaction energy is used. It is expressed as a sum of several components:

$$E_{int}(\mathbf{r}) = E_{es}(\mathbf{r}) + E_{ct}(\mathbf{r}) + E_{ex}(\mathbf{r}) \tag{10.9}$$

where $\mathbf{r}$ specifies the position of the incoming reactant in the vicinity of the organometallic substrate, E_{es}, E_{ct} and E_{ex} being electrostatic, charge-transfer (CT) and exchange-repulsion energy components, respectively (Figure 10.13).

To keep the model computationally simple, it is assumed that selection of the most favourable region(s) of S to be attacked by R occurs at rather large distances, well in advance of the transition state, so that geometrical distortions of S are negligible. This assumption is commonly made when modelling organic reactions using the MEP reactivity index. For organometallic reaction mechanisms, the CT component can be quite important at such distances, and therefore it may lead to a non-negligible E_{ct} energy. This is especially true for electrophilic addition reactions, as exemplified by ferrocene

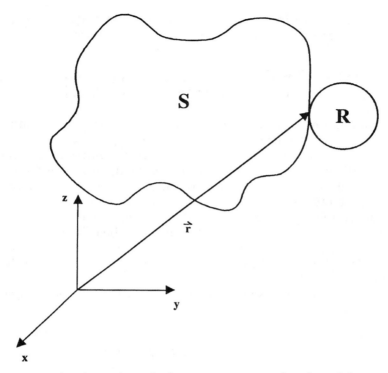

Figure 10.13 Molecular surface of substrate S represented with model reactant R; **r** is the point of the surface where E_{int} is calculated.

where the MEP component alone predicts protonation on the ligand ring while the introduction of E_{ct} completely changes the selectivity and, in agreement with experiment, favours the metal site (see below).

It is important to choose a simple, though realistic, model for the reactant R, as the computer time required to evaluate E_{int} increases rapidly as a function of the complexity of R. To obtain E_{int} values that are dependent on the position of R only in the vicinity of S, and not on its orientation, two spherically symmetrical model reactants have been selected: a proton with a virtual 1s orbital for the electrophile and H^- hydride ion for the nucleophile.

In the case of electrophilic attack, E_{es} is equal to the MEP of substrate S calculated using equation (10.6) and extended Hückel (EH) wavefunctions. In the case of nucleophilic attack, it is assumed that the electrostatic interaction between S and the H^- ion reduces to that between S and a negative point charge, which is obviously correct for rather large S–R distances where the so-called *penetration integrals* vanish. Several test calculations have shown this to be the case when R is at a distance larger than roughly 2.0 Å from any atom of S. For nucleophilic attack, the electrostatic component is, therefore, given by $-V(\mathbf{r})$ (equation (10.6)).

As the calculation of electronic attraction integrals using the EH basis of atomic orbitals is time-consuming because they are of the Slater type, the MEP can be approximated using the neglect of diatomic differential overlap (NDDO) scheme [78], according to which the second right hand term of equation (10.6) becomes:

$$\sum_{\mu}\sum_{\upsilon}P_{\mu\upsilon}\int\frac{\chi_{\mu}(\mathbf{r}')\chi_{\upsilon}(\mathbf{r}')}{|\mathbf{r}-\mathbf{r}'|}d\mathbf{r}' = \sum_{A}\sum_{\mu\in A}\sum_{\upsilon\in A}P_{\mu A\upsilon A}\int\frac{\chi_{\mu A}(\mathbf{r}')\chi_{\upsilon A}(\mathbf{r}')}{|\mathbf{r}-\mathbf{r}'|}d\mathbf{r}' \qquad (10.10)$$

The first summation is over all atoms A of S. However, this requires the evaluation of the reduced density-matrix using orthogonalized AOs.

As to the CT component, it is well known that organometallic substrates are often characterized by bands of closely spaced energy levels in both HOMO and LUMO regions [17]. This renders the use of frontier MOs only, in the evaluation of orbital or charge-transfer effects, as questionable. Instead, Brown et al. [17] suggested replacing the perturbation treatment by a complete EH calculation of the S–R supermolecule; the so-called "orbital interaction energy" is obtained as the difference between the total energies of the supermolecule and those of the separate fragments. The same approach is used in this model.

As EH total energies represent fairly comprehensively the sum of covalent energies within chemical bonds, the S–R charge-transfer energy may be approximated as:

$$E_{ct}(\mathbf{r}) = E^{tot}(S-R,\mathbf{r}) - E^{tot}(S) - E^{tot}(R) \qquad (10.11)$$

where $E^{tot}(X)$ represents the EH total energy of system X calculated as:

$$E^{tot}(X) = \sum_{i}n_i\varepsilon_i \qquad (10.12)$$

n_i and ε_i being occupation number and energy respectively, of the ith MO of X and $E^{tot}(S-R,\mathbf{r})$ being calculated for position $\mathbf{r}$ of reactant R. It is to be noted that in the case of the H^- nucleophile reactant, $E_{ct}(\mathbf{r})$ may be positive or negative as well, as it results from the balance of: (i) attractive 2-electron interactions between the HOMO of R and the unoccupied MOs of S, and (ii) repulsive 4-electron interactions between the HOMO of R and the occupied MOs of S [79]. The valence state ionization energy (VSIE) H_{RR} of the 1s orbital of reactant H^- is therefore a critical parameter for the calculation of the charge-transfer component: moving H_{RR} downwards from a value close to the LUMO of S to more negative energies nearby the HOMO, i.e. transforming a soft nucleophile into a hard one, generally amounts to shifting E_{ct} from negative to positive values.

Let us turn to the exchange-repulsion component. In the case of electrophilic attack, there is no such component in this model, as the reactant has no electrons. For nucleophilic attack, whereas in the case of colour-coded molecular surfaces the reactivity index E_{int} usually comprises an exchange-repulsion component based on the hard sphere approximation [74], this procedure is no longer applicable to isoenergy surfaces as there is no molecular envelope to prevent the reactant collapsing on the nuclei of S. A more elaborate short range exchange-repulsion function is therefore needed, and a parametrized potential of Buckingham type is used [80]:

$$E_{ex}(\mathbf{r}) = 828\,000 \sum_{A \epsilon S} k_{AH^-} \exp\left[\frac{|\mathbf{r} - \mathbf{r}_A|}{R_A + R_{H^-}}\right] \tag{10.13}$$

where the summation runs over all atoms A of S, located in $\mathbf{r}_A$, k_{AH^-} being an energy parameter depending on atom A and H^- ion, R_A and R_{H^-} being the van der Waals radii of the atom A and H^- ion, respectively. In equation (10.13), E_{ex} is expressed in kcal/mol, and the values of the parameters have been taken from Eliel *et al.* [81].

The reactivity index defined by equation (10.9) has been successfully applied to a wide range of reaction mechanisms in organometallic chemistry [18, 61, 74–77]. Indeed, in most cases, the regio- and stereoselectivity of electrophilic and nucleophilic addition mechanisms is adequately predicted as demonstrated by the following examples.

Ferrocene is a classical example of organometallic compound with an iron atom symmetrically bound to two cyclopentadienyl rings. It is known from experimental investigation that ferrocene readily fixes a proton in strong acids to give the reaction intermediate $Fe(C_5H_5)_2H^+$ [82] (Figure 10.14). The protonation site is located on the metal, presumably in the equatorial plane, which is confirmed by recent ion thermochemistry experiments [30]. It is seen in Plate XVI that these experimental results are reproduced by the theoretical model as the five lowest minima of E_{int}, reflecting the five-fold symmetry axis of the molecule, are located in equatorial positions of the molecular surface in regions corresponding to an addition on metal [83]. Furthermore, a closer look at the molecular surface of ferrocene, coloured according to the reactivity index, shows that secondary minima, corresponding to less favourable interaction energies, are found on the exo positions of the ligand rings (i.e. towards the face opposite to the metal of complexed C_5H_5), which is in agreement with gas phase measurements pointing out that ring protonation is also a possible, but less probable, process [30]. Interestingly, reactivity indices based on the MEP only lead to the result that the most stable protonation site lies on the ligands [29, 31], which is an expected result as the electrostatic energy between a formally Fe(II) atom and an incoming proton is strongly

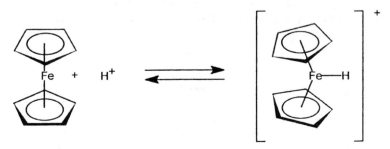

Figure 10.14   Mechanism of protonation reaction of ferrocene.

repulsive. It is therefore only by adding the E_{ct} component to E_{es} that the E_{int} energy can be used in this case as a realistic reactivity index.

Another example of the application of this simple reactivity index is provided by the benzene-chromiumtricarbonyl complex $(C_6H_6)Cr(CO)_3$ (Figure 10.15), whose structure and reactivity has been the subject of intensive investigations in organometallic chemistry [84].

It is generally accepted that metal-benzene bonding leads for this compound to a net intramolecular charge transfer from the ring to the carbonyl groups, with the result that the complex is easily attacked by a nucleophile on the exo-face of the ring. Plate XVII represents the E_{int} reactivity index calculated for the nucleophilic attack of this compound, which shows that, indeed, the most reactive site is located on the exo-face of the ligand ring. For comparison, the same index is also displayed for a free, uncoordinated benzene molecule. It can immediately be seen that, as expected, an important change in reactivity accompanies the complexation of this benzene molecule: whereas for an isolated aromatic ring one observes essentially slightly negative isoenergy envelopes centred on the C6 axis above and below the molecular plane (representative of weak van der Waals interactions), the situation changes dramatically for the coordinated arene. In the latter case, several imbricated surfaces are found at much lower energies, revealing as expected an important activation of the benzene ring through coordination to the electrophilic $Cr(CO)_3$ group. As a consequence, the theoretical results show that the initial nucleophilic attack is likely to occur on the face opposite to the metal of complexed benzene, as has been found by experimentation [85]. The two examples presented here dealing with practical applications of this simple reactivity index indicate that, as reported for several other cases [74–77], it may provide a convenient tool for interpreting, and in some cases predicting, the reactivity of organometallic species.

Among the other molecular properties which can be easily derived from the ground state wavefunction of a given compound and subsequently visualized as a reactivity index, it is worth mentioning the superdelocalizability, which is a measure of how tightly the electrons are held by the molecule at any point [66, 86], and the average local ionization energy, which may be similarly defined as the average energy needed to ionize an electron at any point [68]. Both properties may be used as alternatives to the MEP to correlate with pK_a values [68] or π-aromaticity of hydrocarbon rings [87]. In addition, Parr has extended the concept of hard and soft acids and bases, first introduced by Pearson [88], to reactivity indices based on density functional theory and

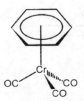

Figure 10.15   Structure of the benzene-chromiumtricarbonyl complex.

defining local softness and local hardness [37]. These properties have recently been applied with success to modelling the reactivity of organic molecules [89, 90]. Though the limitations of such static reactivity indices are real, they may be used as valuable tools to describe the gross features of chemical reaction paths and, as such, they enjoy a growing popularity in CAMD applications.

REFERENCES

1. A.D. McLean and M. Yoshimine *Tables of Linear Molecule Wave Functions, Suppl. IBM J. Res. Dev.* November 1967.
2. L.C. Synder and H. Basch *Molecular Wave Functions and Properties: Tabulated from SCF Calculations in a Gaussian Basis Set*, Wiley, New York, 1972.
3. M.J. Frisch, G.W. Trucks, M. Head-Gordon, P.M.W. Gill, M.W. Wong, J.B. Foresman, B.G. Johnson, H.B. Schlegel, M.A. Robb, E.S. Replogle, R. Gomperts, J.L. Andres, K. Raghavachari, J.S. Binkley, C. Gonzalez, R.L. Martin, D.J. Fox, D.J. Defrees, J. Baker, J.J.P. Stewart and J.A. Pople *Gaussian 92*, Revision A, Gaussian Inc., Pittsburgh PA, 1992.
4. A. Steitwieser and P.H. Owens *Orbital and Electron Density Diagrams: An Application of Computer Graphics*, Macmillan, New York, 1973.
5. W.L. Jorgensen and L. Salem *The Organic Chemist's Book of Orbitals*, Academic Press, New York, 1973.
6. N. Thalmann and J. Weber, *Chimia*, **31:** 1977; 361.
7. A. Schmelzer and E. Haselbach, *Helv. Chim. Acta*, **54:** 1971; 1299.
8. I. Absar and J.R. Van Wazer, *J. Phys. Chem.*, **75:** 1971; 1360.
9. J. Weber in *Computer Aids to Chemistry*, G. Vernin and M. Chanon (eds.), Ellis Horwood, New York, 1986, p. 154 (and references therein).
10. *Spartan 2.0*, Wavefront Inc., 18401 Von Karman, # 210, Irvine CA 92715, USA.
11. R.F. Hout, W.J. Pietro and W.J. Hehre, *A Pictorial Approach to Molecular Structure and Reactivity*, Wiley, New York, 1984.
12. R.B. Woodward and R. Hoffmann, *The Conservation of Orbital Symmetry*, Verlag Chemie, Weinheim, 1970.
13. K. Fukui, T. Yonezawa, C. Nagata and H. Shingu, *J. Chem. Phys.*, **22:** 1954; 1433.
14. I. Fleming, *Frontier Orbitals and Organic Chemical Reactions*, Wiley, Chichester, 1976.
15. G. Mazzara and A. Borgo, *Gaz. Chim. Ital.*, **35:** 1905; II, 20.
16. O. Eisenstein and R. Hoffmann, *J. Am. Chem. Soc.*, **103:** 1981; 4308.
17. D.A. Brown, N.J. Fitzpatrick and M.A. McGinn, *J. Organomet. Chem.*, **293:** 1985; 235.
18. C. Daul, A. Goursot, P.Y. Morgantini and J. Weber, *Int. J. Quant. Chem.*, **38:** 1990; 623.
19. H. Hart, *Organic Chemistry, A Short Course*, 7th edition, Houghton Mifflin, Boston, 1987.
20. J.S. Binkley, J.A. Pople and W.J. Hehre, *J. Am. Chem. Soc.*, **102:** 1980; 939.
21. A.R. Katrizky and J.M. Lagowski, *The Principles of Heterocyclic Chemistry*, Academic Press, New York, 1968.
22. R. Houriet, H. Schwarz, W. Zummack, J.G. Andrade and P.v.R. Schleyer, *Nouv. J. Chim.*, **5:** 1981; 505.
23. J. Catalan and M. Yanez, *J. Am. Chem. Soc.*, **106:** 1984; 421.
24. K.F. Purcell and J.C. Kotz, *Inorganic Chemistry*, Saunders, Philadelphia, 1977.
25. F.A. Cotton and G. Wilkinson *Advanced Inorganic Chemistry. A Comprehensive Text*, 4th edition, Wiley, New York, 1980.

26. Ch. Elschenbroich and A. Salzer *Organometallics, A Concise Introduction*, VCH, New York, 1989.

27. A. St-Amant and D.R. Salahub *Chem. Phys. Lett.*, **169:** 1990; 387.

28. A. Haaland *Acc. Chem. Res.*, **11:** 1979; 415.

29. O. Schaad, M. Roch, H. Chermette and J. Weber *J. Chim. Phys.*, **84:** 1987; 829.

30. M. Meot-Ner *J. Am. Chem. Soc.*, **111:** 1989; 2830.

31. P. Jungwirth, D. Stussi and J. Weber *Chem. Phys. Lett.*, **190:** 1992; 29.

32. J. Weber, A. Goursot, E. Penigault, J.H. Ammeter and J. Bachmann *J. Am. Chem. Soc.*, **104:** 1982; 1491.

33. P.O. Löwdin *Phys. Rev.*, **97:** 1955; 1474.

34. C.W. Kern and M. Karplus *J. Chem. Phys.*, **40:** 1964; 1374.

35. R.F.W. Bader, I. Keaveny and P.E. Cade *J. Chem. Phys.*, **47:** 1967; 3381.

36. R.S. Mulliken and W.C. Ermler *Diatomic Molecules. Results of Ab Initio Calculations*, Academic Press, New York, 1977.

37. R.G. Parr and W. Yang *Density-Functional Theory of Atoms and Molecules*, Oxford University Press, New York, 1989, and references therein.

38. R.F.W. Bader *Atoms in Molecules. A Quantum Theory*, Clarendon Press, Oxford, 1990, and references therein.

39. P. Coppens and E.D. Stevens *Adv. Quant. Chem.*, **10:** 1977; 1.

40. K. Angermund, K.H. Claus, R. Goddard and C. Krüger *Angew. Chem. Int. Ed. Engl.*, **24:** 1985; 237.

41. S.D. Kahn, C.F. Pau, L.E. Overman and W.J. Hehre *J. Am. Chem. Soc.*, **108:** 1986; 7381.

42. J.S. Murray, P. Lane, T. Brinck, P. Politzer and P. Sjoberg *J. Phys. Chem.*, **95:** 1991; 844.

43. M. Roux and R. Daudel *C.R. Acad. Sci. (Paris)*, **240:** 1955; 90.

44. M. Roch, J. Weber and A.F. Williams *Inorg. Chem.*, **23:** 1984; 4571.

45. J. Weber and M. Roch *J. Mol. Graphics*, **4:** 1986; 145.

46. J. Tomasi, in *Chemical Applications of Atomic and Molecular Electrostatic Potentials*, P. Politzer and D.G. Truhlar, (eds.), Plenum, New York, 1981; p. 257.

47. F.L. Pilar *Elementary Quantum Chemistry*, McGraw-Hill, New York, 1968.

48. N.G.J. Richards and J.G. Vinter *J. Comput.-Aided Mol. Des.*, **5:** 1991; 1.

49. R. Bonaccorsi, E. Scrocco and J. Tomasi *J. Chem. Phys.* **52:** 1970; 5270.

50. E. Scrocco and J. Tomasi *Top. Curr. Chem.*, **42:** 1973; 95.

51. R.S. Mulliken *J. Chem. Phys.*, **23:** 1955; 1833.

52. U.C. Singh and P.A. Kollman *J. Comput. Chem.*, **5:** 1984; 129.

53. C.A. Reynolds, G.G. Ferenczy and W.G. Richards *J. Mol. Struct. Theochem*, **256:** 1992; 249.

54. J.R. Rabinowitz, K. Nomboodiri and H. Weinstein *Int. J. Quant. Chem.*, **29:** 1986; 1697.

55. R. Bonaccorsi, E. Scrocco and J. Tomasi, *J. Am. Chem. Soc.*, **98:** 1975; 4049.

56. A. Goldblum, D. Perahia and A. Pullman, *Int. J. Quant. Chem.*, **15:** 1979; 121.

57. H. Weinstein, P. Politzer and S. Srebrenik, *Theor. Chim. Acta*, **38:** 1975; 159.

58. G. Moss and P. Coppens, in *Chemical Applications of Atomic and Molecular Electrostatic Potentials*, P. Politzer and D.G. Truhlar, (eds.), Plenum, New York 1981; p. 427.

59. P.K. Weiner, R. Langridge, J.M. Blaney, R. Schaefer and P.A. Kollman, *Proc. Natl. Acad. Sci. USA*, **79:** 1982; 3754.

60. S.D. Kahn, C.F. Pau and W.J. Hehre, *Int. J. Quant. Chem., Symp.*, **22:** 1988; 75.

61. J. Weber, P.Y. Morgantini, P. Fluekiger and M. Roch, *Comput. & Graphics*, **13:** 1989; 229.

62. J.P. Doucet, S.Y. Yue, J.E. Dubois, M. Roch and J. Weber, *J. Chim. Phys.*, **84:** 1987; 647.

63. R. Bonaccorsi, E. Scrocco, J. Tomasi and A. Pullmann, *Theor. Chim. Acta*, **36:** 1975; 339.

64. UNICHEM 1.0, Cray Research Inc., 655-E Lone Oak Drive, Eagan MN 55121, USA.
65. J.P. Ritchie, *J. Mol. Struct. Theochem*, **255:** 1992; 297.
66. P. Quarendon, C.B. Naylor and W.G. Richards, *J. Mol. Graphics*, **2:** 1984; 4.
67. P. Politzer and J.S. Murray, in *Reviews in Computational Chemistry, vol. 2*, K.B. Lipkowitz and D.B. Boyd, (eds.), VCH, New York (1991); p. 273.
68. J.S. Murray, T. Brinck, M.E. Grice and P. Politzer, *J. Mol. Struct. Theochem.*, **256:** 1992; 29.
69. G.D. Purvis and C. Culberson, *J. Mol. Graphics*, **4:** 1986; 88.
70. M.A. Hermsmeier and T.M. Gund, *J. Mol. Graphics*, **7:** 1989; 150.
71. S.L. Price and N.G.J. Richards, *J. Comput.-Aided Mol. Des.*, **5:** 1991; 41.
72. G.D. Purvis, *J. Comput.-Aided Mol. Des.*, **5:** 1991; 55.
73. H. Moriishi, O. Kikushi, K. Suzuki and G. Klopman, *Theor. Chim. Acta*, **64:** 1984; 319.
74. J. Weber, P. Fluekiger, D. Stussi and P.Y. Morgantini, *J. Mol. Struct. Theochem.*, **227:** 1991; 175.
75. J. Weber, P.Y. Morgantini and O. Eisenstein, *J. Mol. Struct. Theochem.*, **254:** 1992; 343.
76. G. Bernardinelli, A. Cunningham, C. Dupre, E.P. Kündig, D. Stussi and J. Weber, *Chimia*, **46:** 1992; 126.
77. J. Weber, D. Stussi, P. Fluekiger and P.Y. Morgantini, *Comments Inorg. Chem.*, **14:** 1992; 27.
78. J.A. Pople, D.P. Santry and G.A. Segal, *J. Chem. Phys.*, **43:** 1965; S129.
79. P. Sautet, O. Eisenstein and K.M. Nicholas, *Organometallics*, **6:** 1987; 1845.
80. P. Hobza and R. Zahradnik, *Weak Intermolecular Interactions in Chemistry and Biology, Academia*, Prague, 1980.
81. E.L. Eliel, N.L. Allinger, S.J. Angyal and G.A. Morrison, *Conformational Analysis*, Wiley, New York, 1965.
82. T.J. Curphey, J.O. Santer, M. Rosenblum and J.H. Richards, *J. Am. Chem. Soc.*, **82:** 1960; 5249.
83. J. Weber and P.Y. Morgantini, *EPFL Supercomputing Review*, No **2:** 1990; 14.
84. A. Solladie-Cavallo, *Polyhedron*, **4:** 1985; 901.
85. M.F. Semmelhack, H.T. Hall, R. Farina, M. Yoshifuji, G. Clark, T. Bargar, K. Hirotsu and J. Clardy, *J. Am. Chem. Soc.*, **101:** 1979; 3535.
86. K. Fukui, T. Yonezawa and C. Nagata, *Bull. Chem. Soc. Japan*, **27:** 1954; 423.
87. J.S. Murray, J.M. Seminario, P. Politzer and P. Sjoberg, *Int. J. Quant. Chem. Symp.*, **24:** 1990; 645.
88. R.G. Pearson, *J. Am. Chem. Soc.*, **85:** 1963; 3533.
89. C. Lee, W. Yang and R.G. Parr, *J. Mol. Struct. Theochem.*, **163:** 1988; 305.
90. W. Langenaeker, K. Demel and P. Geerlings, *J. Mol. Struct. Theochem.*, **234:** 1991; 329.

11 *Molecular similarity*

Comparison of structural characteristics within a set of molecules that display common features is a frequent problem. It relies on the general assumption that similarity in behaviour implies similarity in structure. Relevant to this well-accepted hypothesis is, at a very simple level of elementary chemistry, the concept of *functional groups*: specific arrangements of some atoms which provide a common property to all molecules where they are embedded. At a more sophisticated level, this also relies on the *pharmacophoric approach* in drug design: a common activity within a set of compounds is due to some similar structural features, which are considered as responsible for that activity. Once these structural features (considered as indispensable for the activity to be present) are identified, some guidelines may be obtained for synthesizing new active drugs. In connection with this first phase of identifying the pharmacophore, the next step in drug design now largely relies on screening databases of 3D coordinates to define new leads that possess the right atoms in the right geometry. After some structural optimization, these leads provide new active substances. The creation of large databases of 3D coordinates, sometimes associated with activity data (see Chapters 4 and 7), prompted increased interest in methods able to determine the maximum of similarity between two molecules or two fragments with respect to their mutual orientation, and also rank candidates according to their similarity with a specific target.

Owing to the multiple facets of chemical behaviour, the search for structural similarity can in fact be processed at various levels of complexity, depending upon the field of application.

Topology and topography

A first set of applications can be efficiently carried out by using only 2D representations. *Topology* is here the main tool for finding *connected* substructures. This involves, for instance, information retrieval from the chemical literature and its very large files, chemical reaction indexing, etc. Identification of 2D structures has also been used widely in spectroscopy (for structural elucidation) and in quantitative structure activity relationships (QSAR).

The success of topological concepts in these fields may appear surprising since molecular properties largely depend upon the true spatial location of atoms, i.e. on *topography*. Nevertheless, in most cases within a given structural area, topology (indication of the nature of the atoms and bonds between them) implicitly reflects topography (i.e. the proximity relationships between atoms): topology is a "rubber sheet geometry" [1].

Comparison of actual 3D representations, on the other hand, is required when interactions between remote groups of stereochemical features are involved. This is, of course, the case in searching for a pharmacophore, or in investigating protein properties. Similar problems also occur in X-ray crystallography, for fitting a structure to a set of peaks in a Fourier map of electron density distribution [2].

Our main concern here being the modelling of chemical behaviour, we only consider the comparisons of 3D structures or fragments.

Levels of 3D comparisons

In 3D comparisons, various levels may be considered. The simplest approach only looks for atom locations. The question is, for example, to compare ligand points on two drugs, or ligand and site points in drug-receptor interactions. Such an approach uses geometrical transformations for superimposing wire-frame representations of the molecular frameworks to be compared. Other applications address the comparison of molecular volumes or surfaces, and use Boolean operators to identify the common parts or the unshared volumes between two or more molecules. Still more complex is the investigation of similarity for electronic properties (MEP, for instance) encoded on the molecular surface or even not directly supported by the molecular framework (location of atomic orbitals, zones of high electrostatic potential, etc.).

Finally, also relevant to this type of problem is the estimation of the *complementarity* between a drug and its receptor.

What similarity?

The various aspects the molecular similiarity concept may address can be illustrated by the following examples. Let us first consider the pair of analgesics, R4238 and R6372, of largely different efficiency (Figure 11.1). Although the structural formulae are quite similar (they differ only by a

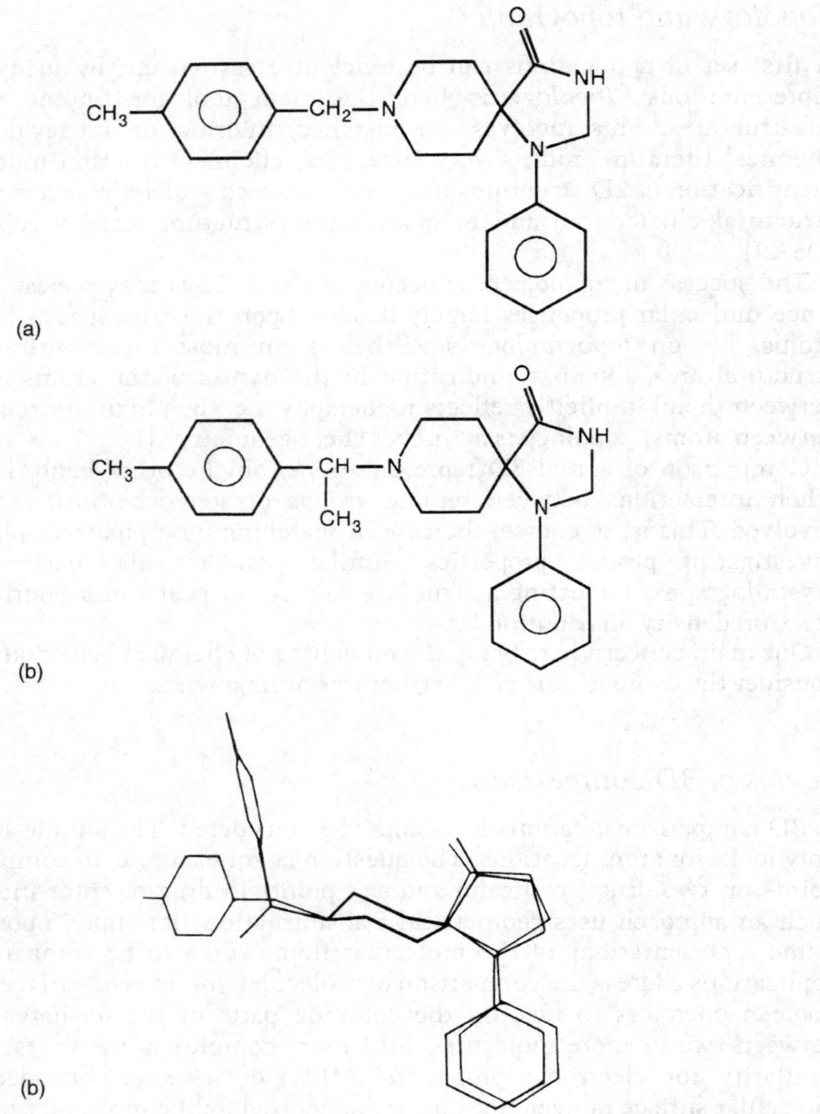

(a)

(b)

(b)

Figure 11.1 Analgesics R4238 (a) and R6372 (b) which have activities of 10 and 180 respectively, with reference to morphine (activity of 1). In (c) the two molecules are superimposed to illustrate their similarity.

methyl group), their geometries are definitely different; the methyl group inducing in R6372 a tilt of the phenyl ring [3, 4].

Conversely, the pteridine rings of dihydrofolic acid and methotrexate look very similar (Figure 11.2). However, there is crystallographic evidence that they bind differently to dihydrofolate reductase: the bound conformation of methotrexate corresponding to a 180° flip of the ring. In such a position, the distributions of the MEP of the two molecules now appear very similar [5–7].

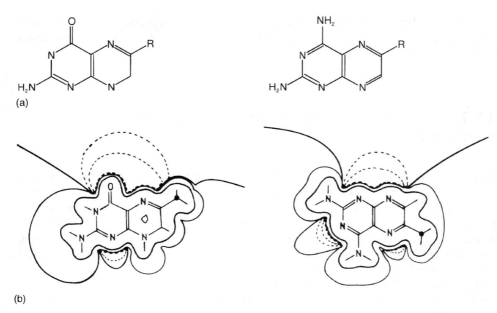

(a)

(b)

Figure 11.2   Model fragments of dihydrofolic acid and methotrexate. MEP pattern for
R = CH$_3$. Dotted lines correspond to negative MEP values regions (attractive for an
electrophile) (from Gund *et al.* with permission [7]).

Computer graphics, with its interactive capabilities, soon emerged as an
attractive tool for displaying molecular shapes, or superimposing active
molecules to detect their common parts. Thanks to the power of analysis of
the human eye, such a visual inspection is very efficient. However, it may
become cumbersome when a large set of molecules has to be examined [8],
and automated detection allowing efficient treatments without any *a priori*
hypothesis is desirable. This relies on the definition of quantitative
similarity indices for evaluating the degree of overlap between given
molecular shapes and the detection of common patterns. A non-visual and (if
possible) quantitative estimation of how one molecule is similar to another
is also highly desirable in view of QSAR applications: for instance, for
determining the best way in which to superimpose two molecules, or for
interpreting how a biological activity common to a set of compounds may be
related to the structural features they share. Such automated approaches are
particularly useful if the molecules to be compared are not closely related. In
such cases, it may be difficult to decide how to superimpose two ligands
with no obvious similar geometry so as to find common binding interactions
to the same receptor pocket. Quantitative similarity indices are also highly
desirable when screening large 3D databases in the search for leads
containing given structural features. Ranking the hits according to their
similarity with a known drug (or a fragment of it) allows one to select those
candidates that are more likely to be good starting points for the design of
new drugs.

Before exploring the more complex problem of comparing structural or property shapes, we first turn to geometrical comparisons, that is comparison of the atom frameworks.

11.1 GEOMETRICAL COMPARISONS: MOLECULAR SUPERIMPOSITION

Comparisons of the atom frameworks in fact encompass different activities. Frequently, there is some evidence about the atoms that correspond in the two structures under comparison. Besides the more naive superimposition of two molecules, where the anchoring points are fixed, the question is mainly one of finding the best *geometrical transformation* in order to superimpose them together as closely as possible and evaluate their *degree of similarity* (in other words, to evaluate how good the fit is).

In more complex cases, such as the comparison of "non-obviously similar" molecules, it can first be necessary to find the *best atom-atom correspondences*.

Also of relevance here, two related problems deserve special attention:

- either searching for the existence of a given fragment (substructure) within a set of compounds: *common substructure search* (CSS);
- or identifying the greatest common substructure: *maximal common substructure search* (MCSS).

Such questions arise, for instance, when screening a 3D database, searching for molecules containing a given pharmacophoric pattern, or conversely, deriving a common atom arrangement (a putative pharmacophore) shared by a set of active molecules.

For such operations, computer programs allow detection of the presence of common patterns, and the evaluation of similarity indices by least squares fitting techniques and interactive computer graphics [9–11].

11.1.1 Crude approach

When the anchoring motif is known, a crude superimposition of two structures can be easily carried out. Given two triplets of corresponding atoms (*a, b, c*) in molecule A and (*r, s, t*) in molecule B, the following steps can be performed (Figure 11.3):

- bring atom *r* on to atom *a* by translation: *the second triplet now becomes a, s′, t′*;
- superimpose planes *a, b, c* and *a, s′, t′*: this superimposes the normals *n, n′* to these planes by rotation along an axis perpendicular to them (intersection of the two planes). The unit vectors along the normal are:

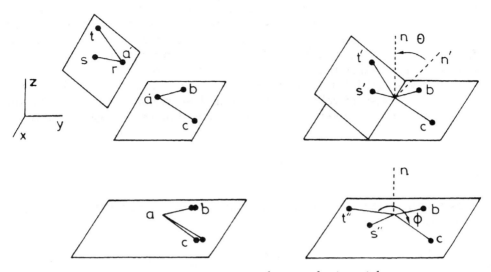

Figure 11.3 Superimposition of two anchoring triples.

$$\mathbf{n} = (ab \times ac)/(ab^2 + ac^2)^{0.5}$$
$$\mathbf{n}' = (as' \times at')/(as'^2 + at'^2)^{0.5}$$

The axis of rotation is given by $\mathbf{n} \times \mathbf{n}'$, and the angle of rotation by:

$$\cos \theta = (\mathbf{n} \cdot \mathbf{n}')/(\mathbf{n}^2 + \mathbf{n}'^2)^{0.5}$$
$$\sin \theta = (\mathbf{n} \times \mathbf{n}')/(\mathbf{n}^2 + \mathbf{n}'^2)^{0.5}$$

Triple a, s', t' now becomes a, s", t";
- superimpose directions *as"* and *ab* by rotation along $\mathbf{n}$. The rotation angle Φ is given by:

$$\cos \Phi = (as'' \cdot ab)/(as''^2 + ab^2)^{0.5}$$
$$\sin \Phi = (as'' \times ab)/(as''^2 + ab^2)^{0.5}$$

In simple examples, this approach seems attractive since it easily allows for visualizing gross comparisons. However, it requires some initial assumptions as to anchoring points for superimposition, and in the preceding form, it privileges first point *a* (*r*), then *b* (*s*) over *c* (*t*), so that one cannot be certain that the best fit is obtained; more sophisticated methods (implying no *a priori* hypothesis) are thus needed. (See Plate XVIII a.)

11.1.2 Finding the best transformation

Comparing the geometrical features of two molecules is often carried out by displaying the atom frameworks superimposed as closely as possible. In contrast to the problems encountered in "common substructure" searching, where the aim is to identify the corresponding atom pairs, here one knows what atom of the first molecule is associated with each atom of the second molecule. Therefore, the problem is to find the rigid body translation and rotation that give the best match between the two structures, that is which minimize the rms distance between corresponding atoms [12–18].

In other words [13–15]: let $\mathbf{a}_n$, $\mathbf{b}_n$, ($n = 1 \ldots N$) be the position vectors of two sets of atoms (respectively molecules A and B). We wish to determine the translation $\mathbf{t}$ and rotation $\mathbf{r}$ that convert the coordinates b_{in} ($i = 1, 2, 3$) into:

$$b'_{in} = \sum_{j}^{3} r_{ij} b_{jn} + t_i$$

so as to minimize the residual (rms difference value in fitting the superimposed atoms):

$$E = \frac{1}{2} \sum_{i,n} w_n (b'_{in} - a_{in})^2$$

w_n being the weight assigned to atom n.

It is easily shown that the best translation $\mathbf{t}$ is that which superimposes the centroids of both molecules. So, we will subsequently consider only rotations between two shapes with a common centroid (supposed to be at the origin of the coordinate system).

The determination of the rotation matrix has been approached in various ways. The first attempts, applying rotation angle procedures, have been superseded by more efficient methods using matrix calculations, Lagrange multipliers or quaternions optimizations. We only give brief guidelines here about the principle of some of these methods (for details the reader is referred to the original papers).

By expanding the preceding equation, we get:

$$E = (A + B) + v$$

where:

$$A = \frac{1}{2} \sum_n w_n a^2_{in} \qquad B = \frac{1}{2} \sum_n w_n b'^2_{in}$$

Since B does not vary, when molecule B rotates around its centroid, these quantities can be evaluated once from the initial data, and the problem becomes the minimizing of the reduced objective function -v:

$$v = \sum_{i,n} w_n b'_{in} a_{in}$$

Matrix treatment

In a first group of methods [13–15], the rotation matrix $\mathbf{R}$ can be derived from two 3×3 matrices $\mathbf{U}$, $\mathbf{V}$ with:

$$V = R\,U$$

as defined as:

$$u_{ij} = \sum_n w_n b_{in} a_{jn} \quad \text{(from initial coordinates)}$$

$$v_{ij} = \sum_n w_n b'_{in} a_{jn}$$

An iterative solution has been proposed [13, 15] by successive rotations. Consider the rotation $\mathbf{r}$ followed by a small additional rotation θ about the direction $\mathbf{l}$. Resolving the vector $\mathbf{b}'$ for any atom before the notation into a component parallel to $\mathbf{l}$: $(\mathbf{l}.\mathbf{b}')\mathbf{l}$ and a perpendicular one $-\mathbf{l}\mathbf{x}(\mathbf{l} \times \mathbf{b}')$, it can be seen that, to the second order, the vector $\mathbf{b}'$ becomes (after rotation θ):

$$\mathbf{b}' + (\Theta \times \mathbf{b}') + 1/2\,\Theta \times (\Theta \times \mathbf{b}') \quad \text{where } \Theta = \mathbf{l}\theta$$

The corresponding change of the reduced objective function is:

$$\delta E = -\mathbf{g}\Theta + 1/2\Theta^t\,T\Theta$$

where vector $\mathbf{g} = \Sigma_n\, w_n\, (\mathbf{b}'_n \times \mathbf{a}_n)$ is the couple that will be produced by a force $w_n\,(\mathbf{a}_n - \mathbf{b}'_n)$ acting at each atom $\mathbf{b}'_n$ as it was attracted by its guide point $\mathbf{a}_n$, and $\Theta =$ column vector and $\Theta^t =$ row vector (transpose).

In terms of $\mathbf{U}$ and $\mathbf{V}$ matrices, the couple $\mathbf{g}$ is given by the antisymmetric part of $\mathbf{V}$ (for instance, $g_1 = V_{23} - V_{32} \ldots$).

$\mathbf{T}$ is the matrix of the second derivatives, with elements T_{ij} given by:

$$T_{ij} = v\,\delta_{ij} - 1/2(V_{ij} + V_{ji})$$
$$\text{with } v = V_{11} + V_{22} + V_{33}$$

In other words, the change in the reduced objective function can be written:

$$\delta E = -G\,\sin\theta + H(1 - \cos\theta)$$

which becomes for small θ

$$\delta E = -G\theta + (1/2)H\theta^2$$

$G = \mathbf{g}.\mathbf{l}$ is the downhill gradient about direction $\mathbf{l}$, and:

$$H = \sum_{ij} l_i T_{ij} l_j \quad \text{is the second derivative.}$$

Searching for the rotation matrix $\mathbf{R}$ (finding the maximum of v) now turns to make G disappear and H be positive for any further direction of motion. This can be carried out by the usual minimization procedures (for instance, conjugate gradient minimization). The minimum along direction **g** is directly obtained by:

$$\sin \theta = G / (G^2 + H^2)^{\frac{1}{2}}$$
$$\cos \theta = H / (G^2 + H^2)^{\frac{1}{2}}$$

In practice, a set of successive rotations is carried out about axes parallel to the couples **g** at the start of each cycle: on typical examples, about 5–8 iterations are needed to obtain the best match (down to a last rotation about 10^{-10} rad.).

Analytical solutions – with diverse variants – have also been proposed, giving the rotation matrix **R** directly.

From the preceding matrices **U**, **V** (with **V = R U**), the solution sought can be written **V = (U^{t}U)$^{1/2}$**, where **U^t** represents the transpose of matrix **U**. Once **V** has been found, rotation matrix **R** is calculated as:

$$\mathbf{R = (U^t U)^{\frac{1}{2}} U^{-1}}$$

McLachlan [14] used a 6×6 partitioned matrix built from **U** and **U^t**, whereas Kabash [12, 12a] introduces Lagrange multipliers to find a direct solution. Another approach uses the two matrices[1]:

$$\mathbf{M_{12} = A^t B} \quad \text{and} \quad \mathbf{M_{21} = B^t A}$$

where **A** and **B** represent the $(n \times 3)$ matrices of the atom coordinates in molecules A and B, and consider matrices $\mathbf{M_{21}M_{12}}$ and $\mathbf{M_{12}M_{21}}$. These two matrices have the same set of eigenvalues. Their diagonalization leads respectively to matrices $\mathbf{D_{21}}$ and $\mathbf{D_{12}}$ of normalized eigenvectors, from which the best orthogonal transformation **R** for superimposition is derived by:

$$\mathbf{R = D_{21}D^t_{12}}$$

Rotation angle method

An efficient algorithm was recently proposed by Zong Jie Liu and van Rapenbusch [16, 16a]. The principle is to express the objective function to minimize in terms of three rotation angles, and simplify its expression by first adjusting the initial positions of the two sets of coordinates.

The reduced objective function v is expressed as:

$$v = \sum \alpha_{ij} \, r_{ji} (\theta, \Phi, \Omega)$$

where r_{ji} (θ, Φ, Ω) defines the rotation sought, and the α_{ij} are constants depending upon the initial positions of A and B. (The $\alpha_{ij} = \sum_{i}^{n} a_{in} \, b_{jn}$ are homologous to the u_{ij} elements with $w_n \equiv 1$.)

The position of B can be adjusted by a preliminary rotation that cancels three

[1]M. Petritjean.

(at least) of the nine α_{ij} coefficients, and consequently reduces v to a simpler expression that can be easily minimized by least squares treatment.

Besides speed, it can be noted that this algorithm is able to find either (or both) the minimum value of the objective function and/or the rotation matrix, without remaining trapped in a local minimum.

Quaternions

Quaternion-based methods have also been proposed [18]: invented by W. Hamilton in 1843, quaternions can be thought of as quadruplets of real numbers, corresponding to a scalar ps and a vector $\mathbf{p}$. They were soon used to parameterize rotations, and after a long spell of oblivion, proved very useful for computer processes: computers work faster with algebraic rather than trigonometric functions, and for quaternions the derivative space is continuous [18].

Given a pair of atom vectors, $\mathbf{b}$ and $\mathbf{b}'$, derived from $\mathbf{b}$ by the rotation described by the quaternion Q, so that:

$$\mathbf{b}' = Q^{-1}\mathbf{b}Q \quad \text{or} \quad Q\mathbf{b}' = \mathbf{b}Q$$

one obtains a set of $3N$ equations (for an N atom comparison) that can be solved by a least squares procedure, leading to the unknowns: the rotation angle θ, and the director cosines l, m, n of the rotation axis [17].

Remark: elements of quaternion algebra

For a rotation θ around an axis (unit vector $\mathbf{d}$, director cosines l, m, n), corresponding vectors are related by:

$$\mathbf{b}' = Q^{-1}\mathbf{b}Q$$

where:
$Q = (\cos \theta/2 + \sin \theta/2)\,\mathbf{d}$ can be expanded along three orthogonal unit vectors $(\mathbf{i}, \mathbf{j}, \mathbf{k})$ as:

$$Q = \cos \theta/2 + l\sin \theta/2\,\mathbf{i} + m \sin \theta/2\,\mathbf{j} + n \sin \theta/2\,\mathbf{k}$$
$$Q^{-1} = \cos \theta/2 - l \sin \theta/2\,\mathbf{i} - m \sin \theta/2\,\mathbf{j} - n \sin \theta/2\,\mathbf{k}$$

For each pair of corresponding atoms (x, y, z), (x', y', z') in A and B, the rotation associating them is defined by $(\mathbf{b}' - \mathbf{b}) = \tan \theta/2\,\mathbf{d} \times (\mathbf{b} + \mathbf{b}')$, or in Cartesian coordinates:

$$(x - x')\mathbf{i} + (y - y')\mathbf{j} + (z - z')\mathbf{k} = \tan \theta/2 \begin{vmatrix} \mathbf{i} & \mathbf{j} & \mathbf{k} \\ l & m & n \\ x + x' & y + y' & z + z' \end{vmatrix}$$

This leads, for N pairs of atoms, to a system of $3N$ equations (with $t = \tan \theta/2$):

$$mt(z + z') - nt(y + y') = x' - x$$
$$-lt(z + z') + nt(x + x') = y' - y$$
$$lt(y + y') - mt(x + x') = z' - z$$

This system can be solved by the usual least squares procedures ($AX = H \rightarrow X = (A^tA)^{-1} A^tH$) giving the unknown lt, mt, nt. Adding the squared values determines t^2. Then l, m, n are easily obtained.

Although such comparisons are generally carried out between a pair of molecules, more complex cases may occur. This happens, for example, when searching for a pharmacophoric pattern within a series of active molecules, the objective here being to superimpose in the best way a set of related molecules to extract common atoms responsible for common building interactions. Obviously, the comparisons can be carried out by choosing a reference structure and mapping on to it the other in a series of pairwise superimpositions. However, this process does not guarantee an optimal fit for any pair of molecules differing from the reference structure. To avoid this stepwise process, Kearsley [18] proposed using quaternions to directly find the optimal superimposition within the family. The residual to minimize here:

$$\sum_{i=1}^{n-1}\sum_{j=i+1}^{n}\sum_{k}^{ns}(Q^i\,\mathbf{x}^i{}_k - Q^j\,\mathbf{x}^j{}_k)^2$$

where Q = rotation operator and $\mathbf{x}$ = atom position.

The first two sums are over the unique pairs of structures (with indices i and j), the third is over the number (ns) of atom-atom associations between the structures.

Flexible fitting

The above approaches are limited to motions of rigid bodies, and consider only geometric criteria to characterize the fit. When some conformational flexibility is allowed for the inspected molecule to map onto a rigid target, one may imagine that the fit obtained using the more stable (favoured) conformer may be improved with a slightly distorted conformation, at the price of a small energetic expense. This point was recently considered by Venkatachalam et al. [19], who proposed introducing, in the objective function, a part related to the energetic cost:

objective function = RMS + a (Energy)
(molecule A, molecule B)

where "Energy" represents the energetic expense for distorting the flexible system. Setting the adjustable coefficient a to zero reduces the problem to the preceding rigid body motions. Conversely, when a is large, it turns to an energy minimization for the flexible molecule.

Atom correspondence between (dissimilar) molecules

When the structures to compare are not close neighbours, the correspondences between atoms are not always obvious. The algorithm described by Danziger and Dean addresses this problem, and proposes the best solution by a systematic search through all possible pairs associating one atom in molecule A and one in B [20]. Candidate solutions are generated by a recursive-descent tree-search. To avoid a combinatorial explosion, partial candidates which cannot lead to a good solution are discarded, and the branches stemming from them are suppressed.

A first obvious way of pruning was to take into account the nature of the atoms. In the example treated, only hydrogen bonding atoms are considered and separated into H-donor, H-acceptor, or both.

The second pruning criterion relies on the rms value Δd of the difference distance matrix (DDM) comparing interatomic distances in molecule A to those in molecule B:

$$\Delta d^2 = 2\left(\sum_{i=1}^{n-1}\sum_{j=i+1}^{n} DDM^2{}_{ij}\right)/n(n-1)$$

where n = number of correspondences and DDM is given by:

$$DDM_{ij} = \left| DM^A{}_{ij} - DM^B{}_{ij} \right|$$

where $DM^A{}_{ij}$ and $DM^B{}_{ij}$ are the elements of the distance matrix for molecule A and B (distance between atoms i, j in A, and between corresponding atoms in B).

Although non-differentiating between mirror images, Δd is here generally preferred to the rms difference value of the corresponding atom positions, (when fitting the superimposed molecules), this criterion being much slower to evaluate. The algorithm also allows for the selection of the number of correspondences (a preset number of null correspondences is permitted) and the tolerance threshold accepted for the fit. Interestingly, a caveat is put on "blind searching" of atom correspondence: a very good superimposition of some atoms can be attained which is of no sense as to a possible way of common binding if steric fitting is poor (Figure 11.4).

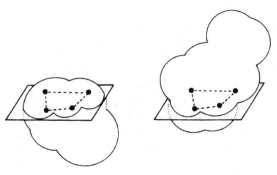

Figure 11.4 A good four-centre correspondence is obtained but steric fit is low. The centroids of the two molecules are not on the same side of the least squares plane through the corresponding atoms.

A similar problem to retrieving corresponding atoms also emerges in the protein field. When two proteins have a similar shape in part of their structure, they can be superimposed by choosing certain sets of atoms as a guide-point for the fit. However, this choice is not necessarily the best one, and a more systematic approach is needed for comparing large scale organization [14]. In the approach of McLachlan, a comparison is carried out between zones of about 50 residues. Once the similarity index ("fitting distance") has been calculated, the problem is updated, deleting the first atom and adding atom $N + 1$. Speed is therefore essential, since comparisons may involve as many as 10^6 least squares fits. Results can be displayed in a plot of distances between α carbons of molecule A and the rotated molecule B for matched lengths of chain ... a representation somewhat similar to an Ooi contact map (but referring here to two distinct proteins) [21].

This method was applied to define chymotrypsin's structure [14]: this enzyme appears as "built from two similar domains, each of which is a hydrogen-bonded barrel". Both domains have highly symmetrical structures, and shape comparison methods strongly suggest that the barrels have evolved by gene duplication: a closely linked dimer of two intertwined half-domains becoming united into one domain. The enzyme then evolved by a second duplication with a second dimer [14].

11.2 COMMON SUBSTRUCTURE SEARCHES

Identifying 3D features common to a set of molecules addresses many research areas: chemical documentation (retrieval of substructures from large files of structures), database exploitation in X-ray crystallography (or in spectroscopy, fitting a structural fragment to spectral signals), pharmacology (in QSAR, or search for a pharacophore). Various approaches have been proposed for automatic identification of a common pattern within a set of molecules, either searching for a given common substructure (CSS) or looking for the maximal common substructure (MCSS).

Automated determination of CSS or MCSS avoids *a priori* hypotheses as to some common atoms. Indeed, these methods specifically generate such hypotheses. Furthermore, the spatial location of atoms is the important parameter with no acknowledgement of topology (i.e. independently of whether the atoms comprising the substructures are connected). The question is similar to the problem of atom correspondences addressed by Dean *et al.* [20], but here one deals with the *recognition of a 3D motif* (given some tolerance about interatomic distances) but with no acknowledgement of any similarity index quantifying the "goodness" of the geometrical fit.

Whereas, in geometrical transformations, Cartesian coordinates are used for interactive molecular superimposition, in automated substructural searches, reasoning on pairwise interatomic distances is generally preferred. This clearly avoids problems due to the orientation of the molecules in respect to the reference axes. It also allows for an easier treatment of tolerance thresholds to cope with small geometrical distortions, and processing is made easier. Note, however, that Lesk's algorithm works in (x, y, z) coordinate space, although this appears to be very time consuming (see below).

We now briefly present some of the most commonly used approaches to CSS search. Note that the examples given below consider only atom positions, without considering their nature. This constitutes the worst case, since in practice examination of atom types (H-bond donor or acceptor character for example) usually reduces the number of possible matches.

11.2.1 Crandell and Smith's algorithm

Crandell and Smith's algorithm [22] uses an iterative process for finding either the maximal common 3D substructure within a set of structures or all the common substructures of a given size.

In the growing step, the algorithm starts with an N-atom common substructure and expands it to identify all common fragments of size $(N + 1)$ atoms. At each step, one checks that any newly generated substructure is common to the whole set (otherwise it is discarded for further phases).

The basic steps of the algorithm can therefore be summarized as:

- creating the initial distance table;
- setting $N = 1$ (or any given common substructure);
- growing substructures of $(N + 1)$ atoms, expanding the N-atom substructures emanating from the preceding iteration;
- obtaining a canonical description of these substructures;
- comparing these, and identifying those which are still common to the whole set of molecules (if none, stop);
- amending the distance table and going back to the second step.

This algorithm deserves some comment:

- To accommodate small geometrical variations, true interatomic distances are gathered in clusters (with a usual tolerance value of about 0.09 Å), and the process is carried out on (integer) codes identifying these clusters for each iteration step. Amending the distance table encodes −1 to all pairwise atom distances, which cannot be elected as common to all molecules in the following steps.
- When various substructures can be generated in the growing step (adding several different atoms), the growth is processed through a breadth-first tree search, since after each step, all the substructures generated are examined and compared with those originating from the other molecules. Only substructures that all molecules have in common survive.

To make these comparisons easier, each substructure generated is given a unique (canonical) name: for each atom pair a triplet is formed, consisting of the two atom types and the relevant entry in the distance table.

11.2.2 Clique-detection algorithm

This approach relies on a topological perception of the molecular structure, and uses graph theoretical procedures to identify common features [23]: a 3D molecule is regarded as a graph, where the nodes and edges are labelled according to atom types and interatomic distances (Figure 11.5).

Given the graphs associated with a chemical structure A (atoms a, b . . .) and a pattern P (atoms p, q . . .), a correspondence graph C is built by the following process:

- Creating the set of all pairs of nodes (one from each of the two graphs A and P) of the same type.
- Building the correspondence graph C. Its nodes are the pairs from the preceding step. Two nodes (A_iP_x), (A_jP_y) are connected in C, if the edges A_iA_j and P_xP_y have similar values (similar distances).

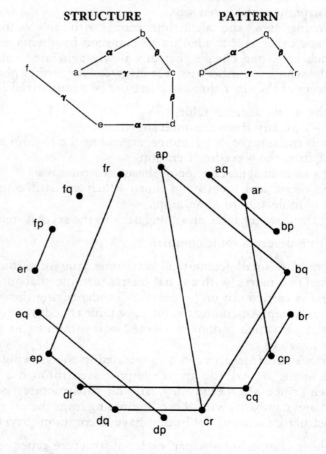

Figure 11.5 Clique-detection. Detection of the clique (ap, bq, cr) in the correspondence graph built from the 6 × 3 possible pairs (structure atom-pattern atom).

The maximum common subgraphs correspond to the cliques of the correspondence graph (clique = a subgraph in which every node is connected to every other node, and which is not contained in any larger subgraph with this property).

Among the various algorithms tested, the most efficient implementation was obtained with the widely used method of Bron and Kerbosch [24]. Applications can be found elsewhere [25, 26].

The clique-detection algorithm allows for the identification of a common substructure between two molecules, and can be used for retrieving substructures consisting of many atoms. However, in its initial implementation, extension to a set of structures, although theoretically feasible, greatly increases the size of the correspondence graph, and the clique-finding procedure appears quite impracticable if more than two molecules of a realistic size are to be processed.

Brint and Willett [27] suggested some enhancements to both the Crandell and Smith and clique-finding algorithms: incorporating in the clique-finding algorithm the concept of "seed pattern" (a starting point for the growth of the Crandell and Smith method) that must be present in the clique, or other user-defined structural constraints, have been proposed as an attractive solution to increasing efficiency [27].

11.2.3 Lesk's algorithm

Lesk's algorithm [2] favours a more geometrical approach, directed rather towards the search for a pre-defined structural moiety (Figure 11.6).

The pattern and structure are each defined by a list of atomic coordinates and identifiers which specify the type of each atom. For an atom in the structure proposed as a candidate to be identified with one centre of the fragment, we check whether its environment in the structure is the same as in the fragment (proper type and distances). In this way, we can determine sets of atoms in the structure candidate for congruence (candidate to superimposition with the fragment through rotation/translation operations). Structure atoms which cannot match to any pattern atom are removed and the distance table of the structure is updated. The process is repeated until no more atoms can be discarded.

Once a set of atoms of the structure has been elected, one has to test these points for congruence with the pattern (within some tolerance).

From the examples given by Lesk, it appears that the required CPU time varies as the fifth power of the number of atoms in the structure (if only the atom type is considered). The score would presumably be better with more discriminating chromatisms.

However this type of method (exhaustive generation, then filtering) is (inherently) subject to problems of combinatorial explosion. Geometrical transformations needed for testing congruence are also a burden on computer time.

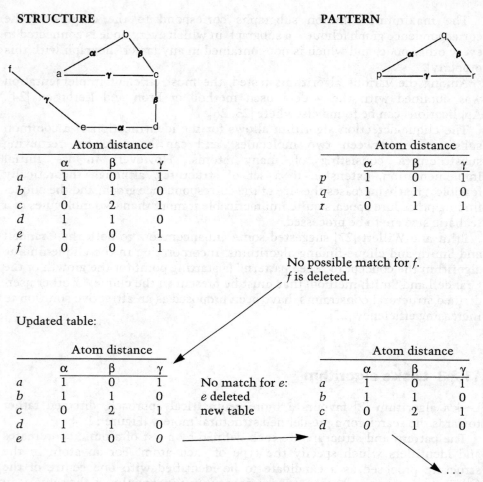

STRUCTURE   PATTERN

Atom distance   Atom distance

	α	β	γ
a	1	0	1
b	1	1	0
c	0	2	1
d	1	1	0
e	1	0	1
f	0	0	1

	α	β	γ
p	1	0	1
q	1	1	0
r	0	1	1

No possible match for *f*.
f is deleted.

Updated table:

Atom distance   Atom distance

	α	β	γ
a	1	0	1
b	1	1	0
c	0	2	1
d	1	1	0
e	1	0	0

No match for *e*:
e deleted
new table

	α	β	γ
a	1	0	1
b	1	1	0
c	0	2	1
d	0	1	0

. . .

Figure 11.6   Lesk's algorithm.

11.2.4   Set-reduction algorithm

This algorithm proceeds by examination of ordered atom pairs [26] and involves the successive elimination of structure atoms thanks to an analysis of their neighbourhood and connectivity.

First, a list of all atom pairs in the pattern is created. To these pairs are associated all the pairs from the structure so that $D(A_i, A_j) = D(P_x, P_y)$, within a tolerance value.

The presence of $A_i A_j$ in the list $P_x P_y$ implies that A_i and A_j are respectively corresponding to P_x and P_y. To confirm this hypothesis, one tests all the lists associated with pairs of P and containing P_x. The only A pairs to be retained are those associating A_i to P_x at least once in each list. This step is repeated until no more elimination can be carried out. As the algorithm works by iteration, no combination problems occur (Figure 11.7).

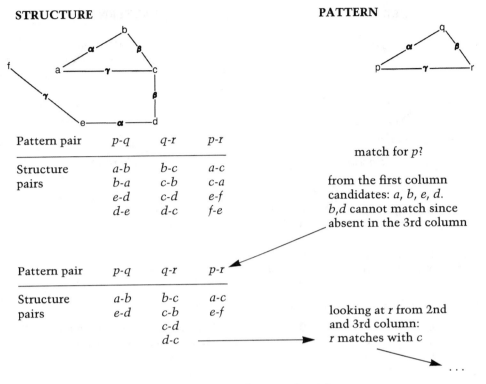

Pattern pair	p-q	q-r	p-r	
				match for p?
Structure	a-b	b-c	a-c	from the first column
pairs	b-a	c-b	c-a	candidates: a, b, e, d.
	e-d	c-d	e-f	b,d cannot match since
	d-e	d-c	f-e	absent in the 3rd column

Pattern pair	p-q	q-r	p-r	
Structure	a-b	b-c	a-c	
pairs	e-d	c-b	e-f	looking at r from 2nd
		c-d		and 3rd column:
		d-c		r matches with c

. . .

Figure 11.7   Set-reduction algorithm.

11.2.5 Ullman's algorithm

The subgraph isomorphism method of Ullman [28] uses distance matrices. Given the two distance matrices, for respectively the molecule A and the pattern P, one builds an association matrix. For two nodes, A_i and P_x, $M_{ix} = 1$ if and only if, for any P_y neighbour to P_x, it exists A_j neighbour to A_i so that:

$$d(A_i, A_j) = d(P_x, P_y)$$

The association matrix M will be modified in successive steps according to the rule:

given P_x and A_i with $M_{ix} = 1$
for any P_z of P so that $d(P_x, P_z) = D$
it exists A_k of A so that
$M_{Ak}, P_z * d(A_k, A_i) = D$
otherwise $M_{ix} = 0$ for the next iteration.

In other words, if A_i has to match P_x, any neighbour to P_x must have an equivalent in A (Figure 11.8).

The final matrix is obtained when each row contains a single 1, and each column contains no more than a single 1.

STRUCTURE   **PATTERN**

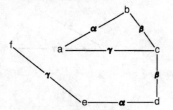   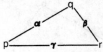

Distance matrices:

	Structure					
	e	d	b	a	c	f
e	0	α	X	X	X	γ
d	α	0	X	X	β	X
b	X	X	0	α	β	X
a	X	X	α	0	γ	X
c	X	β	β	γ	0	X
f	γ	X	X	X	X	0

	Pattern		
	p	q	r
p	0	α	γ
q	α	0	β
r	γ	β	0

Logical matrix of association M:

1	0	0	1	0	0
0	1	1	0	0	0
0	0	0	0	1	0

The 1st line means that node p of the pattern can match nodes e or a of the structure.

Iterations:

i	j	x	D(i, x)	y	M(x, y)	M(i, j)
1	1	2	α	2	1	
		3	γ	6	0	M(1,1) = 0
1	4	2	α	3	1	
		3	γ	5	1	M(1, 4) No change
2	2	1	α	1	0	
		3	β	5	1	M(2, 2) = 0
2	3	1	α	4	1	
		3	β	5	1	M(2, 3) No change
3	5	1	γ	4	1	
		2	β	2, 3	0, 1	M(3, 5) No change

In each step, i, j are first assigned to the rows and column numbers of non-zero M elements. Next x are taken from the pattern matrix before y is found from the structure.

Final matrix:

p	0	0	0	1	0	0
q	0	0	1	0	0	0
r	0	0	0	0	1	0

Figure 11.8 Ullman's algorithm.

This procedure, robust and not suffering from combinatorial problems, seems very efficient to retrieve a given substructure, but was not proposed to search for the maximal common substructure.

In fact, clique-detection, Ullman's and the set reduction algorithms can be viewed as very similar approaches: a link between nodes in the graph of correspondence is equivalent to the association of an atom pair of P with a pair of molecule A in the set reduction. The clique describing the substructure identified is no more than the final table of pairs in the set reduction (Figure 11.9).

Some comparisons about the applicability and efficiency of these varied approaches have been carried out. Crandell and Smith's approach is very demanding both on CPU time and storage requirements. However, it seems faster than the clique-finding algorithm for searching for a small MCS within large molecules.

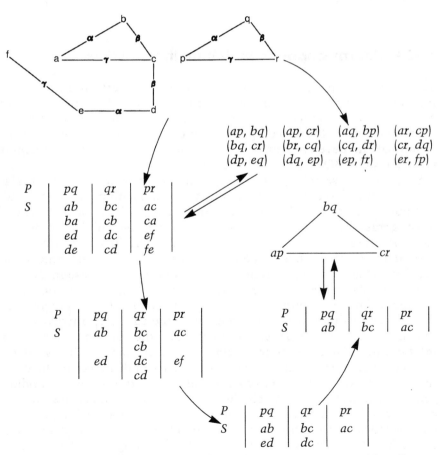

Figure 11.9 Analogies between clique-detection, set-reduction and Ullman's algorithms.

Systematic tests were carried out, searching for the presence of fragments of 3–15 atoms in diverse structures containing from 14–60 non-hydrogen atoms, or for the presence of a given pharmacophoric motif created from a subset of the Cambridge database. According to Brint and Willett [26], "The Ullman's algorithm is by far the quickest and its advantage increases with an increase of the size of the pattern." It appears as an "algorithm of choice for pharmacophoric pattern search." On the other hand, Lesk's method and set-reduction are broadly of comparable efficiency. However, they were sometimes unable to reduce the number of structure atoms to combine. So, when processing large molecules (such as proteins) the clique-detection and Ullman methods are difficult, owing to the size of the distance table, and Lesk's algorithm seems to be the best for substructure searching in macromolecules when it does not suffer from combinational problems [26]. However, note that the clique-detection algorithm has recently been applied by Artymiuk *et al.* to identify patterns of α helices and β strands in 3D space that are common to a pair of proteins [29].

11.2.6 Pharmacophoric matching in large files

The identification of a given pharmacophoric pattern in large files has produced an explosion in interest over the last few years in view of the huge amount of data to process (a problem also encountered when searching for 2D substructures in the large spectroscopic or bibliographical databases), and prompted the development of various substructure searching techniques for the retrieval of 3D fragments [11, 26, 30, 31].

Substructure searching is usually implemented in a two-stage algorithm: first, an initial screen has to eliminate a large number of molecules that obviously cannot contain the query substructure. Then, only those molecules passing this filter are submitted to the complete subgraph-isomorphism search, the very time-consuming step of the search. The first screening is generally carried out via bit-strings encoding the more characteristic features (for details see, for instance the organization of the Cambridge database in Chapter 4).

For 3D databases, the screening criteria usually rely on interatomic distances. To limit computer time, Willett *et al.* [30, 32] proposed adding a third stage in this retrieval methodology: the first step involves a rapid bit-screen matching operation. It looks for the presence (or absence) of interatomic distances within certain preset ranges. These have been previously defined by analysing the occurrence of distances in the database. In the second step, all interatomic distances in the selected candidates are generated to ascertain that all the distances in the query are present in the candidates. The final search, using the set-reduction technique, determines whether these interatomic distances are in the correct geometrical orientation to each other. The examples quoted, searching for some well known pharmacophoric patterns within a subset of the Cambridge database (about 12,000 compounds) indicate that, typically, only about 5% of the file needs the detailed geometrical search (third step). This corresponds to a large decrease in the computational task.

The efficiency of the screening stage, however is lowered when only a few distances are known for the query. To overcome this drawback, Willett *et al.* proposed using "smoothed bounded distances", a technique derived from distance geometry. For the query, the usual distance matrix is built (see Chapter 7) and smoothed (removing inconsistencies) by the triangle inequality. This allows for the determination of possible ranges for some missing distances in the query. Selected tests show that the process may considerably increase the efficiency of search.

Willett *et al.* [33–35] stressed the point that, although the notion of similarity is inherently a little subjective, a useful index would be one in which "the calculated similarities in structure mirror the similarities in activity." They also devoted special interest to the effectiveness and computational requirements when screening large databases. Their approach relies on distance information, the target and the candidate molecule being represented by their interatomic distance matrices. This type of search sets the problem of ranking candidates and therefore quantitatively estimating molecular similarity. The proposed process compares, in a first step, the environment of each atom in the first molecule to that of each atom in the second, to determine the resemblance between each possible pair of atoms. This defines an *atom-match matrix*, the elements of which encode the similarity between pairs of atoms. This matrix is thereafter used to establish what atoms are most alike. The sum of the so established atom-atom similarities gives an evaluation of the overall 3D similarity. Other comparisons were carried out using the frequency distributions of the interatomic distances, possibly separating carbons and heteroatoms, or the count of distances that are identical in the two structures, but they seem to be less effective. The search of the maximum common substructure (MCSS) would be the best method, but it is very demanding on computational requirements, so that atom mapping appears as the most cost-effective technique [33]. Similarity evaluation can be improved by taking into account some atom characteristics: in other words, mapping can only occur for atoms of the same "type". Hydrogen bonding ability and electronegativity, on the one hand, and charge and van der Waals radius on the other, were used to distinguish 18 different atom classes.

A comparison of this 3D search method with a 2D similarity search was performed on a database of ca 4500 structures. It leads to outputs very different from each other, and shows some complementarity of these two substructure searching systems. Some 2D candidates, though containing the same substructures, bear little overall similarity to the query. Conversely, 3D candidates can exhibit strong topographical similarities but look very different from the target in terms of topology (Figure 11.10).

The concept of fuzzy matching (only some of the query features need to be located in the answer), and the use of a similarity index to rank the retrieval, was extended to molecular properties, bearing in mind the exploitation of the Chemical Abstract Service 3D databases [36]. An exploratory approach, the

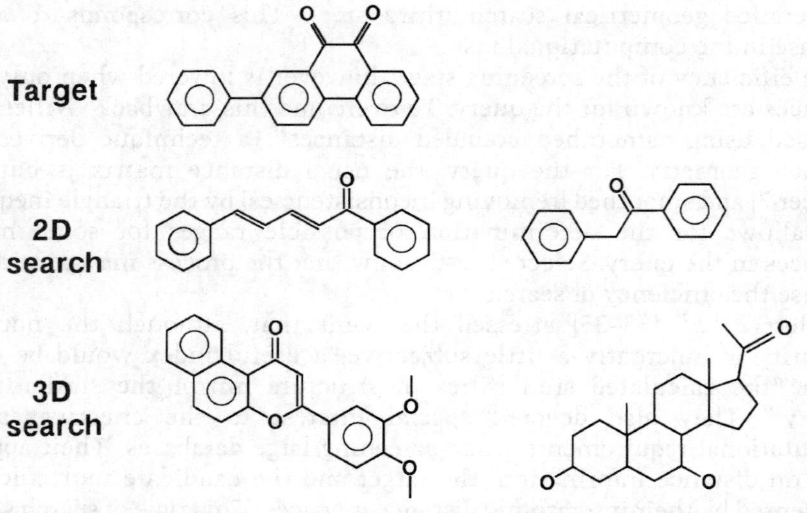

Figure 11.10 Differing first two top hits for the same target in 2D and 3D searches (from Pepperrell *et al.* [34]).

Substance Similarity Search Modeller (SSSM), works on a database comprising about 6000 substances, encoded over 3000 features: 161 molecular property features, 663 2D, 2230 2D/3D and 118 3D features (maximum figures). They comprise global or local molecular properties (heat of formation, ionization potential, partial atom charges, etc.), topological indices, path lengths between atoms, flexibility indices, and so on. The description is completed by atom triangles generated from the 3D coordinates (approximately 2500 per substance, amounting to about 15 million for the whole database). It is hoped that triangles should be better than atom pair distances in detecting the shape and size similarity as they contain more information.

The raw data are reduced (using binning) into "feature definitions" which are given numerical values, and subsequently used by the search system software: for example, atom triangle characteristics (interatomic distances, area, and perimeter counts) are specified within about 1800 features. The profiling of features is automatically performed by the system for the target. Scoring methods rely on the presence/absence criterion or the Tanimoto coefficient [33–35]. Searches can be performed on the entire molecules, or on fragments (substructures) of them. It is also possible on any combination of features: results indicate that fuzzy-match searching on molecular property features detects chemical or isosteric similarity, and that atom triangle features convey important information about shape and size similarity.

11.2.7 Trends and new developments: parallel processing, neural networks

Parallel computers have shown greatly increased speed capabilities which have proved very useful in the field of numerical computation. They also seem likely to offer attractive solutions in the screening of large databases for processing the large amount of non-numerical data stored.

In usual, sequential computing, operations are carried out one after the other, one at a time. It easy to imagine that for certain repetitive operations, for example updating an array, performing the same operation on each of its elements, a large amount of time could be saved if numerous processors were available, each performing the same operation on one element of the array all at the same time, that is, working in parallel. This is (quite naively presented) the principle of multiprocessor computing: the computer is organized with a network of processors. Each one is programmable, and executes its own program while communicating with others over the network.

Attempts at 2D substructure searching [37, 38] have shown that this problem implies inherent parallelism, and can be efficiently speeded up by sharing tasks within a network of processors.

Brint and Willett [39] studied the use of a transputer network for identifying the greatest common substructure for a pair of molecules. First, they use a software simulation, then an actual transputer implementation. The algorithm used was that proposed by Crandell and Smith, the basic modification being that most of the steps can be performed independently for each of the structures being compared (growing and naming substructures, amending the distance table, etc.). The only interaction between the separate processes of each molecule being the comparison phase, which must occur in due time. Encouraging results were obtained, but the authors stated that in practice, "a substantial increase in speed is achieved if, and only if, the MCS is large." Parallel implementation of the atom mapping method is described elsewhere [35].

Neural networks provide other capabilities of non-sequential operations and address learning, classification or combinatorial problems using a holistic process. Remember that in such networks, the status of a given node depends upon the status of all the preceding nodes to which it is connected, and from which it receives inputs. In turn, as output, the status of this node monitors the status of the following nodes connected to it. In the resolution of a given problem, nodes are randomly updated (from treatment of a learning set of data, or minimization of an objective function) until a stable solution is found (Figure 11.11).

A Hopfield's network (a fully interconnected network) has been used to solve the well known Travelling Salesman Problem: find in what order a set of towns can be joined with a minimal trip (and passing through each town only once). Such a problem of correspondence between a series of sites and their order of visit, minimizing the distance criterion, is not far from the problem addressed in finding atom correspondences between two structures

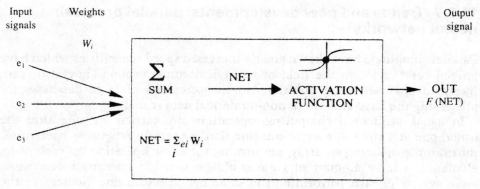

Figure 11.11   Artificial neuron.

with the rms value of the difference distance matrix as an objective function. Indeed, Hopfield's network was used successfully [40] to search for a given 3D (or 2D) pattern within a molecule, or to derive the maximal common substructure between two structures (Figure 11.12).

Using a Boltzmann's machine (a Hopfield's network with a simulated annealing algorithm) avoids becoming trapped in local minima (partial solutions), and so safely limits the number of trials necessary to get the best fit. An example is given in Plate XVIII b.

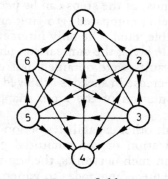

Figure 11.12   Hopfield's network.

11.3 SIMILARITY BETWEEN STRUCTURAL SHAPES

The preceding geometry comparisons work to a large extent on wire-frame representations, where atoms are considered as points. The rms difference between atom locations (once the structures are superimposed) or the difference distance matrix (DDM) offer an efficient way in which to detect to what degree two wire frame structures may be considered as similar. For more realistic comparisons in terms of chemical or biological activity, special interest was devoted to the detection and evaluation of the similarity between "structural shapes" (e.g. volumes) giving more insight into steric requirement and mainly "property shapes" displaying electronic characteristics such as electrostatic potential or electron distribution. Indeed, maximal electronic similarity is often more useful than structural similarity in monitoring molecular reactivity. Note that the MEP evaluation also includes some steric information, since highly positive (repulsive) values result in the vicinity of the atom locations.

New indices therefore had to be defined to quantify the similarity between two structural or electronic shapes. In a pioneering work, as early as 1980 (see below), Carbo *et al.* [41, 42] proposed a "computational" approach, relying on evaluation of the electron density overlap between the two superimposed molecules. However, for electronic properties not easily described through analytical expressions, as in the Carbo definition, or for volume comparisons, numerical (discrete) methods have largely been used. Basically, all these approaches employ the same fundamental procedures [43, 44]: the property is evaluated for two molecules (or the target and the current candidate if screening a database) on the nodes of a 3D lattice structuring the space. A similarity index is calculated. Since molecules, at the beginning, are generally positioned arbitrarily, their relative orientation must be optimized for a maximum of this index. Depending upon the property chosen, or the authors, differences appear in the calculation of the indices, space sampling or method of optimization [43, 44]. Finally, gnomonic projection has also been proposed for surface or property comparisons.

11.3.1 Similarity indices by overlap integral

In the model of Carbo [41, 42], the electron density overlap between the two superimposed structures (quantifying their similarity) is expressed by:

$$R_{AB} = \frac{\int \mu_A \mu_B \, dv}{[\int \mu_A{}^2 dv]^{0.5} [\int \mu_B{}^2 \, dv]^{0.5}}$$

where μ_A, μ_B = electron density of molecule A (resp. B) at the current point.

In this formula, the numerator evaluates the overlap of charge density while the denominator is a normalizing factor, so that the range of R_{AB} variations is 0 to 1 (for identical molecules). Within the same formalism, it is also possible to extend the comparison to other electronic properties such as MEP.

However, it was stated that this index compares *shapes* of electronic distributions rather than *magnitudes*. So setting $\mu_A = \mu_B$ or $\mu_A = n * \mu_B$ leads to the same unit value for R_{AB}. An alternative index was proposed by Hodgkin, which compares both shape and magnitude [45, 46]:

$$H_{AB} = \frac{2 \int \mu_A \mu_B \, dv}{[\int \mu_A{}^2 \, dv] + [\int \mu_B{}^2 \, dv]}$$

In their primitive expression, such definitions involve complex calculations for atomic integrals. Some approximations have been proposed using either semi-empirical CNDO calculations (with rather poor results, at least for electron density comparisons) [47], or Gaussian charge-distributions [48]. This later approximation, consistent with *ab initio* results, is so fast to compute that interactive superimposition processes can be performed: calculation of a similarity index, adjustment of the positions of the molecules to be compared, evaluation of the new similarity index value, etc. Automatic evaluation is also provided by the ASP package [49]. Other practical problems can also appear as to the extent of the comparison: in other words, should the integrals be evaluated over the whole molecule (including the side chains) or restricted to the main framework containing the pharmacophore? Modified molecular and charge-similarity indices have been proposed by Richard [50], with the incorporation of nuclear charges so as to not overestimate the importance of core electrons and better model the effects of the total charge distribution.

11.3.2 Similarity evaluation through space structuralization and Boolean operations

In numerical methods, the molecules to be compared are embedded in a 3D lattice of nodes. We discussed such grid methods for the calculation and display of molecular surfaces and volume in Chapter 8. Suffice it to say here that if the nodes are rationally explored, the description of the molecular property can be reduced to strings of bits encoding whether or not nodes satisfy the criterion selected (for example, inside or outside the molecular volume, MEP higher than a given level). Such a data structure is well suited to easily carrying out Boolean operations. *Union* defines the total imprint or trace of a set of molecules, a theoretical structure embedding all the others; *intersection* selects their common part, the population focus; for two molecules, *symmetrical difference* determines the unshared volume pertaining to one but not to the other, and so on.

Boolean operations can also be performed on the isovalued envelopes associated with electronic properties (MEP, for instance). The problem here is a little more complex, since these envelopes are first derived by extrapolation between nodes on the lattice edges. We developed two approaches: in one method, relevant nodes are first selected through Boolean operations and the resulting isopotential volume is represented thereafter. In the second one, this volume is determined through the intersections of the facets limiting the individual isopotential volumes of each molecule [3, 4].

Grid methods were also the basis of various attempts to quantitatively evaluate (and optimize) the similarity between two molecules regarding a given property. For shape or volume, for example, the goodness of the fit (the similarity criterion) can be easily evaluated with the ratio of the number of nodes (grid points) commonly occupied by both molecules to the total number of nodes occupied (or a mean square of it), or to the number of nodes occupied by the target [51–54]. The method was applied to evaluate shape similarity between linear, branched or cyclic alkyl groups and a phenyl group (group G being assimilated to molecule G–H). Two sets of experiments, dealing with selectivity measurements or biological activity, illustrate potential applications of this similarity index [53]. The complement of the similarity index was also related to the "chiral coefficient" proposed by Gilat [55], and offers a method for its numerical evaluation [51]. Similar indices can be introduced for H-bonding capabilities [52]: ratio of number of commonly formed H-bondings to the number of possible H-bondings in the target molecule. Nodes on the grid are "flagged" as possible H-bonding partners, according to their distance and direction to the H-bonding groups in the target.

Grid methods can also address the problem of evaluating electronic similarity, the integrals intervening in the Carbo and Hodgkin formulae being evaluated by a discrete summation of nodes [52]. But, as claimed by the authors, due to the binary character (yes/no) of the counting, it cannot respond to the gradation of electronic density [51]. For calculating MEP similarity, it was alternatively proposed to introduce, in the numerator of Carbo or Hodgkin indices the differences in the MEP values, and compare it to the maximum value of the MEPs. This was associated with the creation of new software tools, increasing flexibility and allowing for the display of similarity maps, with the possible inclusion of steric penalties [56]. On another hand Manaut *et al.* [5] chose Spearman's rank correlation coefficient to evaluate the similarity of MEP distributions encoded on grid nodes. In this work, the space around the molecule was limited by a shell corresponding to twice the van der Waals atomic radii, internal regions (within the molecular volumes) being excluded to avoid singularities or very high MEP values. Possible areas where the MEP is higher than a preset value are also considered, since such areas may be important to characterize the electrostatic pattern.

The adjustment of the orientation of the two molecules to obtain the better similarity is generally performed by fixing one molecule and moving the other by small translations or rotations. If the calculation of the similarity index is fast enough, adjustment can be performed interactively. In other cases, or when screening a large database, automatic processes are provided, as in ASP [49] or SUPER [54] where the first 20 correspondences for surfaces and charge distributions are selected. In Manaut's treatment [5], maximization of the similarity is carried out by a gradient procedure: one molecule is given translation and rotation displacements around three orthogonal axes, a new grid is calculated and the similarity evaluated. Selected examples (among them the well known case of methotrexate and dihydrofolic acid, where electronic rather than structural similarity has to be sought – see above) testify to the efficiency of the approach. Special interest was devoted in this work [5] to the initial position of the superimposed molecules, since a good starting point may avoid local minima and speed up optimization. A least-squares superimposition is used when the structures to be compared are neighbours. In other cases, molecules are located so that the centroids of their electron distributions coincide and their dipole moments are aligned. Some torsional flexibility for one molecule can be added, optimization being carried out with a SIMPLEX method [57]. Other authors proposed to bypass this cumbersome step of orientation adjustment, taking advantage of elements of similarity (for example, clustering of atoms near symmetry elements) or using weighting factors depending upon the atom type [51].

How fine the grid must be and how far beyond the molecular surface it should extend have been investigated by Richards *et al.* [49]. The best compromise between accuracy and time cost was defined as a 4Å/1Å grid (a grid with increments of 1 Å and extending 4 Å apart from the lowest and highest atom coordinates in the molecule). However, results are less decisive as to the choice of point-charge values obtained through various quantum chemistry methods (MNDO, AM1, STO–3G). In the set investigated, the AM1 approximation was claimed to give satisfactory results. For other examples, see Nakayoma and Richards [58].

A radial-type grid MACRA (Molecular Atom-Centered RAdial grid) was introduced by Richard [59] in place of a cubic grid to evaluate a new quantitative similarity index (MEP-SI) based on a Carbo-type formalism. It was claimed to provide more efficient storage of MEP information and a convenient means to perform local comparisons. Examples of the utility of molecular similarity calculation as a tool for searching 3D databases are given by Good *et al.* [60]. They address the choice of structural modifications of a template or the search for a pharmacophore-matching structure (from 3D database screening) to find a compound that best mimics a target molecule or a lead.

11.3.3 Comparison through gnomonic projection

A general approach, valid for both the molecular surface and a property encoded on it (MEP, for instance), was introduced by Chau and Dean, allowing for the comparison of equivalent surface patches [61]. Such patches may be thought of as representing parts of the molecule able to bind the receptor site. The method relies on *gnomonic projection*, that is a central projection of a property on a point on to a spherical surface (of arbitrary radius), retaining the 3D characteristics of the surface inspected. A ray from the centroid meets the molecular surface at a *pierce point*. The parameter value computed on this point of the surface examined (the "inspection surface") is assigned to the corresponding projection on the sphere (Figure 11.13).

To generate a semi-regular distribution of projection points, an algorithm for

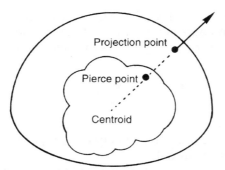

Figure 11.13   Gnomonic projection.

icosahedral tessellation has been proposed, leading typically, to 102 vertices
[61]. In other applications, using an icosahedron plus a dodecahedron with
vertices at the centre of the triangular faces of the icosahedron creates a set of
12 + 20 points to characterize the projection sphere [62] (Figure 11.14).
    Let us suppose we have to compare two molecules. They are first positioned
in their respective matched orientation, and translated so that their centroids
coincide with the origin of the coordinate system. From each projection point
(vertex of the icosahedron approximating the projection sphere) a ray is drawn
towards the centroid and the pierce point (intersection with the inspection
surface) computed. The projection point is encoded with the distance
(centroid/pierce point) for comparing molecular surfaces, or with the MEP
value at the pierce point (if the MEPs on the molecular surfaces have to be
compared). So any shape of surface patch can be compared. The quality of the
match between the two surfaces is determined with Spearman's rank
correlation analysis [63, 64] taken here as:

$$\text{Rank} = 1 - 6 \sum_{k=1}^{n} d_k^2 / (n^3 - n)$$

where d_k = the difference in ranks between the corresponding pairs of surface
parameters, and n the number of data points.

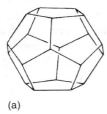

(a)   (b)

Figure 11.14   Generation of surface patches using a dodecahedron (a) or an icosahedron
(b).

Rank correlation is not affected by the actual values of the points, but only by the differences in ranks between paired members of the two sets of points. So, an alternative procedure [63] prefers to consider an error function:

$$\sum_{i=1}^{n} e_i^{\,2}$$

where $e_i = P_{iA} - P_{iB}$ is the difference between property values for molecules A and B at point i (total n points).

In subsequent papers, Dean *et al.* [63, 65] proposed a method for searching for best pattern matches between parameters mapped on to molecular surfaces. The comparison is carried out over a patch of a predefined shape projected on to the first molecule and supposed to characterize the property on this molecule. The second molecule is rotated to find matches within the projected window.

The procedure outlined assumes that we know the patch on one molecule to be matched by rotation of the second molecule. The basic task is therefore to minimize the residual between two faces with one molecule fixed and the other allowed to rotate. Numerous starting positions would have to be examined to get the best match (absolute minimum of an objective function). To save time, the search has two levels. First, partial minimizations define possible solutions for rotational coordinates, gathered around each feasible minimum. Then a cluster analysis of these intermediate results and a complete optimization locates discrete matching orientations.

The problem was generalized to situations where any particular patch to be matched cannot be predefined. In these "blind searching" procedures, the only reference frame is the window through which the molecules will be compared. Each molecule is then allowed to rotate, and the residuals are minimized to find matches, through a six-dimensional cluster analysis [66].

In such comparisons of the spherical parameterized surfaces of two molecules, a large number of starting orientations would have to be considered. Taking advantage of the symmetry properties of the regular icosahedron and dodecahedron largely reduces the part of the conformational space to be considered, and saves computer time [62].

The gnomonic projection method was adapted with the Sperm program [43, 44] to the search in 3D structural databases for molecules showing 3D shape similarity. As in Dean's treatment, the property investigated may be the shape (point inside or outside the molecular body) or any electronic feature (electron density, electrostatic potential). In a first attempt, the property investigated was mapped onto 32 points on the projection sphere (corresponding to the vertices of a dodecahedron and icosahedron), and taking advantage of the icosahedral symmetry for reducing storage requirements and speeding up the scanning of the rotational space. To scan the conformational space, the molecules to be compared are first located with their geometrical centres coincident with the centre of the sphere (assumption justified when looking at similar molecules) and their principal inertial axis aligned. Then a systematic search is carried out with a coarse grid that will subsequently be refined in the interesting regions. The process is therefore slightly different from that of Dean and Chau, which uses minimization from random starting orientations and then clustering. The similarity criterion retained here is the root mean square difference (RMSD):

$$
\mathrm{RMSD} = \left[\sum_{i=1}^{n} (P_{Ai} - P_{Bi})^2 / n \right]^{0.5}
$$

where P_{Ai} and P_{Bi} are properties on molecules A and B at point i, and n = the number of points.

Subsequent studies [44] showed the necessity for tessellation of the polyhedrans to dispose of at least 162 points to obtain reproducible and reliable similarity indices. To reduce the computation time, preliminary screens can be introduced. They are properties which, if matched between target and candidate, are indicative of a high similarity: molecular volume, data on inertial moments, etc. The method was claimed to be able to screen extensive databases (33,000 compounds) in about 7 hours on a VAX station 3100M76. Interestingly, the authors noted that "in addition to molecules showing topological similarity to the query, hits are also obtained showing considerable structural variety: molecules topologically dissimilar but nevertheless resembling closely in space." Note also that in the search for ligands able to bind a common receptor site, SPERM requires only the shape of one ligand to be known. It mainly estimates overall similarity, but allows for local incompatibilities, whereas other programs (such as DOCK, see Chapter 12 require the receptor structure and a shape of the binding site, but allow for ligands extending outside the binding site, provided that bad contacts with the receptor are avoided.

11.3.4 Geometrical complementarity

Close fitting of the molecular surfaces of a macromolecule and a bounded ligand is likely to be an important phenomenon in drug-receptor interactions, since among the main interaction mechanisms are steric requirements (no overlap between atoms) and hydrogen-bond formation, which obeys quite well defined distance constraints. A rapid estimation of how close two molecular surfaces can be is obtained by using a grid of points, let us say $(x_i, y_i, 0)$, projected on to the molecular sufaces along the z direction [67, 68]. A difference matrix D_{ij} can be built, measuring for each projected ray, the difference in z between the two pierce points, corrected by the translation **T**, bringing the surfaces into closest contact at one point (Figure 11.15).

Other aspects of ligand-receptor complementarity, particularly in view of

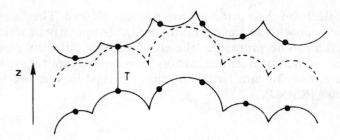

Figure 11.15   2D schematization: dots correspond to the intersection of the projection lines (along z) with the two molecular surfaces. **T** is the translation vector bringing the surfaces into closest contact (from Dean with permission [67]).

designing new active drugs, will be discussed in the next chapter, which is dedicated to the pharmacocophore approach.

REFERENCES

1. B.H. Arnold *Intuitive Concepts in Elementary Topology*, Prentice-Hall, Englewood Cliffs, NJ, 1962; in N.J. Turro, *Angew. Chem. Internal. Ed.*, **25:** 1986; 882–901.
2. A.M. Lesk *Comm. ACM*, **22:** 1979; 219–224.
3. J.E. Dubois, S.Y. Yue and J.P. Doucet *Visual Computer*, **2:** 1986; 367–378.
4. S.Y. Yue Thesis, University of Paris, 1987.
5. F. Manaut, F. Sanz, J. Jose and M. Milesi *J. Comput.-Aided Mol. Des.*, **5:** 1991; 371–380.
6. J.T. Bolin, D.J. Filman, D.A. Matthews and J. Kraut *J. Biol. Chem.*, **257:** 1963; 1982.
7. P. Gund, J.D. Andose, J.B. Rhodes and G.M. Smith *Science*, **208:** 1980; 1425–1431.
8. G.A. Arteca, V.B. Jammal, P.G. Mezey, J.S. Yadav, M.A. Hermsmeier and T.M. Gund *J. Mol. Graph.*, **6:** 1988; 45–53.
9. R.A. Diamond *Acta Cryst.*, **A21:** 1966; 253–266.
10. P.K. Warme, R.W. Tuttle and H.A. Scheraga *Computer Programs in Biomedicine*, **2:** 1972; 248–256.
11. P. Gund, W.T. Wipke and R. Langridge *Proc. Int. Conf. Comput. in Chem. Res. Educ.*, Ljublyana, Yugoslavia, July 1973, **3:** 1974; 5–33.
12. W.A. Kabsch *Acta Cryst.*, **A32:** 1976; 922–923.
12a. W.A. Kabsch *Acta Cryst.*, **A34:** 1978; 827–828.
13. A.D. MacLachlan *Acta Cryst.*, **A28:** 1972; 656–657.
14. A.D. MacLachlan *J. Mol. Biol.*, **128:** 1979; 49–79.
15. A.D. MacLachlan *Acta Cryst.*, **A38:** 1982; 871–873.
16. Zong Jie Liu and R. Van Rapenbusch *J. Comput. Chem.*, **9:** 1988; 596–599.
16a. Zong Jie Liu and R. Van Rapenbusch *Computers Chem.*, **13:** 1989; 5–23.
17. A.L. Mackay *Acta Cryst.*, **A40:** 1984; 165–166.
18. S.K. Kearsley *J. Comput. Chem.*, **11:** 1990; 1187–1192.
19. C.M. Venkatachalam, R. Czerminski and R. Potenzone Jr *Proceedings of the Montreux 1991 International Chemical Information Conference*, Annecy, France. H. Collier (Ed.), Infonortics Ltd, Calne, England, 1991; 209–215.

20. D.J. Danziger and P.M. Dean *J. Theor. Biol.* **116:** 1985; 215–224.
21. K. Mishikawa and T. Ooi *J. Theor. Biol.*, **43:** 1974; 351–374.
22. C.W. Crandell and D.H. Smith *J. Chem. Inf. Comput. Sci.*, **23:** 1983; 186–197.
23. H.G. Barrow and R.M. Burstall *Inf. Proc. Lett.*, **4:** 1976; 83–84.
24. C. Bron and J. Kerbosch *Commun. ACM*, **16:** 1973; 575–577.
25. F.S. Kuhl, G.M. Crippen and D.K. Friesen *J. Comput. Chem.*, **5:** 1984; 24–34.
26. A.T. Brint and P. Willett *J. Mol. Graph.*, **5:** 1987; 49–56.
27. A.T. Brint and P. Willett *J. Chem. Inf. Comput. Sci.*, **27:** 1987; 152–158.
28. J.R. Ullman *J. ACM*, **23:** 1976; 31–42.
29. P.J. Artymiuk, H.M. Grindley, D.W. Rice, E.C. Ujah and P. Willett *Proceedings of the Montreux 1991 International Chemical Information Conference*, Annecy, France. H. Collier (Ed.) Infonortics Ltd, Calne, England, 1991; 91–106.
30. S.E. Jakes, N. Watts, P. Willett, D. Bawden and J.D. Fisher *J. Mol. Graph.*, **5:** 1987; 41–48.
31. R.P. Sheridan, R. Nilakantan, A. Rusinko III, N. Bauman, K.S. Haraki and R. Venkataraghavan *J. Chem. Inf. Comp. Sci.*, **29:** 1989; 255–260.
32. D.E. Clark, P. Willett and P.W. Kenny *J. Mol. Graph.*, **9:** 1991; 157–160.
33. C.A. Pepperrell and P. Willett *J. Comput.-Aided Mol. Des.*, **5:** 1991; 455–474.
34. C.A. Pepperrell, R. Taylor and P. Willett *Tetrah. Comput. Method.*, **3:** 1990; 575–593.
35. P.J. Artymiuk, P.A. Bath, H.M. Grindley, C.A. Pepperrell, A.R. Poirrette, D.W. Rice, D.A. Thorner, D.J. Wild, P. Willett, F. H. Allen and R. Taylor *J. Chem. Inf. Comput. Sci.*, **32:** 1992; 617–630.
36. W. Fisanick, K.P. Cross and A. Rusinko III *J. Chem. Inf. Comput. Sci.*, **32:** 1992; 664–674.
37. W.T. Wipke and D. Rogers *J. Chem. Inf. Comput. Sci.*, **24:** 1984; 255–262.
38. M. Stewart and P. Willett *J. Documentation*, **43:** 1987; 93–111.
39. A.T. Brint and P. Willett *J. Mol. Graph.*, **5:** 1987; 200–207.
40. E. Feuilleaubois, V. Fabant and J.P. Doucet *SAR and USAR in Environmental Research*, **1:** 1993; 97–114.
41. R. Carbo, L. Leyda and M. Arnau *Int. J. Quant. Chem.* **17:** 1980: 1185–1189.
42. R. Carbo and L. Domingo *Int. J. Quant. Chem.*, **32:** 1987, 517–545.
43. V.J. Van Geerestein, N.C. Perry, P.D.J. Grootenhuis and C.A.G. Haasnoot *Tetrah. Comput. Method.*, **3:** 1990; 595–613.
44. N.C. Perry and V.J. van Geerestein *J. Chem. Inf. Comput. Sci.*, **32:** 1992; 607–616.
45. W.G. Richards and E.E. Hodgkin *Chemistry in Britain*, 1988; 1141–1144.
46. E.E. Hodgkin and W.G. Richards *Int. J. Quant. Chem.*, **14:** 1987; 105–110.
47. P.E. Bowen-Jenkins, D.L. Cooper and W.G. Richards *J. Phys. Chem.*, **89:** 1985; 2195–2197.
48. E.E. Hodgkin and W.G. Richards *J. Chem. Soc. Chem. Commun.*, 1986; 1342.
49. C. Burt, W.G. Richards and P. Huxley *J. Comput. Chem.*, **11:** 1990; 1139–1146.
50. A.M. Richard and J.R. Rabinowitz *Int. J. Quantum. Chem.*, **31:** 1987; 309–323.
51. A.Y. Meyer and W.G. Richards *J. Comput.-Aided Mol. Des.*, **5:** 1991; 427–439.
52. Y. Kato, A. Itai, and Y. Iitaka *Tetrahedron*, **43:** 1987; 5229–5236.
53. J.I. Kato, M.M. Ito, M. Tsuyuki, S. Skimizu, Y. Kainami, T. Inakuma, H. Matsuoka, T. Isago, K. Tajima and T. Endo *J. Chem. Soc. Perkin Trans* **2:** 1991; 131–136.
54. R.B. Hermann and D.K. Herron *J. Comput.-Aided Mol. Des.*, **5:** 1991; 511–524.
55. G. Gilat *J. Phys. A: Math. Gen.*, **22:** 1989; L545–550.
56. A.C. Good *J. Mol Graph.*, **10:** 1992; 144–151.
57. C. Burt and W.G. Richards *J. Comput.-Aided Mol. Des.*, **4:** 1990; 231–238.
58. A. Nakayama and W.G. Richards *Quant. Struct.-Act. Relat.*, **6:** 1987; 153–157.
59. A.M. Richard *J. Comput. Chem.* **12:** 1991; 959–969.
60. A.C. Good, E.E. Hodgkin and W.G. Richards *J. Comput.-Aided Mol. Des.*, **6:** 1992; 513–520.
61. P.L. Chau and P.M. Dean *J. Mol. Graph.*, **5:** 1987; 97–100.
62. P. Bladon *J. Mol. Graph.*, **7:** 1989; 130–137.
63. P.M. Dean and P.L. Chau *J. Mol. Graph.*, **5:** 1987; 152–158.

64. S. Namasivayam and P.M. Dean *J. Mol. Graph.*, **4:** 1986; 46–50.
65. P.M. Dean and P. Callow *J. Mol. Graph.*, **5:** 1987; 159–164.
66. P.M. Dean, P. Callow and P.L. Chau *J. Mol. Graph.*, **6:** 1988; 28–34.
67. P.M. Dean *Molecular Foundations of Drug-Receptor Interaction*, Cambridge University Press, 1987; 122.
68. M. Santavy, J. Kypr *J. Mol. Graph.*, **2:** 1984; 47–49.

12 Drug receptor interactions: receptor mapping and pharmacophore approach

Interpreting the mode of action of drugs or designing new active compounds is a fascinating challenge which has prompted immeasurable work. However, until now, in most cases only partial solutions could be proposed. In fact, as stated by Sheridan and Venkataraghavan [1], the design of new drugs is not so advanced as the design of an aircraft or a ship's hull, since the laws governing pharmacological action are not as well known as the basic principles of hydro- or aerodynamics. In the present state-of-the-art, the purpose of computer-assisted drug design (CADD) is mainly to propose models consistent with observations and to suggest new experiments that it is hoped will be the most fruitful [1].

A first step in the development of new drugs is the search for "leads": compounds active in a particular therapeutic area, used as a guide for the synthesis of analogues, which will thereafter be optimized thanks to structure/activity analysis in order to increase activity and decrease toxicity. Until now, leads have frequently been found by screening numerous compounds, or have been discovered by chance. But it may be hoped that molecular modelling can provide more rational and efficient approaches.

A starting point may be the observations of Crum-Brown and Fraser [2]: "... there can be no reasonable doubt but that a relation exists between the physiological action of a substance and its composition and constitution." Important developments arose from Langley's work [3] on the antagonism between pilocarpine and atropine in saliva. He suggested the formation of a complex between exogenic compounds introduced and a material present on the nervous terminations: this was the concept of a receptor.

During the past century, definite and continuous improvement in this field corresponds to the use of quantitative descriptions of structural characteristics: quantitative structure activity relationships (QSAR) give valuable information about the influence of electronic, steric or hydrophobic features, and allow for quantifying their influence upon the biological activity of drug molecules. As shown by Hansch and co-workers, the relative importance of the various mechanisms (steric, hydrophobic) taken into account in the QSAR treatments often offers some insight into the nature of the receptor sites adjacent to the substituent under study [4].

Another important step in the last decade was the use of *computer-assisted molecular modelling*: displaying the three-dimensional structure of molecules is of invaluable help so as to quickly grasp the essential features of complex interactions involving numerous (sometimes hundreds) atoms. Computer graphics appears to be a powerful tool for analyzing and understanding the intimate interaction mechanisms involved in drug action. Evaluating shape or electronic complementarity, taking into account conformational flexibility, and adaptation processes in molecular recognition allow for refining the interpretations, and possibly suggest new solutions.

Today, computer-aided drug design largely takes advantage of the synergy of the two approaches of QSAR and molecular modelling to determine the feasible binding modes of a drug, and derive the subtle energetic and dynamic features of drug-macromolecule interactions. Two main goals are kept in mind: understanding the phenomena, and from this knowledge deriving guidelines for the design of new compounds.

12.1 THE PHARMACOPHORE HYPOTHESIS

Drug action involves very varied (and complex) aspects:

- transport of the drug molecule from the point of administration to the receptor site,
- *in vivo* chemical modifications or free energy changes for flexible drug molecules to adopt the conformation required by the active site,
- energetics of drug (ligand) – receptor (active site of the proteins) interactions,
- and finally, production of the biological response.

As to the drug-receptor interactions, a two-step mechanism is generally accepted, beginning with the formation of a complex between the drug and the receptor. The formation of this complex induces specific conformational changes of the receptor, resulting in variation of its electronic characteristics, able to fire the biological stimulus. Some experiments using fast kinetics methods on acetylcholine agonists and antagonists [5], thermodynamic experiments [6] or even the direct observations of size variations due to complex formation [7] support these hypotheses. In a theoretical approach (studying interaction energy as a function of the distance for simple model compounds), Hall and Kier [8] distinguish a first step where long range interactions are sizeable, without geometrical changes of the drug, these beginning only at shorter distances. Successive phases are thus:

- *Recognition*: (long range interactions). The receptor identifies drugs (in their minimal energy conformation) that are able to bind.
- *Binding*: formation of a complex, energetically favoured and sterically allowed. This implies adequate deformation of the drug (and the receptor).
- *Specific perturbation* of the receptor inducing the biological response.

According to this scheme, *antagonists* are considered either as molecules recognized by the receptor but unable to bind, or alternatively, as molecules able to bind the receptor, but unable to deform it in the expected way. The existence of additional binding sites in the antagonistic structures compared to agonists has recently been illustrated by Hoffmann *et al.* [9] on muscarinic ligands. This can be summarized by the scheme given in Figure 12.1.

Despite this complexity, when looking more precisely at drug-receptor or, in other words, ligand-biomolecule interactions at the molecular level, medicinal chemistry largely relies on the central concept of "pharmacophore" or "3D mimicry". According to the classical "lock and key analogy" [1], drug molecules (keys) exert their effects by binding to receptors (locks). To bind a receptor site, a ligand has to be recognized by it: this implies that the ligand must possess some specific features which are indispensable for recognition, whereas other parts of the molecule may be changed without drastically modifying the affinity or pharmacological action. Such a specific arrangement of essential chemical groups, common to active molecules, constitutes the pharmacophore. This concept of a specific arrangement needed for recognition encompasses two aspects: the geometry of the active drugs (the shape of the key), and the volume of the binding site available for the ligand (the shape of the key/hole). Although more frequently presented in terms of geometry, the pharmacophore concept can involve steric and electronic features for a more realistic description (parts of the space where electron density, molecular electrostatic potential...take given values). This implies supplementary difficulties since, now, identification is not carried out on classical representations (graphs or coordinates), but on more conceptual shapes, not always defined through analytical expressions.

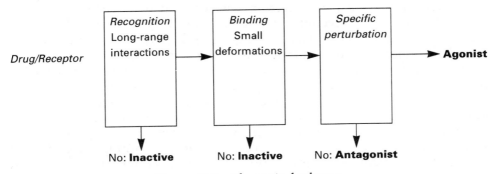

Figure 12.1 Theoretical scheme.

Although this concept of the pharmacophore requires various implicit assumptions (single binding mode, single set of important interacting groups), it proves to be very useful in rationalizing pharmacological data. The problem is to find the ways in which the drug may occupy receptor pockets, giving favourable interactions without overlapping the sterically forbidden regions. Once the pharmacophore is determined or the receptor is known, directly or by complementarity to the pharmacophore, the design of new active compounds becomes possible.

However useful this approach may be, it must be borne in mind that various other aspects, which still cannot be predicted by modelling (transport, metabolism, toxicity *in vivo*), are also likely to modify the activity or the interest of a drug.

Various strategies have been developed, depending upon the information available on the receptor. In the most favourable (but scarce) cases, the nature of the receptor site or, better still, the structure of the complex between the receptor and a bound ligand, is known by X-ray crystallography. This makes investigation of the interaction mechanisms easier, and can give some starting points for the design of other active substances. Sometimes, the receptor structure is still unknown but can be reasonably estimated by homology from analogous biomolecules whose 3D structures are available [10–18]. More ambitious is the prediction of the tertiary structure of a protein, based only on sequence information (even if the secondary structure can be correctly predicted, rules for folding still remain to be found).

More frequently (by far), the receptor is not known. The pharmacophore itself has to be deduced from similarity search within sets of active and closely related inactive molecules [4, 19–23] and the receptor will be thought of as a "negative image" of this pharmacophore. Conformational analysis, examination of various distance constraints in ligands having a different structure or conformational behaviour [24], and the search for a (or the greatest) common substructure constitute the basic tools of this "deductive" approach.

According to Hall and Kier's [8] scheme, the recognition phase involves only long range interactions. Comparison of drug molecules in their *minimal energy conformation* can therefore give valuable information within a series of closely related structures. However, for a better understanding, or when investigating a wider structural scope, it becomes essential to take into account the potential *flexibility* of drug molecules to adapt the receptor shape. This approach is generally carried out by first considering shape complementarity, the energetics of chemical interactions involved being examined afterwards. In the search for new leads (starting proposals for drug design), this sequence saves some computer time in the formidable task of finding a common binding arrangement of essential atoms within all the low-energy conformers allowed for the set of investigated molecules. Once this framework has been attained, the chemist will be able to change the type of atoms, in order to optimize electrostatic, H-bonding or hydrophobic interactions with the receptor for a maximal activity.

12.2 ACTIVE CONFORMATIONS OF A DRUG: FEASIBLE BINDING MODES OF A LIGAND MOLECULE AT THE RECEPTOR SITE

"Molecular docking" explores the binding modes of two interacting molecules, depending upon their topographic features or energy-based considerations [25], and aims to fit them into conformations that lead to favourable interactions. It therefore constitutes an essential step in determining the active conformation of a drug, i.e. its conformation when bound to the receptor.

This is a complex problem, since the active conformation is not necessarily the more stable one, as determined by X-ray crystallography in the condensed phase (where packing effects may intervene), or calculated by theoretical methods for the isolated molecule in the gas phase. The drug can accommodate small conformation changes if they lead to more favourable interactions or avoid the sterically forbidden regions of the receptor pockets.

How to hypothesize the active conformations of a drug? The point is to determine the *atoms* of the receptor where ligands bind on the macromolecule accessible-surface and the ligand *orientation* in the bound state.

12.2.1 The receptor active site is known or can be inferred

Of course, a direct determination of geometry for a ligand- or an inhibitor-bound receptor would give the answer [15]. Indeed, in some privileged situations, the structure of the receptor-bound ligand, or the structures of both the receptor and putative ligands, are known [26]. Such data about receptor geometry can be obtained by X-ray crystallography (although more recently, NMR studies can offer new tools for geometry determinations in solution). For example, Dihydrofolate reductase (DHFR) [10–14], which catalyses the reduction of dihydrofolate to tetrahydrofolate,

$$\text{Dihydrofolate} + \text{NADPH} \rightarrow \text{Tetrahydrofolate} + \text{NADP}$$

has been extensively studied owing to its crucial role in DNA synthesis, and therefore in the growth of any organism where it appears as a key enzyme. Indeed, its inhibitors can be used to control growth in organisms (plant, animal, microorganism) [17]. For instance, they act as antitumoral (methotrexate) or antibacterial agents (trimethoprim). Other results are available on prealbumin/thyroid hormone [27] or tryptophan/tryptophan-repressor complexes [18].

From this X-ray information, the design of possible new ligands can be approached. This is generally carried out as a stepwise construction. First, knowledge of the putative binding sites of the receptor fixes some steric constraints on the geometry of possible ligands. On the other hand, the electronic (and hydrophobic) pattern of the receptor pocket determines the interaction field experienced by the ligand. It gives some restrictions as to the nature of the ligand atoms which can provide the best (steric and chemical) complementarity to the receptor, and so the highest binding affinity. Once the right atoms have been located in the right ligand-point positions, one has to incorporate them into an actual molecular skeleton, and possibly optimize the proposed molecule with appropriate substituents.

Although the structure of an increasing number of macrobiomolecules has now become available thanks to X-ray crystallography (the Brookhaven Database now contains about 3100 structures), it is not always an easy task to crystallize binary or tertiary complexes in order to get X-ray information, and such experiments tell us nothing about quantitative energetics [28]. Complementary computer-driven approaches are therefore clearly needed.

Receptor-based design

A first type of approach corresponds to the so-called "receptor-based design of inhibitors" [14]. Kuyper *et al.*, for example, used a 3D model of the complex formed between Trimethoprim (an antibacterial) and *Escherichia coli* DHFR to design analogues of Trimethoprim with a higher affinity with DHFR, and to derive complementary information on the binding mode (their conclusions were thereafter ascertained by X-ray experiments) (Figure 12.2).

The challenge is to imagine new systems with appropriate substituent groups (carboxylic acid groups in the example quoted) able to ensure enhanced interactions with selected sites of the enzyme (here a guanidinium fragment in an Arg residue), and to optimize their spatial location by adjusting the length of the chain bearing them (Figure 12.3).

In such studies, determining the final bound geometry of a ligand was, to some extent, approached manually [29], adjustment being obtained through the manipulation of 3D coordinates. Appropriate rotations and translations are sought, so as to obtain superimposition of drug atoms and receptor sites and a good steric fit. Novel approaches, where the computer-generated image replaces the usual molecular models try to automate the process of fitting ligands to receptors, avoiding these lengthy coordinate manipulations to find possible binding modes, and making the investigation of the conformational space easier, which may become a formidable task for flexible ligands. This constitutes the aim of "shape matching" methods, which emphasize the complementarity of shapes, thought of as an essential aspect of ligand-receptor interactions.

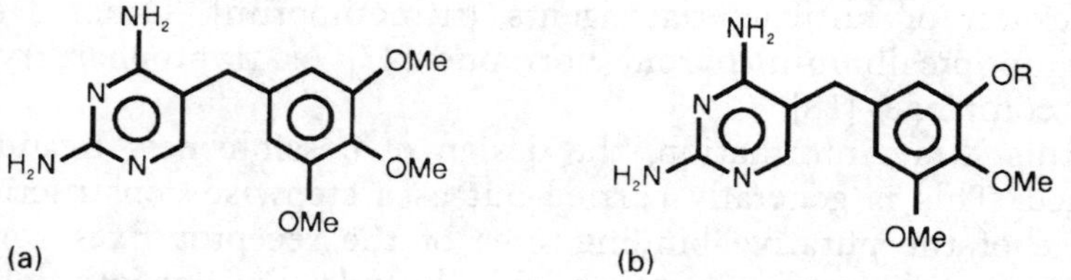

Figure 12.2 (a) Trimethoprin (TMP), (b) active analogue (M) (with R = $(CH_2)_2CO_2H$, affinity is about three times higher than that of TMP).

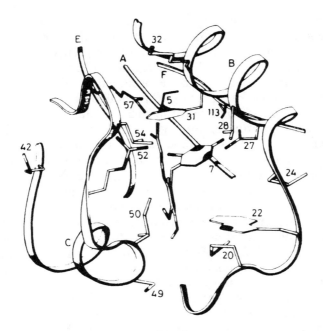

Figure 12.3 Receptor-based design of DHFR inhibitors. Schematic illustration of binding for molecule (M), an analogue of Trimethoprim (TMP) bearing a carbomethoxy chain R = (CH$_2$)$_2$CO$_2$H. Segments A, E and F (parts of β sheets) are located on the rear of the binding cleft. On the left hand side, the cleft is limited by an irregular peptide chain. The right hand side contains an irregular region and a helix (B). As for TMP itself, the pyrimidine ring is bound to Asp-27 (nitrogen atoms are drawn in black), the benzyl moeity is partially enclosed by Phe-31 above, Ile-50 on the left and Leu-28 on the right. To get a better binding affinity, an additional interaction (H-bonding) has been designed thanks to the carboxy group of the side chain introduced, which interacts with the guanidium group of Arg-57 (from Kuyper *et al.* with permission [14]).

Similarly, it may be feared that the search strategy for new drugs largely relies on the designer's knowledge and intuition, both for identifying key features of the receptor pocket and for building molecules under given geometrical and electronic constraints. Danziger and Dean [30] suggested recently that artificial intelligence methods can help to define automated processes, providing a more systematic approach, and avoiding individual bias and the subjective selection of those avenues potentially leading to possible solutions. This is the basis of their "site directed design" (see below).

Finally, binding efficiency is not only a matter of geometry; the energetics of interactions, which may be just as "critically important", must be considered [28, 31, 31a].

To more easily determine the favoured sites where ligands may bind the known active site of a target macromolecule, Goodford *et al.* [31–32] proposed to evaluate the interaction energy on a grid of test points surrounding the target. An empirical energy function was proposed to calculate Lennard–Jones, electrostatic and H-bonding terms (the latter taking into account the nature of the donor and acceptor atoms and their location in space) as a sum of pairwise interaction energies between probe groups and the target. Typical probes include water, methyl, amine nitrogen, carboxy oxygen and hydroxyl. Contour surfaces of negative energy delineate regions of attraction, and allow for specifying the spatial positions to be occupied by the ligand atoms for a favourable interaction with the target. Although, as stated by the authors, evaluation of interaction energy on the basis of additive pairwise terms may only be a crude approximation, good agreement was observed with crystallographic results. This method, which simultaneously considers both shape and energy, correctly retrieves binding sites for water or other ligands in known systems. It therefore appears to be a useful guide for interpretation in docking studies [31–32], but also a helpful tool to be incorporated into programs dedicated to the design of new ligands [33].

In other respects, this study also established that the release of water on binding has an important influence in the hydrophobic effect (entropic terms).

For a "real time" evaluation of energy features, Pattabiraman *et al.* [34] used a similar approach. Prior to docking, a potential, representing electrostatic and van der Waals terms, is calculated on a 3D grid enclosing the receptor. Thanks to this pre-calculation, the interaction in the docking process is quickly evaluated, assuming that the drug atoms always coincide with grid nodes (an approximation which seems satisfactory with a mesh of about 0.5, 0.25 Å). Results appear qualitatively consistent with the usual more elaborate (but slower) methods.

Automated shape matching

Apart from various algorithms to superimpose the atomic framework of molecules or macromolecules (see Chapter 11), docking macromolecules by shape (i.e. taking into account the actual volume occupied by atoms) was considered by Wodak and Janin [35], characterizing the interaction of two proteins with a simplified expression of their interaction energy, and by Santavy and Kypr [37], relying on purely geometrical features.

The algorithms of Kuntz *et al.* [38] allow for fitting small rigid molecules into potentially binding clefts of macromolecular receptors of a known crystal structure. It was further extended to the docking of flexible ligands [36]. The approach relied first only on shape complementarity, without energetics consideration (except a very crude hard sphere test: interatomic distances are compared to the sum of van der Waals radii to rule out all but very small atom overlaps). It has recently been improved with a more refined evaluation of interaction energy (thanks to a molecular mechanics function, taking into account van der Waals and electrostatic terms of the AMBER force field). Computation is made easier by precalculating the receptor-dependent terms in the potential function on a 3D grid mapping the active site [33].

The principle is to determine the surface complementarity between the ligand and the receptor by searching for a fit (or more exactly, some geometrical identity) between the ligand and a "negative image" of the receptor. In other words, the complementarity between the key (drug) and the keyhole (the receptor) is approached by representing the keyhole, in negative, by a virtual shape (featuring the void volume) and looking for the identity of the key(s) and this virtual shape. In the Kuntz method, the molecular surface is used as a starting point to characterize shape (only non-hydrogen atoms with united-atom radii being selected), and both the ligand and the "negative" of the receptor are schematized by two sets of "imaginary" spheres filling, respectively, the key (ligand) and the receptor's negative (the keyhole) (Figure 12.4).

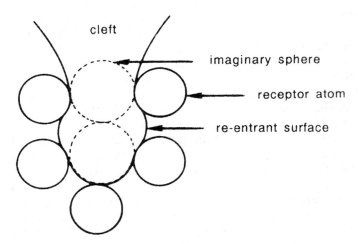

Figure 12.4 The receptor's cleft (the keyhole) is approximated by a set of "imaginary spheres" filling the void volume and defined as tangent to receptor atoms (from Kuntz *et al.* with permission [38]).

First, to build a negative image of the receptor, the program generates from each surface points a set of spheres that fill all pockets and grooves on the surface of the receptor. These spheres are drawn as touching the molecular surface at two points and lying outside the receptor surface (in the void keyhole) (Figure 12.5). Various criteria are introduced to reduce their number: from each surface point, only the smallest sphere is retained since the larger ones

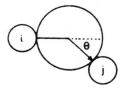

Figure 12.5 Fitting an imaginary sphere (negative of the receptor) to receptor surface atoms *i,j*. The centre of the sphere lies on the surface normal at point *i* (from Kuntz *et al.* with permission [38]).

would intersect the surface. Then among them, spheres with $\theta > 90°$, which are more likely to lie in shallow grooves, or $r > 5\text{Å}$, which extend out of the top of the receptor pocket, are rejected. Similarly, pairs of contact atoms must be at least four consecutive amino acid residues apart (otherwise they would probably be involved in a groove of an α helix). So, the number of imaginary spheres can be reduced to one sphere per atom (the largest sphere from the contact surface points of each receptor atom and the largest from the reentrant surface of each ligand atom). Then, overlapping spheres (considered as belonging to the same site) are gathered into distinct clusters (of typically about 50 units). Over an entire molecule, several clusters are generally found, representing cavities of various sizes and giving a good estimate of the possible binding sites of large biomolecules. The largest set usually corresponds to the binding site observed. Similarly, the ligand is represented by a set of spheres which approximately fill the space it occupies.

Within this approximation of both ligand and receptor shapes by sets of spheres, the problem is to fit the set of ligand spheres within the set of receptor spheres. The first step of the matching algorithm (the more time-consuming one) gathers all fits that are possible on a comparison of internal distances in both ligand and receptor. A systematic distance matching is handled, using distance matrices, without explicitly expressing the internal rotations necessary to actually carry out the real binding. A ligand sphere is paired with a receptor sphere if its internal distances to all the spheres in the ligand set match all the internal distances of the corresponding sphere within the receptor set (some tolerance limit on each distance is permitted). This rule, which allows for the identification of geometrically similar clusters of spheres in the receptor and in the ligand, looks similar to one of the algorithms proposed to identify common 3D substructures (a basic problem in the evaluation of molecular similarity) [39].

The output consists of short lists of pairs of ligands and receptor spheres having all internal distances matching within a tolerance value (1 or 2 Å). At least four pairs of contacts are necessary to ensure a unique docking. The list now has to be examined to resolve handedness, to check that the unmatched ligand spheres do not occupy sterically unacceptable positions, and to locate the ligand within the receptor pocket thanks to coordinate transformation. The final stage explores the suggested pairings: ligand atom coordinates are calculated and the locations of the ligand atoms are optimized to improve the fit (reducing steric interactions and looking for best hydrogen bonding interactions). This algorithm was discussed in examples where ligand and receptor geometries were established by crystallographic measurements: haem-myoglobin interaction in metmyoglobin, binding of thyroxine to prealbumin, etc. It was confirmed that the method produces a sampling of the reasonable geometries (within 1 Å of the geometries derived from X-ray crystallography).

The DOCK package, developed by DesJarlais *et al.* [40], was recently used to study inhibitors of the active site of α-chymotrypsin [41]. Chymotrypsin catalyzes the hydrolysis of peptide amide bonds found on the carboxyl side of an aromatic amino acid residue (Figure 12.6). The binding receptor pocket is

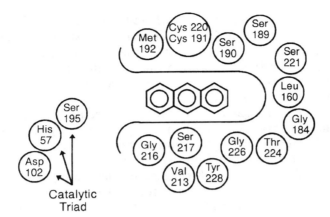

Figure 12.6 Schematic representation of the active site of chymotrypsin. The docked anthracene molecule points out the size of the pocket (from Stewart *et al.* with permission [41]).

known as a hydrophobic cavity with approximately the size of an anthracene ring system. According to the authors, the aim of this study was primarily to estimate, in database searches, the performance of computerized prescreening for the elimination of unequivocally inactive compounds, rather than identifying all active compounds in the case of well known receptor systems.

Each element of a database of 103 putative ligands was systematically docked into the receptor site and evaluated for fit. This receptor site was comprised of 40 amino acid residues identified by X-ray crystallography, but water detected in the site was removed. The binding pocket was filled with 21 spheres from 1–4 Å in diameter and inhibitors docked by sphere-centre/atom-centre matching. The docked orientations were scored from a function approximating a soft van der Waals potential based on the sum of the van der Waals contacts (but up to now ignoring some important interactions such as hydrogen bonding or electrostatic forces). No strict correlation was obtained between known binding strengths and docking scores, but the program ranked eight of the potent inhibitors (acridine or quinoline derivatives) within the top ten best scoring compounds. This can be considered as a quite encouraging agreement between computer predictions and experimental observations.

A fundamental problem in docking strategies is that the orientation space is very large and computation time rapidly becomes enormous. Furthermore, when looking at complexes between two macromolecules, where only a portion of a large ligand is involved in the active interface, many extraneous distances complicate the calculation. To overcome this drawback, Shoicet *et al.* [25] proposed dividing macromolecules into independent, geometrically distinct subsections, which can be matched separately. In this variant, the larger spheres in the clusters modelling the cavity are eliminated so as to create subclusters to be independently considered. Furthermore, various algorithms have been tested to efficiently prune the tree-search matching distances in the receptor and the ligand. The method, tested in 10 protein-ligand systems, including seven cases where the ligand is itself a protein, was able to successfully reproduce the experimentally determined geometries of the ligand in the protein. The interest of the method is to concentrate the search on regions of the orientation space that are likely to have high complementarity. This results in increased speed when compared to searches based on sampling a regular grid.

In the previous approaches, the macromolecule is kept rigid. This assumption does not cause any problem for designing antagonists or inhibitors: if a drug binds tightly to some conformation of the protein, it prevents substrate binding and has an inhibitory effect. However, for agonists, which are generally smaller and more flexible molecules, things are more difficult, since the active conformation of the receptor must be known [40].

So, the approach was later modified by DesJarlais *et al.* [36] to accommodate some flexibility of both ligand and receptor. The ligand is now approximated as a small number of large rigid fragments, such a division allowing for some flexibility at the junction position. For each fragment a match with receptor spheres is sought independently. Then the ligand is recreated joining all fragments and eliminating orientations where fragments cannot be reasonably joined together. The examples presented rely on the binding of methotrexate to dihydrofolate reductase and thyroxine to prealbumin (Figure 12.7).

Encouraging results were obtained, where in each case, ligand binding-geometry was found, very similar to that observed by crystallography. However, interestingly, some other geometries, similar in energy, were sometimes also found. For example, with methotrexate, as well as a solution similar to the X-ray binding mode, another solution was also suggested, with the pteridine ring rotated 180° (which looks very much like that thought to be assumed by folate) [15]. Clearly, several assumptions limiting the use of the

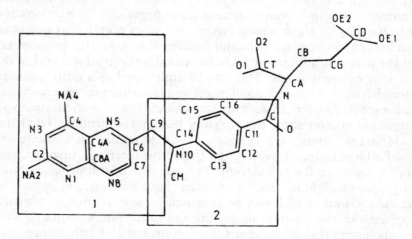

Figure 12.7 Division of the methotrexate molecule into moeities 1 and 2 (from DesJarlais *et al.* with permission [36]).

latter method should to be borne in mind: only shape complementarity is taken into account, without any more detailed discussion of energetics features (strong hydrogen-bonding opposing weak van der Waals contacts, for instance, are neglected), so that one can question the ability to derive predictions. Furthermore, it is also assumed that a receptor site changes only a little when the ligand binds to it. Nevertheless, this shape matching method may constitute an attractive tool for designing novel ligands able to bind a given macromolecular receptor the structure of which is known at the atomic level, by X-ray crystallography or inferred from pharmacophore models (see below) [40].

For the docking of flexible ligands, a recent variant of the DOCK algorithm first determines the orientation of a fragment of the ligand (the "anchor fragment") within the active site [42]. These positions form the basis of an exploration of the conformational space for the other parts of the ligand. The matching procedure includes some information about hydrogen bonding features thanks to additional "hydrogen bonding site points".

In a quite different approach, Goodsell and Olsen [43] proposed a simulated annealing algorithm to take account of conformational flexibility when docking a small molecule to a known receptor site (supposed static). The trial molecule performs a random walk in the space around the receptor. The simulation is broken into a number of cycles, each at a constant temperature and composed of a large number of individual steps. At each step a new substrate conformation is generated by small random displacement in each of its degrees of freedom (translation, rotation, rotatable bonds). The interaction energy is evaluated (using a grid technique) and examined with a Metropolis criterion: if the energy is lower than previously the conformation is automatically accepted; if it is higher it can be accepted with a certain probability, depending upon a user-defined "temperature". At the end of each cycle the temperature is lowered, according to a cooling scheme, and a new cycle is started from the lowest energy conformation of the preceding cycle. The process is able to escape from the local minimum and would, in principle, finally converge to the absolute energy minimum (if the number of steps is sufficient and the cooling scheme slow enough). For more details about annealing methods see Chapter 7.

Flexibility is also tractable with the ellipsoid algorithm, a constrained-optimization method already presented in Chapter 7 for the multiple minimum problem, and which has been adapted to docking studies [44]. Sterically acceptable interactions between a macromolecule and a ligand are discussed as constraints on the distances, rather than in terms of energy evaluation. Upper distance limits between selected atoms, one in the ligand and one in the macromolecule, guide the docking, van der Waals contacts fixing lower distance limits. Examples treated use specific information about probable interatomic distances or consider that only the binding sites of the macromolecule are fixed without prior knowledge of the interaction pattern, so that all possible binding modes have to be investigated. Upper distance limits are used one at a time for a systematic exploration. For this problem of docking (implying only 6 degrees of freedom plus conformational flexibility), the authors insist on the robustness of the ellipsoid algorithm with respect to local minima and relatively small requirements of computer time or memory.

With the generation of several extensive databases of 3D coordinates, a large impetus is now currently given to automated methods, where potential ligands (complementary to the shape of a known macromolecular receptor site) are sought by screening a wide assortment of candidate structures. A first example was presented on two protein receptors: papain (a sulfhydryl proteolytic enzyme with broad substrate specificity) and carbonic anhydrase. Small molecules or fragments were extracted from a shape database built from a subset of the Cambridge Crystallographic Database. They were automatically docked by Kuntz's algorithm, and after geometrical optimization, a scoring routine retained only those proposals fitting best. The approach retrieved known ligands, but also seemed able to suggest reasonable dockings for a large variety of molecular architectures that were able to bind, although generally larger than the known possible ligands [40] (Figure 12.8).

For example, the binding site of papain was identified as an elongated groove with a pocket at each end separated by a ridge. The search retrieves 3-iodophenyl hippurate (molecule (a), a known substrate intentionally added to the database), and proposes other potential ligands.

Some molecules exhibit (like hippurate) two bulky groups separated by a chain, and are likely to bind in the same mode: a bulky group in each pocket and the chain over the ridge (molecule (b)). However, other binding modes are found possible: molecule (c) fits with a pyridinium ring against the wall of the ridge; molecule (d) fits only one pocket (Figure 12.9).

The same strategy (automatic docking of candidates extracted from a database of 3D coordinates with the Kuntz algorithm and sorting according to

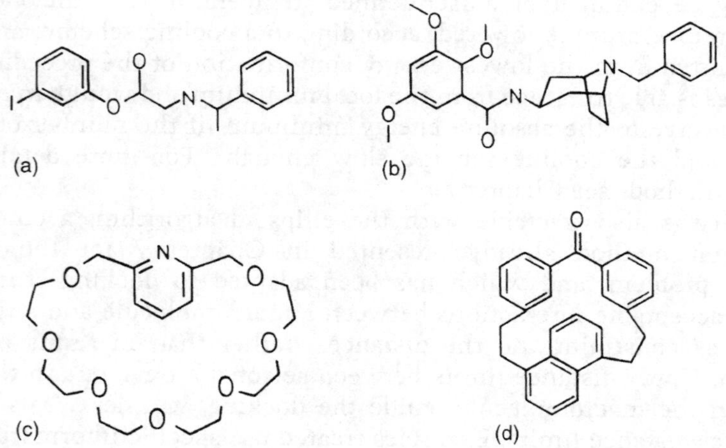

Figure 12.8 Potential ligands of papain. (a) 3-iodophenyl hippurate, (b) N-(8-benzyl-1αH-5αH-nortropan-3β-yl)-2,3,5 trimethoxybenzamide, (c) 2,6 pyrido-24-crown-8, (d) [2,2] (4,4′) benzophenono 2,6 naphtalenophane (from DesJarlais *et al.* with permission [40]).

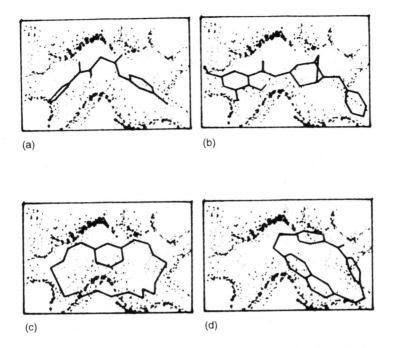

Figure 12.9 Binding of molecules (a–d) into the papain receptor site. The two receptor pockets are located on the left and right sides with a ridge at the centre of the figures (from DesJarlais *et al.* with permission [40]).

their steric fit to the receptor) was applied in a search for compounds that contain a particular pharmacophore [45]. Published examples show that such searches can identify classes of compounds different from the compounds used to derive the pharmacophore but known to have the appropriate activity and may suggest new structural classes for synthesis (Figure 12.10).

According to the authors, it can be hoped that the best candidates, with optimum fitting of the bumps and grooves of the site can serve as precursors. To obtain new drugs, designers then have to modify the candidate so as to ensure steric but also chemical complementarity.

The method was extended [46] to the design of new ligands if a pharmacophore geometry has been established and the receptor-bound conformations of other ligands are known. Now the structure of the receptor itself is not available, and known ligands are used to derive a negative image of the receptor shape (this point will be discussed further). Given enough conformationally constrained active and inactive ligands, one can deduce the pharmacophore geometry. A set of active ligands docked together defines a "minimum binding site volume", space that an active ligand can occupy on the receptor. The hypothetical receptor may be thought of as complementary to this ensemble volume. In other words, the keyhole is approximated by the union of the volumes of the known keys. From a database of molecular shapes, one searches for molecules which can fit inside this minimum binding site

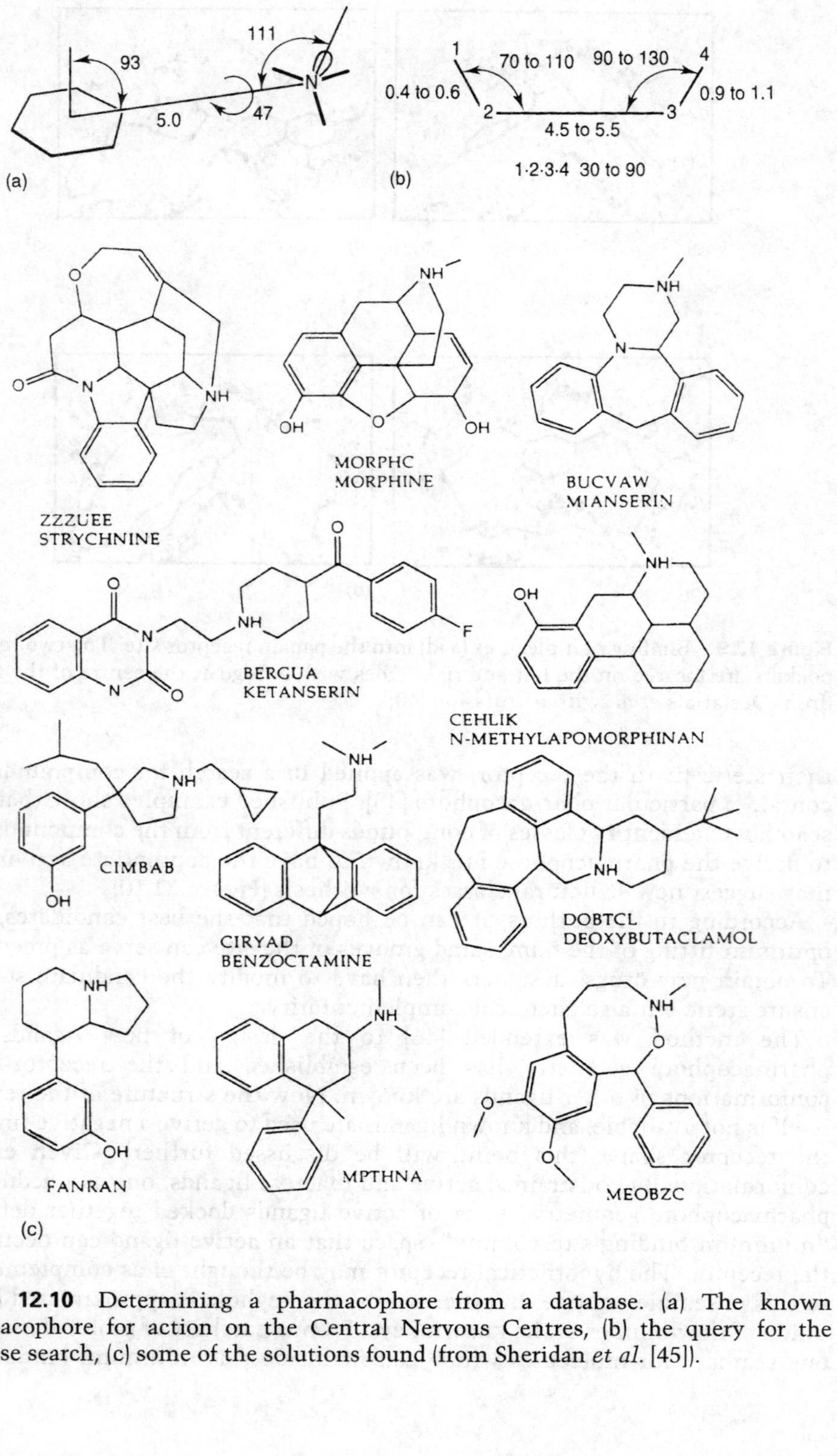

Figure 12.10 Determining a pharmacophore from a database. (a) The known pharmacophore for action on the Central Nervous Centres, (b) the query for the database search, (c) some of the solutions found (from Sheridan *et al.* [45]).

volume and which have interatomic distances compatible with the pharmacophore geometry. This step generates a framework, from which it may be hoped that new drugs can be derived modifying atoms to optimize interactions.

Also relevant to the recognition of compounds meeting geometrical and structural requirements is the ALADDIN system from Martin *et al.* [47, 48]. It constitutes an integrated tool which can be used to perform automated 3D database searches for finding compounds fitting a known binding site (and in other applications receptor mapping). Besides distances, ALADDIN incorporates the steric fit to the binding site (the boundaries of the binding site being specified by dummy atoms), and specification of the structural environment of the atoms involved.

A recent application [49] deals with the design of novel non-peptidic inhibitors of HIV protease, an important therapeutic goal in the treatment of AIDS. The active site of the enzyme was previously determined by X-ray analysis of the crystal structure of HIV-1 protease/inhibitor complexes. Important key interactions, identified from the complex structure, suggest a model of the pharmacophoric pattern: a central hydroxyl group (to interact with carboxylate side chains of Asp 25 and 25'), one or two symmetrically opposed hydrogen-bond donating groups (to interact with carbonyl oxygens of glycine 27 and 27'), a hydrogen-bond accepting moiety (to interact with Ile 50 and 50' and displace the buried bridging water molecule). The set of the corresponding distance ranges fixes the geometric constraints retained. An extensive search with ALADDIN, among a large 3D database (140,000 sets of coordinates) leads to about 30 compounds, obeying steric and geometrical constraints, to be tested for their ability to inhibit HIV-1 protease (Figure 12.11). Among them, three hydroxy-substituted benzophenones exhibit moderate levels of inhibition and may constitute leads for the design of more potent and potentially active HIV-1 protease inhibitors.

Graph theory techniques have also been applied to dock a flexible drug on to a rigid macromolecular receptor on the basis of distance information [50]. Given the binding atoms of the receptor and the candidate atoms of the drug (regarding their H-bonding capabilities), a correspondence graph is drawn associating to each of the receptor sites any drug atom. Then, the point is to find the graph cliques to determine what interactions between pairs of receptor atom/drug atom can occur simultaneously. Clique-detection has already been presented in Chapter 11. Ligand flexibility is introduced by considering upper and lower limits rather than precise distances, and distance geometry is then invoked to generate docked structures.

Artificial intelligence and site-directed design

As just stated, the design of new ligands (leads or new active molecules) can be carried out by analogy: known molecules selected from a database are placed in the known receptor pocket and scored for "goodness-of-fit". Another avenue, complementary to this design by analogy, is "*de novo* design", in which the ligand model is constructed piecewise in the receptor [51]. As already

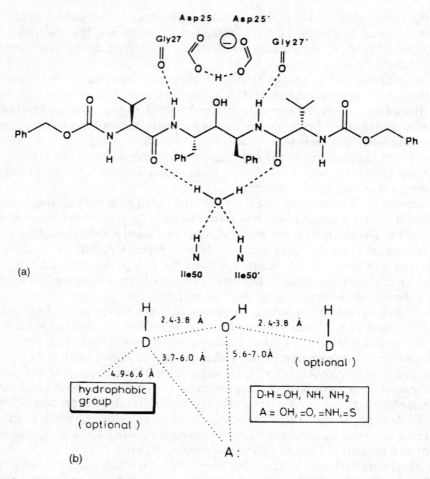

Figure 12.11 (a) 2D representation of inhibitor A-74704 and hydrogen-bonding interactions with HIV-1 and (b) schematic substructure used for the search (from Bures *et al.* with permission [49]).

indicated, this automated design [30] appears as a three-stage process: first, one identifies the interesting points of the receptor pocket; then, some ligand points able to interact with them are located; finally, these atoms are incorporated in a true molecular structure which is refined to meet the steric and electronic constraints. As to the location of relevant ligand points, the authors mainly focused interest on hydrogen bonding, owing to their prime importance in drug-receptor interactions.

From the atomic coordinates of a protein, extracted from the Brookhaven Database, hydrogen bonding groups at the surface of the protein are identified (and classified as either H-donor or -acceptor) (Figure 12.12). Then, in the surroundings of these atoms, the probability of hydrogen-bond formation with good complementarity (acceptor region in the site corresponding to a donor

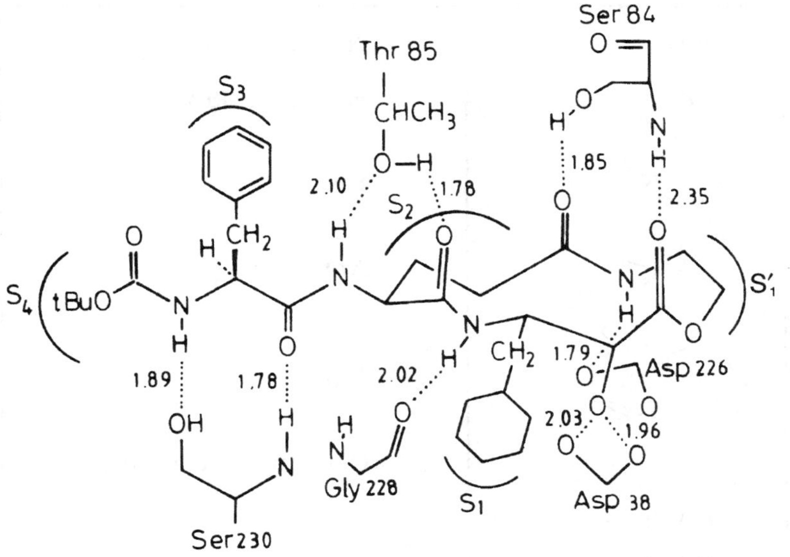

Figure 12.12 Schematic representation of the H-bonding network of a macrocyclic inhibitor in the active site of human renin (from Weber *et al.* with permission [52]).

ligand point, and *vice versa*) is mapped on to a grid and displayed, locating probable hydrogen bonding regions of the ligand (Figure 12.13–12.15).

This approach, which considers regions outside the receptor as possible (virtual) locations for ligand sites, is therefore similar to that of Goodford [31, 31a], although the energetic aspects are considered in less detail. The method takes advantage of the directional properties of the hydrogen bond (as shown by several crystallographic studies). This limits well-defined possible anchoring regions where ligand atoms, able to bind the protein through H-bonds, must be located. Steric hindrance from neighbouring protein atoms and the possibility of intramolecular H-bonds preventing receptor-ligand interactions are also looked for.

The predictive power of the algorithm can be ascertained by the agreement observed in comparisons with some crystallographic results: for example, the position of oxygen atoms of water molecules on hydrated proteins myoglobin and plastocyanin (described in the Brookhaven Database with, respectively, 388 and 44 included water molecules), or the positions of H-bonded atoms in protein-ligand co-crystals such as the enzyme dihydrofolate reductase (DHFR) co-crystallised with methotrexate (MTX) and NADPH and amidinophenyl pyruvate (APPA) complexed with trypsin [30, 53].

Once some ligand points able to interact with a receptor pocket have been defined, the following steps incorporate them in an actual structure. This generation can be divided into two parts: first creating a *molecular graph* spanning the binding site and fitting the ligand points, then going to the *true molecule* able to fit the mould formed by the receptor active site: that is proposing a ligand not intersecting the protein accessible surface, matching

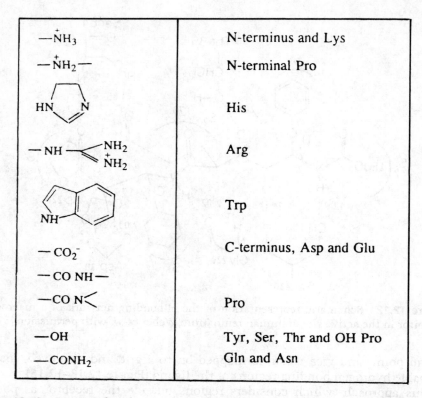

Figure 12.13 Main H-bonding groups in proteins (from Danziger and Dean with permission [30]).

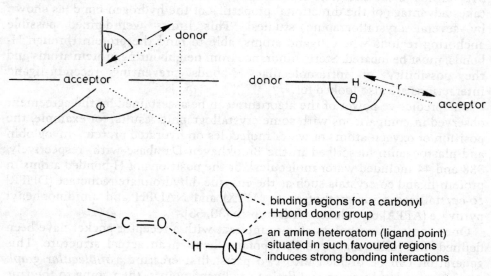

Figure 12.14 Geometrical parameters describing donor or acceptor regions (from Lewis and Dean [56], and Danziger and Dean [30], with permission).

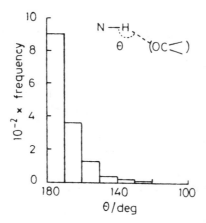

Figure 12.15 Distribution of the H-bond angle θ. Such histograms (derived from analysis of crystallographic data) allow for the determination of the probability of occurrence of a hydrogen bond in a given geometrical location (from Danziger and Dean [30, 53] with permission).

local hydrophobic and electronic patterns, and corresponding to a low energy conformation [30, 53–55].

Manually designing a graph for a fragment rapidly becomes a very difficult process, quite intractable by comprehensive searching for even a moderate sized fragment of, let us say seven atoms. So, rather than a bond-by-bond construction, suffering from complex combinatorial problems, building the molecular graph can be made faster by using some predefined bricks, such as a "building kit", to form links between ligand points [54]. This leads to the concept of a "spacer skeleton": a "topological artefact, composed of vertices and edges, which is able to model several distinct classes of compounds (ring systems) within just one graph and hence reduces the number of different structures that must be fitted to the defined ligand points" [54] (Figure 12.16).

For the sake of simplicity, the problem was first tackled on 2D representations (avoiding conformational flexibility), a not unreasonable assumption since many molecular structures of biological interest contain cyclic parts.

From a screening of the Cambridge database, an average geometry was defined for 20 different planar (or nearly planar) ring arrangements (spacer skeletons), designed to model many different fragments. These moieties provide useful frameworks from which molecular templates and then actual molecules (putative ligands) can be derived [56]. From the convenient spacer, fitting to ligand points is carried out through a distance matrix method, which examines all the possible correspondences between spacer vertices and ligand points, and optimizes the fit. Then, clipping (erasing certain parts lying beneath the accessible surface of the ligand-binding site) leads to a so-called "molecular template" which is still an (abstract) topological graph. Specifying the nature of its atoms and links converts it to a real molecule.

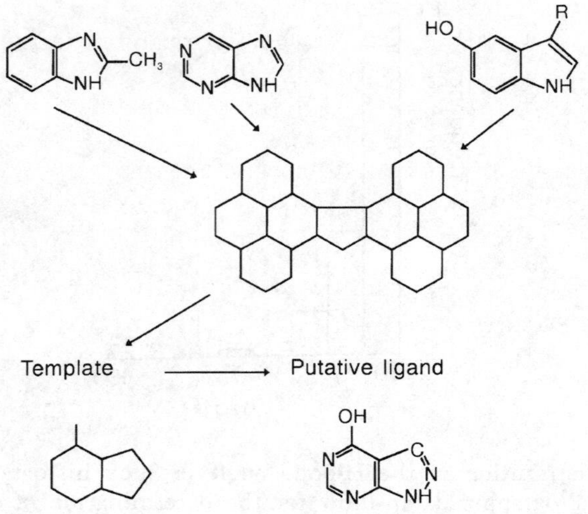

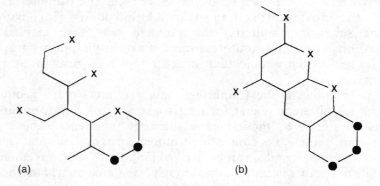

Figure 12.16 Many distinct compounds can be represented from the same spacer skeleton (top). From a subset of a spacer fitted to the receptor binding site, a molecular template is produced and then converted to a putative ligand (bottom) (adapted from Lewis and Dean with permission [54]).

Note, however, that erasing some links or modifying hybridization of some atoms, going from the spacer to the molecule, can induce changes in steric or geometrical characteristics, and this point has to be considered when deriving a molecular structure from a spacer and a template. The method has been tested at two binding clefts: the pteridine binding site in DHFR and the amidino phenyl pyruvate site of trypsin, using a spacer formed by a regular honeycomb of 11 hexagons. The graphs automatically generated showed strong similarity to those of real ligands, determined by crystallography [56] (Figure 12.17).

Figure 12.17 Templates fitting the binding sites on dihydrofolate reductase (derived from a honeycomb of regular hexagons). Crosses mark the atoms corresponding to ligand points. Side chains can be added to seed atoms (dark circles). Template (b) is similar to that of the pteridine group of methotrexate (from Lewis and Dean with permission [56]).

The approach was later extended to 3D through the generation of connecting chains. An efficient method consists of generating an alkyl chain between ligand points: Lewis [55, 57] suggested as a spacer skeleton the union of alkyl chains in the diamond lattice. An optimization process is proposed, selecting atoms of the lattice slightly overshooting the target atom be to linked so as to give some flexibility for a better adjustment through energy minimization/ flexible fitting (Figure 12.18).

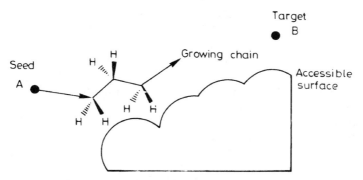

Figure 12.18 A chain is grown along the surface of the receptor, avoiding steric interactions with it. Spare valencies are filled with hydrogen atoms (from Lewis with permission [57]).

In the actual state, only steric requirements are examined for the chain generated crawling on to the receptor surface. It seems possible in future to also take into account hydrophobic or electrostatic effects (Figure 12.19).

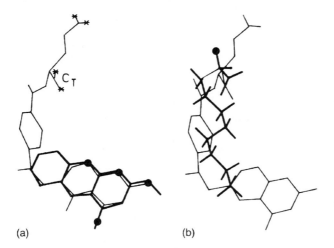

(a)   (b)

Figure 12.19 (a) Fitting a 2D template (bold) to the pteridine ring of methotrexate, and (b) a 3D template on the chain C_6–C_T. Ligand points and H-bonding atoms are marked as dots or asterisks (from Lewis with permission [57]).

Artificial intelligence techniques for an automated structure generation from template joining have also been developed by Gillet *et al.* [58, 58a] for the design of new compounds based on 3D criteria. For this constrained structure generation, the same approach was retained: creating a molecular graph to fit designated ligand points, then converting the graph into a true molecular structure. To form "skeleton structures", i.e. approximate structures that satisfy the primary constraints (steric contacts with a boundary or binding constraints) building blocks (corresponding to commonly occurring substructures) are pieced together in all possible ways to prevent the construction of identical skeletons. Different conformations are considered, since steric effects in the structure can sometimes favour conformations that are not the low energy ones for the templates (Figure 12.20).

Heuristics were developed to restrict the combinatorial explosion when joining templates. For example, grouping templates in similarity classes ("super templates") avoids unnecessary processing (if, in a class, the template of minimum steric bulk violates steric constraints, it is no longer necessary to try using the rest of the class). Tests were presented on the active site of an enzyme known by X-ray crystallography: the APPA (p-amidinophenyl pyruvate) binding site of trypsin (Figure 12.21). The program generates skeletons very similar to that of APPA, but also suggests possible novel

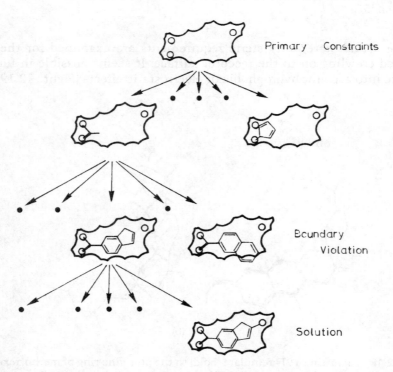

Figure 12.20   Progressive structure generation by incorporating templates (from Gillet *et al.* with permission [58a]).

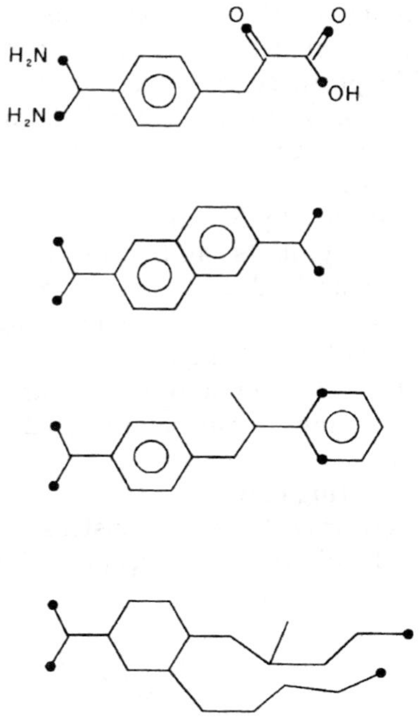

Figure 12.21   Generation of structures mimicking APPA (p-amidinophenyl pyruvate). Skeletons A and B are very similar to APPA. Although looking quite different, skeleton C also meets the geometrical constraints. Corresponding atoms are indicated by dots (from Gillet *et al.* with permission [58]).

solutions, looking quite different but maintaining a good overlap with the binding site. For other examples, see [58a].

In recent developments (the SPROUT progam [58a]) additional functions are included: *structure evaluation,* since the proposed structures must often satisfy other conditions: to be synthetically accessible, to present required transport properties, etc., *organization of the results* (clustering, ranking, etc.) since such programs of structure generation can produce a very large number of candidates. It can be noted also that in SPROUT, structure generation encompasses two parts: *generation* of molecules that satisfy the geometrical and steric constraints, and *possible substitution* of atoms in the skeletons to get better electrostatic and hydrophobic properties.

For building up peptide or peptide-like ligands binding known active sites, the GROW program of Moon and Howe [51] uses a library of low-energy conformations of amino acids (and some chain-terminator fragments or non-hydrolyzable chain inserts) as templates. From a user-defined seed point, fragments are gradually joined together within the active site, and a scoring function is evaluated considering non-bonded interactions between receptor and ligand, desolvation penalties and internal strain. At each stage, all templates stored in the library are considered and the n (let us say 10) highest scoring are retained. The efficiency of the program was demonstrated by reproducing the known bound conformation of an inhibitor of the aspartyl protease rhizopuspepsin and HIV-1 protease. An interesting application is the combined used of *de novo* design and screening a 3D database: computer-generated structures can constitute templates to search for new classes of compounds able to fit the receptor site.

A similar approach was also developed by Böhm with LUDI [59, 59a], which exploits libraries of about 600 fragments and uses empirical rules or information extracted from the Cambridge database to locate hydrogen bonding and hydrophobic groups in the binding site. These fragments are then connected to form a molecule thanks to bridge fragments from a second library. Applications were presented for the crystal packing of benzoic acid and the enzymes dihydrofolate reductase and trypsin (Figure 12.22).

Automated detection of receptor binding regions:
Also relevant to site-point directed drug design is, upstream, the automated determination at the atomic level of the receptor binding regions.

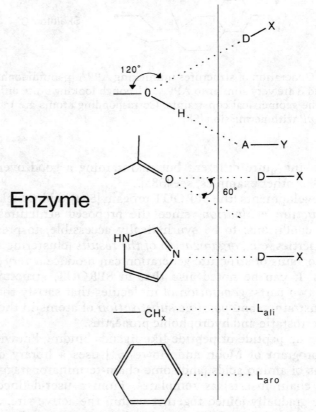

Figure 12.22   Some rules of LUDI to generate interaction sites (L_{ali}, L_{aro} = lipophilic -aliphatic, -aromatic; A, D = hydrogen acceptor, donor) (from Böhm with permission [59]).

Frequently, these active regions correspond to clefts or dimples. Identification of such features from the atom-coordinate file of a protein can therefore give interesting clues as to its putative binding sites. The method proposed by Lewis relies on a Voronoi tessellation of the molecular surface [60]. It finds the complete molecular surface in a planar slice through the receptor, and also locates clefts and dimples in this surface. Recall that a Voronoi tessellation divides space into domains such as points, within a certain domain, lie closer to the centre of that domain than to any other centre.

Atoms intervening in the section of the molecular surface intersect the plane along circles (the radii of which depend upon the van der Waals radii and the position of the atom centres with respect to the slice plane). In the simplest case, the trace of the molecular surface in that plane is described by a closed polygon through atomic centre projections. This convex polygon can be drawn using the "gift wrapping algorithm": starting from the atom with the lowest y coordinate, for instance, the remaining atomic points are scanned counterclockwise. The surface is drawn as wrapping a paper around a gift. The same is applied to Voronoi tessellation.

Given two atom centres, the surface is closed (and tessellation can be carried out) only if their distance (d) is less than $2r_p + R_1 + R_2$, where R_1 and R_2 are the radii associated with atoms 1 and 2, and r_p the radius of the probe rolling on to the molecular surface. Otherwise (dimple region) the probe can slip between atoms 1 and 2. Then another vertex, a common neighbour to atoms 1 and 2, has to be found sweeping counterclockwise. For larger clefts, the surface cannot be completed (the distance to two common neighbours is too large to prevent the passage of the probe). A simple expedient can then be used: placing a "traffic island" allows for directing the search algorithm around the correct path. This adds to the surface tessellation a *dummy tile*, which marks the cleft and allows for its identification (for large clefts, a clump of adjacent dummy tiles is similarly built) (Figure 12.23).

Putative receptor sites

In some cases, only the amino acid sequence of the target protein, but not its three-dimensional structure, is known. As stated by Marshall, "the tertiary information resides in the sequence, but the translation rules (to derive 3D structure) have defied definition", especially for turn prediction. However, some (approximate) heuristic approaches have been developed to propose tentative 3D structures and derive models for the receptor site [4]. Valuable clues are sometimes provided by X-ray data on proteins next to the target protein itself. Computer modelling (performing side chain replacement and subsequent geometry optimization) can then more safely propose a reasonable 3D structure for the target protein itself [61].

These approaches of the receptor site by inference or homology rely more closely on protein engineering and will be discussed in Chapter 13. For some examples, see Marshall [4].

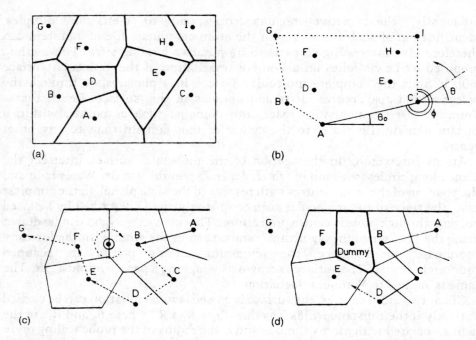

(a)   (b)

(c)   (d)

Figure 12.23   Detection of dimples (receptor binding regions). (a) Voronoi tessellation of a set of points within a rectangular window, (b) "gift wrapping algorithm" in the construction of the convex hull ACIGB. From the lowest point A, a line is swept around counterclockwise (from the x axis) until the next vertex C is found. The same process is repeated from C, finding I ..., (c) distances between B and F or the common neighbour E are too large to prevent the passage of the solvent probe. A "traffic island" added to the set of atoms redirects the drawing to the path B-C-D-E-F, (d) a "dummy tile" is added indicating a dimple in that region (from Lewis with permission [60]).

12.2.2 Unknown receptor site: deducing the pharmacophore (and the receptor) from binding drugs

In most cases, one can only infer the target (the receptor) from drugs binding to it. Some approaches for finding the pharmacophore geometry are discussed elsewhere [62, 63], and will be detailed below. The main steps can be schematized as follows. First, one has to select the essential groups assumed to constitute the pharmacophore. Then, from examination of low-energy conformers for the set of drugs investigated, one has to find a *common arrangement* of the groups *appearing in each molecule*. It corresponds to the receptor-bound conformation. The choice may be modified or refined by supplementary constraints until only one common solution is possible [1]. Distance constraints of equivalent atoms in different ligands, having different structure or conformational behaviour, give some clues regarding the receptor pockets accommodating these ligand atoms [24]. Investigation of strongly or weakly bound ligands have also been proposed [19, 22, 23].

In the "receptor mapping technique", all receptor-bound conformers are docked together with essential groups superimposed. The union of their volumes defines a minimal available space which is sterically allowed for the binding site. Even with good pharmacophore geometry, a molecule may be still inactive if it protudes beyond this allowed space, resulting in negative steric interactions, which preclude binding. The pharmacophoric hypothesis must be thereafter checked by examining compounds bearing the assumed pharmacophore but devoid of activity, and searching for any interpretation (differing metabolism, steric interactions, etc.).

Pharmacophores were often derived from the examination of conformationally more rigid ligands (usually antagonists) for which the problem is simpler, the agonists (more flexible) being considered afterwards. Indeed, conformational flexibility sets a major problem in pharmacophoric pattern search and receptor mapping techniques. First, when only flexible or semi-rigid agonists are used, a conformational search method is necessary to find conformations for which the superimposition of essential groups is possible. Second, ligands bind the receptor in a conformation corresponding to a minimal energy for the complex, but the free energy of association generally outweighs the energy of a conformational change of the ligand [22, 23]. Although various studies consider only minimum energy structures, there is some evidence of systems where the actual bound conformer lies above the energy minimum by about 3 kcal/mol (and corresponds, therefore, for the isolated molecule, to species not easily detected under usual conditions because of that very low abundance) [4].

There is no doubt that minimal energy conformations may be of value in the search for a pharmacophore, since, at least for closely related molecules, similarity in the ground state is likely to reflect some similarity for the capability of adjustment to a common receptor. However, some caveat must be given. An illuminating example in the field is given in a study by Cohen [64] on some β-lactam antibiotics: a comparison of 3D features of active Δ^3-cephalosporin (Figure 12.24a) and inactive Δ^2-cephalosporin (Figure 12.24b), extracted from the Cambridge database, indicates a clear geometry difference, suggesting an easy criterion for biological activity. However, active penicillin G (Figure 12.24c) does not match this pleasant picture, since its shape looks like that of inactive compounds.

This puzzling situation comes from the fact that the geometry attributed to penicillin G was that determined by X-ray crystallography (and in the more

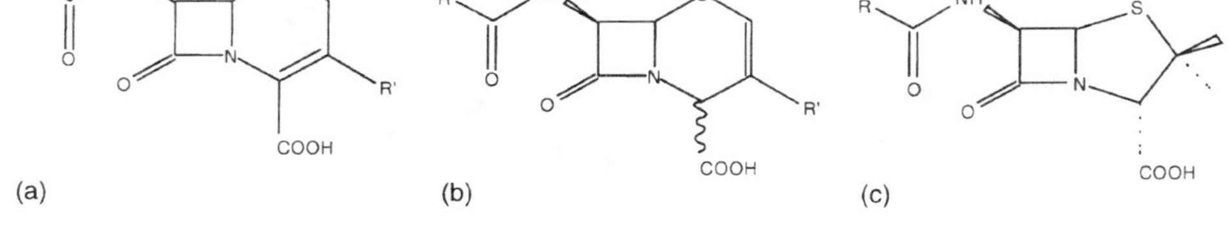

(a)   (b)   (c)

Figure 12.24   β lactam antibiotics example (from Cohen [64] with permission).

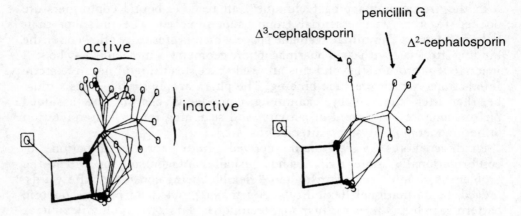

Figure 12.25 3D features from X-ray data (from Cohen with permission [64]).

stable crystal phase) (Figure 12.25). However, this molecule can suffer an easy pseudo-rotation of the penam nucleus, leading to a conformation able to mimic the active compounds and only slightly hampered on energy considerations (probably less than 1 kcal/mol above the more stable conformation) (Figure 12.26). Adaptation of the ligand to better fit the enzyme cavity gives an attractive interpretation of this apparent discrepancy.

Many efforts concentrated on conformational aspects to take into account some flexibility of the ligand and of the receptor during the complexation process. In other words, in place of the common image of "lock and key", the model of "hand and glove" was substituted, both able to undergo some deformation for a better mutual adaptation. Within this framework various approaches have been proposed. They mainly work on interatomic distances, which appear more convenient than dihedral angles (although some redundancy is involved) to traduce the geometrical information, particularly for flexible rings.

Active analogue approach

In this first type of approach, proposed by Marshall *et al.* [4, 19–21], active molecules are superimposed in one of their possible conformations, so that important corresponding groups (the pharmacophore) coincide. To cope with

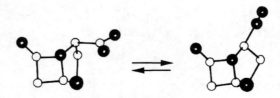

Figure 12.26 Adaptation by pseudo-rotation (from Cohen with permission [64]).

conformational flexibility, each rotatable bond is rotated by a fixed increment and distances between atoms of the pharmacophore are systematically recorded for each conformation allowed (i.e. conformations not suffering from atom-atom overlap). Sets of interatomic distances that can be achieved by all active molecules represent the possible pharmacophore geometry.

This approach requires compounds homologous enough so that superimposition is unambiguous. Furthermore the conclusions are mainly qualitative (active vs. inactive), since no account is made of energetic factors (relative affinities, etc.).

To make the derivation of the pharmacologically active conformations of acetylcholine and other agonists interacting with the muscarinic receptor easier, Schulman [65] proposed a novel model defining geometry with new parameters (rather than the usual dihedral angle). So, dummy atoms are introduced, representing the location of a carboxylate oxygen of the receptor and an electrophilic site (such as a hydrogen bonding atom) located at the point of minimum electrostatic potential of the ester oxygen (at about 1.2 Å of the oxygen atom). These dummy points constitute some characteristic invariants of the receptor site, and the dihedral angle PNOQ an invariant of the possible drug receptor complexes (Figures 12.27).

The same superimposition scheme is the basis of the model of Simon *et al.* [66]. Although less sophisticated in the treatment of steric effects, it (roughly) takes into account quantitative molecule-site interactions, mainly in terms of

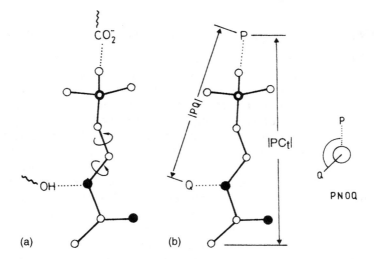

Figure 12.27 Symbolic structural descriptors. (a) Acetylcholine interacting with the receptor's carboxylate oxygen and an electrophilic group (here a hydrogen-bonding proton), (b) symbolic descriptors: P corresponds to the oxygen location, Q to the minimum of the electrostatic potential near the ester oxygen, and the dihedral angle PNOQ is an invariant descriptor of the receptor-bound acetylcholine system (N atoms are depicted as, open **circles**, O atoms as closed circles) (from Schulman *et al.* with permission [65]).

steric accessibility (hydrogen-bonding regions, hydrophobic pockets are not considered). Some other attempts to quantify the degree of molecular similarity, and to take into account energetic factors, have been published. So Hopfinger [22], in his *molecular shape analysis* approach, proposes to introduce the *volume overlap* (between the superimposed molecules) as a supplementary parameter in Hansch-type QSAR. In a later work [23], he calculated, via a molecular mechanics approach, an intermolecular potential field around a drug molecule, and used it as an indication of how it may interact with the site. This concept of an intermolecular potential in the surroundings of a macromolecule, approached here by means of the drug, is somewhat similar to that used by Goodford [31, 31a] to delineate favourable binding sites directly from the study of a known receptor.

A similar concern appears in Kato *et al.*'s work, trying to obtain some information about the important regions of the receptor cavity from superimposition of a set of active compounds. Their RECEPS system [67, 68] emphasizes the fact that a superimposition scheme based on properties rather than on atom positions may be more fruitful, and allows for deriving a reasonable model of the active site, as exemplified by the enzyme dihydrofolate reductase. It is expected that such computer-originated models for describing the receptor environment would be of help for the chemist in modelling the most likely bound conformations and designing other active ligands able to bind the receptor cavity.

Crippen's distance geometry and the ensemble approach

We have already introduced the distance geometry approach of Crippen (see Chapter 7) as an alternative avenue for traversal of the conformational space. It corresponds to a Monte Carlo sampling within the constraints of the distance limits. Distance geometry is able to propose the possible conformers for an isolated molecule. In a set of molecules, it can also be used to determine upper and lower bounds for the distances between the pharmacophore groups over all conformations allowed. From these common upper and lower bounds, a 3D arrangement of site points can be generated, giving a binding site model which can be used to dock additional molecules and rationalize binding data.

Complementing Crippen's earlier programs, an efficient algorithm has been proposed [69] within the framework of distance geometry. Rather than examining all sterically allowed conformations to derive the upper and lower distance matrices, one only uses a finite number of selected situations which can be viewed as discrete points in the conformational space. They are chosen by incrementing the conformational variables at regular intervals (sufficiently large so as not to multiply the number of points to examine). The distance matrix is now calculated from two successive conformations, which are both sterically or energetically allowed, and a "grace" value (minimum distance limit) is introduced for the superimposition. The advantage is that missing feasible binding modes is highly unlikely while maintaining reasonable computational tasks.

The ensemble approach:

In the "ensemble approach", developed by Sheridan *et al.* [70] for determining the feasible binding modes of flexible ligands on the receptor site, the classical distance geometry of Crippen is modified so that two or more molecules are treated simultaneously as a single "ensemble". In Marshall's and the original Crippen methods, individual molecules are first considered before examining the possibility of superimposition. In the ensemble approach, the superimposition of groups is assigned directly, since all molecules are treated together within an "ensemble". This allows for incorporating additional constraints, such as "excluded volume" information. Also, computation time is independent of the number of rotatable bonds.

The approach can generate, in one step, coordinates for a set of molecules in their "active conformation", i.e. conformations such that their essential points can be superimposed. Among the main features of the method, we can note that all the coordinates for the various molecules investigated are included in the same distance matrix. However, additional conditions are introduced. So, whereas the lower limit L_{ij} is set to the sum of the van der Waals radii for two atoms (i and j) in the same molecule, on the contrary $L_{ij} = 0$ when i and j are paired positions (from two different molecules) that can be superimposed. In that case, the upper U_{ij} is chosen equal to a small tolerance parameter (typically 0.3 Å). That is, in the final structure, superimposed atoms may not be further apart than this tolerance value. Then the standard algorithm is followed.

This approach has been illustrated for the vertebrate nicotinic acetylcholine receptor. Nicotinic agonists induce an open-channel form of the receptor. Antagonists bind to the receptor but do not open the channel. A common pharmacophore was extracted from four semi-rigid nicotinic agonists: (−)-nicotine, (−)-cytisine, (−)-ferruginine methiodide and (−)-muscarone (Figure 12.28).

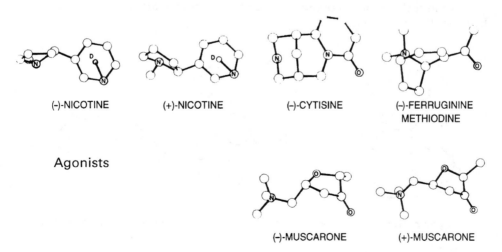

(−)-NICOTINE   (+)-NICOTINE   (−)-CYTISINE   (−)-FERRUGININE METHIODINE

Agonists

(−)-MUSCARONE   (+)-MUSCARONE

Figure 12.28   The ensemble approach (from Sheridan *et al.* with permission [70]).

Three essential groups were chosen: a cationic centre A (quaternary nitrogen or protonatable nitrogen), an electronegative atom or a centre which may act as a hydrogen bond acceptor B, and a third atom C (carbonyl carbon or dummy atom at the centre of a phenyl ring) forming a dipole with B, and defining a direction along which a hydrogen bond is likely to form (Figure 12.29). After generating several sets of superimposed structures, the only possible pharmacophore is derived, where points A, B, C form a triangle with sides 4.8 Å ± 0.3 (AB), 4.0 Å ± 0.3 (AC) and 1.2 Å (BC).

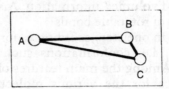

Figure 12.29 Deriving the only possible pharmacophore (from Sheridan *et al.* with permission [70]).

Then for each individual agonist, conformations obeying the pharmacophore requirements are generated. The antagonists strychnine, trimethaphan, etc. also meet this triangle (Figure 12.30).

Once the active conformations have been selected, docking them together (by superimposition of the essential atoms on to an ideal pharmacophore triangle) defines, by union of their volumes, that part of space that any agonist may occupy and so defines the handedness needed for fitting the cavity of the receptor and activating it (Figure 12.31).

With such information about the volume that an agonist may occupy on the receptor and the pharmacophore geometry, one may start to design new active drugs that fit the model.

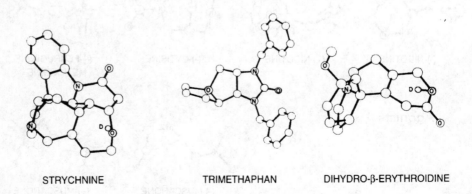

STRYCHNINE   TRIMETHAPHAN   DIHYDRO-β-ERYTHROIDINE

Figure 12.30   Antagonists (from Sheridan *et al.* with permission [70]).

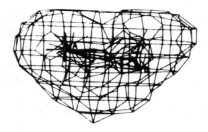

Figure 12.31 Combined volume of agonists (from Sheridan *et al.* with permission [70]).

Site modelling

One of the important applications of distance geometry is in receptor site modelling, particularly regarding the binding of small ligand molecules to sites on proteins or macromolecules. Compared to other approaches developed in QSAR, using purely empirical correlations, the treatment proposed by Crippen emphasizes the spatial aspects of the problem [71]. It proposes a rationalization of the data thanks to an *understanding of the actual mode of binding*. A reasonable picture of the geometry of the site and plausible deductions as to the chemical nature of the binding sites can be so attained, allowing for heuristic indications in the design of new, more active inhibitors or drugs. Furthermore, as previously stressed, using a distance matrix (invariant under translation or rotation) avoids geometrical displacements involved in the usual docking studies.

The approach relies on the following assumptions:

- Binding is observed to occur on a single site of a receptor protein.
- Each ligand has a given chemical structure and stereochemistry, but suffers from some flexibility due to rotation about single bonds.
- No chemical modification of the ligand occurs during binding, although its conformation may change to better fit the receptor sites.
- Such conformational changes involve small energy variations (as compared to binding free energy).
- The observed free energy of binding may be calculated from additive interaction energies for all contacts between parts of the ligand molecule and parts of the receptor sites.
- A slight flexibility is allowed for the receptor site, without energetic cost.

From known drug structures, one constructs a geometry for the receptor and, given experimental free energies of binding, deduces possible binding sites in terms of geometry and the chemical character of the various parts of the site. Subsequent evaluation of interaction energy as a sum of pairwise terms then proposes the best binding mode.

The ligand is represented by a collection of representative *ligand points* in space. These may be atoms (presumably involved in the binding process according to the chemist's intuition) or dummy atoms which are likely to properly describe the interactions of parts of the drug molecule with the receptor macromolecule. The centre of a benzene ring may be a good descriptor for the location of the ring if binding is not specific of a given orientation of that moiety of the ligand, for instance, in a large receptor pocket.

Once the ligand points have been chosen, the molecule is defined by a distance matrix gathering all distances between pairs of ligand points. To cope with conformational flexibility, the upper and lower triangles of this distance matrix represent, respectively, the upper and the lower limits of the distances for the different conformers investigated. Although some problems of correlated flexibility with cycles have been reported, this approximate description is quite convenient to condense conformational flexibility, a major problem in pharmacophoric pattern-search and receptor-mapping techniques. Note that from this matrix, usual distance geometry techniques allow for determining the coordinates of the points.

Similarly, a binding site is proposed as a series of *site points* whose positions are specified in a fixed distance matrix (only small variations on site point distances are allowed). The number of these site points depends upon the details required for describing the receptor part. Site points are called *filled* (sterically blocked position, forbidden for any ligand point on binding) or *empty* (vacant position where a ligand point may come during binding). These are best thought of as corresponding in the real receptor to the locations of pockets of various types (accommodating a phenyl ring, an ethyl group, able to form a hydrogen-bond, and so on).

Thanks to the description of both ligand and receptor by a set of discrete points, the binding mode is represented by a list of which ligand points coincide with which empty site points. It can be determined in a simple combinatorial way, searching for energetically favoured interactions and enforcing geometrical constraints.

Two approaches have been developed. The first, if the number of points in each ligand is small, consists of automatically finding the simplest binding site consistent with the binding data. An exhaustive search of all combinations of the number of site points, their types, their distance matrix and their interaction energy matrix is carried out. In the second approach (more convenient in the usual case of more complicated ligands), one proposes a binding site and then the computer fits the data in an interactive "cut and try" fashion.

Binding energy is calculated by a summation of contributions from all contacts between site and ligand points. These terms correspond to the free energy variation ΔG for the process:

$$\begin{array}{c} \text{solvated ligand point} \\ + \\ \text{solvated site point} \end{array} \quad \rightarrow \quad \text{occupied site point}$$

The values are taken from a table where rows and columns correspond respectively to the type of ligand points and site points. In a first step, this table is proposed by the user according to reasonable assumptions (and will be refined further as part of the fitting process). For example, a hydrophobic pocket may be attractive for a phenyl ring, but not so strongly attractive to an ionized group; some filled points may be highly repulsive for a t-Bu group, but mildly attractive to a methyl [71]. The experimental free energy of binding can be derived from the measured I_{50} (millimolar concentration of an inhibitor required to produce 50% inhibition) thanks to the usual Michaelis–Menten model.

Determining how complicated a site must be, in order to explain the experimental results, is not easy. Ordinarily, one begins by arranging some empty site points to match some common feature of the ligand structures "base group". Ligand points of the base group corresponding to a common geometrical arrangement (within upper and lower bounds) can be thought of as forming a pharmacophoric pattern, necessary for a ligand to be recognized and to bind to the receptor site. Other surrounding points are then introduced to accommodate the occurrence of various substituents on the ligands, i.e. groups occurring one or more times each in some, but not all, of the ligands. They often represent functional groups associated with only one ligand point, such as NH_2-, S-. Rough interaction energies are used to ensure reasonable binding. Then the process is refined for an objective calculation of interaction energies.

The key step in these two approaches is finding the optimal mode of binding. This is carried out by the following sequence:

- All possible modes of binding are sought by systematically generating all possible combinations of contacts between ligand and site points (unused ligand or empty site points are allowed).
- Various pruning steps on the proposed solutions eliminate contacts with unfavourable interaction energy or geometrically forbidden situations (distances between site points and between ligand points must match according to a given tolerance for flexibility). "Forced contacts" and chirality conditions are then examined. They correspond to contacts necessarily resulting from certain combinations of the contacts already chosen. When four points of a chiral quartet are involved in a contact combination, chirality of the ligand points and the site points must also match.
- When a proposal has passed all the above tests, the corresponding binding energy is calculated. The solution retained corresponds to the minimal (more favourable) binding energy. Underlying this approach, therefore, is the concept that the *ligand binds at the site in whatever conformation and orientation minimizes its free energy of binding.*

A least squares fit of the experimental free energy to a sum of interaction contributions is performed. If the binding scheme is correct, this should lead to a small residual error. Refined values of interaction energies between ligand and site points are then proposed. For outliers, additional hypotheses can be interactively introduced, imagining different binding modes to refine the model. Poor agreement may arise from a choice of ligand points which does not adequately represent significant features of their structure.

Once the binding mode is identified (fixing site points for each of the base and substituent sets), calculation of the coordinates of the associated site points is straightforward, giving a 3D representation of the corresponding part of the receptor. The exhaustive algorithm for deducing the site geometry and the interaction energy matrix and the interactive binding algorithm are described by Crippen [71].

The example of substituted phenoxyacetones $R–C_6H_5–OCH_2–CO–CH_3$ (chymotrypsin inhibitors) allows us to give some details about the approach [71]. In a first step, the molecules can be represented with three ligand points corresponding to the carbonyl carbon, the ether oxygen and the centre of the phenyl ring (Figure 12.32).

Empty site points will be associated with these ligand points, but other site points are necessary to interpret the fact that the binding site can accommodate an m-methyl, but not two m,m'-methyl, groups (Figure 12.33). A first additional binding site (4) is provided for the m-methyl. To treat the low

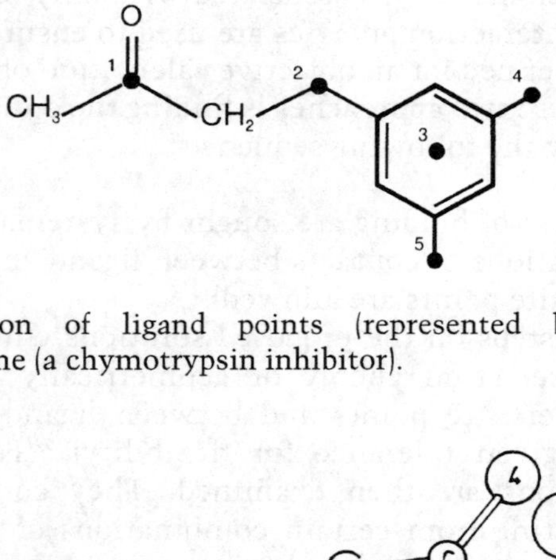

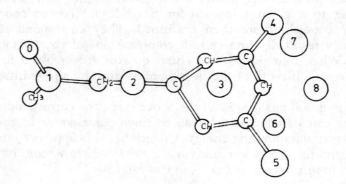

Figure 12.32 Selection of ligand points (represented by dots) for m,m'-dimethylphenoxyacetone (a chymotrypsin inhibitor).

Figure 12.33 Binding site for chymotrypsin inhibitors. Site points are represented by large spheres, ligand atoms by small spheres. Site points (1) and (2) coincide respectively with the carbonyl carbon and the ether oxygen, and site points (4) and (5) to the methyl groups (see text). This molecular orientation would be sterically unfavourable (adapted from Crippen with permission [71]).

affinity of the m,m′ derivative, it is necessary to include a repulsive site point (in (5)), but this is not sufficient, since the phenyl would then be able to rotate so as to avoid this unfavourable site point (5). So three additional repulsive site points (in (6–8)) are introduced, 3 Å apart from the phenyl site.

Interestingly, from the refined values of binding contributions, some insight can be gained as to the nature of the interactions. Point (1) is presumably associated with an H-bond donor to the carbonyl, point (2) corresponds to a polar pocket, and point (3) to a strongly hydrophobic region for binding the phenyl group. It may be thought as "the center of a structured pocket surrounded by sterically repulsive regions (5–8) and a small pocket (4)" which can accommodate one methyl group. Furthermore, Crippen was able to tentatively locate these binding points by respect to the chymotrypsin residues, in a proposal consistent with some X-ray data [71, 72] (Figure 12.34).

As a concluding remark, one can note that Crippen's approach gives information not only about the geometrical requirements, but also about interaction energy, and focuses on building a discrete model of the receptor site which allows a more thorough and realistic examination of all modes of interaction with each drug molecule.

A major problem with respect to the general applicability of the method arises from the type of data needed [71]: the data set must contain various classes of structurally diverse ligands to give significant indications on the geometry of the site. It must also be large enough to allow statistically sound derivation of the interaction contributions. Dissociation constants of the

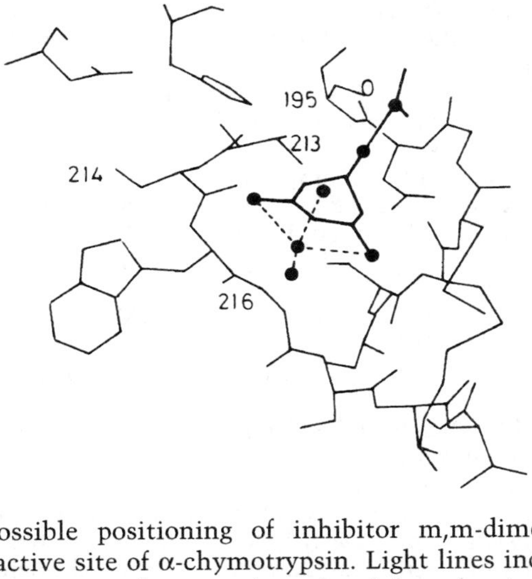

Figure 12.34 A possible positioning of inhibitor m,m-dimethylphenoxyacetone (heavy lines) in the active site of α-chymotrypsin. Light lines indicate residues of the active site (some sequence numbers are shown) and dots the site points. The ligand's carbonyl is near the side chain oxygen of serine 195 in the foreground. Points (4) and (5) are in the plane of the ligand phenyl ring (respectively at the right and left sides). Site points (6) and (7) are above and below this plane, and point (8) is in the background in the plane (dashed lines are drawn only to convey a sense of depth) (from Crippen with permission [71]).

ligand-receptor complex must also be available to derive the free energy of binding.

Finally, as for any other approaches (except, of course, direct X-ray measurements of the ligand-receptor complex), this receptor mapping procedure only generates a *possible model* for the binding site of a receptor protein, corresponding to necessary (but not sufficient) conditions for an allowed binding mode.

REFERENCES

1. R.P. Sheridan and R. Venkataraghavan *Acc. Chem. Res.*, **20:** 1987; 322–329.
2. A. Crum-Brown and T.R. Fraser *On the Connection between Chemical Constitution and Physiological Action*, Royal Society of Edinburgh. Neil and Co, 1869.
3. J.N. Langley *J. Physiol. (Lond)*, **1:** 1878; 339–369.
4. G.R. Marshall *Ann. Rev. Pharmacol. Toxicol.* **27:** 1987; 193–213.
5. H.H. Grunhagen, M. Iwatsubo and J.P. Changeux *Eur. J. Biochem.*, **80:** 1977; 225.
6. L.J. Pike *Mol. Pharmacol.*, **14:** 1978; 370.
7. J.T. Harmon, E.S. Kempner and C.R. Kahn *J. Biol. Chem.* **256:** 1981; 7719–7722.
8. L.H. Hall and L.B. Kier *J. Theor. Biol.*, **58:** 1976; 177–195.
9. R. Hoffmann, J.J. Bourguignon and C.G. Wermuth *QSAR: Rational Approaches to the Design of Bioactive Compounds*, C. Silipo and A. Vittoria (eds). Elsevier, 1991; 283–292.
10. D.A. Matthews, R.A. Alden, J.T. Bolin, S.T. Freer, R. Hamlin, N. Xuong, J. Kraut, M. Poo, M. Williams and K. Hoogsteen *Science*, **197:** 1977; 452–455.
11. D.A. Matthews, R.A. Alden, S.T. Freer, N. Xuong and J. Kraut *J. Biol. Chem.*, **254:** 1979; 4144–4151.
12. D.A. Matthews, R.A. Alden, J.T. Bolin, D.A. Filman, S.T. Freer, R. Hamlin, W.G.J. Hol, R.L. Kisliuk, E.J. Pastore, L.T. Plante, N. Xuong and J. Karut *J. Biol. Chem.*, **253:** 1978; 6946–6954.
13. D.J. Baker, C.R. Beddell, J.N. Champness, P.J. Goodford, F.E.A. Norrington, D.R. Smith and D.K. Stammers *FEBS Lett.*, **126:** 1981; 49–52.
14. L.F. Kuyper, B. Roth, D.P. Baccanari, R. Ferone, C.R. Beddell, J.N. Champness, D.K. Stammers, J.G. Dann, F.E.A. Norrington, D.J. Baker and P.J. Goodford *J. Med. Chem.*, **25:** 1982; 1120–1122.
15. J.T. Bolin, D.J. Filman, D.A. Matthews, R.C. Hamlin and J. Kraut *J. Biol. Chem.*, **257:** 1982; 13650–13662.
16. D.J. Filman, J.T. Bolin, D.A. Matthews and J. Kraut *J. Biol. Chem.*, **257:** 1982; 13663–13672.
17. C. Hansch, R. Li, J.M. Blaney and R. Langridge *J. Med. Chem.*, **25:** 1982; 777–784.
18. R.W. Schevitz, Z. Otwinowski, A. Joachimiak, C.L. Lawson and P.B. Sigler *Nature (Lond)*, **317:** 1985; 782–786.
19. G.R. Marshall, C.D. Barry, H.E. Bosshard, R.A. Dammkoehler and D.A. Dunn *Computer Assisted Drug Design*, ACS symposium, Ser. No 112, E.C. Olson and R.E. Christoffersen (Eds). American Chemical Society. Washington DC, 1979, pp. 205–226.
20. C. Humblet and G.R. Marshall *Ann. Rep. Med. Chem.*, **15:** 1980; 267–276.
21. C. Humblet and G.R. Marshall *Drug Dev. Res.*, **1:** 1981; 409–434.
22. A.J. Hopfinger *J. Am. Chem. Soc.*, **102:** 1980; 7196–7206.
23. A.J. Hopfinger *J. Med. Chem.*, **26:** 1983; 990–996.
24. A.K. Ghose and G.M. Crippen *J. Med. Chem.*, **25:** 1982; 892–899.

25. B.K. Shoichet, D.L. Bodian and I.D. Kuntz *J. Comput. Chem.*, **13:** 1992; 380–397.
26. C.R. Beddell, P.J. Goodford, F.E.A. Norrington, S. Wilkinson and R. Wootton *Br. J. Pharmacol.*, **57:** 1976; 201–209.
27. J.M. Blaney, E.C. Jorgensen, M.L. Connolly, T.E. Ferrin, R. Langridge, S.J. Oatley, J.M. Burridge and C.C.F. Blake *J. Med. Chem.*, **25:** 1982; 785–790.
28. A.K. Ghose and G.M. Crippen *J. Med. Chem.*, **27:** 1984; 901–914.
29. C. Freundenreich, J.P. Samana and J.F. Biellmann *J. Am. Chem. Soc.*, **106:** 1984; 3344–3353.
30. D.J. Danziger and P.M. Dean *Proc. Roy. Soc. Lond., B*, **236:** 1989; 101–113.
31. P.J. Goodford *J. Med. Chem.*, **27:** 1984; 557–564.
31a. P.J. Goodford *J. Med. Chem.*, **28:** 1985; 849–857.
32. D.N.A. Boobbyer, P.J. Goodford, P.M. McWhinnie and R.C. Wade *J. Med. Chem.*, **32:** 1989; 1083–1094.
33. E.C. Meng, B.K. Shoichet and I.D. Kuntz *J. Comput. Chem.*, **13:** 1992; 505–524.
34. N. Pattabiraman, M. Levitt, T.E. Ferrin and R. Langridge *J. Comput. Chem.*, **6:** 1985; 432–436.
35. S.J. Wodak and J. Janin *J. Mol. Biol.*, **124:** 1978; 323–342.
36. R.L. DesJarlais, R.P. Sheridan, J.C. Dixon, I.D. Kuntz and R. Venkataraghavan *J. Med. Chem.*, **29:** 1986; 2149–2153.
37. M. Santavy and J. Kypr *J. Mol. Graphics*, **2:** 1984; 47–49.
38. I.D. Kuntz, J.M. Blaney, S.J. Oatley, R. Langridge and T.E. Ferrin *J. Mol. Biol.*, **161:** 1982; 269–288.
39. A.M. Lesk *Comm. ACM*, **22:** 1979; 219–224.
40. R.L. DesJarlais, R.P. Sheridan, G.L. Seibel, J.C. Dixon, I.D. Kuntz and R. Venkataraghavan *J. Med. Chem.*, **31:** 1988; 722–729.
41. K.D. Stewart, J.A. Bentley and M. Cory *Tetrahedron Computer Methodology*, **3:** 1990; 713–722.
42. A. Leach and I.D. Kuntz *J. Comput. Chem.*, **13:** 1992; 730–748.
43. D.S. Goodsell and A.J. Olsen *Proteins, Struct. Funct. Gen.*, **8:** 1990; 195–202.
44. M. Billeter, T.F. Havel and I.D. Kuntz *Biopolymers*, **26:** 1987; 777–793.
45. R.P. Sheridan, A. Rusinko III, R. Nilakantan and R. Venkataraghavan *Proc. Natl. Acad. Sci. USA*, **86:** 1989; 8165–8169.
46. R.P. Sheridan and R. Venkataraghavan *J. Comput.-Aided Mol. Design*, **1:** 1987; 243–256.
47. J.H. van Drie, D. Weininger and Y.C. Martin *J. Comput.-Aided Mol. Design*, **3:** 1989; 225–251.
48. Y.C. Martin *J. Med. Chem.*, **35:** 1992; 2145–2154.
49. M.G. Bures, C.W. Hutchins, M. Maus, W. Kohlbrenner, S. Kadam and J.W. Erickson *Tetrahedron Computer Methodology*, **3:** 1990; 673–680.
50. A.S. Smellie, G.M. Crippen and W.G. Richards *J. Chem. Inf. Comput. Sci.*, **31:** 1991; 386–392.
51. J.B. Moon and W.J. Howe *Tetrahedron Computer Methodology*, **3:** 1990; 697–711.
52. A.E. Weber, T.A. Halgren, J.J. Doyle, R.J. Lynch, P.K.S. Siegl, W.H. Parsons, W.J. Greenlee and A.A. Patchett *J. Med. Chem.*, **34:** 1991; 2692–2701.
53. D.J. Danziger and P.M. Dean *Proc. Roy. Soc. Lond. B*, **236:** 1989; 115–124.
54. R.A. Lewis and P.M. Dean *Proc. Roy. Soc. Lond. B*, **236:** 1989; 125–140.
55. R.A. Lewis *J. Comput.-Aided Mol. Design*, **4:** 1990; 205–210.
56. R.A. Lewis and P.M. Dean *Proc. Roy. Soc. Lond. B*, **236:** 1989; 141–162.
57. R.A. Lewis in *Proceedings of the Sixth European Seminar and Exhibition on Computer Aided Molecular Design*, London, October 1989.
58. V.J. Gillet, A.P. Johnson, P. Mata and S. Sike *Tetrahedron Computer Methodology*, **3:** 1990; 681–696.
58a. V.J. Gillet, W. Newell, P. Mata, G. Myatt, S. Sike, Z, Zsoldos and A.P. Johnson *J. Chem. Inf. Comput. Sci.*, **34:** 1994; 207–217.
59. H.J. Böhm *J. Comput.-Aided Mol. Design*, **6:** 1992; 61–78.
59a. H.J. Böhm *J. Comput.-Aided Mol. Design*, **6:** 1992; 593–606.

60. R.A. Lewis *J. Comput.-Aided Mol. Design*, **3:** 1989; 133–147.
61. D.W. Cushman, H.S. Cheung, E.F. Sabo and M.A. Ondetti *Biochemistry*, **25:** 1977; 5484–5491.
62. L. Motoc, R.A. Dammkoehler, D. Mayer and J. Labanowski *J. Quant. Struct. Relat. Pharmacol. Chem. Biol.*, **5:** 1986; 99–105.
63. G.R. Marshall *Ann. N.Y. Acad. Sci.*, **439:** 1985; 162–169.
64. N.C. Cohen *J. Med. Chem.*, **26:** 1983; 259–264.
65. J.M. Schulman, M.L. Sabio and R.L. Disch *J. Med. Chem.*, **26:** 1983; 817–823.
66. Z. Simon, I. Badilescu and T. Racovitan *J. Theor. Biol.*, **66:** 1977; 485–495.
67. Y. Kato, A. Itai and Y. Iitaka *Tetrahedron*, **43:** 1987; 5229–5236.
68. Y. Kato, A. Inoue, M. Yamada, N. Tomioka and A. Itai *J. Comput.-Aided Mol. Design*, **6:** 1992; 475–486.
69. A.K. Ghose and G.M. Crippen *J. Comput. Chem.*, **6:** 1985; 350–359.
70. R.P. Sheridan, R. Nilakantan, J.S. Dixon and R. Venkataraghavan *J. Med. Chem.*, **29:** 1986; 899–906.
71. G.M. Crippen *J. Med. Chem.*, **22:** 1979; 988–997.
72. M.R. Linschoten, T. Bultsma, A.P. Ijzerman and H. Timmerman *J. Med. Chem.*, **29:** 1986; 278–286.

13 *Modelling proteins*

Proteins are the focus of constant and prominent interest in molecular modelling applications, in view of their key role in nearly all biological processes. Among their main functions one can note enzymatic catalysis, the transport and storage of small molecules or ions, transmission of nerve impulses, control of growth and differentiation, immune protection (antibodies), hormones, repressors, coordinated motion, mechanical support, etc.

Proteins are built from a basic set of 20 (naturally occurring) amino acids linked together by peptide bonds to form large polypeptide chains (typically from 100 to about 1000 amino acids). A peptide bond links the carbonyl of one amino acid and the amino group of the next one. These amino acids are identified by either a three- or one-letter code (see Table 13.1), and by convention the polypeptide chain is written starting from the amino end towards the carboxyl terminal residue. Despite this small number of

Table 13.1 The 20 naturally occurring amino acids found in proteins (HOOC–CHR–NH$_2$).

Structure of R	Name	Abbreviation	
R is alkyl			
- H	Glycine	Gly G	
- CH$_3$	Alanine	Ala A	
- CH(CH$_3$)$_2$	Valine	Val V	■
- CH$_2$CH(CH$_3$)$_2$	Leucine	Leu L	■
- CH(CH$_3$)CH$_2$CH$_3$	Isoleucine	Ile I	■
R contains an OH group			
- CH$_2$OH	Serine	Ser S	
- CH(CH$_3$)OH	Threonine	Thr T	
- CH$_2$—⬡—OH	Tyrosine	Tyr Y	
R contains sulphur			
- CH$_2$SH	Cysteine	Cys C	
- CH$_2$CH$_2$SCH$_3$	Methionine	Met M	■
R contains a carboxyl or amido group			
- CH$_2$CO$_2$H	Aspartic Acid	Asp D	*i*
- CH$_2$CH$_2$CO$_2$H	Glutamic Acid	Glu E	*i*
- CH$_2$CONH$_2$	Asparagine	Asn N	
- CH$_2$CH$_2$CONH$_2$	Glutamine	Gln Q	*i*
R contains a basic amino group			
- CH$_2$CH$_2$CH$_2$NHC(NH)NH$_2$	Arginine	Arg R	*i*
- CH$_2$CH$_2$CH$_2$CH$_2$NH$_2$	Lysine	Lys K	*i*
- CH$_2$ (imidazole ring)	Histidine	His H	*i*
R contains an aromatic group			
- CH$_2$—⬡	Phenyl Alanine	Phe F	■
- CH$_2$ (indole ring)	Tryptophane	Trp W	
see above	Tyrosine	Tyr Y	
imino acid (complete structure)			
(proline ring)—CO$_2$H	Proline	Pro P	

■ largely hydrophobic / *i* largely hydrophilic. They are respectively almost always found on the internal part (■) or the external part (*i*) of globular proteins.

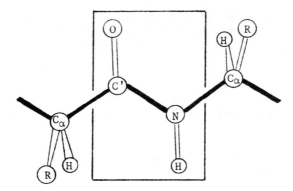

Figure 13.1   The peptide bond: a planar and rigid arrangement. Bond lengths (Å): C'O 1.24, C'N 1.32, C'C$_\alpha$ 1.53, NC$_\alpha$ 1.47; bond angles (°): C$_\alpha$–C'–N 113, C'–N–C$_\alpha$ 123, N–C$_\alpha$–C' 110 (from Stryer with permission [1] p 29).

elementary bricks, there is a huge diversity in assembling them in a polypeptide chain to obtain an enormous number of structures, largely differing in shape and function: each protein corresponds to a specific arrangement, which is genetically determined (Figure 13.1).

Except for glycine, all intervening amino acids are chiral molecules, but only *L*-enantiomers are found in proteins (Figure 13.2). Some proteins also contain special amino acids formed by modification of naturally occurring amino acids. For basic features and properties on proteins, see elsewhere [1–5].

All amino acid residues share a common CO–CHR–NH moiety; but they can be subdivided according to the nature of the side chain R: hydrophobic alkyl or aryl groups; highly polar groups either neutral (hydroxyl, amide) or ionized (acidic: carboxyl, basic: ammonium); planar rigid aromatic groups, sulphur containing. Some of these characteristics of side chains are reflected in the ^{1}H NMR spectrum: for instance, protons attached to an aromatic ring and amide NH appear at low field. Resonance signals of methyl groups are generally shifted upfield as they are subject to local ring current fields.

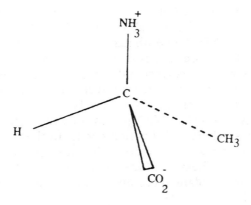

Figure 13.2   Absolute configuration of *L*-Alanine.

Owing to the complexity of such systems, various levels of simplification have been introduced, for both structural description and display, depending upon the nature and precision required for a selected application.

13.1 STRUCTURAL ANALYSIS

Determining the 3D structure of a protein can be roughly presented as hierarchically solving:

- *Primary structure*: identification of the amino acids present (intra-residue assignments) and determination of their sequence in the molecule (inter-residue or "sequential" assignment). Chemical methods exist, but we will mention here only spectroscopic approaches mainly based on NMR.
- *Secondary structure*: recognition of rigid regular regions such as α-helices, β-sheets, etc.
- *Tertiary structure*: specifying the relative locations of these motifs and looking at the overall 3D appearance.

Hydrogen-bonds, between the carbonyl group of one residue and the amino group of another, play a preeminent role in the building of regular structures in proteins, with α-helices and β-sheets as the most characteristic patterns:

$$C = O \cdots\cdots H - N$$

In an *α-helix*, the polypeptide chain is tightly coiled in a helical array with the side chains of the constituting residues directed outwards, the screw being right handed (clockwise). The structure is stabilized by a net of H-bonds between the carbonyl groups and the NH groups situated four residues ahead, all the CO and NH of the polypeptide chain being bonded. The pitch of the helix is 5.4 Å or 3.6 amino acid units, each residue being related to the next one by a rotation of 100° and a 1.5 Å translation along the axis. Starting from the N terminal end (supposed at the bottom of the scheme), CO groups are oriented upwards (towards N atoms situated above) and H-bonds are roughly parallel to the helix axis (Figure 13.3). The 3_{10} helix (sometimes found) corresponds to a variant of α helix with 3 residues per turn, instead of 3.6.

Another regular structure corresponds to β *pleated sheets*: the chains are almost fully extended (with a distance between following residues of 3.5 Å) and gathered in sheets of (generally) 3–5 strands. Stabilization is ensured by hydrogen-bonds between NH and CO groups from different polypeptide chains (contrary to α-helices, where stabilization occurs in the same chain). Adjacent strands can run in the same direction (parallel β-sheet (βp) or opposite directions (antiparallel β-sheet (β)). Side chains are, alternatively, up and down with respect to the mean plane (possibly leading to steric repulsions between chains) (Figure 13.4).

Collagen, present in skin, bone and cartilage, involves repeated sequences of

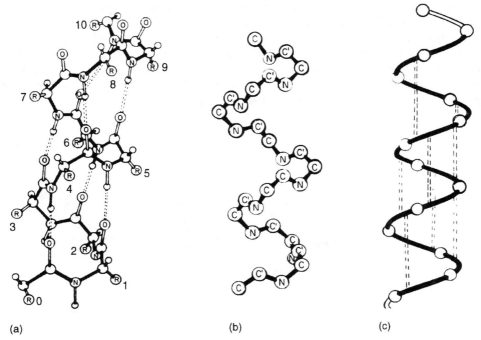

Figure 13.3 α-helix representation (a) and schematic models (b) and (c) indicating only the main-chain atoms (b) or locating the α-carbons on a helical thread (c). In (a) and (c) broken lines identify hydrogen-bonding between NH and CO four residues apart (adapted from Stryer with permission [1] p 29).

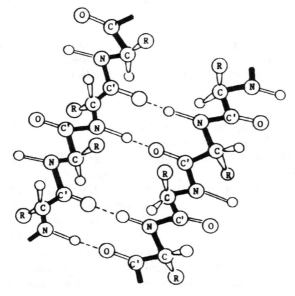

Figure 13.4 Schematic structure of an antiparallel β-sheet. Broken lines symbolize hydrogen-bonds. R groups are alternately above or below the mean plane of the sheet (adapted from Stryer with permission [1] p 31).

Gly-Pro-Pro arranged in another helical motif (called a type II trans helix). Stabilization is here ensured by mutual repulsion of the pyrrolidone rings of the proline residues. In collagen, three helices wind around each other, leading to the constitution of fibres of remarkable strength. Glycine (one for each three-residue unit), with its side chain limited to a hydrogen atom, is necessary for reducing steric interactions inside this triple stranded helical cable. The pitch of this superhelix is about 3.3 residues per turn (corresponding to 2.9 Å) (Figure 13.5).

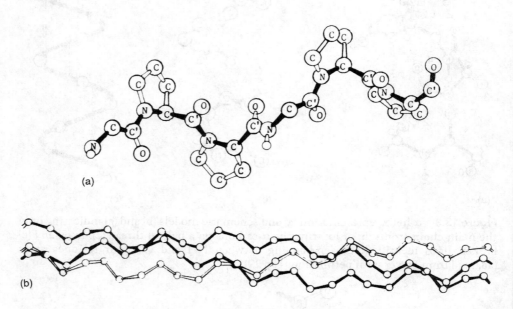

(a)

(b)

Figure 13.5 Collagen structure. (a) Single strand helix (dark line shows the main chain ···NC$_\alpha$C′··· in the sequence -Gly-Pro-Pro-Gly-Pro-Pro-); (b) model of the triple-stranded collagen cable (α-carbons only) (adapted from Stryer with permission [1] p 188).

Frequently, chain direction can be abruptly reversed in a *β-turn*, corresponding to an H-bond formation between a CO group and an NH group three residues apart. β-turns occur between four residues when the distance C$_{\alpha i}$–C$_{\alpha i+3}$ is less than 7 Å and O$_i$–N$_{i+3}$ < 3.5 Å (Figure 13.6).

They are more frequently observed with D,G,S,P, at the end of α-helices, at the folding of a β-sheet, in the junction between helices or between a helix and a β-sheet, or in disulphide-bridge stabilized structures. According to the torsional angles for residues *i*, *i+1* and *i+2*, eleven types of β-turn have been identified. Proline has a high probability of being involved in such turns.

Tertiary structure refers to the overall shape of the protein. This division between secondary and tertiary structure may appear somewhat artificial. It corresponds to the fact that tertiary structure more directly relies on the interactions between residues far apart in the sequence (H-bonding and

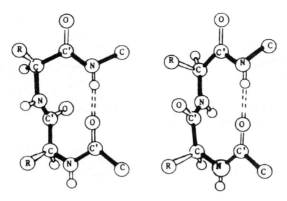

Figure 13.6 Two examples of β-turns (adapted from Cantor and Schimmel with permission [2] p 95).

covalent disulphide bonds resulting from the oxidation of SH groups of cysteine) (Figure 13.7).

The overall shape distinguishes fibrous from globular proteins. Fibrous proteins such as keratins (from skin or hair) comprise an important proportion of α-helices joined by disulphide bonds between chains thanks to cysteine units. The particular structure of collagen has been indicated above. Silks (such as fibroin from a spider's web or a cocoon) corresponds to stacked β-sheets, leading to flexible but poorly extensible structures. Globular proteins (water soluble), involving enzymes, hormones, transport or storage proteins, have a roughly spherical shape.

Finally, *quaternary structure* refers to aggregates that more easily keep the non-polar parts out of the aqueous cellular environment (Figure 13.8).

13.2 REPRESENTATION

For such large systems, it is very difficult to find a simple way in which to express the global shape. Solid models (space filling or wire-frame models), and more recently computerized images, give attractive representations, allowing for a rapid perception of essential features, but such representations cannot be easily input into quantitative models.

Owing to the size of such systems, it would sometimes be difficult to display all of the atoms, as is commonly done for simpler molecules (see Chapter 3) in the usual CPK representations. Of course, for specific purposes a part of the protein can be represented at the atomic level: for instance, one can represent the subset of atoms constituting the active site in order for studying the electronic and steric requirements when binding a drug. However, in many cases, the point is only to express the overall shape and demonstrate relationships between families. In such cases, schematic representations, using as graphical primitives symbolic icons which are simpler to generate, are

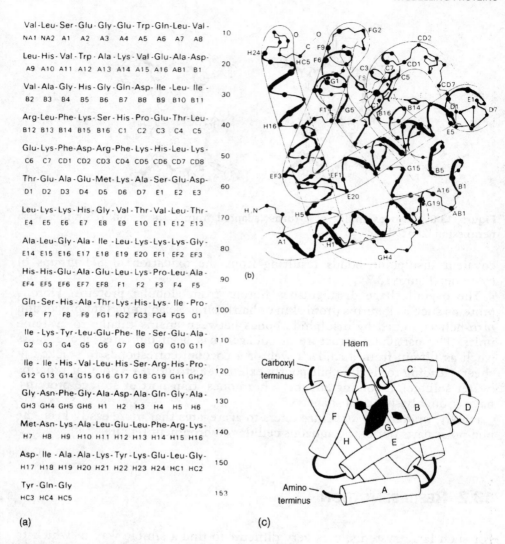

Val-Leu-Ser-Glu-Gly-Glu-Trp-Gln-Leu-Val-
NA1 NA2 A1 A2 A3 A4 A5 A6 A7 A8   10

Leu-His-Val-Trp-Ala-Lys-Val-Glu-Ala-Asp-
A9 A10 A11 A12 A13 A14 A15 A16 AB1 B1   20

Val-Ala-Gly-His-Gly-Gln-Asp-Ile-Leu-Ile-
B2 B3 B4 B5 B6 B7 B8 B9 B10 B11   30

Arg-Leu-Phe-Lys-Ser-His-Pro-Glu-Thr-Leu-
B12 B13 B14 B15 B16 C1 C2 C3 C4 C5   40

Glu-Lys-Phe-Asp-Arg-Phe-Lys-His-Leu-Lys-
C6 C7 CD1 CD2 CD3 CD4 CD5 CD6 CD7 CD8   50

Thr-Glu-Ala-Glu-Met-Lys-Ala-Ser-Glu-Asp-
D1 D2 D3 D4 D5 D6 D7 E1 E2 E3   60

Leu-Lys-Lys-His-Gly-Val-Thr-Val-Leu-Thr-
E4 E5 E6 E7 E8 E9 E10 E11 E12 E13   70

Ala-Leu-Gly-Ala-Ile-Leu-Lys-Lys-Lys-Gly-
E14 E15 E16 E17 E18 E19 E20 EF1 EF2 EF3   80

His-His-Glu-Ala-Glu-Leu-Lys-Pro-Leu-Ala-
EF4 EF5 EF6 EF7 EF8 F1 F2 F3 F4 F5   90

Gln-Ser-His-Ala-Thr-Lys-His-Lys-Ile-Pro-
F6 F7 F8 F9 FG1 FG2 FG3 FG4 FG5 G1   100

Ile-Lys-Tyr-Leu-Glu-Phe-Ile-Ser-Glu-Ala-
G2 G3 G4 G5 G6 G7 G8 G9 G10 G11   110

Ile-Ile-His-Val-Leu-His-Ser-Arg-His-Pro-
G12 G13 G14 G15 G16 G17 G18 G19 GH1 GH2   120

Gly-Asn-Phe-Gly-Ala-Asp-Ala-Gln-Gly-Ala-
GH3 GH4 GH5 GH6 H1 H2 H3 H4 H5 H6   130

Met-Asn-Lys-Ala-Leu-Glu-Leu-Phe-Arg-Lys-
H7 H8 H9 H10 H11 H12 H13 H14 H15 H16   140

Asp-Ile-Ala-Ala-Lys-Tyr-Lys-Glu-Leu-Gly-
H17 H18 H19 H20 H21 H22 H23 H24 HC1 HC2   150

Tyr-Gln-Gly
HC3 HC4 HC5   153

(a)

(b)

(c)

Figure 13.7   Representation of myoglobin (oxygen carrier to muscles), a protein of 153 residues. The molecule comprises eight segments of α-helices (with proline in four turns). The external region is the most polar one, the internal part the most hydrophobic, except for the polar groups of two histidines near the active site. The haema group (oxygen binding site) is bonded to histidine nitrogen. A second histidine non-bonded to central iron lies nearby. (a) Sequence of amino acids (the label below each residue refers to its location in an α-helical region A–H or in a non-helical region between two helices) after Edmundson with permission [128] and Watson with permission [129]; (b) only α-carbons are shown after Dickerson with permission [130] (from Stryer with permission [1] p 49 and Martin *et al.* [4] p 47); (c) schematic representation of the eight α-helices by cylinders (modified from Widom and Edelstein with permission [5] p 555).

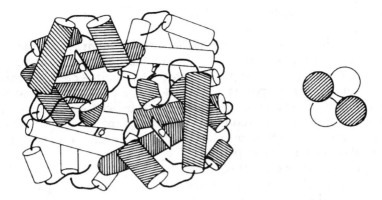

Figure 13.8 Haemoglobin. Two α-polypeptides with 141 amino acids and two β-polypeptides with 146 residues are assembled in a tetrahedron (from Widom and Edelstein with permission [5] p 556).

very useful for a rapid perception of essential features of the secondary and tertiary structures, the hydrophobic character of various surface areas, and so on (see Plate XIX).

In the most simple representations, only *α-carbons* are considered. The structure is schematized by a succession of line segments connecting the α-carbons in the sequence order: these segments are called *virtual bonds*. α-carbons are also the basic elements of the model-builder of Toma: virtual C_α–C_α bonds are drawn with a fixed length of 3.8 Å between neighbouring atoms (except in some cis peptide bonds). The geometrical adjustment is carried out via the angles formed by three successive C_α–C_α bonds. These angles, which are allowed to vary from 60°–160°, are good indicators of the secondary structure, and are easier to adjust than actual dihedral angles [6]. A crude volume representation can then be achieved by colour encoding spheres centred on α-carbons and assumed to represent the various amino acids. The formal chain joining the α-carbons is also the basis of the rapid automatic procedure for recognition of a user-defined pattern proposed by Brint *et al.* [7].

More refined graphic primitives were also used for an easier perception of regular arrangements: in such displays α-helices are generally represented as cylinders or regular coils, and β-strands as arrows. Random coils are drawn as single line virtual bonds, constrained to meet the end points of icons representing the regular patterns.

A drawing program working on a calligraphic device, for real time interactivity, has been proposed by Burridge and Todd [8]. For cylinders or regular coils representing α-helices, the axis is placed using the mean coordinates of the first and third residues as the starting point, and of the third from last and final ones as the end point. The radius is chosen as 2.5 Å. For connection to other moieties, the starting and end points are located on radial vectors joining the axis to the starting and end carbons, with a length adapted to the diameter. The pitch of the coil is 3.6 residues per turn, but other values can be selected.

β-strands are shown as broad arrows. For each sequential peptide, edges are defined initially by the α-carbon, carbonyl carbon and amide nitrogen, which fix the orientation of the planar peptide unit. Then points are smoothed with a simple formula ("average points i and $i-1$ and i and $i+1$, then average the averages"). Virtual bonds, used for randomly coiled chains, are constrained to meet the end points of the α-coil, the axis end point for α-cylinders and the centre points of the arrowhead or tail for β-strands. Similar programs have been implemented on raster devices [9, 10].

Other representations consist of ribbon models [11] adapted from the model of Richardson [12]. The ribbon is composed of approximately parallel smooth threads running along its length (for instance, 3 Å in width for a secondary structure, 1 Å for turns and random coils) (Figure 13.9). These threads are drawn thanks to β-spline functions. A series of closely spaced guide points is generated in the peptide planes, the basis of the construction. They lie on a line halfway between the α-carbons, and are equally spaced along the ribbon width desired. Things are easier when splines are available as built-in functions of the graphical device (as offered by various workstations), but an algorithm has also been proposed to generate them by software [13]. From the adjacent β-splines, a network of quadrilaterals can be drawn and rendered as filled areas leading to a solid surface model on a raster device. The method was extended to nucleic acids, with guide points based on a vector centred at the phosphate's position and directed towards the O_5 atom of the phosphate-sugar backbone.

Also relevant to the simplification of 2D representation, making it possible to see the essential features of the whole molecule at a glance, is the map projection of protein charge distribution [14]. Whereas, in a usual screen

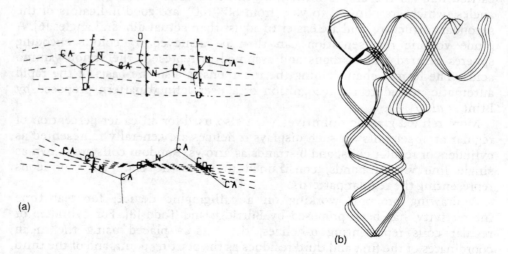

(a)

(b)

Figure 13.9 Ribbon model. (a) Creation thanks to β-splines (the guide points are located on a line in the peptide plane half way between the α-carbons); (b) resulting display (from Carson [13]).

representation, it is impossible to view the entire surface in a single figure (for the entire globe as for a globular protein), one half of the shape being hidden, it is possible to represent the whole surface on a single continuous projection map. Hammer projection with eight equal area projections was selected, as it suffers from less distortion. For example, such maps can be used to identify surface hydrophobic patches by displaying the distribution of polar/apolar groups on the protein's surface. Hammer projections (x,y) are calculated from the polar coordinates (r,θ,Φ) from:

$$x = [2.2^{1/2}\, r \cos\theta \cdot \sin(\Phi/2)] / [1 + \cos\theta \cdot \cos(\Phi/2)]^{1/2}$$
$$y = [2^{1/2}\, r \sin\theta] / [1 + \cos\theta \cdot \cos(\Phi/2)]^{1/2}$$

where r is the radius of the globe being generated (Figure 13.10).

13.3 DETERMINATION OF GEOMETRICAL DATA: 2D NMR IN PROTEIN STRUCTURE ANALYSIS

The Brookhaven Database constitutes the main source of 3D coordinates for proteins, with about 3500 entries (October 1995), originating mainly from

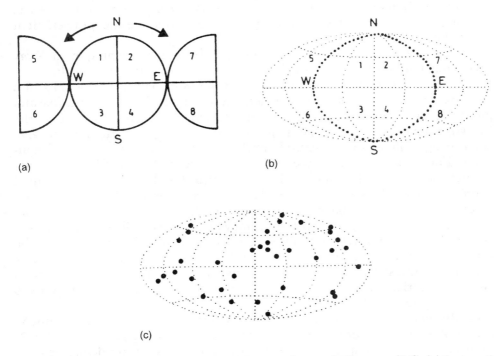

Figure 13.10 Hammer map projection (from Barlow and Thornton [14]). (a) Segments of a generating globe (parts 5–8 correspond to the rear region) and (b) the resulting map; (c) representation of the distribution of the charged groups in Ferredoxin.

crystallographic studies. Further information on the Protein Database (PDB), which has become a reference and invaluable tool in the field of protein modelling, is detailed in Chapter 4. However, NMR, with its newly extended capabilities, coupled with computer simulation, is becoming a very useful and widespread tool in the field.

For smaller polypeptide systems (< 100 amino acids), NMR techniques are now able to completely assign the resonance peaks and propose 3D structures within a 2.0–2.5 Å accuracy, comparable to the resolution obtained in crystallographic studies. We have already cited the work of Havel and Wüthrich [15] on basic pancreatic trypsin inhibitor (a small protein of 58 residues), where NMR data and a distance geometry approach lead to a model reproducing the crystal structure with a good accuracy.

With larger systems, however, it was not feasible until recently to assign even about one half of the amino acids present without the help of X-rays. Large spectral overlap (due to the number of residues) generally makes it possible to only assign resonances from the extremities of the spectrum, taking advantage of ring current effects in the vicinity of aromatic systems, changes induced by pH variations (indicators of neighbouring ionizable groups), and so on. Although structure determination for large systems (15–20 kDa MW) still remains a formidable challenge, improved resolution by increasing the dimensionality of the spectrum (thanks to heteronuclear 3D or 4D NMR) overcomes some of the problems arising from shift overlap, degeneracy and linewidth effects, and provides an attractive and quite efficient methodology.

On the one hand, NMR is a useful complementary technique to X-ray crystallography for proteins which do not easily form crystals. On the other hand (and this is perhaps the more important point), NMR provides information for liquid samples, which more closely look like the natural physiological environment. Such measurements also allow for the investigation of rational changes in pH, temperature or ionic strength, and give some insight about the structural similarities between crystalline and non-crystalline states. For instance, it appears that NMR coupling constants predicted from crystal geometry (through Karplus laws) often fit the measured couplings well, showing that side chains in the interior of the protein are blocked in a rigid spatial orientation. For surface residues, however, rapid averaging between several orientations may be observed [16]. Similarly, it was shown that "the α-amylase inhibitor Tendamistat (with 74 residues) has the same globular architecture in crystals and solution, but localized differences appear near the surface, including the active site." In other cases, differences are more drastic, including most of the cooperative bonds linking metal ions to the polypeptide chain [17, 18].

Finally, changing the conventional picture of a static system derived from X-ray crystallography, some dynamic information is available through NMR, regarding both low frequency processes (exchange of labile amide protons, for example, loss of tertiary structure by denaturing agents) and high frequency motions such as those occurring during libration of flexible side chains, isotropic tumbling of globular proteins, flipping of aromatic rings, etc. [16].

Apart from complete structure determination, an important application area relies on the study of protein-substrate interactions in a solution state. Suffice it here to have some assigned protons at various points of the structure and look at the changes of their chemical shifts under selective perturbations (salt, temperature effects, presence of inhibitors of enzyme action): any change indicates that the corresponding residue is likely to be involved in the interaction process [16].

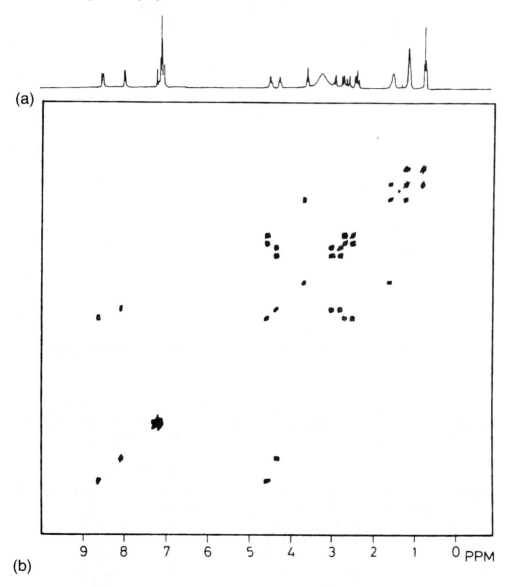

(a)

(b)

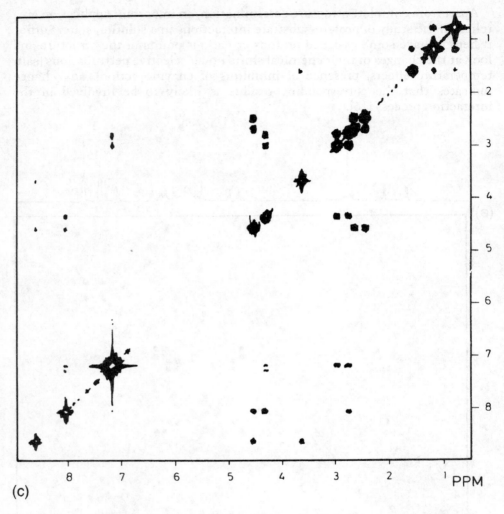

(c)

Figure 13.11 ^{1}H NMR spectra of amide tripeptide NLe-Asp-Phe-NH$_2$ at 270 MHz in DMSO. (a) 1D spectrum; (b) 2D COSY. (c) 2D NOESY. (NLe nor-Leucine, a natural, non-essential amino acid, with a R = n-Bu side chain) (J. Gharbi with permission).

The main tools currently used for such analyses come from *2D homonuclear* (and more recently, 3D heteronuclear) *NMR*, which specifies sites correlated through space or through scalar coupling. The basic principles of such experiments have already been presented in Chapter 4. Indeed, Nuclear Overhauser Enhancement is the source of excellent and varied information: within a single amino acid (to determine its nature, in complement to COSY data), between neighbour amino acids (to fix their sequencing), and between distant residues (*i, i+2* or more) to identify regular 3D motifs in the secondary structure (Figure 13.11). Particularly important are NOEs between residues that are far apart in the sequence but close together in space, to detect folding of the backbone or side chains. *Scalar couplings* related to torsion angles, specific effects on selected *chemical shifts* or the examination of *hydrogen bonds* can also be of help.

To mimic the physiological conditions, experiments are generally carried out in water solutions (about 1–5 mM), but this requires the elimination of the exceedingly strong solvent signal which can mask some signals of the substrate. Solvent-peak suppression can be achieved using various techniques. Otherwise, use of D_2O avoids such strong solvent signals, and furthermore, simplifies the spectrum. For example, spectra run in D_2O exhibit a clearer low field region: labile amide protons are exchanged, and the suppression of couplings to α-NH simplifies the α-CH and β-CH$_a$H$_b$ signals.

13.3.1 Sequence assignment

In the typical case of smaller systems, the first task is of course to determine *what amino acids* constitute the polypeptide chain. From spectra in H_2O, the spin system of the present residues NH C$_\alpha$H C$_\beta$H$_a$(H$_b$) can be identified. For the 20 essential amino acids, COSY and RELAY connectivity peaks gather in 14 characteristic fingerprints (among which four are for aromatic protons) [19, 20], easily identified thanks to coupling between α, β and γ protons. Information is gained through COSY spectra for C$_\alpha$H/NH correlation and C$_\beta$H/NH in RELAY spectra (Figure 13.12). This point was developed in Chapter 4. For a random chain organization, tables specify proton chemical shifts (at pH = 7) and $^3J_{HN\alpha}$ or $^3J_{\alpha\beta}$ coupling constants [21]. These values, providing a significant reduction of assignment possibilities, are of definite help in protein analysis (Table 13.2).

Once the constituent amino acids are known, the next problem is to determine in what order they are linked together. This is the *sequence specific assignment*, which is mainly performed from the NH and Hα peaks in COSY and NOESY spectra. COSY and RELAY spectra indicate connectivities within the amino acid residues, via scalar coupling constants between protons separated by no more than three bonds (COSY) or via couplings relayed by a common proton (RELAY). The NOESY spectrum specifies dipolar correlation between neighbours that are close in space rather than in sequence. Determining such neighbourhood relations allows for specification of the relative position of a residue within the chain. Besides intra-residue correlation (also detected on COSY spectra), between NH$_i$ and αCH$_i$ (d$_{N\alpha(i,i)}$), the main NOESY information is here given by inter-residue cross peaks N$\alpha_{(i,i+1)}$ between the αCH of residue i and NH on residue $i+1$ (d$_{\alpha N}$); and NN$_{(i,i+1)}$, between NH$_i$ and NH$_{i+1}$ (d$_{NN}$) (see Chapter 4). Starting with one characteristic peak, intra-residue correlation identifies the NH peak of a given residue and inter-residue analysis determines which residues are sequentially related. Such relations between protons in neighbouring residues are sufficient to match a dipeptide motif, if this dipeptide segment is contained only once in the amino acid sequence. Otherwise, tri- or tetra-peptide segments are studied (such motifs seldom occur more than once in globular proteins).

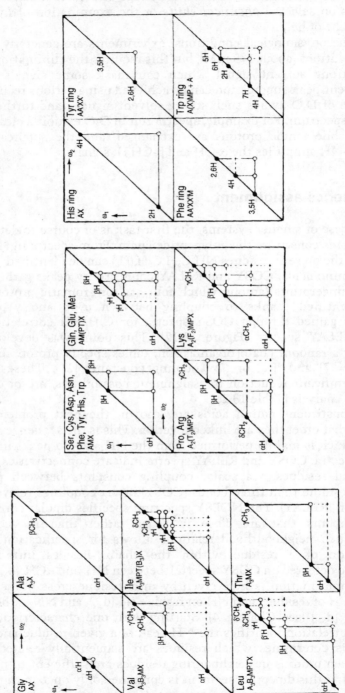

Figure 13.12 Characteristics of COSY (O) and RELAY (+) connectivity patterns for non-labile protons in common amino acid residues (from Abraham *et al.* with permission [20]).

Table 13.2 ^{1}H NMR parameters of the 20 common L-amino acid residues in the linear tetrapeptides H-Gly-Gly-X-L-Ala-OH (from Bundi and Wüthrich [21]).

Amino acid residue	Chemical shifts δ (from TSP ± 0.002 ppm)					Spin-spin coupling constant J (± 0.5 Hz)	
	NH$_\alpha$	CH$_\alpha$	CH$_\beta$	others		^{3}J$_{H\alpha NH}$	^{3}J$_{\alpha\beta}$
Gly	8.391	3.972				5.6	
Ala	8.249	4.349	1.395			6.5	7.0
Val	8.436	4.184	2.130	γCH$_3$	0.969	7.0	6.9
					0.942		
Ile	8.195	4.224	1.894	γCH$_2$	1.478	7.0	7.6
					1.190		
				γCH$_3$	0.943		
				δCH$_3$	0.885		
Leu	8.423	4.385	(1.649)	γCH	(1.649)	6.5	7.2
			(1.649)	δCH$_3$	0.943		7.2
					0.899		
Ser^b	8.380	4.498	3.885			6.5	5.1
			3.885				5.1
Thr	8.236	4.346	4.220	γCH$_3$	1.232	6.9	5.0
Asp^b	8.410	4.765	2.837			7.0	5.7
			2.753				8.3
Glu	8.368	4.295	2.092	γCH$_2$	2.314	7.0	4.6
			1.969		2.283		9.5
Lys	8.408	4.358	1.870	γCH$_2$	(1.471)	6.5	5.6
			1.747	δCH$_2$	(1.708)		7.8
				εCH$_2$	3.023		
				εNH$_3^+$	7.519^a		
Arg	8.274	4.396	1.915	γCH$_2$	(1.719)	6.9	5.5
			1.796	δCH$_2$	3.312		7.6
				NH	6.622^a		
					7.166^a		
Asn^b	8.747	4.755	2.831	γNH$_2$	6.912^a	7.5	5.8
			2.755		7.591^a		8.3
Gln	8.411	4.373	2.131	γCH$_2$	2.379	6.0	5.0
			2.010	δNH$_2$	6.875^a		8.8
					7.594^a		
Met	8.418	4.513	(2.164)	γCH$_2$	(2.633)		5.7
			(2.000)		(2.633)		8.6
				εCH$_3$	2.128		
Cys^c	8.312	4.686	3.278			7.7	4.0
			2.958				9.6
Trp	8.094	4.702	3.322	*Ring Protons*			6.0
			3.195	C2H	7.244		7.8
				C4H	7.649		
				C5H	7.167		
				C6H	7.244		
				C7H	7.504		
				NH	10.220^a		
Phe	8.228	4.663	3.223		(7.339)	9.4	5.6
			2.991				10.3
Tyr	8.183	4.604	3.127	C3,5H	6.857	6.8	5.6
			2.922	C2,6H	7.149		9.0
His	8.415	4.630	3.263	C2H	8.120	8.0	6.0
			3.198	C4H	7.140		6.9
Pro^{b,d}		4.471	(2.295)	γCH$_2$	(2.030)		8.8
			(1.981)	δCH$_2$	3.653		5.0

^{a}Conditions: solvent D$_2$O (H$_2$O for labile H), pD 7.0, t=35°C.
^{b}Data from the protected peptides CF$_3$CO-Gly-Gly-X-Ala-OCH$_3$.
^{c}Measured in Z-Gly-Gly-Cys-L-Ala-OH where Z = carbobenzoxy protecting group.
^{d}Only the parameters for *trans* Pro are given.

For larger proteins (MW from 10,000 to 20,000), however, many CH_α and amide protons have shifts that are very close, limiting such sequential assignment techniques. Isotope labelling then becomes a possible solution.

13.3.2 Secondary structure determination

Once the sequence (primary structure) is known, we now have to determine the spatial location of atoms, that is the secondary structure. Karplus laws for coupling constants and NOESY experiments are the main tools in this field.

Torsional angles

According to IUPAC conventions, the conformation of an amino acid in a polypeptide chain is defined by four angles (Figure 13.13, Tables 13.3, 13.4).

The torsion angle $H-N-C-H_\alpha$ ($\Theta = \Phi - 60°$ for L-amino acids) and the corresponding ^3J obey a Karplus law [23–26][1] from which it can be deduced that $^3J_{HN\alpha} < 5.5$ Hz corresponds to Φ between $-60°$ and $40°$; whereas $^3J_{HN\alpha} > 8$ Hz corresponds to Φ between $-80°$ and $-160°$ [27, 28].

Only three values (60°, $-180°$, $-60°$) are possible for χ_1, due to energetic reasons. Similarly, a Karplus relationship relates χ_1 to the coupling constant $^3J_{\alpha\beta}$, itself related to the percentage of the three more stable rotamers around the $C_\alpha-C_\beta$ bond [29].

The ψ angle is not directly related to any H–H coupling. However, possible ψ values can be derived from sequential NOE $d_{\alpha N(i,i+1)}$ and $d_{\beta N(i,i+1)}$. Finally, d_{NN} values are related to the pair of angles Φ, ψ as indicated in a Ramachandran

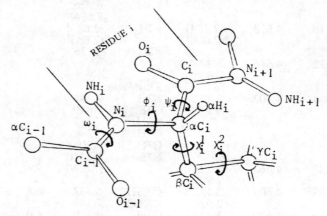

Figure 13.13 Four torsion angle scheme. ω is 0° for a cis arrangement of the main chain and 180° for a trans one. Φ and Ψ define the orientations around the $N-C_\alpha$ and $C_\alpha-C,0$ bonds (and take a 180° value for an extended transconfiguration). The χ angles specify the orientation of the side chain with respect to the backbone and within the side chain.

[1]$^3J_{HN\alpha} = A \cos^2\theta - B \cos\theta + C$; coefficients proposed for proteins: $A = 6.4$, $B = 1.4$, $C = 1.9$ Hz.

Table 13.3 Angles of rotation in the backbone (in polypeptides of L-amino acids). Adapted from Scheraga [22], assuming $\Phi = \psi = 180°$ for a fully extended chain.

Φ (rotation around N–C$_\alpha$)	ψ (rotation around C$_\alpha$–C)
180° C$_\alpha$–C′ cis to N–H	180° C$_\alpha$–N cis to C–O
–60° C$_\alpha$–R cis to N–H	–60° C$_\alpha$–H cis to C–O
60° C$_\alpha$–H cis to N–H	60° C$_\alpha$–R cis to C–O

Table 13.4 Angles of rotation for some regular structures. Adapted from Scheraga [22], assuming $\Phi = \psi = 180°$ for a fully extended chain.

	Φ	ψ
Fully extended chain	180°	180°
Right handed α-helix	–58°	–47°
Parallel pleated sheet	–119°	113°
Antiparallel pleated sheet	–139°	135°
Collagen	≈–60°	≈160°

map [30][2]. In such a map, energy contours delineate the values allowed for angles Φ and ψ so as to avoid steric interactions for rotations along C$_i$ C$_{i\alpha}$ (ψ) and C$_{i\alpha}$N$_i$ (Φ). With an amide group fixed in a trans form, such rotations are interdependent in a given residue but are sterically independent of rotations within neighbouring residues (Figure 13.14).

Regular motifs

An important aspect of the determination of the secondary structure is the recognition of regular rigid domains, mainly α-helices and β-sheets, leading to characteristic patterns. NOE is particularly useful for looking for residues that are close in space rather than neighbours in the sequence.

In α-helices (comprising about 3.6 residues per turn), residues in position i and $i+3$ are approximately 4 Å apart, and are therefore able to exhibit NOE effects, whereas in an extended form they would largely be too distant (more than 10 Å apart) for any NOE enhancement. Characteristic values are summarized in Tables 13.4 and 13.5.

Short sequential distances d_{NN}, small successive vicinal couplings $^3J_{NH\alpha}$ (about 4 Hz), and slowed exchange rates of amide protons have been reported as other indicators of α-helices. Corresponding patterns are also available for other regular structures; β-sheets, tight turns, etc. [17, 19]. For instance, in β-sheets, $^3J_{HN\alpha}$ are large (about 10 Hz for parallel β-sheets, and 9 Hz for anti-parallel ones) and sequential NOEs largely differ from those noted in helices (Figure 13.15).

It must be recalled, however, that NOE is directly dependent upon the ratio of the dipolar interaction vs. all other relaxation processes. So, caution is necessary as to a direct interpretation of the experimental results if these other

[2]We will not detail the particular case of proline, where the cyclic structure imposes a bent direction to the main chain.

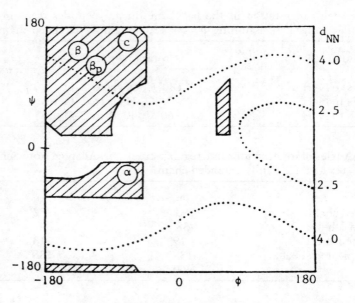

Figure 13.14 The map is drawn for an *L*-alanyl residue in a polypeptide chain. Hatched areas correspond to sterically allowed geometries. Φ, ψ values for α-helix, β- and β-parallel sheets, collagen are shown. Dotted lines indicate NN isodistance contours (from Ramachandran and Sasisekharan with permission [30]).

Table 13.5 (a) Interproton distances (Å) in regular motifs: α-helix (H) and β-sheets (S); (b) interstrand distances in β-sheets.

(a)

	$d_{\alpha N}$	d_{NN}	$d_{NN(i,i+2)}$	$D_{\alpha N(i,i+3)}$	$d_{\alpha\beta(i,i+3)}$	$d_{\alpha N(i,i+4)}$
H	3.5	2.8	4.2	3.4	2.5–4.4	4.2
S	2.2	4.3	No NOE between fragments i, $i+2$ or more			
			$d_{\beta N}$ 3.2–4.7			

(b)

	$d_{\alpha\alpha(i,j)}$	$d_{\alpha N(i,j)}$	$d_{NN(i,j)}$
β	2.3	3.2	3.3
β-p	4.8	3.0	4.0

mechanisms are important. For instance, in β-sheets the most reliable data concern CH_α and NH_α.

Indirect clues may also be used: so, H_α shifts depend upon the environment, and particularly on anisotropic effects from carbonyls or phenyl rings. Displacements from the values quoted for random structures thus give some indices about the secondary structure. A difference greater than 0.4 ppm is an indicator of β-sheets; an α-helix is probable if the difference is about –0.4 ppm [21].

Among other helpful data, amide exchange rates can give some insight to the secondary structure thanks to information about the formation of hydrogen-bonding between backbone amide protons and carbonyl oxygen. In

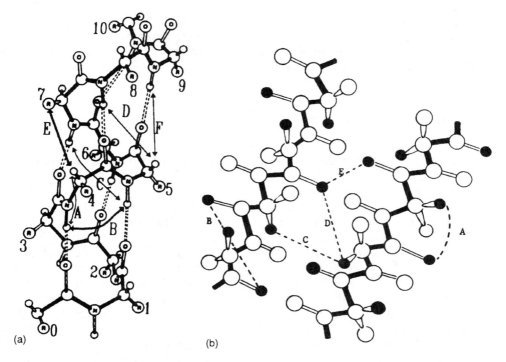

Figure 13.15 Characteristic atom-atom distances involved for NOE in an α-helix and β-sheet. For the sake of clarity, only one example is shown for each NOE type (adapted from Wüthrich [19]). (a) α-helix: sequential NOE: (A) $d_{\alpha N} = 3.5$ Å; (B) $d_{NN} = 2.8$ Å; longer range NOE (C) $d_{NN(i,i+2)} = 4.2$ Å; (D) $d_{\alpha N(i,i+3)} = 3.4$ Å; (E) $d_{\alpha\beta(i,i+3)} = 2.5–4.4$ Å; (F) $d_{\alpha N(i,i+4)} = 4.2$ Å; (b) antiparallel β-sheet. Sequential NOE (A) $d_{\alpha N} = 2.2$ Å; (B) $d_{NN} = 4.3$ Å. Interstrand NOE (C) $d_{\alpha\alpha(i,j)} = 2.3$ Å; (D) $d_{\alpha N(i,j)} = 3.2$ Å; (E) $d_{NN(i,j)} = 3.3$Å.

α-helices, hydrogen bonds appear between $CO_{(i)}$ and $NH_{(i+4)}$. In β-sheets, they form a dense network. In D_2O solutions, labile amide protons exchange rapidly and the corresponding ^1H signal disappears. However, the rate of decay is slow for protons involved in H-bonds or buried in the internal part of the molecule.

13.3.3 3D and 4D NMR

For larger systems, determining the three dimensional structure becomes a formidable challenge, owing to chemical shift degeneracy, peak overlap and increased linewidths. The new techniques of heteronuclear 3D and 4D NMR, carried out on uniformly labelled (> 95%) ^{15}N and/or ^{13}C substrates, provide an attractive solution for proteins in the 15–30 kDa range. The basic principles of this approach have been briefly summarized in Chapter 4. Suffice it here to recall that as a 2D experiment, the aim is to identify NOE connectivities between adjacent residues involving NH, $C_\alpha H$ and $C_\beta H$ protons. However, in a

3D experiment, resolution is improved by increasing the dimensionality of the spectrum, since the proton/proton cross-peaks are now separated into different planes according to the shifts of bonded heteronuclei (^{13}C or ^{15}N). This leads to a massive simplification of the analysis.

13.4 COMPUTER BUILDING

In complement to experimental access to geometrical characteristics, computer-aided modelling plays an important role in structure determination of proteins. It intervenes in fact at several levels.

As already pointed out, the three-dimensional structure from X-ray crystallography is now available only for a few proteins, for which results are collected in the Brookhaven Database (PDB) see Chapter 4. In some other cases, experimental data (from NMR, for example) are available, but not sufficient, if considered alone, to completely define the structure: they can only be viewed as constraints regarding the spatial location of some atoms. Computer modelling is therefore very useful to generate structures consistent with these constraints. A similar help can be also sought when an X-ray structure is difficult to solve.

At a more ambitious level, when experimental determination of the geometry is not possible (for example, when good crystals are not easily obtained), a computer-based structure prediction would be very valuable for understanding for example structure-activity relationships. Modelling will be also of important help in future to define guidelines for rational engineering and building of novel proteins with specific functions. However, at the present time, we are still far from an *ex nihilo* generation of a possible 3D structure. Work has generally to be carried out by comparison with similar systems, and the introduction of limited structural changes.

A particular emphasis will be put here on two major aspects:

- The prediction of the 3D structure (secondary or tertiary) from the sequence of the amino acids, an important point since sequence-assignment is much faster than 3D structure determination, and the gap between known sequences and known structures is widening owing to the development of rapid sequencing techniques [31].
- The detection and quantification of structural similarity or homology at the 3D- or the sequence-level.

It is now quite clear that the sequence of the amino acids constituting a protein is of enormous importance for both its 3D structure and functional properties. However, up to now, the relationships between these different facets of protein behaviour still remain unknown. They would of course be of paramount importance since for many systems the residue sequence is known well before the 3D structure is solved. Such knowledge would also give powerful clues for designing, in the near future, proteins dedicated to selected applications.

Indeed, the function of a protein may be dramatically sensitive to the constitution of its sequence. Pincus and Scheraga [40] quoted that changing a single amino acid at position 12 of a P21 protein (molecular weight 21,000) suffices to cause the protein to become oncogenic (i.e., to induce malignant transformation of normal cells) (Figure 13.16).

From another point of view, prediction of protein structures can be considered as relying on two main approaches (in fact complementary and which can be used in conjunction with each other):

1. *Energy-based* calculations through theoretical models and energy minimization, or
2. *Knowledge-based* models combining sequence data to other information such as the known structure of a homologous protein.

Energy-based structure prediction mainly relies on energy minimization and molecular mechanics or dynamics. Nevertheless, it still often remains reduced to limited optimization of strain induced by changes in side chains or small insertions or deletions. These methods are less successful for loop regions or when large changes take place. Up to now, they seem more useful for refining a structure than for predicting it. Their successes depend, of course, on the quality of the potential functions used. Nevertheless, basically, these methods are faced with the problem of a large number of possible multiple minima, making the traversal of the conformational space difficult, and the detection of the real energy minimum uncertain. These difficulties are

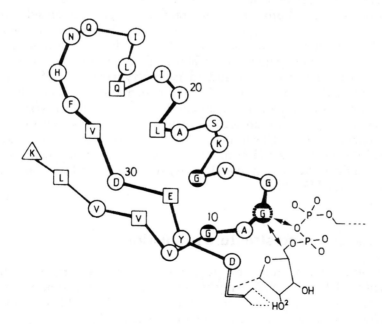

Figure 13.16 Mutation of glycine 12 modifies the postulated mode of binding of a nucleotide on the P21 protein (from Hol with permission [41]).

however somewhat relieved by the alternative approach based on distance geometry. In fact, systematic conformational search for a protein of about 100 residues as it is constitutes a huge problem. Looking only at angles Φ and ψ, and even assuming that for most residues, there are roughly only three regions allowed in the (Φ, ψ) conformational map, one would have to examine about 3^{100} (i.e., 10^{48}) different conformations [2] p. 299.

13.4.1 Refinement of a structure under constraints

Once the main structural patterns of the 3D structure have been experimentally determined, for example from NMR data, the next step deals with the elaboration of a 3D representation, using the techniques of computer modelling. Generally, starting with standard bond lengths and angles, one proposes a model consistent with the data (constraints derived from NOE measurements, torsion angles, selective non-bonded atom-atom distances, etc.) and refines it by energy minimization programs. Among the first solutions proposed, the CONFOR package [17, 32] starting from NMR data, visualizes violations of the geometrical constraints and allows for interactively relieving them by modification of torsion angles. This interactive construction was, however, limited to small motifs. Another pioneering package [17, 33], was limited to a size of about 10 residues and mainly applied to lipid-bound polypeptide hormones.

An extreme example corresponds to situations where the starting backbone geometry of the desired protein is constrained to reproduce closely the X-ray structure of a homologous protein before side chains are introduced, as illustrated in a study of angiogenin (a 123 residue protein) based on its homology to bovine pancreatic ribonuclease of known structure. A preliminary step generated a standard geometry structure of ribonuclease consistent with X-ray data. Small L-peptide segments of about 30 residues were first energy-minimized along dihedral angles and then juxtaposed, with 10 residues overlapping, for an initial fitting of the X-ray structure of RNase. Then the backbone of angiogenin was built from that of RNase, taking into account residue deletions and minimizing energy (including a penalty term to constrain the fit of the two chains). Finally, side chains and backbone dihedral angles were refined [50–52].

13.4.2 Energy-based structure prediction

In earlier attempts, a global reconstruction of the tertiary structure was approached by Scheraga *et al.* using chosen atomic distances, and considering only two or three degrees of internal freedom (backbone dihedral angles) for each residue, owing to computational requirements [40].

An empirical force field program (ECEPP: Empirical Conformational Energy for Polypeptides and Proteins) [42, 43] is first used to determine the energy minima for the naturally occurring amino acids considered individually (single energy minima). The process is then progressively and systematically extended to the preferred conformations of oligopeptides (di- to hexa-) and polysaccharides. Starting from the first two residues of the sequence, all possible single residue minima are combined and subject to energy minimization. Dipeptide conformations whose energy lies within a given cut-off energy value above the lowest minimum are retained and combined with all the single residue minima of the next amino acid, and the process is repeated [44, 45]. The procedure was applied, for instance, to gramicidin (acyclic decapeptide with symmetry) and a model of collagen poly(Gly-Pro-Pro) [40].

Nevertheless, for an extension to larger systems without symmetry or perfectly repeating units it is necessary to limit the number of low energy conformations to be combined. This can be achieved thanks to the concept of *non-degenerate minima* corresponding to different backbone conformations (eliminating redundant local minima with the same backbone geometry but differing by the orientation of their side-chains) [46, 47]. This method was applied to a number of membrane polypeptides and proteins with a particular interest devoted to leader sequences (sequences of about 20 non-polar residues on their amino terminal) that cause translocation of the protein chain across membranes and are supposed to act as Chain Folding Initiation Sites (CFIS) [46]. It was also used to determine the preferred structures of the transformation region of the P21 protein to obtain some information about the structural basis of oncogenesis [48].

This progressive building-up procedure stresses the general principle that protein shape and folding can be successfully approached using successive approximations: at a first level, short range interactions limit the conformations possible for residues in the final structure. Medium range interactions (such as those occurring between residues four positions apart in helices) select subsets of these possible conformations. Finally, long range interactions (packing of non-polar residues, side chains interactions, etc.) must be introduced to enable the molecule to achieve its overall shape; the whole structure then being subjected to an energy minimization.

Alternatively, simulated annealing methods have been proposed to locate the global minimum energy conformations, and were proved to work successfully for the "dipeptide models" of the 20 natural amino acids as well as for polyalanines (up to Ala_{80}) [49].

For larger systems, limiting the conformational space to be traversed and reducing computer time involved in the building-up process can be efficiently achieved by introducing constraints (for example distance constraints from NOE data) in a penalty term.

13.4.3 Including knowledge from solved structures

Energy-based methods can be efficiently improved by the addition of some amount of knowledge extracted from proteins of known structure [53].

Reconstruction of the tertiary structure of flavodoxin from only the crystallographic coordinates of its α-carbons constitutes an illuminating example to check the accuracy of such an approach. This protein of moderate size (138 residues) whose structure is known can be considered as a typical test case since it contains a mixture of helix, sheet, turn and coil conformations. Main chain atoms were here generated from a dictionary of backbone structures for polyalanine short segments (4–7 residues). Then side chains were initially set according to statistical distribution criteria (deduced from an analysis of side chain torsional angles as a function of secondary structure). To solve clashes, a base of 190,000 pairwise side chain interactions was built. Finally, energy minimization was carried out. The deviation with respect to the native structure is quite satisfactory (RMSD 1.7 Å across all atoms) but, as the authors noted, about 40% only of the χ_1 torsional angles (for side chains) are predicted correctly.

Generally speaking, the accuracy of such a methodology is often difficult to assess. In the preceding study, special attention was paid to the search for various parameters that can be used as a possible caveat from incorrect modelling. Surface areas, internal cavity volume, close contact distances, hydrogen-bonding patterns, energy, etc. were examined and their values compared to statistical averages observed in known structures. Although a discrepancy would indicate no more than a deviation with respect to "usual behaviour", using these indices can greatly assist the modelling process.

13.4.4 Tertiary structure calculation through distance geometry

Distance geometry was then proposed as a more general approach, requiring massive computation but able to cope with large systems [34]. The DISGEO program [15, 35–37] (working with up to 100 residues) enabled the first structure determination from NMR data for a globular protein (bull seminal proteinase inhibitor IIA (BUSI IIA), including 57 residues).

It must be recalled, however, that distance geometry, using experimental NMR constraints, works only on upper limits rather than specific distances [17], often representing only a (largely) incomplete set of distances. As a result, the trial structures generated from random distance selection generally lead to similar (but not exactly identical) solutions, and correspond rather to groups of neighbouring conformers. Although largely used for polypeptides and proteins, distance geometry was also extended to DNA structure determination by Tinoco *et al.* [38].

Distance geometry was also incorporated in tertiary structure prediction methodology (see below [39]). For more details about distance geometry methods, see Chapter 7. Note also that molecular dynamics (see Chapter 6) constitutes an alternative avenue for refining structures obtained by interactive modelling (and possibly the distance geometry approach).

Kuntz *et al.* [39] have extended the distance geometry method, relying on upper and lower limit matrices, to incorporate varied theoretical or experimental constraints. The aim was to predict hairpin turns and specify a set of long-range contacts. As the method uses some "knowledge" (a turn predictor), it appears as intermediate between energy-based and knowledge-based approaches. For large protein systems, calculation cannot be carried out at the atomic level and for the sake of simplification, each amino acid is here represented just by a sphere centred on the α-carbon. The originality of the treatment is that besides mathematical or geometrical constraints (triangle inequality for distances, known angles or bond lengths, distance of closest approach for non-bonded atoms, etc.) other long range constraints may be introduced: disulphide bonds, proximity of specific residues, partial data from NMR or X-rays, and so on.

The method then generates a partial contact/no-contact list. The first step proposes a residue pairing scheme on the basis of hydrophobic interactions (evaluated from hydrophobic pairing functions, a coefficient roughly proportional to the surface area of the side chain, and expressing the energy of hydrophobic bonding). In the second step, pairs "geometrically inconsistent" with other contacts are eliminated. For the purpose of generating a list of contacts, a turn predictor model relying on hydrophobic interactions was used (giving better results than the Chou–Fasman predictor of secondary structure (see below)). This "firehose" model assumes that the chain is made of linear fragments (featuring helices and extended strands) and turns and that all secondary structures except turns have the same spatial advance per residue. Possible interacting pairs are generated, eliminating residues which are too near or too far apart (very large loops). From weighted hydrophobic interactions calculated around each residue pair i,j (with $6 < j-i < 25$), hairpin turns permitting the maximum overlap of hydrophobic residues are predicted. According to the location of these turns, a screening algorithm determines pairs that can be reasonably in contact.

After these manipulations of the boundary matrices comes the generation of 3D structures meeting the imposed constraints, using for example optimization or Monte Carlo procedures [54, 54a]. In the example of the small protein pancreatic trypsin inhibitor, this methodology correctly locates the hairpin turns and predicts plausible long range hydrophobic contacts. It leads to a proposed backbone conformation with errors of 4–8 Å compared to the native structure.

13.5 KNOWLEDGE-BASED PREDICTION: MODEL BUILDING FROM HOMOLOGY

Most of knowledge-based model building relies largely on the concepts of similarity and homology. As pointed out by Boswell and Lesk [55, 56] speaking of homologous sequences would strictly imply that they are derived from a common ancestor. However, in a broader sense (as we adopt here), "homologous" is often taken as being synonymous with "similar".

In knowledge-based modelling, a database of proteins for which structural and sequence information is available is used to predict structural features for proteins of neighbouring sequences. From proteins of known structures, a comparison of sequence and 3D geometry makes it possible to derive rules or parameters that will subsequently enable us to determine the probability for a given fragment to arrange into a particular regular structure. As for tertiary structures, they also allow to define sequence templates for families of protein structures that adopt a common fold. If a sequence of an unknown structure can be matched with a template, a model of the fold can be built by analogy, as in the COMPOSER program [57]. However, up to now, much more confidence can be placed on a comparison with homologous proteins rather than on a prediction from knowledge bases.

The initial premise is that functionally analogous proteins with homologous seqeuences will have closely related structures with, in particular, common tertiary folding patterns [58]. As can be expected, when sequence homology with a known protein is high, modelling of an unknown structure by comparison can be carried out with reasonable faith [59]. However, (although this is of no use in modelling by homology where sequence similarity is a prerequisite [60, 61]) it can be noted that structural homology may remain significant even if sequence homology is low (as for example in cytochrome c3, or globins where eleven proteins with very different amino acid sequences exhibit remarkably similar secondary and tertiary structures) [62]. In other words, 3D structure seems better conserved than the residue sequence. From a comparison of 32 pairs of homologous proteins [59], it was established that pairs with a sequence homology > 50% have 90% or more of the residues within a structurally conserved common core (with the same fold). When similarity drops to about 20%, the common core contains between 42% and 98% of the residues of the individual proteins.

Schematically, for predicting structures of unknown proteins, model building proceeds through three main alternate approaches [58]:

1. Starting from the sequence knowledge.
2. Assembling fragments from different, known homologous structures.
3. Carrying out limited structural changes from a known neighbouring protein.

13.5.1 Prediction of secondary or tertiary structures from the sequence

These approaches, which have prompted considerable work, have so far shown only limited reliability (less than 70% accuracy). Prediction from sequence works better at the secondary structure level or within a class of proteins, but for tertiary structure or for molecules outside the class, reliability is often poor [63], although over the years, these approaches have gained slow but continuous improvements. We will only present here the most classical approache. 'ecent refinements (although they do not seem to drastically modify the situation, see Lesk [56] and Garnier et al. [64].

Generally speaking, these methods were first tackled at the secondary structure level, applying simple physicochemical constraints to short chains of amino acids [65–67a]. Two aspects were considered: the propensity of individual residues to adopt a conformation favouring a given arrangement, and the influence of neighbouring residues. Generally, hydrophobic patterns over several neighbouring residues play an outstanding role in these evaluations. The principle of these methods can be traced back to the observation of Wu and Kabat [68], that in three-peptide fragments XYZ, the conformation of a specific residue Y is largely determined by local interactions with X and Z, and so is likely to be similar to that observed in fragment XYZ of another protein. Various methods have been proposed, relying on information theory, statistics, artificial intelligence or neural networks.

As indicated by Garnier *et al.* [64] prediction of the secondary structure can be of some help for the alignment of homologous proteins, and can guide the choice for mutation of an amino acid with respect to the conservation of the structure.

Information theory and structure prediction

The method proposed by Garnier *et al.* [69] basically aims to evaluate the information that residues R_1, R_2, R_3... in the sequence carry on the conformational state S_j (say $S_j = H$ helix..., sheet...) of any given residue j:

$$I\,(S_j = H, R_1, R_2, \ldots\ldots R_{last})$$

This is expressed, in a simple form, by the summation of separate terms characterizing the information on residue j (under investigation) carried by its neighbours, i.e. residues separated by no more than eight positions in the sequence:

$$I\,(S_j; R_1, R_2, \ldots\ldots R_{last}) \approx \sum_{m=-8}^{m=+8} I\,(S_j; R_{j+m})$$

The model proposes a four state prediction (*Helix, Extended chain, reverse Turn, Coil*). For the 20 natural amino acids, and 17 positions (from j–8 to j+8), a table of 20×17×4 parameters is needed. Their values were obtained from a statistical analysis of 25 proteins of known sequence and structure. For each residue, the information is calculated for each of its four conformational states, and corrected from a subtractive decision constant (that is specific to the state (H,E,T,C) and empirically adjusted to get better predictions). The conformation with the highest information content is elected. An overall accuracy of about 63% is attained (for a four state prediction, where random probability would be 25%). This seems a quite general limit for several methods of secondary structure prediction, since long range interactions are omitted. However, prediction can be improved by using supplementary information, for example classifying proteins into α-helix-rich proteins, β-pleated sheet-rich proteins and other proteins, classes for which long range interactions differ. Circular dichroism, which gives some indications about the content of helices or β-sheets, can be used for that purpose.

In the absence of any evaluation of long range interactions, well defined subassemblies of local secondary structures (which dominate the tertiary folds of many globular proteins) reflect intermediate range contacts between sequentially adjacent units of a secondary structure, and can help in solving the secondary structure [70]. The method (based on recognition of supersecondary structures) was exemplified on a βαβ-unit prediction, leading to an improvement of about 7.5% over the original prediction. The point was to find the best positions along the sequence for a scaled ideal βαβ template derived from frequency histograms of secondary structure occurrence. This was carried out by adding to the probability calculated for helices, extended or coil forms (according to the Garnier or Chou methods) additional information on the distribution of hydrophobic residues specifically required for βαβ units. Then the secondary structure prediction was refined (Figure 13.17).

Statistical approach

The model of Chou and Fasman [65–67a] relies on the observation, derived from a wide number of crystal structures of proteins, that some amino acids are more likely than others to be found in a type of secondary motif, for example an α-helix. From the frequency of occurrence of each of the 20 amino acids in regular arrangements (α-helix, β-sheet, β-turn) observed in the crystal structures of 29 proteins, the authors have calculated the probability (called a structural or conformational parameter) of finding a given amino acid in a helix, a sheet, etc. (see Table 13.6).

To predict the secondary structure, a mean value of these structural parameters is evaluated for four residues all along the protein. Four residues among six with $P\alpha > 1.06$ can initialize a helix and for the residues involved, $<P\alpha>$ must be greater than 1.03. The helix stops if $<P\alpha> < 1.0$ for four residues. Three residues among five with $P\beta > 1.05$ may initialize a β-sheet. The sheet stops if $<P\beta> < 1.0$ for four residues. If a sheet is predicted, $<P\beta>$ must be greater than $<P\alpha>$ (and the reverse for the prediction of a helix). Similarly, β-turn prediction can be carried out from the structural parameter of residue i and the

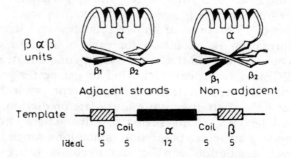

Figure 13.17 Idealized templates for βαβ units (from Taylor and Thornton with permission [70]).

Table 13.6 Conformational parameters of Chou–Fassman–Prevelige [65, 65a].

Amino acids	α-helix $P\alpha$	β-sheet $P\beta$	β-turn Pt	β-turn			
				f_i	f_{i+1}	f_{i+2}	f_{i+3}
A	1.42 H	0.83 i	0.660	0.060	0.076	0.035	0.058
C	0.70 i	1.19 h	1.190	0.149	0.053	0.117	0.128
D	1.01 I	0.54 B	1.460	0.147	0.110	0.179	0.081
E	1.51 H	0.37 B	0.740	0.056	0.060	0.077	0.064
F	1.13 h	1.38 h	0.600	0.059	0.041	0.065	0.065
G	0.57 B	0.75 b	1.560	0.102	0.085	0.190	0.152
H	1.00 I	0.87 i	0.950	0.140	0.047	0.093	0.054
I	1.08 h	1.60 H	0.470	0.043	0.034	0.013	0.056
K	1.16 h	0.74 b	1.010	0.055	0.115	0.072	0.095
L	1.21 H	1.30 h	0.590	0.061	0.025	0.036	0.070
M	1.45 H	1.05 h	0.600	0.068	0.082	0.014	0.055
N	0.67 b	0.89 i	1.560	0.161	0.083	0.191	0.091
P	0.57 B	0.55 B	1.520	0.102	0.301	0.034	0.068
Q	1.11 h	1.10 h	0.980	0.074	0.098	0.037	0.098
R	0.98 i	0.93 i	0.950	0.070	0.106	0.099	0.085
S	0.77 i	0.75 b	1.430	0.120	0.139	0.125	0.106
T	0.83 i	1.19 h	0.960	0.086	0.108	0.065	0.079
V	1.06 h	1.70 H	0.500	0.062	0.048	0.028	0.053
W	1.08 h	1.37 h	0.960	0.077	0.013	0.064	0.167
Y	0.69 b	1.47 H	1.140	0.082	0.065	0.114	0.125

H, h, I = strong, medium or weak former character for helices or sheets, i = indifferent, B, b = breaker character (strong, weak).
P values are derived from fractional occurrences. For example, for an α-helix, the ratio $f_\alpha = n_\alpha/n$ of the occurrences of a given residue within helices versus the total occurrence of this residue in the learning set gives the fractional occurrence of the residue in the helices. $P = f_\alpha/\langle f_\alpha \rangle$, where $\langle f_\alpha \rangle$, the average value of f_α, is simply the sum of f_α values divided by 20, the number of different residues.
P = 1 means that the residue adopts that conformation at the same frequency on the "average" residue.

fractional occurrences of residues i, $i+1$, $i+2$, $i+3$. Given amino acid i, and $\langle p_t \rangle$ = $f_i * f_{i+1} * f_{i+2} * f_{i+3}$ a turn is likely to start in position i if $\langle p_t \rangle > 0.75 * 10^{-4}$, $\langle Pt \rangle$ > 1.0 and $\langle Pt \rangle$ > $\langle P\alpha \rangle$ and $\langle P\beta \rangle$.

Other models introduced rules developed from an *a priori* stereochemical theory of the secondary structure of globular proteins. The prediction is based on the hydrophobic character and size of amino acids within short fragments (typically five residues): for example, large hydrophobics at i, $i+1$, $i+4$ or i, $i+3$, $i+4$ are good clues for a helix [63].

Neural networks and structure prediction

In various chemical research areas, neural networks have shown their ability to acquire complex mapping between input and output data sets, and to establish relationships even before the underlying mechanisms are understood. Neural networks are connected structures of summing nodes in which each node is linked to other nodes via weighted connections. Allowing the strength of the connections to change under the control of a learning algorithm makes such networks able to self-organize in a supervised learning environment. (For a discussion of supervised and unsupervised learning from neural networks, see Zupan and Gasteiger [71].)

Various network architectures have been proposed to address the problem of predicting protein secondary and tertiary structures from sequence alone, or classifying sequences into structurally and functionally defined families [71–76]. Neural networks have also been used for the analysis of nucleic acid sequences [77]. Although the various approaches use different types of network, an important set of studies devoted to the prediction of the secondary structure, focuses on a feed-forward, layered organization. The learning step is carried out thanks to the algorithm of back propagation of errors in the learning stage: that is, information propagates forward (from input to output) while error correction is performed backwards (from output to input) (Figure 13.18).

Nodes in a layer only communicate with immediately adjacent layers, according to connections provided with suitable weights w_{ji}. At each node i, the weighted sum of all outputs O_j from the nodes j of the preceding layer connected to i (possibly with a bias B_j) gives the activation A_i. The output from node i is then calculated applying to A_i a squashing function, frequently the sigmoidal function:

$$O_i = 1/(1 + e^{-A_i})$$

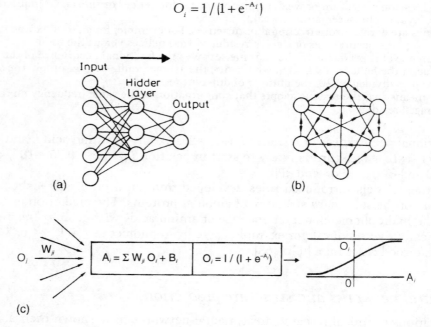

(a)   (b)

(c)

Figure 13.18 (a) Three-layer and (b) Hopefield-type networks; (c) neuron computation and sigmoidal squashing function. A_i = activation of unit i (bias B_i); O_i = output of unit i.

This output is then transmitted to the nodes of the next layer connected to node i. The input layer receives structural information from the environment of the position considered and sends it to the next layer. Results are read from the output layer. One or more hidden layers can be placed between the input and output layers.

In applications to proteins, inputs are typically characteristic of the sequence and outputs are secondary-structure features or distances between α-carbons. In the learning stage, a set of known input/output pairs are fed to the network iteratively, and the system adjusts the weights of the connections and the biases so as to retrieve the correct values at output. This is done by back propagation of errors: starting from the differences between observed and calculated values, weights are modified from the output layer toward the preceding layer, and so on. The system is now trained and ready to deal with new compounds. As to inputs, in the studies quoted, the nature of the amino acids is described within a window centred on the current residue (that under scrutiny) and progressively moved along the sequence. For each amino acid of the window, its nature is specified by binary encoding within a 20- or 21-bit group of inputs (one for each possible amino acid and one spacer to cope with windows overlapping the end points of the protein sequence). Within each group of 21 inputs, the input corresponding to the amino acid present is set to 1 and the others to zero. The window width was chosen as 13 amino acids (six neighbours before and after the one investigated) [72, 74], or 17 [75], leading to a high number of input neurons (say $13 \times 21 = 273$). In [72], the output is constituted by three neurons for the three possible states considered (α-helix, β-sheet, coil), but variants have been investigated: two outputs (helix and sheet) for Holley and Karplus [75], whereas Bohr et al. [73] used separate networks for the decisions on helix, sheet or coil. Various topologies (as to the number of hidden units) were also tried, but the number of hidden units does not seem to significantly modify the results. The predictive accuracy is about 63%, and so comparable to other approaches. However, subdividing proteins into structural classes (α, β, α/β) or adding supplementary information (hydrophobicity) significantly improves the results.

Beyond the recognition of secondary structure patterns, neural networks were also used to predict the tertiary structure. Bohr et al. [76] still used a feed-forward network, but extended the input window to 61 residues. Outputs specify the secondary structure assignment (helix, sheet, coil) and tertiary structure information that is treated in a binary coding of the C_α distances (0 if the distance between the two residues in question is less than an 8 Å threshold value, 1 otherwise). Thirty (binary) distance constraints for each C_α are so provided as a starting point for an energy minimization process (with a steepest descent approach) to generate a folded conformation of the protein backbone (Figure 13.19).

The approach of Wilcox et al. [78] is similar, but the network is trained to reproduce a real distance matrix (output), consistent with the data extracted from the Brookhaven Database. Indeed, a distance matrix can be used for classification and preserves sufficient information for predicting the 3D structure. Sequences given as inputs are now encoded by Leibman hydrophobicity parameters (scaled to +1/−1) [79], rather than by their nature, as

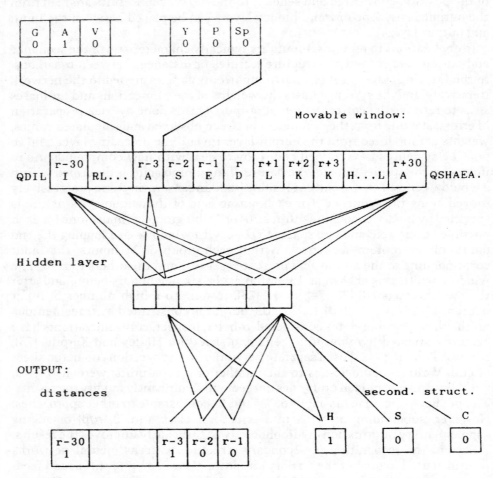

Figure 13.19 Schematic representation of the three layer feed-forward network used by Bohr *et al.* [76] for predicting 3D protein structure. The input is constituted by a movable window of 61 residues centred on the residue under scrutiny (r in the scheme). Each residue is identified by binary coding in a 20 letter descriptor (one per amino acid). For example, in this scheme, residue r–3 is alanine. Outputs indicate the secondary structure (here r belongs to an α helix), and distances between r and its preceding neighbours; residues r–30, r–3 are further apart from r than the threshold distance (8 Å). Typically, the network comprises 20 × 61 input units, 300 hidden units and 30 + 3 units as output. Each unit is connected to all units of the preceding and following layers. Only some connections are shown here for the sake of clarity.

in the previous applications. The network used in their study comprises an input layer of 140 units, and one or two hidden layers can be managed (up to 240 nodes for the one layer model, 2 × 140 hidden nodes when two hidden layers are considered). A possibility of direct connection between input and output was also provided. The output is formed by a normalized (+1/−1) distance matrix (140 × 140) for α-carbons. Such an organization corresponds to about ten million connections executing up to five million updates per second, which were carried out on a Cray supercomputer.

The training set was formed of 15 heterologous proteins of less than 133 residues each, inputs being determined by means of a movable window of 10–20 residues along the sequence. As to the possible prediction of a simple helix or strand for novel sequences, a reliability for the prediction of about 65% is achieved. This is only roughly similar to the results of Chou *et al.* [65–67a], although given data are learned to 98% recall fidelity, with an output distance matrix within 0.3–1.5% RMS deviation. According to the authors, the training set obviously seems too small for useful generalization (although encouraging results are obtained for pairs of homologous structures, confirming that a trained network is able to predict structure within a homologous family). Increasing the size of the training window and the complexity of the description, it is hoped, should lead to better results.

Interesting remarks about the way in which neural simulators work also emerge from that study. Networks with direct input-output connections learn faster but perform as simple associative memory, demonstrating little generalizing ability. Hidden layers forcing the formation of global associations are more able to generalize.

Prediction of folding has recently been tackled [80] with another network structure: a Boltzmann machine. Rather than taking into account the identity of some fragments with learned sequences, the system minimizes a parameterized energy function. The Boltzmann machine (already presented in Chapter 11) is a feedback, completely connected network with a binary threshold and symmetric connections. Each state of the system has an energy that is a function of the state of the units (the nodes) and connections. The probability of a state is related to its energy by Boltzmann's probability law. In classical use, learning adjusts the connections so that the probability distributions of the values of the units are as close as possible to some desired probability distribution. Here the system has to adjust free parameters (thought of as representing physical quantities such as hydrophobicity) rather than adjusting connections. Learning is carried out by adjusting the probability of finding each amino acid residue in a given conformation. Configuration is simply encoded by indicating whether the C_α carbon considered is in an α-helix. The system learns the probability of the configuration of segments (or the whole) of the protein. To fit this probability distribution with that which it calculates during the learning stage, the system adjusts the energy parameters involved in the energy function. This energy function is calculated according to the helix propensity parameter of the individual amino acids (free energy of the helix state for each amino acid). It involves consideration of residues i, $i+1$ and $i+2$ (that must be in an α-helix configuration to allow for the formation of a hydrogen-bond network between the carbonyl of residue $i-1$ and NH group of residue $i+3$ [2]. Another energy parameter also intervenes in representing the H-bond energy between residues (situations where amino acids i and $i+3$ are both hydrophobic, hydrophilic, or one of each type being distinguished). The results obtained on a training set of 110 protein chains (mean dimension = 165 residues) compare well with other approaches.

The framework proposed by Friedrichs and Wolynes [81] to tackle the problem of protein folding relies on associative memory. The formalism is similar to that used in spin associative memories, where interactions are given by the spin correlation function over the memory set. Here the Hamiltonian expressing the interaction potential between residues is evaluated by a "charge density" correlation function, comparing "charges" (in fact a binary encoding of hydrophobicity) over the database. According to the authors, the system has a large capacity for recall, and seems able to recognize tertiary structures for moderately variant sequences.

Artificial intelligence technique

Knowledge engineering was also recently introduced by Clark *et al.* [82] for protein sequence analysis and structure prediction. Identification of analogies in secondary structure, domains or interactions with ligands was at the centre of numerous studies, with the common aim of deriving rules for structure prediction. For example, Blundell *et al.* [83] proposed extracting loop conformation from a database. In attempts towards automated procedures [84, 84a], rules were proposed for the conformation of side chains when replacing residues in homologous proteins.

Common processes and decisions involved in the analysis of a sequence for structural prediction are summarized in Taylor's flowchart of "possible paths to follow in the prediction of structure" [85]. This flowchart represents one possible strategy, but it may be argued that it imposes a defined order in process execution (although many operations involved could be used at diverse stages), and possible extensions cannot be easily incorporated (Figure 13.20).

However, in a parallel approach, it was established that the processes used in these flowcharts can be expressed as a set of rules [86], making it possible to include them in a knowledge-based system. Such systems now offer opportunities to incorporate rules, facts and hypotheses in a new logical approach. One of their main characteristics is the clear separation and explicit representation of descriptive knowledge (facts, hypotheses) and strategic knowledge (how to use the descriptive knowledge to solve problems) combined by logic inference [87]. In fact, very often, most of the rules used in knowledge-based systems, and generally expressed in the simple formalism

"IF CONDITION ... IS FULFILLED, THEN ACTION ... RESULTS"

could be incorporated into traditional procedural programming (written in common languages such as Fortran). However, the declarative structure of knowledge-based systems offers important advantages as compared to traditional programming:

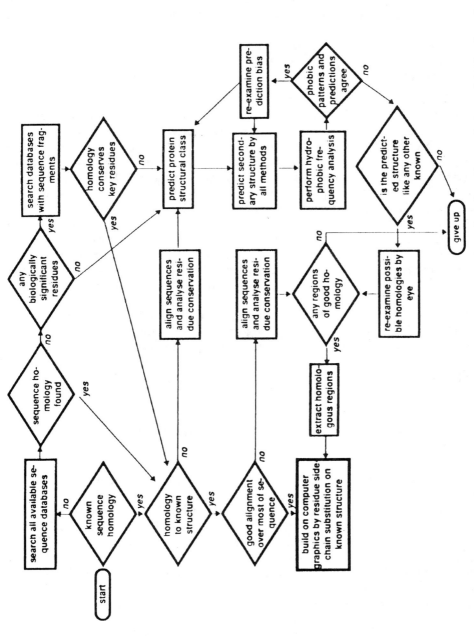

Figure 13.20 Possible paths in predicting a protein structure according to Taylor's flowchart (from Clark *et al.* [82] and Taylor [85]).

- modularity to permit extension,
- flexibility (opportunity to integrate diverse knowledge from various sources of information, and coherently represent it with the same framework... easy addition of new knowledge as more problems are solved),
- consistency and truth maintenance of facts and hypotheses,
- robustness and capability to accommodate conflicts,
- ability for reasoning about high level relationships, etc.

These qualities are offered in the proposed system, which is based on a network architecture (according to the blackboard model in expert systems) where nodes represent entities and links represent processes. The prototype system presented consists of 29 nodes and over 100 links. Knowledge comprises sequence homology, biologically significant residues, results of biochemical or biophysical assays (proteolytic cleavage, NMR data). Entities (storing knowledge) comprise structural description, identifiers, results of assays or of database queries. Links represent relationships between entities. These relationships can be constraints (consistency requirements between entities), and minimal preconditions associated with processes relating entities (for example, similar sequences are a precondition for alignment, but not sufficient alone, since additional preconditions are necessary). When the required conditions are met, it is possible to execute the process linking the two entities.

Besides secondary structure prediction, tertiary structure can be attained, thanks to Cohen's combinatorial algorithm using disulphide linkage and structural class folding rules. However, the authors indicate that prediction still suffers from non-standardization of some of the knowledge, and a lack of a coherent functional classification of non-enzyme proteins [88].

13.5.2 Combining structural elements from homologous proteins

Proteins have been classified into families based upon sequence homology; globins, cytochromes and serine proteases are some of these. Within a family, homologous amino acid sequences have virtually identical folding patterns. Such similarities in the 3D organization suggest that the structure of an unknown protein can be attained from known members of the family by comparative model building [60, 61, 61a, 89]. In fact, structure, presumably determined by the requirement to preserve the functionally correct fold, has evolved more slowly and is better conserved than the sequence.

The example of serine proteases (the structures of which are known) can be used to present the methodology of Greer [61, 61a]. The basic observation is that known serine proteases have Structurally Conserved Regions (SCRs) (with high sequence homology) and Structurally Variable Regions (SVRs) (including additions and deletions), usually corresponding to external loops. For regions of high sequence homology, construction of the SCR is straightforward, whereas building SVRs remains a more challenging task. Greer proposes to model SVRs from *parts* selected from several *different* known structures. For example, when modelling chymotrypsin from trypsin and elastase, trypsin is a good model for the 97–101 loop and elastase for the 203–206 loop (Figure 13.21). In other words, the method "capitalizes upon the availability" of several experimentally known homologous proteins.

However, linking together fragments extracted from different proteins remains somewhat hazardous, since this approach is unaware of environmental effects. The active site of a protein is generally composed of residues from different strands. It therefore seems difficult to mimic this structural organization by simply assembling fragments from various sources.

In a recent approach, Haneef *et al.* [90] proposed building the framework from an average (regularized) structure that retains important interactions and is derived from several known structures. Then a distance geometry method is used (formation of a metric matrix, extraction of its three largest eigenvalues) before an optimization to relieve distance constraints. The method is applicable to both proteins and nucleic acids.

13.5.3 Modelling by modification of a known homologous protein

The most promising methodology to date relies on modifications of a closely related, functionally analogous, known homologous molecule. This is the basis of *homology modelling* for deriving a putative 3D structure: from a known 3D structure, residues are changed in the sequence with minimal disturbance to the geometry. Then an energy minimization process optimizes

```
      80       85          90          95          100
-KI' |Q K I K i A |K v F K n S |K Y N S| L T I - - N N |d i t  CHT
GNq  |Q F I S a S |K s I V H P |S Y N S| N T L - - N n |d i m  TRP
GTe  |Q Y v G v Q |K I V V h P |Y w N T| O O V A A g Y |d i a  ELA

      200        205        210       215        220
CHT   v c K|  K N - - G A |W t l v g i v s W g|  s S - T c S T
TRP   v c S|  - - - - - - |G K l q g i v s W g|  S G - - c a Q
ELA   h c L|  V N - - G Q |Y a v h g v t s F v|  s R L g c N V
```

Figure 13.21 Sequences of chymotrypsin (CHT), trypsin (TRP) and elastase (ELA) aligned by comparison of their 3D structures (partial view). The structurally conserved regions are shown in boxes. Lower case letters correspond to buried residues (low solvent accessibility of the side chain). When modelling chymotrypsin, trypsin is a good model for the 97–101 loop and elastase for the 203–206 loop (from Greer with permission [61]).

the mutated structure [58]. However, one must bear in mind an important caveat: major changes in the conformation of the external part cannot be predicted.

One of the first attempts in that field, as quoted by Blundell *et al.* [83], seems to be the construction of a model of α-lactalbumin (Figure 13.22) from the X-ray structure of lysozyme [91]. This was achieved, at that time, on physical models, but the approach now using computerized images, with their extended capabilities, is similar. For example Kretsinger *et al.* [92, 93] predicted the structure of the calcium-binding region of the protein troponin C, thanks to the known structure of another Ca-binding protein, parvalbumin. Subsequent determination of the crystal structure showed that the prediction was correct for the binding site and supersecondary structure (helix, loop, helix). Similarly, the structure of antigen-binding domains of an antibody (immunoglobulin D1.3) could be predicted by comparative analysis of known antibodies or conformational energy calculations [94].

Determining to what extent sequence homology can be useful for structure prediction of an unknown protein is a huge problem. On the one hand, Chothia and Lesk [60] investigated the influence of evolution on pairs of homologous (strictly speaking) proteins. From this study, it appears that the degree of success in predicting structure from the sequence by comparison to a known homologous structure is good when sequence homology is > 50%. However, if sequence homology is about 20%, large and unforeseeable structural

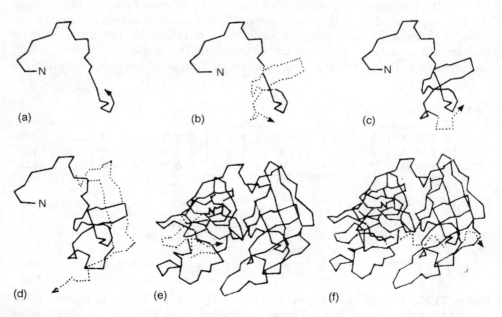

(a)　　　　　　　　　　(b)　　　　　　　　　　(c)

(d)　　　　　　　　(e)　　　　　　　　(f)

Figure 13.22 Modelling using multiple structures. Construction of tissue plasminogen activator (f) from 70 segments of other known proteins on the basis of sequence homology. The first steps are shown in (a)–(d), the last one in (e)–(f) (from Blundell *et al.* with permission [83]).

differences are to be expected. However, the structure of the active site itself may be not heavily changed.

On the other hand, if some rules for side chain substitution have been derived and graphical visualization avoids crass errors (a buried, charged side chain that is not hydrogen bonded is almost certainly incorrect) [83], things are not always as simple as expected. Great care must be taken when carrying out side chain substitution. This was elegantly demonstrated by Karplus *et al.* [89]. Given two proteins of a similar overall shape, *Themiste dyscritum* haemerythrin and mouse myeloma immunoglobulin (respectively, all –α or – β), they constructed two patently incorrectly folded structures: haemerythrin side chains were substituted into the immunoglobulin, and *vice versa*. Surprisingly, energy minimization gives, for these incorrect structures, energies quite comparable to those of the correct proteins (Figure 13.23). In other words, avoiding bad non-bonded contacts, though necessary, is not sufficient to ascertain that the correct solution has been found. Other criteria must be looked for. So, it was shown that the incorrect structures have less stabilizing H-bonding, electrostatic and van der Waals interactions (differences are still more pronounced when the influence of the solvent on the latter two terms is evaluated). The incorrect structures also have a larger solvent-accessible surface, and a greater fraction of non-polar side chain atoms exposed to the solvent.

13.5.4 Recent applications of homology modelling

Among recent applications, Toma *et al.* [95] developed the construction of an atomic model of protease inhibitor regions in APPI (amyloid β-protein precursor, intervening in Alzheimer's disease) based on the structure of bovine pancreatic trypsin inhibitor (BPTI). After the alignment of the two sequences, BPTI side chains were changed with best fitting amino acid residues and steric hindrance removed through energy minimization (AMBER force field) in a 58 amino acid fragment (from residues 287 to 345). To complete the information gained from only the inhibitor model, enzymatic subsite models for serine proteases were also built. Site mapping then provides more insight into target enzyme specificity of the inhibitory activity.

The same methodology (mutation of residues from a homologous structure and energy minimization) was used in building a model of the active site of cytochrome P-450 nifedipine oxidase (P-450$_{NF}$) on the basis of sequence homology with cytochrome P-450$_{CAM}$, with the design of new selective inhibitors [96] in mind. Cytochrome P-450 is a largely widespread mono-oxygenase enzyme, which catalyses oxidation of endogenous and exogenous compounds in the body, with, for example, among its varied roles, detoxification of drugs. Its structure still remains unknown because of the lack of suitable samples for crystallographic studies, except for P-450$_{CAM}$. The latter protein provides a suitable model for building P-450$_{NF}$, for which the sequence is known and exhibits significant sequence homology. Similarity between

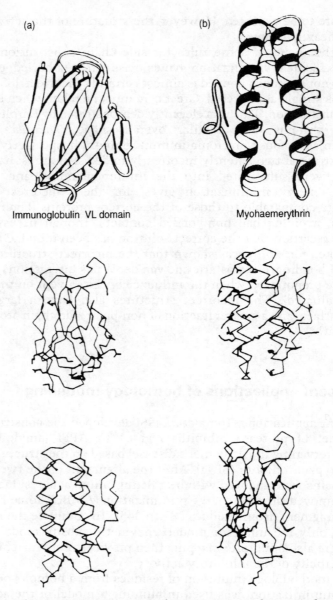

(a)   (b)

Immunoglobulin VL domain   Myohaemerythrin

Figure 13.23 Schematic representation of MCPC 603 myeloma VL domain (a) and *T. dyscritum* haemerythrin (b). Polypeptide chain for correctly and incorrectly folded haemerythrin (right) and VL (left). Correct structures are shown in the upper part of the figure. The chain is represented by α-carbons only. Bold lines correspond to residues lysine, arginine, aspartic and glutamic acid (from Novotny *et al.* with permission [89]).

P-450$_{CAM}$ and P-450$_{NF}$ was assessed using the global alignment method of Needleman and Wunsch and local alignment (see below) [97, 98]. From this study, it appears that the necessary changes in residues maintain the overall hydrophobicity and a similar environment for the haem group, whereas space available for the substrate is increased. Successful docking in the postulated binding pocket with known substrates of very varied types supports the model of active site proposed. From energy minimization it was suggested that van der Waals terms are larger than electrostatic ones in the enzyme-substrate interaction (Figure 13.24).

13.6 EVALUATING SIMILARITY

Rapid comparison of protein structures regarding the sequence of amino acids or their actual 3D organization is a problem of considerable importance. Identifying residues essential for maintaining the structure or functionality is a determining step to understand the properties of known molecules and predict the behaviour of newly isolated or synthesized structures.

In simpler cases, for closely related molecules, a simple, qualitative sequence inspection is sufficient to detect significant similarity. In more complex situations, it can be necessary to quantify the degree of similarity, for example to construct an evolutionary tree (Figure 13.25), or even to explicitly build the best "alignment" (residue-residue correspondence) between two structures. Finally, for proteins having diverged long ago, any similarity could be hard to recognize on sequences, although related functions and 3D structures may be largely maintained: in such cases, comparisons must be carried out directly on the spatial location of amino acids.

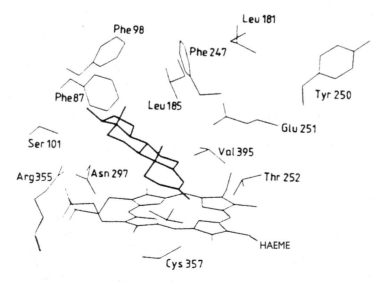

Figure 13.24 Fitting testosterone (bold lines) to the modelled active site of P-450 nifedipine oxidase (from Ferenczy and Morris [96]).

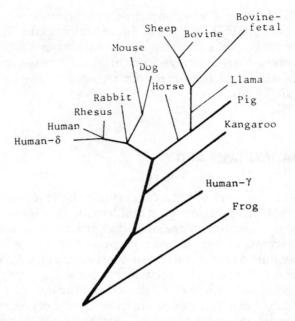

Figure 13.25 An evolutionary (phylogenetic) tree obtained by comparing sequences of β-haemoglobin chains and related variants in various species. (The length of each branch is proportional to the number of point mutations to pass from one sequence to the next one.) (From Cantor and Schimmel [2], p 79, adapted from Dayhoff *et al.* with permission [103].)

These various aspects can be summarized by the three following typical examples, which are extracted, among others, from Lesk [55]:

1. The high similarity observed between proinsulin chains of man, pig, cow, etc. indicating an evolutionary relationship is confirmed by their identical function. Pig insulin is used clinically in the treatment of diabetes in man.
2. Sperm whale myoglobin and lupin leghaemoglobin (with only 15% common residues) have similar secondary and tertiary structures, and are distantly related proteins.
3. On the contrary, chymotrypsin and subtilisin (although a common proteolytic function and a common catalytic mechanism) are not similar nor related. They correspond to a convergent evolution.

In the protein field, this problem of evaluating similarity between two molecules ("where, why, and to what extent, two protein sequences are similar" [98]), was tackled at different levels of complexity, depending upon the knowledge available or the degree of precision and quantification desired.

A first type of comparison involves only the sequence (the order in which the amino acids are linked in the chain). This is a challenge of prime interest, since the sequence is known well before the 3D structure is solved, and sequence comparisons are at the basis of many interpretative or predictive studies. More sophisticated approaches deal with 3D coordinates to specify spatial similarities. From a theoretical point of view, such studies were necessary to determine to what extent sequence similarity can confidently reflect 3D similarity. At a more practical level, comparisons of 3D structures are useful for understanding the differences in properties (say binding ability, for example) of neighbouring proteins, or for defining templates to be used in protein modelling by homology.

13.6.1 Sequence similarity

Looking first at 2D comparisons (sequence level), two approaches can be considered:

- sequence comparison to detect common features,
- sequence alignment which defines the best one-to-one correspondence between amino acid sequences of two proteins.

A very basic tool in these comparisons (providing a simple approach and an attractive method for visualizing the results) is a dot-plot similarity matrix. The sequences are recorded along the two axes of a 2D graph. In this graph, each point corresponds to one residue (of the first protein) along the vertical axis and one (of the second protein) along the horizontal axis (in 3D applications, for the sake of simplicity, points would correspond to α-carbons, to which each residue is reduced, ignoring details of side chain arrangements) [99]. Points are encoded according to any predefined criterion of similarity, and are marked (as dots, for example) if this similarity criterion is met. Parts of the compared sequences which are similar are clearly identified as a succession of consecutive dots parallel to the diagonal. Insertions or deletions are detected as skips to a parallel segment (Figure 13.26).

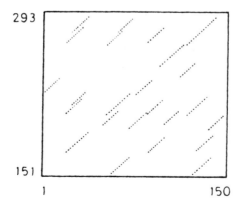

Figure 13.26 Diagonal plot of fragment similarities between two domains of rhodanese (respectively, residues 1–150 and 151–293). Each dot corresponds to a pair of residues (from Vriens and Sander with permission [100]).

In their simpler form, homology matrices indicate only the identity between the ith and jth residue in the two sequences (scored as 1 or 0). A similar binary coding is the rule for nucleotides in DNA. However, for proteins, various other properties can be encoded, such as the nature of the residue, charge, hydrophobicity, bulkiness, or propensity to form α or β secondary structures. We will develop this point later. For example, Greer proposes gathering residues into classes of equivalence (depending upon bulk, size, charge, polarity, etc.), with identity quoted as 1, and intra-class equivalence as 0.5 [61, 61a], according to the scheme of McLachlan [101, 102].

Classes of equivalence of naturally occurring amino acids are:

(D,E,K,R) (G,A,V) (A,V,L,I) (V,L,I,M)
(F,Y,W) (S,T) (Q,N)...(G,P, for turns)

Another widely used scoring process (particularly when distant proteins are to be compared) is the MDM78 matrix of Dayhoff. This matrix (empirically derived from evolutionary patterns) reflects the amino acid mutations per 100 residues, and scores them from a logarithmic scale of probability for residue replacement [103, 104] (Table 13.7).

Table 13.7 Dayhoff mutational substitution matrix (from Gray [99]).

	A	R	N	D	C	Q	E	G	H	I	L	K	M	F	P	S	T	W	Y	V	B	Z	X
A	2	-2	0	0	-2	0	0	1	-1	-1	-2	-1	-1	-4	1	1	1	-6	-3	0	0	0	0
R		6	0	-1	-4	1	-1	-3	2	-2	-3	3	0	-4	0	0	-1	2	-4	-2	-1	0	0
N			2	2	-4	1	1	0	2	-2	-3	1	-2	-4	-1	1	0	-4	-2	-2	2	1	0
D				4	-5	2	3	1	1	-2	-4	0	-3	-6	-1	0	0	-7	-4	-2	3	3	0
C					12	-5	-5	-3	-3	-2	-6	-5	-5	-4	-3	0	-2	-8	0	-2	-4	-5	0
Q						4	2	-1	3	-2	-2	1	-1	-5	0	-1	-1	-5	-4	-2	1	3	0
E							4	0	1	-2	-3	0	-2	-5	-1	0	0	-7	-4	-2	2	3	0
G								5	-2	-3	-4	-2	-3	-5	-1	1	0	-7	-5	-1	0	-1	0
H									6	-2	-2	0	-2	-2	0	-1	-1	-3	0	-2	1	2	0
I										5	2	-2	2	1	-2	-1	0	-5	-1	4	-2	-2	0
L											6	-3	4	2	-3	-3	-2	-2	-1	2	-3	-3	0
K												5	0	-5	-1	0	0	-3	-4	-2	1	0	0
M													6	0	-2	-2	-1	-4	-2	2	-2	-2	0
F														9	-5	-3	-3	0	7	-1	-5	-5	0
P															6	1	0	-6	-5	-1	-1	0	0
S																2	1	-2	-3	-1	0	0	0
T																	3	-5	-3	0	0	-1	0
W																		17	0	-6	-5	-6	0
Y																			10	-2	-3	-4	0
V																				4	-2	-2	0
B																					2	2	0
Z																						3	0
X																							0

Value quoted at position (i,j) represents the likelihood for residue i (with respect to j) to replace residue j (with respect to i) during the process of evolution.

Comparisons can, of course, be carried out by looking directly at the characteristics encoded for individual residues i and j. However, sometimes one prefers to plot at point i information corresponding to a local window and representing a weighted score for a short segment centred on position i.

13.6.2 Optimal alignment

More detailed analysis requires the best alignment (one-to-one correspondence of residues) to be determined, that is the highest scoring correspondence on the entire sequences (summing individual pairwise correspondence scores) and accommodating the possibility of insertions and deletions, i.e. gaps in the correspondences. A basic and widely used solution is the algorithm proposed by Needleman and Wunsch [97], which relies on dynamic programming, the optimal alignment being obtained by induction. The best score for an alignment of sequences ending at residue i through a pair of residues i (of the first protein) and j (of the second one) is obtained by adding the score of the pair (i,j) to the best score for the preceding optimal alignment (that of a sequence ending at $i-1$). A matrix of scores is progressively computed, then the overall alignment is obtained by tracing back along the path of induction that gives the highest score.

A gap in the correspondence between sequences decreases the similarity, and is assigned a penalty. Differing values distinguish gap start and gap extension so as to reduce the number of unmatched residues and gather them in few segments. A small penalty will lead to many (unrealistic) gaps. Too large a penalty limits the number of possible solutions examined. A good compromise seems to be fixing the penalty at two or three times the score for matching identical residues (Figure 13.27).

Global alignment over the entire sequence does not detect local similarities over short stretches which do not contribute to the global optimal alignment. To extract these regional similarities, various modifications of the Needleman–Wunsch algorithm have been proposed. Sellers [106] goes over the path of induction through the sequence in the forward and reverse directions and selects the common parts, whereas Boswell and McLachlan [107]

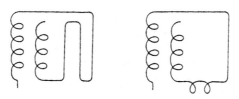

Figure 13.27 Two similar proteins with a deletion from the left molecule and insertion into the right one (from Rossmann and Argos with permission [105a]). For the other parts, the topology (fold) is identical in both structures.

introduced a damping factor reducing the contribution of distant parts of the alignment. Comparison of fixed length segments was also investigated [101, 102, 108, 108a], but suffers from interruption by insertions or deletions.

A similar approach was developed by Morris (the RELATE program) [98]. A window from one protein is displaced along the second structure and the similarity is scored. Then the window is displaced along the first structure and the comparison is repeated, and so on. The best score is stored with the position of the similar segments. A display where each amino acid is represented as a colour bar in the sequence allows for an immediate perception of similarity. Colour groups can be based on hydrophobicity, charge or polarity, acidic character, bulk, propensity to be involved in a secondary structure, etc. Alternatively, a dot matrix analysis offers an easy way in which to rapidly visualize the degree of similarity between homologous proteins.

Identifying similar sequences in two proteins would, in principle, require a high number of comparisons of residues. However, the computational task can be reduced by substantially avoiding pairwise comparisons, and using a look-up table indicating for each of the 20 amino acids its diverse locations along the chain of the first protein [109]. Equivalences for the successive residues of the second chain (with their position offset with respect to the first chain) are readily identified, and the similar sequences (residues identical to those of the first protein with the same offset) are detected and displayed in a dot matrix homology plot. Groups of identities can also be sought in view of increased speed.

Although the problem of optimal alignment has prompted numerous studies, simultaneously aligning more than two structures has so far remained an open challenge [55].

13.6.3 3D similarity

Looking at 3D similarity poses the problems of identifying common 3D features, and possibly superimposing parts of the molecular framework. These problems have already been discussed in Chapter 11, and here we just recall the essential points.

Evaluating 3D similarity

When looking for structural analogy, a rapid geometrical search in protein tertiary structures may be very useful. An early solution in the protein field was proposed by Lesk for the automatic identification of a *user-defined* pattern of atoms in a chain represented by the coordinates of the α-carbons [110]. However, this algorithm, although appropriate for searching for patterns of atoms in proteins, generates a large number of atom combinations and suffers from a need for considerable computational requirements. A two-stage procedure using Lesk's algorithm as a precursor of Ullman's subgraph isomorphism method [111] leads to a definite improvement [7]. As to the geometrical transformations to be performed for superimposition of molecular frameworks, the algorithm of McLachlan is widely used (see Chapter 11) [112].

In an earlier attempt to detect similarity in the 3D structure of proteins, Remington and Matthews proposed dividing the two structures into overlapping segments of a predetermined length and fitting all pairs of segments by a least-squares procedure, the RMS deviations being plotted in a comparison matrix [113].

Rossmann and Argos [105–105b] developed procedures for identifying structural and topological similarities thanks to a probability function with two parameters. The first one indicates the spatial proximity between pairs of residues i and j, and the second the relative orientation of successive residues. A weighting factor allows for giving more importance to topology or spatial equivalence, or both. This function can also skip over portions which are inserted in one molecule. For any given orientation of the two molecules, equivalences are first determined (starting from highest P_{ij} values, then extending the sequential equivalences). Systematic rotations of one of the two proteins are carried out in a search on a grid of the three Eulerian angles [114]. The number of equivalent residues for each node of the search grid indicates the rotation which best relates the two molecules.

Barton and Sternberg [115] address the problem of comparing loops, which may be of different lengths, and extracting rules as to loop formation to be input into a knowledge-based system. They extended to 3D comparisons on C_α coordinates the algorithm of Needleman and Wunsch, originally developed for sequence comparison, and proposed a method able to align equivalent regions automatically. This method seems particularly suited to the comparison of short variable regions bounded by structures of high similarity. These regions (conserved cores of α-helices and β-strands) are first superimposed, and for the intermediate part linking the regular regions (loops) a similarity matrix is built from distances between residues. The best alignment calculated is stored as a list of vectors joining α-carbons for display (Figure 13.28).

Comparison of distant proteins

Comparison of secondary-structure features in the analysis of proteins with different sequences but common folding motifs enlightens new facets of

Figure 13.28   Alignment of loop regions. A model for a β-turn of chymosin (dotted line) from the known structure of endothiapepsin (bold line) and deletion of a single residue (from Blundell *et al.* with permission [83]).

structural similarity [116, 117]. As a first step, a system is built for retrieving secondary structure patterns from 3D coordinates (extracted, for instance, from the Brookhaven Database). Although a detailed search can be carried out from C_α coordinates [79], a simple vectorial representation was preferred for characterizing β-strands and helices.

According to the method of Kabsch and Sander [118], secondary structural elements are recognized by calculating the hydrogen-bond pattern of atoms from their X-ray coordinates. An alternative approach tries to match a template built from the coordinates of four α-carbons (with ideal angles for helices or strand) to each four-residue section of the protein. From the starting and ending residues of the patterns recognized, the program calculates the axes of α-helices and strands. Then it evaluates the angles between each pair of axes, distances between their mid-points and their closest approach distance. These parameters are stored in a matrix representing the protein. This matrix will be used in the POSSUM program (Protein On-line Substructure Searching–Ullman Method) to recognize a user-defined secondary pattern [116–117].

The protein is considered as a labelled graph: nodes represent secondary structures (either helix or strands) expressed in the linear representation (vector) developed, and edges correspond to the relationships between them (inter-vector angles and distances). The query is also described in terms of a similar graph according to the pattern to be recognized. Ullman subgraph isomorphism (in a slightly modified version, introducing some tolerance values and allowing for the retrieval of all occurrences of matching substructures) is then used to determine whether the query graph is contained within the graph representing the protein. Tests with the β-structural motifs of Richardson [119] show the general effectiveness of the matching procedure, with the advantage that it is not necessary to specify all the possible distances and angles in the search pattern [116] [Figure 13.29].

Thanks to this methodology, a comparison of the tertiary fold in the *Salmonella typhimurium* Che Y chemotaxis protein and that of a GDP binding domain of *Escherichia coli* elongation factor Tu (EF Tu) demonstrates definite similarity, far beyond the previously recognized resemblances of each protein's

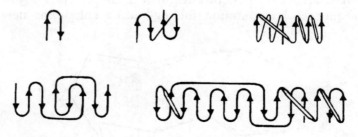

Figure 13.29 Some of the 37 Richardson β-sheet patterns. Drawings correspond to patterns 1, 5, 21, 29, 37 (from Richardson with permission [119]). Arrows point toward the C-terminus. Connections between strands are drawn as double lines (if they occur above the plane or as single lines (below the plane).

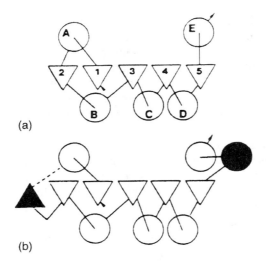

(a)

(b)

Figure 13.30 Schematic representation of (a) Che Y and (b) EF Tu (GDP binding domain) according to the topological representation of Sternberg and Thornton [120]. Circles = α-helices, triangles = β-strands (apex down indicates that the strand is viewed from the C-terminus; shaded parts represent the extra strand and helix in EF Tu) (from Artymiuk *et al.* with permission [116]).

fold to that of a generic nucleotide-binding domain (Figure 13.30). This intriguing result, between two structures lacking significant sequence homology, may reflect a particularly stable folding motif or an evolutionary relationship, but one that is so remote that sequence homology has been lost, or finally, binding to receptors with related structures. According to the authors, it may possibly reveal some previously unsuspected link between families of signal-transduction proteins, playing crucial roles in the regulation and control of cell functions in prokaryotic and eukaryotic cells, respectively [116, 117] (Figure 13.31).

Representation of 3D features

As for displaying inter-sequence relationships, dot maps also offer an efficient way in which to visualize, in easily understandable pictures, essential features

```
         21                                                          50
CheY:    V  R  N  L  L  K  E  L  G  F  N  N  V  E  E  A  E  D  G  V  D  A  L  N  K  L  Q  A  G  G
         *  *  :  *  *     :     :  *              :     :     *  *  :     *  :  :  :
EF Tu:   V  R  E  L  L  S  Q  Y  D  F  P  G  D  D  T  P  I  V  R  G  S  A  L  K  A  L  E  G  D  A
         154                                                        183
```

Figure 13.31 Poor homology in the sequences of Che Y and EF Tu (*represents identity, : represents potentially conservative substitutions) between stretches appearing in structurally unrelated parts of the respective proteins (from Artymiuk *et al.* with permission [116]).

related to either the geometrical organization of a single protein or some similarity (according to a user-defined criterion) between two macromolecules.

A *distance map* [121, 122] provides a simple way in which to understand the 3D folding of a protein in a 2D representation. It is built by plotting the relative distances r_{ij} between all pairs of α-carbons for residues i and j, and colouring areas according to given ranges of d_{ij} (with the advantage of a representation independent of the coordinate system). Such a map allows for easy identification of neighbouring residues (darker regions in the scheme) (Figure 13.32). Proximities occur most frequently for residue neighbours in the sequence, i.e. those appearing near to the diagonal. Other dark regions indicate residues which are neighbours in space but apart in the sequence. They correspond to a folding of the polypeptide chain, which is clearly demonstrated. Characteristic patterns can also be associated with the basic features of the secondary structure (helices, sheets, etc.).

Various other analyses can be carried out from a distance map. For example, the number of residues within a fixed distance (say 10 Å) of a given residue can be evaluated, giving some insight into the solvent accessibility of this residue. Similarly, a difference map between an observed and a computed conformation visualizes the degree of similarity between them. Interaction energies between residues or the distribution of hydrogen bonds can also be displayed. Distance

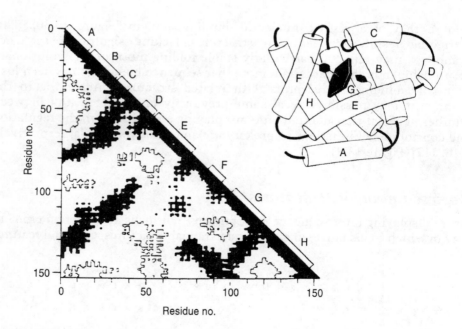

Figure 13.32 Distance map for the native structure of myoglobin. Dark regions 0 Å < r_{ij} < 15 Å; close areas r_{ij} > 30 Å; blank areas 15 Å < r_{ij} < 30 Å. Dark regions perpendicular to the diagonal correspond to folding in antiparallel mode, and to a parallel mode for arrangements parallel to the diagonal (from Ooi and Nishikawa [121]) and adapted from Widom and Edelstein [5] p 555 for the scheme.

maps are also useful for detecting homology in sequences of proteins which have the same (or a similar) function. They allow one to examine whether folding of polypeptides having common residues gives rise to the same tertiary structure. In such a way, distance maps constitute a first rapid approach that can later be complemented by a visual superposition method for more detailed comparisons.

Structure prediction from sequences can at most have a limited success, but *hydrophobicity profiles* (which are also directly available from the amino acid sequence) can give valuable complementary information about the 3D structure. Given an amino acid, hydrophobicity (for which various scales have been proposed) can be viewed as its relative preference for a non-aqueous medium (non-polar solvent or interior region of proteins) rather than the aqueous external region. In fact, for a protein, one cannot look only at the hydrophobicity of each amino acid in the chain, but rather plot a value averaged on a moving window centred on the residue vs. its position in the sequence. Such profiles can be used to detect segments rich in residues with hydrophobic side chains, which tend to bury themselves in the interior part, avoiding the aqueous surrounding. This gives a method by which to distinguish interior or exterior regions. The example of lysozyme shows quite a fair agreement of this method (with a nine-residue window) with solvent accessibility computed from the 3D structure [123, 124] (Figure 13.33).

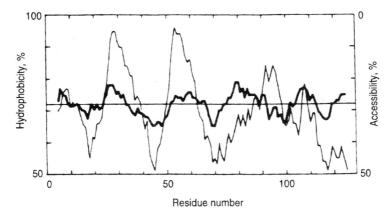

Figure 13.33 Hydrophobicity (bold line) and computed fractional accessibility (thin line) profiles for lysozyme using a nine-residue window. Pronounced minima of the hydrophobicity profile correspond to major solvent-accessible regions, and maxima to buried parts (from Rose *et al.* with permission [124]).

APPENDIX: SELECTED PARAMETERS FOR COMMON AMINO ACID RESIDUES

	A_0	$\langle A\rangle$	f	V	Log P	A_{10}	A_{50}	A_{80}	Cav.	Hy.P.	F
ALA	118.1	31.5	0.74	92	-1.52	43.4	67.7	82.0	24.7	-0.5	0
ARG	256.0	93.8	0.64	202	-2.84	17.9	59.7	90.6	50.4	3.0	0
ASN	165.5	62.2	0.63	135	-2.41	19.9	56.0	84.1	36.7	0.2	0
ASP	158.7	60.9	0.62	125	-2.60	20.8	49.8	77.6	37.1	3.0	0
CYS	146.1	13.9	0.91	118	-0.29	73.1	92.3	97.3	28.6	-1.0	5
GLN	193.2	74.0	0.62	161	-2.05	21.4	54.9	87.0	41.9	0.2	0
GLU	186.2	72.3	0.62	155	-2.47	17.3	45.3	76.4	40.2	3.0	0
GLY	88.1	25.2	0.72	66	-1.83	xxx	xxx	xxx	18.1	0.0	0
HIS	202.5	46.7	0.78	167	-1.70	46.0	72.9	93.3	43.0	-0.5	0
ILE	181.0	23.0	0.88	169	-0.03	61.6	87.8	97.1	48.9	-1.8	5
LEU	193.1	29.0	0.85	168	-0.13	61.2	89.2	96.2	51.0	-1.8	5
LYS	225.8	110.3	0.52	171	-2.82	12.9	41.6	81.8	38.4	3.0	0
MET	203.4	30.5	0.85	171	-0.60	62.5	84.2	96.7	50.5	-1.3	2
PHE	222.8	28.7	0.88	203	-0.04	60.1	89.0	97.4	63.5	-2.5	8
PRO	146.8	53.7	0.64	129	-1.34	27.4	64.6	94.3	27.4	0.0	2
SER	129.8	44.2	0.66	99	-1.87	26.7	52.2	76.2	26.8	0.3	0
THR	152.5	46.0	0.70	122	-1.57	28.1	63.7	90.1	34.0	-0.4	0
TRP	266.3	41.7	0.85	238	0.42	48.6	88.4	98.9	73.0	-3.4	9
TYR	236.8	59.1	0.76	204	-0.87	40.7	85.2	97.0	58.3	-2.3	7
VAL	164.5	23.5	0.86	142	-0.61	59.9	85.8	96.7	40.5	-1.5	2

A_0 ($Å^2$) standard state surface accessible area; $\langle A\rangle$ ($Å^2$) mean solvent accessible surface area; f mean fractional area loss ($=1-\langle A\rangle/A_0$) [124]. V ($Å^3$) average buried volume of residues [125]. Log P partition coefficient [for amino acids] [126, 127]. A_{10}, A_{50}, A_{80}, relative side chain accessibility; A_{10}, A_{50}, A_{80} refer to accessibility wells up to 10%, 50%, 80% (ref. GLY-X-GLY) [53]. Cav. cavity volume: internal unfilled volume not accessible to solvent ($Å^3$) [53]. Hy.P. hydrophobicity parameter [79]. F contribution to hydrophobic pairing function $P_i = F_i + F_j$ [39].

REFERENCES

1. L. Stryer *Biochemistry*, W.H. Freeman and Co., San Francisco, 1981, p 189.
2. C.R. Cantor and P.R. Schimmel *Biophysical Chemistry. Part II. Techniques for the Study of Biological Structures and Functions*, W.H. Freeman and Co. New York, 1980.
3. A.L. Lehninger *Principles of Biochemistry*, Worth Publishers Inc., New York, 1982.
4. D.W. Martin Jr, P.A. Mayes and V.W. Rodwell *Harper's Review of Biochemistry*, Lange Medical Publications, Los Altos, 1983.
5. J.M. Widom and S.J. Edelstein *Chemistry*, W.H. Freeman and Co., San Francisco, 1981, pp 555–556.
6. K. Toma *J. Mol. Graph.*, **5**: 1987; 101–102.
7. A.T. Brint, H.M. Davies, E.M. Mitchell and P. Willett *J. Mol. Graph.*, **7**: 1989; 48–53.
8. J.M. Burridge and S.J.P. Todd *J. Mol. Graph.*, **4**: 1986; 220–221.
9. A.M. Lesk and K.D. Hardman *Science*, **216**: 1982; 539–540.
10. P. Quarendon *J. Mol. Graph.*, **2**: 1984; 91–95.
11. M. Carson and C.E. Bugg *J. Mol. Graph.*, **4**: 1986; 121–122.
12. J.S. Richardson *Adv. Prot. Chem.*, **34**: 1981; 167–339.
13. M. Carson *J. Mol. Graph.*, **5**: 1987; 103–106.
14. D.J. Barlow and J.M. Thornton *J. Mol. Graph.*, **4**: 1986; 97–100.
15. T.F. Havel and K. Wüthrich *Bull. Mathematical Biology*, **46**: 1984; 673–698.
16. J.S. Cohen, L.J. Hughes and J.B. Wooten in *Magnetic Resonance in Biology*, J.S. Cohen (Ed.) J. Wiley, 1983; vol 2, 130–247.
17. K. Wüthrich *Acc. Chem. Res.* **22**: 1989; 36–44.
18. A.D. Kline, W. Braun and K. Wüthrich *J. Mol. Biol.*, **189**: 1986; 377–382.
19. K. Wüthrich *NMR of Proteins and Nucleic Acids*, Wiley, New York, 1986.
20. R.J. Abraham, J. Fisher and P. Loftus *Introduction to NMR Spectroscopy*, John Wiley and Sons, Chichester, 1988; 145–171 and 217–232.
21. A. Bundi and K. Wüthrich *Biopolymers*, **18**: 1979; 285–297.
22. H.A. Scheraga in *Advances in Physical Organic Chemistry*, V. Gold (Ed.), Academic Press, London, 1968; vol. 6, 103–184.
23. A. Pardi, M. Billeter and K. Wüthrich *J. Mol. Biol.*, **180**: 1984; 741–751.
24. G.N. Ramachandran, R. Chandrasekaran and K.D. Kopple *Biopolymers*, **10**: 1971; 2113–2131.
25. V.F. Bystrov, V.T. Ivanov, S.L. Portonova, T.A. Balashova and Y.A. Ovchinnikov *Tetrahedron*, **29**: 1973; 873–877.
26. M. Cung, M. Marraud and J. Neel *Macromolecules*, **7**: 1974; 606–613.
27. D. Marion and K. Wüthrich *Biochem. Biophys. Res. Commun.*, **113**: 1983; 967–974.
28. J.M. Moore, D.A. Case, W.J. Chazin, G.P. Gippert, T.F. Havel, R. Powls and P.E. Wright *Science*, **240**: 1988; 314–317.
29. K.G.R. Pachler *Spectrochemica Acta* **20**: 1964; 581–587.
30. G.N. Ramachandran and V. Sasisekharan *Adv. Protein Chem.*, **23**: 1968; 283–438.
31. B. Wittman-Liebold (ed.) *Methods in Protein Sequence Analysis*, Springer Verlag, Heidelberg, 1989.
32. M. Billeter, M. Engeli and K. Wüthrich *J. Mol. Graph.* **3**: 1985; 79–83.
33. W. Braun, C. Bosch, L.R. Brown, N. Go and K. Wüthrich *Biochim. Biophys. Acta.*, **667**: 1981; 377–396.
34. G.M. Crippen *J. Comput. Phys.*, **24**: 1977; 96–107.
35. M.P. Williamson, T.F. Havel and K. Wüthrich *J. Mol. Biol.*, **182**: 1985; 295–315.
36. DISGEO Program: T.F. Havel, QCPE, 1985; 507.
37. T.F. Havel, G.M. Crippen and I.D. Kuntz *Bull. Math. Biol.*, **45**: 1983; 665–720.
38. I. Tinoco, O. Uhlenbeck and M. Levine *Nature*, **230**: 1971; 362–366.

39. I.D. Kuntz, G.M. Crippen and P.A. Kollman *Biopolymers*, 1979; 939–957.
40. M.R. Pincus and H.A. Scheraga *Acc. Chem. Res.*, **18:** 1985; 372–379.
41. W.G.J. Hol *Angew. Chem. Int. Ed. Engl.*, **25:** 1986; 767–778.
42. F.A. Momany, R.F. McGuire, A.W. Burgess and H.A. Scheraga *J. Phys. Chem.*, **79:** 1975; 2361–2381.
43. G. Nemethy, M.S. Pottle and H.A. Scheraga *J. Phys. Chem.*, **87:** 1983; 1882–1887.
44. H.A. Scheraga *Biopolymers*, **22:** 1983; 1–14.
45. M. Vasquez and H.A. Scheraga *Biopolymers*, **24:** 1985; 1437–1447.
46. M.R. Pincus and R.D. Klausner *Proc. Natl. Acad. Sci. USA*, **79:** 1982; 3413–3417.
47. M.R. Pincus, R.D. Klausner and H.A. Scheraga *Proc. Natl. Acad. Sci. USA*, **79:** 1982; 5107–5110.
48. M.R. Pincus, J.V. Renswoude, J.B. Harford, E.H. Chang, R.P. Carty and R.D. Klausner *Proc. Natl. Acad. Sci. USA*, **80:** 1983; 5253–5257.
49. S.R. Wilson and W. Cui *Biopolymers*, **29:** 1990; 225–235.
50. K.A. Palmer, H.A. Scheraga, J.F. Riordan and B.L. Vallee *Proc. Natl. Acad. Sci. USA*, **83:** 1986; 1965–1969.
51. S. Tanaka and H.A. Scheraga *Macromolecules*, **9:** 1976; 945–950.
52. H.A. Scheraga and G.H. Paine *Ann. N.Y. Acad. Sci.*, **482:** 1986; 60–68.
53. L.S. Reid and J.M. Thornton *Proteins*, **5:** 1989; 170–182.
54. G.M. Crippen *Macromolecules*, **10:** 1977; 21–25.
54a. G.M. Crippen *Macromolecules*, **10:** 1977; 25–28.
55. D.R. Boswell and A.M. Lesk in *Computational Molecular Biology*, A.M. Lesk (Ed.), Oxford Uni. Press, 1988; pp 161–178.
56. A.M. Lesk in *Computational Molecular Biology*, A.M. Lesk (Ed.), Oxford Uni. Press, 1988; p 192–197.
57. C.M. Topham, P. Thomas, J.P. Overington, M.S. Johnson, F. Eiesenmenger and T.L. Blundell *Protein Structure, Prediction and Design*, Biochem. Soc. Symp.
58. J. Kay, G.G. Lunt and D.J. Osguthorpe (Eds.), London, The Biochemical Society, 1990; pp 1–9.
59. D.E. Stewart, P.K. Weiner and J.E. Wampler *J. Mol. Graph.*, **5:** 1987: 133–140.
60. C. Chothia and A.M. Lesk *EMBO J.*, **5:** 1986; 823–826.
61. J. Greer *J. Mol. Biol.*, **153:** 1981; 1027–1042.
61a. J. Greer *J. Mol. Biol.*, **153:** 1981; 1043–1053.
62. A.M. Lesk and C. Chothia *J. Mol. Biol.*, **136:** 1980; 225–270.
63. V.I. Lim *J. Mol. Biol.*, **88:** 1974; 873–894.
64. J. Garnier, J.M. Levin, J.F. Gibrat and V. Biou *Protein Structure, Prediction and Design*, Biochem. Soc. Symp., 57. J. Kay, G.G. Lunt and D.J. Osguthorpe (Eds.), London, The Biochemical Society, 1990; pp 11–24.
65. P.Y. Chou and G.D. Fasman *Biochemistry*, **13:** 1974; 211–222.
65a. P.Y. Chou and G.D. Fasman *Biochemistry*, **13:** 1974; 223–245.
65b. P.Y. Chou and G.D. Fasman *Adv. Enzymol.*, **47:** 1978; 45–148.
66. P. Prevelige Jr and G.D. Fasman *Prediction of Protein Structure and the Principles of Protein Conformation*, G.D. Fasman (Ed.) Plenum Press, New York, 1989; 391.
67. G.D. Fasman *ibid*, p 193.
67a. P.Y. Chou *ibid*, p 549.
68. T.T. Wu and E.A. Kabat *J. Mol. Biol.*, **75:** 1973; 13–31.
69. J. Garnier, D.J. Osguthorpe and B.J. Robson *J. Mol. Biol.*, **120:** 1978; 97–120.
70. W.R. Taylor and J.M. Thornton *Nature*, **301:** 1983; 540–542.
71. J. Zupan and J. Gasteiger *Anal. Chim. Acta*, **248:** 1991; 1–30.
72. N. Qian and T.J. Sejnowski *J. Mol. Biol.*, **202:** 1988; 865–884.
73. H. Bohr, J. Bohr, S. Brunak, R.M.J. Cotterill, B. Lautrup, L. Norskov, O.H. Olsen and S.B. Petersen *FEBS Lett.*, **241:** 1988; 223–228.
74. D.G. Kneller, F.E. Cohen and R. Langridge *J. Mol. Biol.*, **214:** 1990; 171–182.
75. L.H. Holley and M. Karplus *Proc. Natl. Acad. Sci. USA*, **86:** 1989; 152–156.

76. H. Bohr, J. Bohr, S. Brunak, R.M.J. Cotterill, H. Fredholm, B. Lautrup and S.B. Petersen *FEBS Lett.*, **261:** 1990; 43–46.

77. E.W. Steeg *Neural Network Algorithms for the Prediction of RNA Secondary Structure*, Computer Science Dept, Univ. Toronto, Ontario, Canada, 1988.

78. G.L. Wilcox, M. Poliac and M.N. Liebman *Tetrahedron Comp. Method.*, **3:** 1990; 191–211.

79. M.N. Liebman, C.A. Venanzi and H. Weinstein *Biopolymers*, **24:** 1985; 1721–1758.

80. J.D. Bryngelson, J.J. Hopfield and S.N. Southard Jr *Tetrahedron Comput. Method.*, **3:** 1990; 129–141.

81. M.S. Friedrichs and P.G. Wolynes *Science*, **246:** 1989; 371–373.

82. D.A. Clark, G.J. Barton and C.J. Rawlings *J. Mol. Graph.*, **8:** 1990; 94–107.

83. T.L. Blundell, B.L. Sibanda, M.J.E. Sternberg and J.M. Thornton *Nature*, **326:** 1987; 347–352.

84. M.J. Sutcliffe, F.R.F. Hayes and T.L. Blundell *Prot. Eng.*, **1:** 1987; 385–392.

84a. M.J. Sutcliffe, I. Haneef, D. Carney and T.L. Blundell *Prot. Eng.*, **1:** 1987; 377–384.

85. W.R. Taylor in *Nucleic Acids and Protein Sequence Analysis, a Practical Approach*, M.J. Bishop and C.J. Rawlings (Ed.) IRL Press, Oxford, UK, 1987.

86. C.J. Rawlings *Artificial Intelligence and Protein Structure Prediction*, Proceedings of Biotechnology Information '86. IRL Press Oxford, UK, 1987; 59–77.

87. C.J. Rawlings, W.R. Taylor, J. Nyakairu, J. Fox and M.J.E. Sternberg *J. Mol. Graph.*, **3:** 1985; 151–157.

88. F.E. Cohen and I.D. Kuntz *Proteins, Structure, Function and Genetics*, **2:** 1987; 162–166.

89. J. Novotny, R.E. Bruccoleri and M. Karplus *J. Mol. Biol.*, 1984; 787–818.

90. I. Haneef, J. Talbot and P.G. Stockley *J. Mol. Graph.*, **7:** 1989; 186–195.

91. W.J. Browne, A.C.T. North, D.C. Phillips, K. Brew, T.C. Vanaman and R.L. Hill *J. Mol. Biol.*, **42:** 1969; 65–86.

92. R.H. Kretsinger and C.D. Barry *Biochim. Biophys. Acta*, **405:** 1975; 40–52.

93. R.M. Tufty and R.H. Kretsinger *Science*, **187:** 1975; 167–169.

94. C. Chothia, A.M. Lesk, M. Levitt, A.G. Amit, R.A. Mariuzza, S.E.V. Phillips and R.J. Poljak *Science*, **233:** 1986; 755–758.

95. K. Toma, N. Kitaguchi and H. Ito *J. Mol. Graph.*, **7:** 1989; 202–205.

96. G.G. Ferenczy and G.M. Morris *J. Mol. Graph.*, **7:** 1989; 206–211.

97. S.B. Needleman and C.D. Wunsch *J. Mol. Biol.*, **48:** 1970; 443–453.

98. G.M. Morris *J. Mol. Graph.*, **6:** 1988; 135–140.

99. N. Gray *J. Mol. Graph.*, **8:** 1990; 11–15.

100. G. Vriend and C. Sander *Proteins, Structure, Function and Genetics*, **11:** 1991; 52–58.

101. A.D. McLachlan *J. Mol. Biol.*, **61:** 1971; 409–424.

102. A.D. McLachlan *J. Mol. Biol.*, **64:** 1972; 417–437.

103. M.O. Dayhoff, R.M. Schwartz and B.C. Orcutt *Atlas of Protein Sequence and Structure*, National Biomedical Research Foundation, Washington DC, USA, 1979; vol 5, suppl. 3, 345–362.

104. P.A. Argos *J. Mol. Biol.*, **193:** 1987; 385–396.

105. M.G. Rossmann and P.A. Argos *J. Biol. Chem.*, **250:** 1975; 7525–7532.

105a. M.G. Rossmann and P.A. Argos *J. Mol. Biol.*, **105:** 1976; 75–95.

105b. M.G. Rossmann and P.A. Argos *J. Mol. Biol.*, **109:** 1977; 99–129.

106. P.H. Sellers *J. Algorithms*, **1:** 1980; 359–373.

107. D.R. Boswell and A.D. McLachlan *Nucl. Acids Res.*, **12:** 1984; 457–464.

108. W.M. Fitch *J. Mol. Biol.*, **16:** 1966; 9–16.

108a. W.M. Fitch *J. Mol. Biol.*, **16:** 1966; 17–27.

109. D.J. Lipman and W.R. Pearsons *Science*, **227:** 1985; 1435–1441.

110. A.M. Lesk *Comm. ACM.* **22:** 1979; 219–224.

111. J.R. Ullman *J. ACM*, **16:** 1976; 31–42.

112. A.D. McLachlan *J. Mol. Biol.*, **128:** 1979; 49–79.
113. S.J. Remington and B.W. Matthews *Proc. Natl. Acad. Sci. USA*, **75:** 1978; 2180–2184.
114. M.G. Rossmann and D.M. Blow *Acta Crystallographica*, **15:** 1962; 24–31.
115. G.J. Barton and M.J.E. Sternberg *J. Mol. Graph.*, **6:** 1988; 190–196.
116. P.J. Artymiuk, D.W. Rice, E.M. Mitchell and P. Willett *Protein Engineering*, **4:** 1990; 39–43.
117. E.M. Mitchell, P.J. Artymiuk, D.W. Rice and P. Willett *J. Mol. Biol.*, **212:** 1990; 151–166.
118. W. Kabsch and C. Sander *Biopolymers*, **22:** 1983; 2577–2637.
119. J.S. Richardson *Nature (Lond.)*, **268:** 1977; 495–500.
120. M.J.E. Sternberg and J.M. Thornton *J. Mol. Biol.*, **105:** 1975; 367–382.
121. T. Ooi and K. Nishikawa *Conformation of Biological Molecules and Polymers*, E.D. Bergmann and B. Pullman (Ed.) Jerusalem, 1973, Israel Academy of Sciences and Humanities, 1973; pp 173–187.
122. K. Nishikawa and T. Ooi *J. Theor. Biol.*, **43:** 1974; 351–374.
123. G.D. Rose in *Computational Molecular Biology*, A.M. Lesk (Ed.) Oxford University Press, 1988; pp 198–204.
124. G.D. Rose, A.R. Geselowitz, G.J. Lesser, R.H. Lee and M.H. Zehfus, *Science*, **229:** 1985; 834–838.
125. C. Chothia *Nature*, **254:** 1975; 304–308.
126. P. Furet, A. Sele and N.C. Cohen *J. Mol. Graph.*, **6:** 1988; 182–189.
127. J.L. Fauchere and V. Pliska *Eur. J. Med. Chem.–Chim. Ther.*, **18:** 1983; 369–375.
128. A.C. Edmundson *Nature*, **205:** 1965; 883.
129. H.C. Watson *Prog. Stereochem. 4*, 1969; 299–333.
130. R.E. Dickerson *The Proteins*, H. Neurath (ed.), Vol 2, Academic Press, 1964, p. 634.
131. J. Weber, P.Y. Mongantini, P. Fluebiger and A. Gounsot. *The Visual Computer* **7:** 1991; 158–169.

Subject Index

Author Index

DATE DUE